D1192161

FLORIDA STATE
UNIVERSITY LIBRARIES

JUL 6 1995

Tallahassee, Florida

Probability and its Applications

A Series of the Applied Probability Trust

Editors: J. Gani, C.C. Heyde, T.E. Kurtz

Probability and its Applications

Richard F. Bass

Probabilistic Techniques in Analysis

With 12 Illustrations

Springer-Verlag
New York Berlin Heidelberg London Paris
Tokyo Hong Kong Barcelona Budapest

SCI
QA
300
B33
1995

Richard F. Bass
Department of Mathematics
University of Washington
Seattle, WA 98195
USA

Series Editors

J. Gani
Stochastic Analysis
  Group, CMA
Australian National
  University
Canberra ACT 0200
Australia

C.C. Heyde
Stochastic Analysis
  Group, CMA
Australian National
  University
Canberra ACT 0200
Australia

T.E. Kurtz
Department of
  Mathematics
University of Wisconsin
480 Lincoln Drive
Madison, WI 53706
USA

Mathematics Subject Classifications (1991): 60J65, 42A61, 31C15

Library of Congress Cataloging-in-Publication Data
Bass, Richard F.
    Probabilistic techniques in analysis / Richard F. Bass.
      p.   cm. — (Probability and its applications)
    Includes bibliographical references and index.
    ISBN 0-387-94387-0
    1. Mathematical analysis.   2. Probabilities.   I. Title
    II. Series: Probability and its applications.
    QA300.B33   1995
    515—dc20   94-34721

Printed on acid-free paper.

© 1995 by the Applied Probability Trust.
All rights reserved. This work may not be translated or copied in whole or in part without the
written permission of the publisher (Springer-Verlag New York, Inc., 175 Fifth Avenue, New
York, NY 10010, USA), except for brief excerpts in connection with reviews or scholarly
analysis. Use in connection with any form of information storage and retrieval, electronic
adaptation, computer software, or by similar or dissimilar methodology now known or hereaf-
ter developed is forbidden.
The use of general descriptive names, trade names, trademarks, etc., in this publication, even
if the former are not especially identified, is not to be taken as a sign that such names, as
understood by the Trade Marks and Merchandise Marks Act, may accordingly be used freely
by anyone.

Production managed by Bill Imbornoni; manufacturing supervised by Genieve Shaw.
Photocomposed pages prepared from the author's TEX files using Springer-Verlag's
"svplain.sty" macro.
Printed and bound by Edwards Brothers, Inc., Ann Arbor, MI.
Printed in the United States of America.

9 8 7 6 5 4 3 2 1

ISBN 0-387-94387-0 Springer-Verlag New York Berlin Heidelberg

To Meredith, Benjamin, and Peter

PREFACE

A glance at almost any probability book shows that there has been a large flow of ideas from analysis to probability theory. This book is concerned with the flow of ideas in the opposite direction. The topics covered are those branches of analysis to which probability has contributed something, in new results, new proofs, or new insights.

I had two audiences in mind when I wrote this, the first being students and researchers in probability and the second being students and researchers in analysis. Because of the varied backgrounds one would expect from these two audiences, I have tried to make the book as nearly self-contained as possible. Only the standard material covered in any first-year course in real analysis is needed for the first four chapters. For the fifth chapter, the reader should also be acquainted with what is covered in an undergraduate complex variables course. The basic facts needed from a first-year probability course are presented in the first section of the first chapter.

Chapter I is a minicourse in stochastic analysis. By the end of Chapter I the reader will have seen all the probability theory that is needed in the rest of the book. After a preliminary section that gives some of the very basic concepts from probability, Section 2 gives the definition and construction of Brownian motion. This stochastic process provides the basis for the relationship between probability and analysis. There are two main reasons for this: Brownian motion is a Markov process, and Brownian motion is a martingale. The Markov structure is described in Section 3, including the strong Markov property and a brief account of semigroups. On the martingale side, definitions and basic theorems such as Doob's optional stopping theorem, Doob's inequalities, and the decomposition of supermartingales, are given in Section 4. One of the interesting aspects of continuous time martingales such as Brownian motion is that one can define a stochastic integral with respect to them, even though their paths are not of bounded variation. These stochastic integrals, and the crucial change of variables formula (Itô's formula), are constructed in Section 5. Section 6 is devoted to applications of stochastic integrals. These include change of measure formulas, some very useful inequalities, a brief introduction to stochastic differential equations, and local times. Finally, Section 7 is an account of weak convergence of probability measures.

Chapter II is on harmonic functions and potential theory. I have tried to cover the major results in sufficient generality for most purposes while aiming the material toward applications in later chapters. The first section describes the basics of harmonic functions, Poisson kernels for the ball and the upper half-space, and the solution of the Dirichlet problem. The second section is a short section giving Choquet's capacitability theorem and its use to prove that hitting times of Borel sets are measurable. In Section 3 Newtonian potentials are described, as is the definition of Green functions of a domain. Section 4 proves the symmetry of the transition densities of Brownian motion killed on leaving an arbitrary set and describes the solution to the heat equation by means of eigenvalue expansions of the transition densities of Brownian motion. Section 5 is devoted to a description of the Newtonian capacity of a set and applications. Section 6 is about excessive functions; included is a proof of the Riesz decomposition of superharmonic functions. The last section, Section 7, gives a construction of the Martin boundary; a complete proof of Choquet's theorem on representations in terms of extremal elements is also provided.

The material in the first two chapters is mostly classical. What is covered in the last three chapters is mostly results that have been proved since the late 1970s and that for the most part have appeared only in journal articles.

Chapter III is concerned with the behavior of harmonic functions in Lipschitz domains. This topic is considered fundamental to harmonic analysis, and there is a strong interplay here between probability and analysis. First, a technical tool, the boundary Harnack principle, is proved in two different ways. In Section 2, it is shown how the boundary Harnack principle implies that the Martin boundary for a Lipschitz domain may be identified with the Euclidean boundary. Section 3 is a description of the conditional lifetime problem and its close relative, the conditional gauge theorem. Fatou theorems for Lipschitz domains are proved in Section 4; again the boundary Harnack principle is key. Section 5 is a description of the support of harmonic measure in Lipschitz domains; as part of the proofs an account of Gehring's inequality and reverse Hölder inequalities is provided.

The subject of Chapter IV is topics related to singular integrals. I have tried to concentrate primarily on those aspects that have a probability component. Section 1 is about maximal functions, while Section 2 describes the Hilbert transform, Riesz' theorem, and the weak (1-1) inequality. Section 3 describes how to extend the methods used in the Hilbert transform case to higher dimensions and gives applications: Riesz transforms, Sobolev inequalities, and the spaces C^α. Much more general singular integral operators are possible; the necessary tools such as the Littlewood-Paley inequalities are provided in Section 4, while Section 5 gives the applications to singular integral operators and Fourier multiplier operators. The last two sections are concerned with the spaces H^1 and BMO, respectively: Section 6 is devoted to various characterizations of the space H^1, and Section 7 proves the H^1–BMO duality and some sharp inequalities regarding singular integral operators.

The final chapter, Chapter V, is on analytic functions. The first section proves Lévy's theorem and gives several applications, including the Phragmén-Lindelöf theorem and the Riemann mapping theorem. Section 2 is concerned with giving probabilistic proofs of two important theorems, Picard's little theorem and the Koebe distortion theorem. Section 3 describes the boundary behavior of analytic functions; included are Privalov's theorem, Plessner's theorem, and Makarov's law of the iterated logarithm. Closely related is Section 4; there probabilistic proofs of Beurling's projection theorem and the Hall projection theorem are provided, as is Makarov's description of the Hausdorff dimension of the support of harmonic measure. The angular derivative problem and an account of Burdzy's theorem is the subject of Section 5. Finally, in Section 6 Varopoulos' probabilistic proof of the famous corona problem is described.

In almost every section of the last three chapters are a number of interesting open problems. Many of these are quite hard and would provide good topics of research. More in line with the needs of the beginning student are the exercises that are provided at the end of each chapter. These vary from routine to quite difficult. Also at the end of each chapter are notes that describe where I obtained the material. These are not meant to be complete, and the references mentioned in the notes need to be consulted to get the full story.

Much of the material in this book has been covered, some of it several times, in courses at the University of Washington. My thanks go to the students who patiently helped me work through the details. Special thanks go to several people who read parts of the manuscript and made valuable suggestions: Siva Athreya, Rodrigo Bañuelos, Martin Barlow, Krzysztof Burdzy, Michael Cranston, Ping Gao, Davar Khoshnevisan, Youngmee Kwon, Joseph Leighly, and Carlos Tolmasky. Partial support for this project has been provided by the National Science Foundation grant DMS91-00244.

A few words about notation

There are a number of different definitions of the Fourier transform. We use

$$\widehat{f}(\xi) = \int_{\mathbb{R}^d} e^{i\xi \cdot x} f(x) \, dx$$

in this text. $B(x, r)$ denotes the open ball of radius r centered at x. We use $|\cdot|$ to denote either Lebesgue measure in $\mathbb{R}^d$ or, when the meaning is clear, surface measure on the boundary of either a ball or a half-space. We write elements of $\mathbb{R}^d$ as $x = (x^1, \ldots, x^d)$. Although there is a danger of confusion with powers, this makes writing the ith component of the point x_n easy. The letter c, with or without subscripts, signifies a constant whose value is unimportant and which may change from location to location, even within a line. So the reader should not be alarmed at equations like $cx + 2cx = cx$.

CONTENTS

I
PROBABILITY

1 Preliminaries

In this section we give some preliminaries on probabilistic terminology, independence, Gaussian random variables, and conditional expectation. The first and fourth of these are the most important for what follows. The reader with some familiarity with probability theory can safely skip this section.

Probabilistic terminology

Given a space Ω and a σ-field (or σ-algebra) $\mathcal{F}$ on it, a probability measure, or just a probability, is a positive finite measure with total mass 1. $(\Omega, \mathcal{F}, \mathbb{P})$ is called a probability space. Elements of $\mathcal{F}$ are called events. Measurable functions from Ω to $\mathbb{R}$ are called random variables and are usually denoted X or Y instead of f or g. The integral of X with respect to $\mathbb{P}$ is called the expectation of X or the expected value of X, and $\int X(\omega)\mathbb{P}(d\omega)$ is often written $\mathbb{E}\,X$, while $\int_A X(\omega)\mathbb{P}(d\omega)$ is often written $\mathbb{E}\,[X; A]$. If an event occurs with probability one, one says "almost surely" instead of "almost everywhere" and writes a.s. What analysts call the characteristic function of a set probabilists call the indicator of the set; in this book we will use the notation 1_A for the function or random variable that is 1 on A and 0 on the complement. The complement of A is denoted A^c.

If A_n is a sequence of sets, $(A_n$ i.o.), read "infinitely often," is defined to be $\cap_{j=1}^{\infty} \cup_{n=j}^{\infty} A_n$. The following easy fact is called the Borel-Cantelli lemma or sometimes the first half of the Borel-Cantelli lemma.

(1.1) Proposition. *If $\sum_{n=1}^{\infty} \mathbb{P}(A_n) < \infty$, then $\mathbb{P}(A_n \text{ i.o.}) = 0$.*

Proof. Note $\mathbb{P}(A_n \text{ i.o.}) = \lim_{j \to \infty} \mathbb{P}(\cup_{n=j}^{\infty} A_n)$. If $\sum_{n=1}^{\infty} \mathbb{P}(A_n) < \infty$, then $\mathbb{P}(\cup_{n=j}^{\infty} A_n) \leq \sum_{n=j}^{\infty} \mathbb{P}(A_n) \to 0$ as $j \to \infty$. $\square$

Chebyshev's inequality is the following inequality.

(1.2) Proposition. *If $X \geq 0$, a.s., then*

$$\mathbb{P}(X \geq a) \leq EX/a.$$

Proof. This follows from

$$\mathbb{P}(X \geq a) = \mathbb{E}\left[1_{(X \geq a)}\right] \leq \mathbb{E}\left[X/a; X \geq a\right] \leq \mathbb{E} X/a.$$

$\square$

Jensen's inequality is the following.

(1.3) Proposition. *If g is convex and X and $g(X)$ are integrable, then*

$$\mathbb{E} g(X) \geq g(\mathbb{E} X).$$

Proof. If g is convex, then g lies above all its tangent lines. So for each x_0, there exists c such that $g(X) \geq g(x_0) + c(X - x_0)$. Letting $x_0 = \mathbb{E} X$ and taking expectations on both sides, we obtain our result. $\square$

The law or distribution of X is the probability measure $\mathbb{P}_X$ on $\mathbb{R}$, given by

(1.1) $\mathbb{P}_X(A) = \mathbb{P}(X \in A).$

Given any measure μ on $\mathbb{R}$ with total mass 1, we can construct a random variable X such that $\mathbb{P}_X = \mu$: let $\Omega = [0, 1]$ and let $\mathbb{P}$ be Lebesgue measure on $[0, 1]$. For $\omega \in [0, 1]$, let

(1.2) $X(\omega) = \inf\{t : \mu((-\infty, t]) \geq \omega\}.$

Then $\mathbb{P}_X((-\infty, a]) = \mathbb{P}(X \leq a) = \mu(-\infty, a]$ for each a, hence $\mathbb{P}_X = \mu$.

(1.4) Proposition. *If $f \geq 0$ or f is bounded,*

$$\mathbb{E} f(X) = \int f(x)\mathbb{P}_X(dx).$$

Proof. If $f = 1_{(-\infty, a]}$, then

(1.3) $\mathbb{E} f(X) = \mathbb{P}(X \leq a) = \mathbb{P}_X((-\infty, a]) = \int f(x)\mathbb{P}_X(dx).$

By linearity we obtain our result for simple functions f. Taking limits then gives us our result for f bounded or f nonnegative. □

A random vector is just a measurable map from Ω to $\mathbb{R}^d$, and the definition of law and Proposition 1.4 extend to this case.

We will frequently use the following equality.

(1.5) Proposition. *If $X \geq 0$,*

$$(1.4) \qquad \mathbb{E}\, X^p = \int_0^\infty p\lambda^{p-1}\mathbb{P}(X > \lambda)d\lambda.$$

Proof. By Fubini's theorem, the right-hand side is equal to

$$\mathbb{E} \int_0^\infty p\lambda^{p-1}1_{(X>\lambda)}d\lambda = \mathbb{E} \int_0^X p\lambda^{p-1}d\lambda,$$

which is equal to the left-hand side. □

Independence

Two events A, B are independent if $\mathbb{P}(A \cap B) = \mathbb{P}(A)\mathbb{P}(B)$. This definition generalizes to n events: $A_1, \ldots, A_n$ are independent if $\mathbb{P}(\cap_{i=1}^n A_i) = \prod_{i=1}^n \mathbb{P}(A_i)$. A σ-field $\mathcal{F}$ is independent of a σ-field $\mathcal{G}$ if each $A \in \mathcal{F}$ is independent of each $B \in \mathcal{G}$. The σ-field generated by X, denoted $\sigma(X)$, is the collection $\{(X \in A); A \text{ Borel}\}$. Two random variables are independent if the σ-fields generated by X, Y are independent. The notion of an event and a random variable being independent, or a random variable and a σ-field being independent are defined in the obvious way. Note that if X and Y are independent and f and g are Borel measurable functions, then $f(X)$ and $g(Y)$ are independent.

An example of independent random variables is to let $\Omega = [0, 1]^2$, $\mathbb{P}$ Lebesgue measure, X a function of just the first variable, and Y a function of just the second variable. In fact, it can be shown that independent random variables can always be represented by means of some suitable product space.

(1.6) Proposition. *If X, Y, and XY are integrable and X and Y are independent, then $\mathbb{E}\, XY = (\mathbb{E}\, X)(\mathbb{E}\, Y)$.*

Proof. If X is of the form $\sum_{i=1}^I a_i 1_{A_i}$, Y is of the form $\sum_{j=1}^J b_j 1_{B_j}$ and X and Y are independent, then by linearity and the definition of independence, $\mathbb{E}\, XY = (\mathbb{E}\, X)(\mathbb{E}\, Y)$. Approximating nonnegative X and Y by simple random variables of this form, we obtain our result by monotone convergence. The case of general X and Y then follows by linearity. □

The characteristic function of a random variable X is the Fourier transform of its law: $\int e^{iux}\mathbb{P}_X(dx) = \mathbb{E}\,e^{iuX}$ (we are using Proposition 1.4 here). If X and Y are independent, so are e^{iuX} and e^{ivY}, and hence by Proposition 1.6, $\mathbb{E}\,e^{i(uX+vY)} = \mathbb{E}\,e^{iuX}\mathbb{E}\,e^{ivY}$. Thus when X and Y are independent, the joint characteristic function of X and Y factors into the product of the respective characteristic functions.

The converse also holds.

(1.7) Proposition. *If $\mathbb{E}\,e^{i(uX+vY)} = \mathbb{E}\,e^{iuX}\mathbb{E}\,e^{ivY}$ for all u and v, then X and Y are independent random variables.*

Proof. Let X' be a random variable with the same law as X, Y' one with the same law as Y, and X', Y' independent. (We let $\Omega = [0,1]^2$, $\mathbb{P}$ Lebesgue measure, X' a function of the first variable, and Y' a function of the second variable defined as in (1.2).) Then $\mathbb{E}\,e^{i(uX'+vY')} = \mathbb{E}\,e^{iuX'}\mathbb{E}\,e^{ivY'}$. Since X, X' have the same law, they have the same characteristic function, and similarly for Y, Y'. Therefore (X', Y') has the same joint characteristic function as (X, Y). By the uniqueness of the Fourier transform, (X', Y') has the same joint law as (X, Y), which is easily seen to imply that X and Y are independent. $\qquad\square$

The second half of the Borel-Cantelli lemma is the following assertion.

(1.8) Proposition. *If A_n is a sequence of independent events and $\sum_1^\infty \mathbb{P}(A_n) = \infty$, then $\mathbb{P}(A_n \text{ i.o.}) = 1$.*

Proof. Note

$$\mathbb{P}(\cup_{n=j}^N A_n) = 1 - \mathbb{P}(\cap_{n=j}^N A_n^c) = 1 - \prod_{n=j}^N \mathbb{P}(A_n^c)$$

$$= 1 - \prod_{n=j}^N (1 - \mathbb{P}(A_n)) \geq 1 - \exp\left(-\sum_{n=j}^N \mathbb{P}(A_n)\right) \to 1$$

as $N \to \infty$. So $\mathbb{P}(A_n \text{ i.o.}) = \lim_{j\to\infty}(\cup_{n=j}^\infty A_n) = 1$. $\qquad\square$

Gaussian random variables

A mean zero, variance one, Gaussian or normal random variable is one where

(1.5) $$\mathbb{P}(X \in A) = \int_A \frac{1}{\sqrt{2\pi}} e^{-x^2/2} dx, \qquad A \text{ Borel}.$$

We also describe such an X as having an $\mathcal{N}(0,1)$ distribution or law, read as a "normal 0,1" law. It is routine to check that such an X has mean or expectation 0 and variance $\mathbb{E}(X - \mathbb{E}X)^2 = 1$. It is standard (see Durrett [2], pp. 66-67) that $\mathbb{E}e^{iuX} = e^{-u^2/2}$. Such random variables exist by the construction in (1.2), with

$$\mu(dx) = (2\pi)^{-1/2}e^{-x^2/2}dx.$$

X has an $\mathcal{N}(a, b^2)$ distribution if $X = bZ + a$ for some Z having a $\mathcal{N}(0,1)$ distribution.

A sequence of random variables $X_1, \ldots, X_n$ is said to be jointly normal if there exists a sequence of independent $\mathcal{N}(0,1)$ random variables $Z_1, \ldots, Z_m$ and constants b_{ij} and a_i such that $X_i = \sum_{j=1}^{m} b_{ij}Z_j + a_i$, $i = 1, \ldots, n$. In matrix notation, $X = BZ + A$. For simplicity, in what follows let us take $A = 0$; the modifications for the general case are easy. The covariance of two random variables X and Y is defined to be $\mathbb{E}[(X - \mathbb{E}X)(Y - \mathbb{E}Y)]$. Since we are assuming our normal random variables are mean 0, we can omit the centering at expectations. Given a sequence of mean 0 random variables, we can talk about the covariance matrix, which is $\text{Cov}(X) = \mathbb{E}XX^t$, where X^t denotes the transpose of the vector X. In the above case, we see $\text{Cov}(X) = \mathbb{E}[(BZ)(BZ)^t] = \mathbb{E}[BZZ^tB^t] = BB^t$, since $\mathbb{E}ZZ^t = I$, the identity.

Let us compute the joint characteristic function $\mathbb{E}e^{iu^tX}$ of the vector X, where u is an n-dimensional vector. First, if v is an m-dimensional vector,

$$\mathbb{E}e^{iv^tZ} = \mathbb{E}\prod_{j=1}^{m}e^{iv_jZ_j} = \prod_{j=1}^{m}\mathbb{E}e^{iv_jZ_j} = \prod_{j=1}^{m}e^{-v_j^2/2} = e^{-v^tv/2}$$

using the independence of the Zs. So

(1.6) $$\mathbb{E}e^{iu^tX} = \mathbb{E}e^{iu^tBZ} = e^{-u^tBB^tu/2}.$$

By taking $u = (0, \ldots, 0, a, 0, \ldots, 0)$ to be a constant times the unit vector in the jth coordinate direction, we deduce that each of the Xs is indeed normal. Taking $m = 2$, $a \in \mathbb{R}$, $B = \begin{pmatrix} b_1 & 0 \\ 0 & b_2 \end{pmatrix}$, and $u = (a, a)$, we see that the sum of an $\mathcal{N}(0, b_1^2)$ and an independent $\mathcal{N}(0, b_2^2)$ is an $\mathcal{N}(0, b_1^2 + b_2^2)$. If $\text{Cov}(X) = BB^t$ is a diagonal matrix, then the joint characteristic function of the Xs factors, and so by Proposition 1.7, the Xs would in this case be independent.

If X_n are normal random variables converging in probability (i.e., in measure) to a random variable X, then X is also normal. This follows since $\mathbb{E}e^{iuX_n} \to \mathbb{E}e^{iuX}$ by dominated convergence. The analogue for random vectors is also true.

Finally we will use the following estimate.

(1.9) Proposition. *If Z is an $\mathcal{N}(0,1)$ and $a > 1$,*

(1.7) $$\mathbb{P}(Z \geq a) \leq e^{-a^2/2}.$$

Proof. This follows from

$$\mathbb{P}(Z \geq a) \leq \int_a^\infty e^{-x^2/2}dx \leq \int_a^\infty (x/a)e^{-x^2/2}dx.$$

$\square$

Conditional expectation

A concept that will be used extensively in the rest of this book will be that of conditional expectation.

(1.10) Definition. *If $\mathcal{F} \subseteq \mathcal{G}$ are two σ-fields and X is an integrable $\mathcal{G}$ measurable random variable, the conditional expectation of X given $\mathcal{F}$, written $\mathbb{E}[X|\mathcal{F}]$ and read as "the expectation (or expected value) of X given $\mathcal{F}$," is any $\mathcal{F}$ measurable random variable Y such that $\mathbb{E}[Y; A] = \mathbb{E}[X; A]$ for every $A \in \mathcal{F}$.*

If Y_1, Y_2 are two $\mathcal{F}$ measurable random variables with $\mathbb{E}[Y_1; A] = \mathbb{E}[Y_2; A]$ for all $A \in \mathcal{F}$, then $Y_1 = Y_2$, a.s., or conditional expectation is unique up to a.s. equivalence.

In the case X is already $\mathcal{F}$ measurable, $\mathbb{E}[X|\mathcal{F}] = X$. If X is independent of $\mathcal{F}$, $\mathbb{E}[X|\mathcal{F}] = \mathbb{E}X$. Both of these facts follow immediately from the definition. For another example, which ties this definition with the one used in elementary probability courses, if $\{A_i\}$ is a finite collection of disjoint sets whose union is Ω, $\mathbb{P}(A_i) > 0$ for all i, and $\mathcal{F}$ is the σ–field generated by the A_is, then

(1.8) $$\mathbb{E}[X|\mathcal{F}] = \sum_i \frac{\mathbb{E}[X; A_i]}{\mathbb{P}(A_i)} 1_{A_i}.$$

This follows since the right-hand side is $\mathcal{F}$ measurable and its expectation over any set A_i is $\mathbb{E}[X; A_i]$.

It is easy to check that limit theorems such as monotone convergence and dominated convergence have conditional expectation versions, as do inequalities like Jensen's and Chebyshev's inequalities. Thus, for example, we have the following.

(1.11) Proposition. (Jensen's inequality for conditional expectations) *If g is convex and X and $g(X)$ are integrable,*

(1.9) $$\mathbb{E}[g(X)|\mathcal{F}] \geq g(\mathbb{E}[X|\mathcal{F}]), \quad \text{a.s.}$$

A key fact is the following.

(1.12) Proposition. *If X and XY are integrable and Y is measurable with respect to $\mathcal{F}$, then*

$$(1.10) \qquad\qquad \mathbb{E}\left[XY|\mathcal{F}\right] = Y\mathbb{E}\left[X|\mathcal{F}\right].$$

Proof. If $A \in \mathcal{F}$, then for any $B \in \mathcal{F}$,

$$\mathbb{E}\left[1_A\mathbb{E}\left[X|\mathcal{F}\right]; B\right] = \mathbb{E}\left[\mathbb{E}\left[X|\mathcal{F}\right]; A \cap B\right] = \mathbb{E}\left[X; A \cap B\right] = \mathbb{E}\left[1_A X; B\right].$$

Since $1_A\mathbb{E}\left[X|\mathcal{F}\right]$ is $\mathcal{F}$ measurable, this shows that (1.10) holds when $Y = 1_A$ and $A \in \mathcal{F}$. Using linearity and taking limits shows that (1.10) holds whenever Y is $\mathcal{F}$ measurable and X and XY are integrable. $\qquad\square$

Two other equalities follow.

(1.13) Proposition. *If $\mathcal{E} \subseteq \mathcal{F} \subseteq \mathcal{G}$, then*

$$(1.11) \qquad\quad \mathbb{E}\left[\mathbb{E}\left[X|\mathcal{F}\right]|\mathcal{E}\right] = \mathbb{E}\left[X|\mathcal{E}\right] = \mathbb{E}\left[\mathbb{E}\left[X|\mathcal{E}\right]|\mathcal{F}\right].$$

Proof. The right equality holds because $\mathbb{E}\left[X|\mathcal{E}\right]$ is $\mathcal{E}$ measurable, hence $\mathcal{F}$ measurable. To show the left equality, let $A \in \mathcal{E}$. Then since A is also in $\mathcal{F}$,

$$\mathbb{E}\left[\mathbb{E}\left[\mathbb{E}\left[X|\mathcal{F}\right]|\mathcal{E}\right]; A\right] = \mathbb{E}\left[\mathbb{E}\left[X|\mathcal{F}\right]; A\right] = \mathbb{E}\left[X; A\right] = \mathbb{E}\left[\mathbb{E}\left[X|\mathcal{E}\right]; A\right].$$

Since both sides are $\mathcal{E}$ measurable, the equality follows. $\qquad\square$

To show the existence of $\mathbb{E}\left[X|\mathcal{F}\right]$, we proceed as follows.

(1.14) Proposition. *If X is integrable, then $\mathbb{E}\left[X|\mathcal{F}\right]$ exists.*

Proof. Using linearity, we need only consider $X \geq 0$. Define a measure $\mathbb{Q}$ on $\mathcal{F}$ by $\mathbb{Q}(A) = \mathbb{E}\left[X; A\right]$ for $A \in \mathcal{F}$. This is trivially absolutely continuous with respect to $\mathbb{P}|_{\mathcal{F}}$, the restriction of $\mathbb{P}$ to $\mathcal{F}$. Let $\mathbb{E}\left[X|\mathcal{F}\right]$ be the Radon-Nikodym derivative of $\mathbb{Q}$ with respect to $\mathbb{P}|_{\mathcal{F}}$. The Radon-Nikodym derivative is $\mathcal{F}$ measurable by construction and so provides the desired random variable. $\qquad\square$

Equation (1.10) shows that if Y is $\mathcal{F}$ measurable, then $\mathbb{E}\left[Y(X - \mathbb{E}\left[X|\mathcal{F}\right])\right] = \mathbb{E}\,XY - \mathbb{E}\left[Y\mathbb{E}\left[X|\mathcal{F}\right]\right] = 0$. So one way to view conditional expectation is that if $X \in L^2(\mathbb{P})$, then $\mathbb{E}\left[X|\mathcal{F}\right]$ is the projection of X onto the subspace of $L^2(\mathbb{P})$ generated by the $\mathcal{F}$ measurable functions, or the conditional expectation operator on $L^2(\mathbb{P})$ is the projection operator onto this subspace. The conditional expectation operator on $L^1(\mathbb{P})$ is then the (unique) extension of this projection operator.

When $\mathcal{F} = \sigma(Y)$, one usually writes $\mathbb{E}[X|Y]$ for $\mathbb{E}[X|\mathcal{F}]$. Notation that is commonly used (however, we will use it only very occasionally and only for heuristic purposes) is $\mathbb{E}[X|Y = y]$. The definition is as follows. If $A \in \sigma(Y)$, then $A = (Y \in B)$ for some Borel set B by the definition of $\sigma(Y)$, or $1_A = 1_B(Y)$. By linearity and taking limits, if Z is $\sigma(Y)$ measurable, $Z = f(Y)$ for some Borel measurable function f. Set $Z = \mathbb{E}[X|Y]$ and choose f Borel measurable so that $Z = f(Y)$. Then $\mathbb{E}[X|Y = y]$ is defined to be $f(y)$.

2 Brownian motion

In this section we construct Brownian motion and examine a few of its easiest properties. The reader who is not very familiar with probability should probably skip the subsection on the construction of Brownian motion for the first reading.

Definitions

Let $(\Omega, \mathcal{F}, \mathbb{P})$ be a probability space and let $\mathcal{B}$ be the Borel σ-field on $[0, \infty)$. A stochastic process, denoted $X(t, \omega)$ or $X_t(\omega)$ or just X_t, is a map from $[0, \infty) \times \Omega$ to $\mathbb{R}$ that is measurable with respect to the product σ-field of $\mathcal{B}$ and $\mathcal{F}$.

(2.1) Definition. *A stochastic process X_t is a one-dimensional Brownian motion started at 0 if*
(a) $X_0 = 0$, a.s.;
(b) for all $s \le t$, $X_t - X_s$ is a mean zero Gaussian random variable with variance $t - s$;
(c) for all $s < t$, $X_t - X_s$ is independent of $\sigma(X_r; r \le s)$;
(d) with probability 1 the map $t \to X_t(\omega)$ is continuous.

$\sigma(X_r; r \le s)$ is the smallest σ–field with respect to which each X_r, $r \le s$, is measurable.

Our Brownian motion is one dimensional. To define d-dimensional Brownian motion, let $X_t^1, \ldots, X_t^d$ be independent one-dimensional Brownian motions. Then

$$X_t = (X_t^1, \ldots, X_t^d)$$

is d-dimensional Brownian motion.

We have $X_0 = 0$, or our Brownian motion is started at 0. We also want to consider Brownian motion started at $x \in \mathbb{R}^d$. We can get this just by looking at $x + X_t$.

There is, however, another point of view that is more useful. Instead of having $\mathbb{P}$ and $X_t^x = x + X_t$, i.e., one probability measure and many processes, it is usually more convenient to have one process and many probability measures.

Let Z_t be d-dimensional Brownian motion as defined above. Let Ω be the set of continuous functions from $[0, \infty)$ to $\mathbb{R}^d$. Each element ω in Ω is thus a continuous function. (We do not require that $\omega(0) = 0$ here.) Define

$$(2.1) \qquad X_t(\omega) = \omega(t).$$

This will be our process. Let $\mathcal{F} = \sigma(X_r; r < \infty)$. Now define $\mathbb{P}^x$ to be the probability measure on $(\Omega, \mathcal{F})$ given by

$$(2.2) \qquad \mathbb{P}^x(X. \in A) = \mathbb{P}(x + Z. \in A), \qquad x \in \mathbb{R}^d, A \in \mathcal{F}.$$

So, for example,

$$\mathbb{P}^x(|X_1| \in [1, 2], \sup_{h \leq 1} |X_{3+h} - X_3| \leq 1)$$
$$= \mathbb{P}(|x + Z_1| \in [1, 2], \sup_{h \leq 1} |Z_{3+h} - Z_3| \leq 1).$$

We call the pair $(\mathbb{P}^x, X_t)$, $x \in \mathbb{R}$, $t \geq 0$, a Brownian motion.

The paths of Brownian motion are very irregular. See Exercise 19, for example.

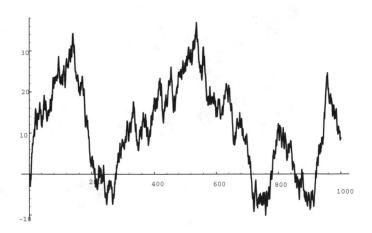

FIGURE 2.1. Simulation of the graph of a one-dimensional Brownian motion X_t plotted versus time.

Brownian motion has a scaling property that is extremely useful.

(2.2) Proposition. *If* $(\mathbb{P}^x, X_t)$ *is a Brownian motion starting from* x *and* $a > 0$, *then* $(\mathbb{P}^{x/a}, a^{-1}X_{a^2t})$ *is a Brownian motion started from* x/a.

Proof. We do the one-dimensional case; the higher-dimensional case is exactly similar. Since $\mathbb{P}^x$ is defined by translation, it is enough to consider the case $x = 0$. $a^{-1}X_{a^2t}$ is continuous in t, and

$$\mathrm{Cov}\,(a^{-1}X_{a^2t}, a^{-1}X_{a^2s}) = a^{-2}(a^2t \wedge a^2s) = s \wedge t.$$

Using (1.6), if $t_1 \leq t_2 \leq \ldots \leq t_n$, the law of $(a^{-1}X_{a^2t_1}, \ldots, a^{-1}X_{a^2t_n})$ is determined by the covariance matrix, and hence agrees with that of a Brownian motion. Since both processes are continuous, $a^{-1}X_{a^2t}$ must be a Brownian motion. $\qquad\square$

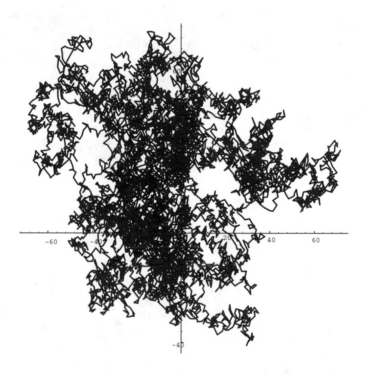

FIGURE 2.2. Simulation of the graph of a two-dimensional Brownian motion (X_t^1, X_t^2).

Brownian motion is also rotationally invariant.

(2.3) Proposition. *If* A *is an orthogonal matrix, then* $(\mathbb{P}^{Ax}, AX_t)$ *is a Brownian motion starting from* Ax.

Proof. AX_t still has continuous paths and independent increments. Since $\mathrm{Cov}\,(X_t - X_s) = (t - s)I$, where I is the identity matrix, then $\mathrm{Cov}\,(AX_t -$

$AX_s) = (t - s)A^tA = (t - s)I$. By the remarks following (1.6), this shows the components of $AX_t - AX_s$ are independent and each component is a Brownian motion.   □

Construction of Brownian motion

Let us show that there exists a process satisfying Definition 2.1. First we construct Brownian motion for $t \in [0, 1]$. We give the Haar function construction, which is one of the quickest ways to the construction of Brownian motion.

For $i = 1, 2, \ldots, j = 1, 2, \ldots, 2^{i-1}$, let φ_{ij} be the function on $[0, 1]$ that takes the value $2^{(i-1)/2}$ if $x \in [(2j-2)/2^i, (2j-1)/2^i)$, the value $-2^{(i-1)/2}$ if $x \in [(2j-1)/2^i, 2j/2^i)$, and 0 otherwise. Let φ_{00} be the function that is identically 1. The φ_{ij} are called the Haar functions. If $\langle \cdot, \cdot \rangle$ denotes the inner product in $L^2([0,1])$, that is, $\langle f, g \rangle = \int_0^1 f(x)\overline{g}(x)dx$, note the φ_{ij} are orthogonal and have norm 1. It is also easy to see that they form a complete orthonormal system for L^2: $\varphi_{00} \equiv 1$; $1_{[0,1/2)}$ and $1_{[1/2,1)}$ are both linear combinations of φ_{00} and φ_{11}; $1_{[0,1/4)}$ and $1_{[1/4,1/2)}$ are both linear combinations of $1_{[0,1/2)}$, φ_{21}, and φ_{22}. Continuing in this way, we see that $1_{[k/2^n,(k+1)/2^n)}$ is a linear combination of the φ_{ij}s for each n and each $k \leq 2^n$. Since any continuous function can be uniformly approximated by step functions whose jumps are at the dyadic rationals, linear combinations of the Haar functions are dense in the set of continuous functions, which in turn is dense in $L^2([0, 1])$.

Let $\psi_{ij}(t) = \int_0^t \varphi_{ij}(s)ds$. Let Y_{ij} be a sequence of independent identically distributed mean zero Gaussian random variables with variance 1. Let

$$V_0(t) = Y_{00}\psi_{00}(t), \qquad V_i(t) = \sum_{j=1}^{2^{i-1}} Y_{ij}\psi_{ij}(t), \qquad i \geq 1.$$

(2.4) Theorem. $\sum_{i=0}^{\infty} V_i(t)$ converges uniformly in t, a.s. If we call the sum X_t, then X_t is a Brownian motion started at 0.

Proof. We start with the proof of convergence. We will show

$$\sum_{i=1}^{\infty} \mathbb{P}(|V_i(t)| > i^{-2} \text{ for some } t \in [0, 1]) < \infty.$$

Then by the Borel–Cantelli lemma (Proposition 1.1), with probability 1, from some i on (depending on ω), $\sup_t |V_i(t)| \leq i^{-2}$. This will show $\sum_{i=0}^{I} V_i(t)$ converges as $I \to \infty$, uniformly over $t \in [0, 1]$. Moreover, since each $\psi_{ij}(t)$ is continuous in t, then so is each $V_i(t)$, and we thus deduce that X_t is continuous in t.

Now for $i \geq 1$, each ψ_{ij} has a graph that looks like a little tent, and for fixed i, $\psi_{ij_1}(t)$ and $\psi_{ij_2}(t)$ are not both nonzero if $j_1 \neq j_2$. Also, the maximum value of ψ_{ij} is $2^{-(i+1)/2}$. Hence

$$\mathbb{P}(|V_i(t)| > i^{-2} \text{ for some } t \in [0,1])$$
$$\leq \mathbb{P}(|Y_{ij}|\psi_{ij}(t) > i^{-2} \text{ for some } t \in [0,1], \text{ some } 0 \leq j \leq 2^{i-1})$$
$$\leq \mathbb{P}(|Y_{ij}|2^{-(i+1)/2} > i^{-2} \text{ for some } 0 \leq j \leq 2^{i-1})$$
$$\leq (2^{i-1}+1)\mathbb{P}(|Z| > 2^{(i+1)/2}i^{-2})$$

where Z is a standard mean zero variance one Gaussian random variable. By Proposition 1.9, we bound the right-hand side by $2^i \exp(-2^i/i^4)$, which is easily checked to be summable in i.

This was the hard part. The V_js are Gaussian and independent, so the X_t are also Gaussian. It is obvious that they have mean zero. So we need only check the covariances and variances. If $f \in L^2([0,1])$, Parseval's identity says that $\langle f, f \rangle = \sum_{i,j}\langle \varphi_{ij}, f \rangle^2$. Since $\psi_{ij}(t) - \psi_{ij}(s) = \langle \varphi_{ij}, 1_{[s,t]} \rangle$, then $X_t - X_s = \sum_{i,j} Y_{ij}\langle \varphi_{ij}, 1_{[s,t]} \rangle$. Since the Y_{ij} are independent with mean 0 and variance 1,

$$\mathbb{E}[X_t - X_s]^2 = \mathbb{E}\left[\sum_{i,j} Y_{ij}\langle \varphi_{ij}, 1_{[s,t]}\rangle\right]^2$$
$$= \sum_{i,j}\langle \varphi_{ij}, 1_{[s,t]}\rangle^2$$
$$= \langle 1_{[s,t]}, 1_{[s,t]}\rangle = t - s.$$

Similarly, $f = \sum_{i,j}\langle \varphi_{ij}, f \rangle \varphi_{ij}$, hence $\langle f, g \rangle = \sum_{i,j}\langle \varphi_{ij}, f \rangle\langle \varphi_{ij}, g \rangle$, and therefore

$$\mathbb{E}[X_t - X_s)(X_r - X_q)]$$
$$= \mathbb{E}\left[\left(\sum_{i,j} Y_{ij}\langle \varphi_{ij}, 1_{[s,t]}\rangle\right)\left(\sum_{k,\ell} Y_{k\ell}\langle \varphi_{k\ell}, 1_{[q,r]}\rangle\right)\right]$$
$$= \sum_{i,j}\langle \varphi_{ij}, 1_{[s,t]}\rangle\langle \varphi_{ij}, 1_{[q,r]}\rangle$$
$$= \langle 1_{[s,t]}, 1_{[q,r]}\rangle.$$

If $r \leq s$, this last inner product is 0, and $X_t - X_s$ is a mean zero variance $t - s$ Gaussian random variable independent of $X_{r_{i+1}} - X_{r_i}$ if $r_1 \leq r_2 \leq \cdots \leq r_n \leq s$. This proves the theorem.   □

We have constructed Brownian motion on $[0,1]$, but we also want it on $[0,\infty)$. We obtain it through a trick called time inversion. Let $Y_t = tX_{1/t}$, $t \in [1,\infty)$. Clearly Y_t is mean 0, Gaussian, and continuous in t. Let us check the covariances of Y.

$$\text{Cov}\,(Y_t, Y_s) = st\text{Cov}\,(X_{1/s}, X_{1/t}) = st((1/t) \wedge (1/s)).$$

This equals $s \wedge t$. So Y_t is a Brownian motion for $t \geq 1$. Now let X_t' be an independent Brownian motion on $[0, 1]$, and let

$$Z_t = \begin{cases} X_t' & \text{if } t \leq 1; \\ X_1' + Y_t - Y_1 & \text{if } t > 1. \end{cases}$$

To check that Z_t is a Brownian motion is routine and is left to the reader.

Stopping times

We next want to talk about stopping times. Suppose we have a stochastic process X_t (think of Brownian motion as a good example) and a filtration $\mathcal{F}_t$. This is an increasing collection of σ-fields. We suppose each $\mathcal{F}_t$ is right continuous (i.e., $\mathcal{F}_t = \mathcal{F}_{t+}$ for each t, where $\mathcal{F}_{t+} = \bigcap_{\varepsilon>0} \mathcal{F}_{t+\varepsilon}$). We suppose that X_t is adapted to $\mathcal{F}_t$: for each t, X_t is $\mathcal{F}_t$ measurable.

(2.5) Definition. *A random mapping T from Ω to $[0, \infty)$ is called a stopping time if for each t, $(T < t) \in \mathcal{F}_t$.*

A stopping time is also called an optional time in the Markov theory literature.

The intuition is that the process knows whether T has happened by time t by looking at $\mathcal{F}_t$. Suppose some motorists are told to drive north on Highway 99 in Seattle and stop at the first motorcycle shop past the second realtor after the city limits. So they drive north, pass the city limits, pass two realtors, and come to the next motorcycle shop, and stop. That is a stopping time. If they are instead told to stop at the third stop light before the city limits (and they had not been there before), they would need to drive to the city limits, then turn around and return past three stop lights. That is not a stopping time, because they have to go ahead of where they wanted to stop to know to stop there.

(2.6) Proposition.
(a) T is a stopping time if and only if $(T \leq t) \in \mathcal{F}_t$ for all t.
(b) Fixed times t are stopping times.
(c) If S and T are stopping times, then so are $S \wedge T$ and $S \vee T$.
(d) If T_n is a nondecreasing sequence of stopping times, then so is $T = \sup_n T_n$.
(e) If T_n is a nonincreasing sequence of stopping times, then so is $T = \inf_n T_n$.
(f) If S is a stopping time, then so is $S + t$.

Proof. These follow from the definitions. We will do (a), leaving the rest as Exercise 8. If T is a stopping time, then $(T \leq t) = \cap_{n>N}(T < t + 1/n) \in$

$\cap_{n>N} \mathcal{F}_{t+1/n}$ for each N. So $(T \le t) \in \mathcal{F}_{t+} = \mathcal{F}_t$. On the other hand, if $(T \le t) \in \mathcal{F}_t$ for all t, then $(T < t) = \cup_{n=1}^{\infty} (T \le t - 1/n) \in \mathcal{F}_t$, since $\mathcal{F}_t$ is increasing.   □

Among the kinds of stopping times we will be interested in are the first times that X_t hits a set A or leaves a set A (the second is the same as hitting A^c). In the notation we will use throughout the rest of the book, let

$$(2.3) \qquad\qquad T_A = \inf\{t > 0 : X_t \in A\}$$

and

$$(2.4) \qquad\qquad \tau_A = \inf\{t > 0 : X_t \notin A\}.$$

Of course, $T_A = \tau_{A^c}$ and $\tau_A = T_{A^c}$.

It will turn out that T_A and τ_A are stopping times for all Borel measurable sets A, but that is a hard result, and we only need it in that generality for Chap. II. We leave the general result to Sect. II.2; here we only prove the result we need for the other chapters, namely, the cases when A is either open or closed.

(2.7) Proposition.
(a) If A is an open set, T_A is a stopping time.
(b) If A is a closed set, T_A is a stopping time.

Proof. Suppose A is open. If $T < t$, then for some s less than t, we have $X_s \in A$. Since the paths are continuous (right continuity would do), there is a rational $q < t$ with $X_q \in A$. Hence

$$(T < t) = \bigcup_{q < t, q \text{ rational}} (X_q \in A) \in \mathcal{F}_t,$$

which proves (a).

If A is closed, let $A_n = \{x : \text{dist}\,(x, A) < 1/n\}$. The A_ns are open, so T_{A_n} is an increasing sequence of stopping times, hence by Proposition 2.6, $T = \sup_n T_{A_n}$ is a stopping time. Since $T_A \ge T_{A_n}$ for each n, then $T_A \ge T$. If $T = \infty$, then T_A must also equal ∞. If $T < \infty$, by the continuity of paths, $X_T = \lim X(T_{A_n})$. Since $X(T_{A_n})$ is in the closure of A_m if $n \ge m$, it follows that X_T is also, for each m. That implies $X_T \in A$, or $T_A \le T$. In either case, $T_A = T$, which proves (b).   □

(2.8) Proposition. $\mathbb{P}^x(X_{\tau(B(x,r))} \in dy)$ *is normalized surface measure on* $\partial B(x, r)$.

Proof. By the rotational invariance of Brownian motion (Proposition 2.3), if C is any Borel subset of $\partial B(0, r)$ and A any orthogonal matrix,

$$\mathbb{P}^0(X(\tau_{B(0,r)}) \in C) = \mathbb{P}^0(AX(\tau_{B(0,r)}) \in C)$$
$$= \mathbb{P}^0(X(\tau_{B(0,r)}) \in A^{-1}C).$$

Therefore $\mathbb{P}^0(X(\tau_{B(0,r)}) \in dy)$ must be surface measure on $\partial B(0,r)$, normalized to have total mass 1. The result follows by translation invariance.

$\square$

It follows that if f is a function defined on the boundary of $B(x,r)$, then

$$\mathbb{E}^x f(X_{\tau(B(x,r))}) = \int_{\partial B(x,r)} f(y)\, \sigma(dy),$$

where σ is normalized surface measure on the boundary of $B(x,r)$. We will use this fact in the next chapter as part of our solution of the Dirichlet problem.

3 Markov properties

The Markov property

Before starting out on the Markov property and strong Markov properties, let us first introduce some notation. Define $\mathcal{F}_t^{00} = \sigma(X_s; s \le t)$, $t \in [0, \infty]$. We let $\mathcal{F}_t^0$ be the completion of $\mathcal{F}_t^{00}$, but we need to be careful what we mean by completion here, because we have more than one probability measure floating around. Let $\mathcal{F}_t^0$ be the σ-field generated by $\mathcal{F}_t^{00}$ and $\mathcal{N}$, where $\mathcal{N}$ is the collection of sets that are $\mathbb{P}^x$ null for every $x \in \mathbb{R}^d$. Finally, let $\mathcal{F}_t = \mathcal{F}_{t+}^0 = \cap_{\varepsilon > 0} \mathcal{F}_{t+\varepsilon}^0$. Ultimately, we will work only with $\mathcal{F}_t$, but we need the other two σ-fields at intermediate stages. The reason for worrying about which filtrations to use is that $\mathcal{F}_t^{00}$ is too small to include many interesting sets (such as those arising in the law of the iterated logarithm), while if the filtration is too large, the Markov property will not hold for that filtration.

We will denote expectation with respect to $\mathbb{P}^x$ by $\mathbb{E}^x$. Let us explain the notation $\mathbb{E}^{X_s} Y$. This is the random variable $\varphi(X_s)$, where $\varphi(y) = \mathbb{E}^y Y$. Thus, for example, $\mathbb{E}^x \mathbb{E}^{X_s} Y = \mathbb{E}^x \varphi(X_s)$.

Let

(3.1) $p(t,x,y) = (2\pi t)^{-d/2} e^{-|x-y|^2/2t}, \qquad x, y \in \mathbb{R}^d, \quad t > 0.$

The corresponding operator P_t on functions is given by

(3.2) $$P_t f(x) = \int f(y) p(t,x,y) dy.$$

Note that if f is bounded and $t > 0$, then $P_t f(x)$ is continuous in x by the continuity of $p(t,x,y)$ and dominated convergence.

If Ω is the set of continuous functions from $[0, \infty)$ to $\mathbb{R}^d$, we define the shift operators $\theta_t : \Omega \to \Omega$ by

$$\theta(\omega)(s) = \omega(t+s).$$

Then
$$X_s \circ \theta_t(\omega) = X_s(\theta_t \omega) = \theta_t \omega(s) = \omega(s+t) = X_{t+s}(\omega)$$

if the X_s process is given by coordinate maps as in (2.1). Even if we are not in this canonical setup, we will suppose there exist shift operators mapping Ω into itself so that

(3.3) $$X_s \circ \theta_t = X_{s+t}.$$

We can now state the Markov property, at least the one for $\mathcal{F}_t^{00}$.

(3.1) Proposition. *If Y is bounded and $\mathcal{F}_\infty^{00}$ measurable and $(\mathbb{P}^x, X_t)$ is a Brownian motion, then*

$$\mathbb{E}^x[Y \circ \theta_s | \mathcal{F}_s^{00}] = \mathbb{E}^{X_s} Y, \qquad \text{a.s. } (\mathbb{P}^x).$$

For the first time through for this and the next subsection, it would be a good idea to read the statements of the results and the examples but to skip the proofs.

Before proving this proposition, let us first give some examples. For example, if $Y = f(X_t)$, Proposition 3.1 says that

$$\mathbb{E}^x[f(X_{t+s})|\mathcal{F}_s^{00}] = \mathbb{E}^x[f(X_t) \circ \theta_s | \mathcal{F}_s^{00}] = \mathbb{E}^{X_s} f(X_t).$$

As another example,

$$\mathbb{P}(X_t \in [4,5] \text{ for some } t \in [3,4] \,|\, \mathcal{F}_2^{00}]$$
$$= \mathbb{P}^{X_2}[X_t \in [4,5] \text{ for some } t \in [1,2]).$$

We start the proof of Proposition 3.1 by proving a lemma, which is just the first example given above.

(3.2) Lemma. *If f is bounded and Borel measurable,*

$$\mathbb{E}^x[f(X_t) \circ \theta_s | \mathcal{F}_s^{00}] = \mathbb{E}^{X_s} f(X_t), \qquad \text{a.s. } (\mathbb{P}^x).$$

Proof. It will suffice to prove $\mathbb{E}^x[f(X_{t+s})|\mathcal{F}_s^{00}] = \mathbb{E}^{X_s} f(X_t)$ when $f(x) = e^{iu \cdot x}$, where $u \cdot x$ is the inner product in $\mathbb{R}^d$, for we can then take linear combinations and limits. Using independent increments, the fact that $X_{t+s} - X_s$ is a $\mathcal{N}(0,t)$, and (1.6), we have

$$\mathbb{E}^x\left[e^{iu \cdot X_{t+s}}|\mathcal{F}_s^{00}\right] = \mathbb{E}^x\left[e^{iu \cdot (X_{t+s}-X_s)}|\mathcal{F}_s^{00}\right]e^{iu \cdot X_s}$$
$$= \mathbb{E}^x\left[e^{iu \cdot (X_{t+s}-X_s)}\right]e^{iu \cdot X_s}$$
$$= e^{-|u|^2 t/2}e^{iu \cdot X_s}.$$

On the other hand, for any y,

$$\mathbb{E}^{y}e^{iu\cdot X_t} = \mathbb{E}^{0}e^{iu\cdot X_t}e^{iu\cdot y} = e^{-|u|^2 t/2}e^{iu\cdot y}.$$

Replacing y by X_s gives the desired equality. $\qquad\square$

Proof of Proposition 3.1. By linearity and limits, it suffices to show the equality when $Y = \prod_{i=1}^{n} f_i(X_{t_i})$, where the f_i are bounded and $s \leq t_1 \leq \ldots \leq t_n$. We will prove it for such Y by induction on n. The case $n = 1$ was Lemma 3.2, so we suppose the equality for n and prove it for $n+1$.

Let $V = \prod_{j=2}^{n+1} f_j(X_{t_j-t_1})$, $h(y) = \mathbb{E}^{y}V$. By the induction hypothesis,

$$\mathbb{E}^{x}\left[\prod_{j=1}^{n+1} f_j(X_{t_j})\Big|\mathcal{F}_s^{00}\right] = \mathbb{E}^{x}\left[\mathbb{E}^{x}[V\circ\theta_{t_1}|\mathcal{F}_{t_1}^{00}]f_1(X_{t_1})\Big|\mathcal{F}_s^{00}\right]$$

$$= \mathbb{E}^{x}\left[(\mathbb{E}^{X_{t_1}}V)f_1(X_{t_1})|\mathcal{F}_s^{00}\right]$$

$$= \mathbb{E}^{x}[(hf_1)(X_{t_1})|\mathcal{F}_s^{00}].$$

By Lemma 3.2, this is $\mathbb{E}^{X_s}[(hf_1)(X_{t_1-s})]$. For any y,

$$\mathbb{E}^{y}[(hf_1)(X_{t_1-s})] = \mathbb{E}^{y}[(\mathbb{E}^{X_{t_1-s}}V)f_1(X_{t_1-s})]$$

$$= \mathbb{E}^{y}\left[\mathbb{E}^{y}[V\circ\theta_{t_1-s}|\mathcal{F}_{t_1-s}^{00}]f_1(X_{t_1-s})\right]$$

$$= \mathbb{E}^{y}[(V\circ\theta_{t_1-s})f_1(X_{t_1-s})].$$

If we replace V by its definition, replace y by X_s, and use the definition of θ_{t_1-s}, we get the desired equality for $n+1$ and hence the induction step and the proposition. $\qquad\square$

A moment's thought shows that adding null sets does not hurt anything, so we have the following corollary.

(3.3) Corollary. *If Y is bounded and is $\mathcal{F}_\infty^0$ measurable, then*

$$\mathbb{E}^{x}[Y\circ\theta_t|\mathcal{F}_t^0] = \mathbb{E}^{X_t}Y, \qquad \text{a.s. } (\mathbb{P}^x).$$

The next step is to go from $\mathcal{F}_t^0$ to $\mathcal{F}_t$.

(3.4) Theorem. *If Y is bounded and $\mathcal{F}_\infty$ measurable and $(\mathbb{P}^x, X_t)$ is a Brownian motion,*

$$\mathbb{E}^{x}[Y\circ\theta_t|\mathcal{F}_t] = \mathbb{E}^{X_t}Y.$$

Proof. Just as in the proof of Proposition 3.1, it suffices to prove the result for Y of the form $f(X_t)$ and then use induction. By a limiting process, we may also assume f is continuous.

If $A \in \mathcal{F}_s = \mathcal{F}_{s+}^0$, then $A \in \mathcal{F}_{s+\varepsilon}^0$ for every $\varepsilon > 0$. So by the Markov property with respect to $\mathcal{F}_{s+\varepsilon}^0$,

$$\mathbb{E}^x[f(X_{t+s+\varepsilon}); A] = \mathbb{E}^x[P_t f(X_{s+\varepsilon}); A].$$

If we let $\varepsilon \to 0$, the left-hand side converges to $\mathbb{E}^x[f(X_{t+s}); A]$ by dominated convergence and the continuity of f and X. As we noted following (3.2), $P_t f$ is continuous. The right-hand side converges to $\mathbb{E}^x[P_t f(X_s); A]$ by dominated convergence and the continuity of $P_t f$ and X. Sorting out the notation, we see we have exactly what we wanted. □

Zero–one laws

We show that going from $\mathcal{F}_s^0$ to $\mathcal{F}_s$ did not really change anything essential.

(3.5) Proposition. $\mathcal{F}_s = \mathcal{F}_s^0$.

Proof. Let $Y = \prod_{j=1}^n f_j(X_{t_j})$. If $t_j \leq s < t_{j+1}$, let $Y_1 = \prod_{\{j : t_j \leq s\}} f_j(X_{t_j})$ and $Y_2 = \prod_{\{j : t_j > s\}} f_j(X_{t_j})$. Then

$$\mathbb{E}^x[Y | \mathcal{F}_{s+}^0] = Y_1 \mathbb{E}^{X_s} Y_2,$$

which is $\mathcal{F}_s^0$ measurable.

By linearity and limits, $\mathbb{E}^x[Y | \mathcal{F}_{s+}^0]$ is $\mathcal{F}_s^0$ measurable whenever $Y \in \mathcal{F}_\infty^0$. If $A \in \mathcal{F}_s = \mathcal{F}_{s+}^0$, letting $Y = 1_A$ shows that $A \in \mathcal{F}_s^0$. □

(3.6) Corollary. (Blumenthal zero-one law) *If $A \in \mathcal{F}_0$, then $\mathbb{P}^x(A)$ equals either 0 or 1.*

Proof. If $A \in \mathcal{F}_0$, then under $\mathbb{P}^x$,

$$1_A = \mathbb{E}^x[1_A | \mathcal{F}_{0+}] = \mathbb{E}^{X_0} 1_A = \mathbb{P}^x(A).$$

Hence $\mathbb{P}^x(A) = 1_A \in \{0, 1\}$. □

As an example of the use of the zero-one law, we show $\mathbb{P}^0(T_{(0,\infty)} = 0) = 1$, i.e., Brownian motion immediately enters the positive real axis. Since by symmetry it also enters the negative reals immediately, the behavior of Brownian motion is rather peculiar. The explanation is that Brownian motion oscillates between positive and negative values in every neighborhood of the origin on the time axis. For a more precise statement, see the law of the iterated logarithm (Exercise 4). To show the assertion about $T_{(0,\infty)}$, for any t, $\mathbb{P}^0(T_{(0,\infty)} \leq t) \geq \mathbb{P}^0(X_t > 0)$, which by the symmetry of the $\mathcal{N}(0, t)$ distribution is $1/2$. Letting $t \downarrow 0$, $\mathbb{P}^0(T_{(0,\infty)} = 0) \geq 1/2$. By the zero-one law the probability, then, must be one.

As another example, let $\varphi(t)$ be any nondecreasing function with $\varphi(0) = 0$. Then the event $(\limsup_{t \to 0} X_t / \varphi(t) > a)$ is in $\mathcal{F}_0$ and hence has probability 0 or 1 for each a. Therefore $\limsup_{t \to 0} X_t / \varphi(t)$ is constant, a.s. (the constant might be 0 or ∞).

Semigroups

The Markov property says that for Brownian motion $\mathbb{E}^x[\mathbb{E}^{X_s}f(X_t)] = \mathbb{E}^x f(X_{s+t})$. This can be rewritten as $P_{s+t}f(x) = P_s(P_t f)(x)$ or

(3.4) $$P_{s+t} = P_s P_t.$$

Hence the operators form a semigroup (in the functional analysis sense). Since the paths of Brownian motion are continuous, it is easy to see that the semigroup is strongly continuous with respect to the space of continuous functions (that is, $P_t f \to P_s f$ uniformly if f is continuous and $t \downarrow s$). Given such a semigroup, it is common to compute the infinitesimal generator

$$\lim_{t \downarrow 0} \frac{P_t f - f}{t}$$

for some collection of fs and where the limit is in some appropriate norm. In Exercise 6, we show that the infinitesimal generator of Brownian motion in $\mathbb{R}^d$ has the bounded C^2 functions as its domain, and on such functions it is $(1/2)\Delta$, one half the Laplacian.

If we let $U^\lambda f(x) = \int_0^\infty e^{-\lambda t} P_t f(x)dt$, then

$$U^\lambda U^\beta f(x) = \int_0^\infty e^{-\lambda t} \int_0^\infty e^{-\beta s} P_t P_s f(x)\, ds\, dt.$$

Using the semigroup property and a change of variables, this is

$$\int_0^\infty e^{(\beta - \lambda)t} \int_t^\infty e^{-\beta s} P_s f(x)\, ds\, dt$$
$$= \int_0^\infty P_s f(x) e^{-\beta s} \int_0^s e^{(\beta - \lambda)t}\, dt\, ds$$
$$= \frac{U^\lambda f(x) - U^\beta f(x)}{\beta - \lambda}.$$

The equation

(3.5) $$(\beta - \lambda)U^\lambda U^\beta = U^\lambda - U^\beta$$

is called the resolvent equation.

Suppose f is smooth, say C^∞ with compact support. By translation invariance, $\partial P_t f(x)/\partial x^i = P_t(\partial f/\partial x^i)$, or $P_t f$ is also smooth, hence so is $U^\lambda f$. Let g be C^∞ with compact support.

Applying Green's first identity (see Sect. II.1) on $B(0, N)$, and letting $N \to \infty$, we see

$$-(1/2)\int_{\mathbb{R}^d} g(x)(\Delta U^\lambda f(x))dx = (1/2)\int_{\mathbb{R}^d} \nabla U^\lambda f(x) \cdot \nabla g(x)dx.$$

If we write $\mathcal{E}(g, h)$ for $1/2 \int_{\mathbb{R}^d} \nabla g(x) \cdot \nabla h(x)dx$, we have

(3.6) $$\mathcal{E}(g, U^\lambda f) = -(1/2) \int_{\mathbb{R}^d} g(x)(\Delta U^\lambda f(x))dx.$$

$\mathcal{E}$ is an example of a Dirichlet form and (3.6) shows that the Dirichlet form corresponding to Brownian motion is $\mathcal{E}(f, f) = (1/2) \int |\nabla f|^2(x)dx$. One can show that the domain of $\mathcal{E}$ is $\{f : \int |\nabla f|^2(x)dx < \infty, f \to 0 \text{ as } |x| \to \infty\}$.

Strong Markov property

In the previous section we defined stopping times. Given a stopping time T, the σ–field of events known up to time T is defined to be

(3.7) $$\mathcal{F}_T = \{A \in \mathcal{F}_\infty : A \cap (T \leq t) \in \mathcal{F}_t \text{ for all } t > 0\}$$

(see Exercise 9). We define θ_T by $\theta_T(\omega)(t) = \omega(T(\omega) + t)$. So, for example, $X_t \circ \theta_T(\omega) = X_{T(\omega)+t}(\omega)$ and $X_T(\omega) = X_{T(\omega)}(\omega)$.

Now that we have these definitions (admittedly a bit opaque at this stage—be patient until we reach the examples), we can state the strong Markov property. On the first reading, skip the proof of Theorem 3.7.

(3.7) Theorem. *If Y is bounded and $\mathcal{F}_\infty$ measurable and $(\mathbb{P}^x, X_t)$ is a Brownian motion, then*

$$\mathbb{E}^x[Y \circ \theta_T | \mathcal{F}_T] = \mathbb{E}^{X_T} Y, \qquad a.s. \quad on \ (T < \infty).$$

Proof. Following the proofs of Proposition 3.1 and Theorem 3.4, it is enough to prove

$$\mathbb{E}^x[f(X_{T+t})|\mathcal{F}_T] = \mathbb{E}^{X_T} f(X_t)$$

for f bounded and continuous. Define T_n by $T_n(\omega) = k/2^n$ if $T(\omega) \in [(k-1)/2^n, k/2^n)$. Then T_n are stopping times decreasing to T on the set $(T < \infty)$.

If $A \in \mathcal{F}_T$, then $A \in \mathcal{F}_{T_n}$. Therefore $A \cap (T_n = k/2^n) \in \mathcal{F}_{k/2^n}$ and we have

$$\begin{aligned}
\mathbb{E}^x[f(X_{T_n+t}); A \cap (T_n = k/2^n)] &= \mathbb{E}^x[f(X_{t+k/2^n}); A \cap (T = k/2^n)] \\
&= \mathbb{E}^x[\mathbb{E}^{X_{k/2^n}} f(X_t); A \cap (T_n = k/2^n)] \\
&= \mathbb{E}^x[\mathbb{E}^{X_{T_n}} f(X_t); A \cap (T_n = k/2^n)].
\end{aligned}$$

Then

$$\begin{aligned}
\mathbb{E}^x[f(X_{T_n+t}); A \cap (T < \infty)] &= \sum_{k=1}^\infty \mathbb{E}^x[f(X_{T_n+t}); A \cap (T_n = k/2^n)] \\
&= \sum_{k=1}^\infty \mathbb{E}^x[\mathbb{E}^{X_{T_n}} f(X_t); A \cap (T_n = k/2^n)] \\
&= \mathbb{E}^x[\mathbb{E}^{X_{T_n}} f(X_t); A \cap (T < \infty)].
\end{aligned}$$

Now let $n \to \infty$. $\mathbb{E}^x[f(X_{T_n+t}); A \cap (T < \infty)] \to \mathbb{E}^x[f(X_T); A \cap (T < \infty)]$ by dominated convergence and the continuity of f and X_t. On the other hand, $\mathbb{E}^{X_{T_n}} f(X_t) = P_t f(X_{T_n}) \to P_t f(X_T) = \mathbb{E}^{X_T} f(X_t)$ since $P_t f$ is, as we saw following (3.2), a continuous function. $\qquad \square$

The first example we give is the key to the probabilistic solution to the Dirichlet problem. Let D be an open domain, $\tau_D = \inf\{t : X_t \notin D\}$, f a bounded function on ∂D, and $u(x) = \mathbb{E}^x f(X_{\tau_D})$. Let $x \in D$, $\delta <$ dist $(x, \partial D)$, and $S = \inf\{t : X_t \notin B(x, \delta)\}$. If ω is a continuous curve starting at x and going until the boundary of D, then $\theta_S(\omega)$ is the portion of that same curve starting from the first time ω hits the ball of radius δ about x and continuing until the boundary of D. So the location where the longer curve and shorter piece of the curve hit the boundary of D will be the same place, and hence $X_{\tau_D} \circ \theta_S = X_{\tau_D}$. (See Fig. 3.1.) Although we do not need it here, the time for the longer curve to hit the boundary is equal to the time for the curve to first hit the ball plus the time for the shorter piece of the curve to go from the ball to the boundary of D, or $\tau_D = S + \tau_D \circ \theta_S$. We then have

$$(3.8) \qquad u(x) = \mathbb{E}^x f(X_{\tau_D}) = \mathbb{E}^x \big[\mathbb{E}^x[f(X_{\tau_D}) \circ \theta_S | \mathcal{F}_S] \big]$$
$$= \mathbb{E}^x \mathbb{E}^{X_S} f(X_{\tau_D}) = \mathbb{E}^x u(X_S).$$

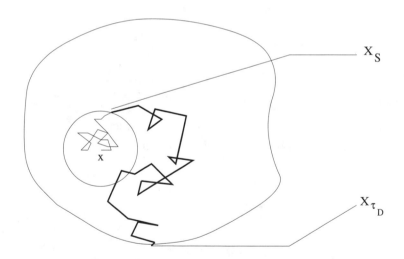

FIGURE 3.1. Graph of Brownian motion until exiting D.

(To anticipate Chap. II slightly, the connection with the Dirichlet problem is that since the distribution of X_S is uniform on the surface of $B(x, \delta)$, (3.8) says that $u(x)$ is the average of its values on the surfaces of small balls centered at x. This implies that u is harmonic.)

For a second example, let D be a domain that has the property that there exists $c > 0$ such that

$$\inf_{x \in D} \mathbb{P}^x(X(\tau_{B(x,2)}) \notin D) \geq c.$$

The strip $[0,1] \times \mathbb{R}^{d-1}$ has this property, for example, because the $\mathbb{P}^x$ law of $X(\tau_{B(x,2)})$ is uniform on $\partial B(x,2)$. Let S_1 be the first time X_t moves a distance more than 2 from its starting point, S_2 the first time after S_1 that X_t moves a distance at least 2 from X_{S_1}, and so on. We can write $S_1 = \inf\{t > 0 : |X_t - X_0| \geq 2\}$ and $S_2 = \inf\{t > S_1 : |X_t - X_{S_1}| \geq 2\}$, and so on, but a more compact way to write this is to define S_1 as we just did and then define $S_2 = S_1 + S_1 \circ \theta_{S_1}$, $S_3 = S_2 + S_1 \circ \theta_{S_2}$, etc. To see this, note that the length of time $S_2 - S_1$ for the X_t curve to move a distance 2 is the same as the time for $X_t \circ \theta_{S_1}$, the portion of the curve with the first S_1 units discarded, to move a distance 2.

The probability that the process can move a distance 2 at least k times before exiting D can be bounded by

$$\mathbb{P}^x(S_k < \tau_D) \leq \mathbb{P}^x(S_{k-1} < \tau_D, S_1 \circ \theta_{S_{k-1}} < \tau_D \circ \theta_{S_{k-1}})$$
$$= \mathbb{E}^x\big[\mathbb{P}^x(S_1 \circ \theta_{S_{k-1}} < \tau_D \circ \theta_{S_{k-1}}|\mathcal{F}_{S_{k-1}}); S_{k-1} < \tau_D\big]$$

since on $(S_{k-1} < \tau_D)$, $\tau_D = S_{k-1} + \tau_D \circ \theta_{S_{k-1}}$ as in the first example, and $(S_{k-1} < \tau_D)$ is $\mathcal{F}_{S_{k-1}}$ measurable.

By the strong Markov property and the fact that $X_{S_{k-1}} \in D$ on $(S_{k-1} < \tau_D)$,

$$\mathbb{P}^x(S_k < \tau_D) \leq \mathbb{E}^x\big[\mathbb{P}^{X(S_{k-1})}(S_1 < \tau_D); S_{k-1} < \tau_D\big]$$
$$\leq (1-c)\mathbb{P}^x(S_{k-1} < \tau_D).$$

Induction then gives $\mathbb{P}^x(S_k < \tau_D) \leq (1-c)^k$.

A third example is to let $f : B(0,R) \to [0,\infty)$, let $S = \tau_{B(0,R)}$, and let $A = \sup_{x \in B(0,R)} \mathbb{E}^x \int_0^S f(X_s)\,ds$. Assume $A < \infty$. We want to show

$$(3.9) \qquad \sup_{x \in B(0,R)} \mathbb{P}^x\bigg(\int_0^S f(X_s)ds \geq 2kA\bigg) \leq 2^{-k}.$$

Let $U_1 = \inf\{t : \int_0^{t \wedge S} f(X_s)ds \geq 2A\}$, and again $U_{i+1} = U_i + U_1 \circ \theta_{U_i}$. The event $\mathbb{P}^x(\int_0^S f(X_s)ds \geq 2kA)$ is bounded by

$$\mathbb{P}^x(U_k \leq S) \leq \mathbb{P}^x(U_{k-1} \leq S, U_1 \circ \theta_{U_{k-1}} \leq S \circ \theta_{U_{k-1}})$$
$$= \mathbb{E}^x\big[\mathbb{P}^x(U_1 \circ \theta_{U_{k-1}} \leq S \circ \theta_{U_{k-1}}|\mathcal{F}_{U_{k-1}}); U_{k-1} \leq S\big]$$
$$= \mathbb{E}^x\big[\mathbb{P}^{X(U_{k-1})}(U_1 \leq S); U_{k-1} \leq S\big].$$

If $U_{k-1} \leq S$, then $X_{U_{k-1}} \in B(0,R)$. If $y \in B(0,R)$,

$$\mathbb{P}^y(U_1 \le S) \le \mathbb{P}^y\left(\int_0^S f(X_s)ds \ge 2A\right) \le \frac{\mathbb{E}^y \int_0^S f(X_s)ds}{2A} \le 1/2$$

by Chebyshev's inequality. So then

$$\mathbb{P}^x(U_k \le S) \le (1/2)\mathbb{P}^x(U_{k-1} \le S)$$

and (3.9) follows by induction.

A fourth example: Suppose A and B are closed sets and $\inf_{x \in A} \mathbb{P}^x(T_B < \infty) \ge c$. (Since d-dimensional Brownian motion tends to ∞ as $t \to \infty$ if $d \ge 3$, it is possible $\sup_{x \in A} \mathbb{P}^x(T_B < \infty) < 1$.) Then for any y,

$$\mathbb{P}^y(T_B < \infty) \ge c\mathbb{P}^y(T_A < \infty).$$

This follows because if $T_A < T_B$, then $T_B = T_A + T_B \circ \theta_{T_A}$. Since $X_{T_A} \in A$,

$$\begin{aligned}
\mathbb{P}^y(T_B < \infty) &\ge \mathbb{P}^y(T_A < T_B < \infty) \ge \mathbb{P}^y(T_A < \infty, T_B \circ \theta_{T_A} < \infty) \\
&= \mathbb{E}^y\left[\mathbb{P}^y(T_B \circ \theta_{T_A} < \infty | \mathcal{F}_{T_A}); T_A < \infty\right] \\
&= \mathbb{E}^y\left[\mathbb{P}^{X(T_A)}(T_B < \infty); T_A < \infty\right] \\
&\ge c\mathbb{P}^y(T_A < \infty).
\end{aligned}$$

Reflection principle

Let X_t be one-dimensional Brownian motion and $M_t = \sup_{s \le t} X_s$. Another use of the Markov property is to calculate the joint law of X_t and M_t.

(3.8) Theorem. *If* $a \ge b$,

(3.10) $$\mathbb{P}^0(M_t \ge a, X_t < b) = \mathbb{P}^0(X_t > 2a - b)$$

and

(3.11) $$\mathbb{P}^0(M_t \ge a) = 2\mathbb{P}^0(X_t > a).$$

Proof. Let $S_a = \inf\{s : X_s \ge a\}$. Then

(3.12) $$\begin{aligned}
\mathbb{P}^0(M_t \ge a, X_t < b) &= \mathbb{P}^0(S_a \le t, X_t < b) \\
&= \int_0^t \mathbb{P}^0(X_t < b, S_a \in ds) \\
&= \int_0^t \mathbb{P}^0(X_t - X_s < b - a, S_a \in ds).
\end{aligned}$$

We condition on $\mathcal{F}_s$. Using the independence of $X_t - X_s$ from $\mathcal{F}_s$ and the fact that the distribution of $X_t - X_s$ is the same as that of $-(X_t - X_s)$, this is

$$(3.13) \qquad \int_0^t \mathbb{P}^0(X_s - X_t < b - a, S_a \in ds)$$

$$= \int_0^t \mathbb{P}^0(X_t > 2a - b, S_a \in ds)$$

$$= \mathbb{P}^0(X_t > 2a - b, M_t \geq a).$$

If $X_t > 2a - b$, then $M_t \geq a$, and we have (3.10).

To obtain (3.11), set $a = b$:

$$\mathbb{P}^0(M_t \geq a, X_t < a) = \mathbb{P}^0(X_t > a).$$

Since $\mathbb{P}^0(M_t \geq a, X_t > a) = \mathbb{P}^0(X_t > a)$ (if the value of the Brownian motion at time t is larger than a, the maximum must automatically also be larger than a), summing gives (3.11). □

Killed processes

We will often want to consider Brownian motion only up until the time τ_D of exiting some set D, and there is a rather morbid terminology associated with this. Let $\zeta = \tau_D$. ζ is called the "lifetime" of the process; one defines a "cemetery" state by adding on an isolated point Δ; one speaks of "killing" the process at time ζ; and the killed process $\widehat{X}_t$ is defined by

$$(3.14) \qquad \widehat{X}_t = \begin{cases} X_t & t < \zeta; \\ \Delta & t \geq \zeta. \end{cases}$$

Every function f is extended to be 0 at Δ. The lifetime need not be the time of exiting a set. Another common occurrence is to let $\zeta = S$, where S is a random variable independent of X_t with an exponential distribution with parameter λ, i.e., $\mathbb{P}(S > t) = e^{-\lambda t}$. The crucial property of ζ is that it be a terminal time:

$$(3.15) \qquad \zeta = s + \zeta \circ \theta_s \qquad \text{if } s < \zeta.$$

(3.9) Proposition. *If (3.15) holds, $(\mathbb{P}^x, \widehat{X}_t)$ satisfies the Markov and strong Markov properties.*

Proof. As in Proposition 3.1 and Theorem 3.4, we need to show

$$\mathbb{E}^x[f(\widehat{X}_t) \circ \theta_T | \mathcal{F}_T] = \mathbb{E}^{\widehat{X}_T} f(\widehat{X}_t), \qquad \text{a.s. } (\mathbb{P}^x).$$

If $A \in \mathcal{F}_T$,

$$\mathbb{E}^x[f(\widehat{X}_t) \circ \theta_T; A] = \mathbb{E}^x[f(X_{t+T}); A \cap (T + t < \zeta)].$$

On the other hand,

$$\mathbb{E}^{\widehat{X}_T}f(\widehat{X}_t) = \mathbb{E}^{X_T}[f(X_t); t < \zeta]1_{(T<\zeta)}$$
$$= \mathbb{E}^x[f(X_t) \circ \theta_T; t \circ \theta_T < \zeta \circ \theta_T | \mathcal{F}_T]1_{(T<\zeta)}$$
$$= \mathbb{E}^x[f(X_{t+T}); T + t \circ \theta_T < T + \zeta \circ \theta_T, T < \zeta | \mathcal{F}_T]$$
$$= \mathbb{E}^x[f(X_{t+T}); T + t < \zeta | \mathcal{F}_T],$$

since $T + t \circ \theta_T = T + t$ and $T + \zeta \circ \theta_T = \zeta$ on $(T < \zeta)$. Hence $\mathbb{E}^x[\mathbb{E}^{\widehat{X}_T}f(\widehat{X}_t); A] = \mathbb{E}^x[f(X_{t+T}); (T + t < \zeta) \cap A]$, as required. $\square$

Second construction of Brownian motion

It is sometimes useful to be able to start with operators P_t satisfying (3.4) and construct a process, or even to start with the infinitesimal generator or Dirichlet form to get a process. There is a well-developed procedure for going from the infinitesimal generator to P_t centered around the Hille-Yosida theorem (see Loève [1]). To see how to obtain P_t from Dirichlet forms, see Fukushima [1]. Because some of the techniques will be useful later, we show how to construct Brownian motion given the P_t. (This subsection could be safely omitted on the first reading.) There are two main theorems involved.

(3.10) Theorem. (Kolmogorov consistency theorem) *Suppose for each n we have a probability measure μ_n on $\mathbb{R}^n$. Suppose the μ_n are consistent: if $A \subseteq \mathbb{R}^n$, then $\mu_{n+1}(A \times \mathbb{R}) = \mu_n(A)$. Then there exists a probability measure μ on $\mathbb{R}^\infty$ such that $\mu(A \times \mathbb{R}^\infty) = \mu_n(A)$ for all $A \subseteq \mathbb{R}^n$.*

We use the σ-field on $\mathbb{R}^\infty$ generated by the cylindrical sets $\{A \times \mathbb{R}^\infty, A$ a Borel subset of $\mathbb{R}^n$ for some $n\}$.

Proof. Define μ on cylindrical sets by $\mu(A \times \mathbb{R}^\infty) = \mu_n(A)$ if $A \subseteq \mathbb{R}^n$. By the consistency assumption, μ is well defined. By the Carathéodory extension theorem, we can extend μ to the σ-field generated by the cylindrical sets provided we show:

(3.16) *If A_n are cylindrical sets decreasing to $\emptyset$, then $\mu(A_n) \to 0$.*

Suppose this did not hold. Then we would have cylindrical sets $A_n \downarrow \emptyset$ with $\mu(A_n) > \varepsilon > 0$. We will obtain a contradiction. Let $A_1' = A_2' = \cdots = A_{i_1}' = \mathbb{R}^\infty$, $A_{i_1+1}' = \cdots = A_{i_2}' = A_1$, $A_{i_2+1}' = \cdots = A_{i_3}' = A_2$, etc., where the $i_1, i_2, \ldots$ are chosen large enough so that for each n, $A_n' = \widetilde{A}_{m_n} \times \mathbb{R}^\infty$ for some $\widetilde{A}_{m_n} \subseteq \mathbb{R}^{m_n}$ and $m_n \leq n$. There is no loss of generality in working with the A_n' sequence instead of the A_n sequence, so we do that and drop the primes. Since $A_n = \widetilde{A}_{m_n} \times \mathbb{R} \times \cdots \times \mathbb{R} \times \mathbb{R}^\infty$, where the factor $\mathbb{R}$ is repeated $n - m_n$ times, we may assume each $A_n = \widetilde{A}_n \times \mathbb{R}^\infty$, where $\widetilde{A}_n \subseteq \mathbb{R}^n$.

For each n, choose $\tilde{B}_n \subseteq \tilde{A}_n$ so that $\tilde{B}_n$ is compact and $\mu(\tilde{A}_n - \tilde{B}_n) \leq \varepsilon/2^{n+1}$. Let $B_n = \tilde{B}_n \times \mathbb{R}^\infty$ and let $C_n = B_1 \cap \ldots \cap B_n$. Hence $C_n \subseteq B_n \subseteq A_n$, and $C_n \downarrow \emptyset$, but

$$\mu(C_n) \geq \mu(A_n) - \sum_{i=1}^{n} \mu(A_n - B_n) \geq \varepsilon/2,$$

and the projection of C_n onto $\mathbb{R}^n$, say $\tilde{C}_n$, is compact.

We will find $x = (x^1, \ldots, x^n, \ldots) \in \cap_n C_n$ and obtain our contradiction. Let y_n be any point in C_n. The first coordinates of y_n, namely, y_n^1, form a sequence contained in $\tilde{C}_1$, which is compact, hence there is a convergent subsequence. Let x^1 be the limit point. The first and second coordinates of the y_ns in this convergent sequence form a sequence contained in the compact set $\tilde{C}_2$, so a further subsequence converges, say to $x' \in \tilde{C}_2$. Since this second subsequence is a subsequence of the first, the first coordinate of x' is x^1. Suppose $x' = (x^1, x^2)$. We continue this procedure to obtain $x = (x^1, x^2, \ldots, x^n, \ldots)$. By our construction, $(x^1, \ldots, x^n) \in \tilde{C}_n$ for each n, hence $x \in C_n$ for each n, or $x \in \cap_n C_n$, a contradiction. $\square$

The other key theorem is Kolmogorov's continuity criterion. Let D denote the dyadic rationals in $[0, 1]$. Let $D_n = \{k/2^n : k \leq 2^n\}$, so that $D = \cup_n D_n$.

(3.11) Theorem. *Suppose there exist c, ε, and $p > 0$ such that*

$$\mathbb{E}\,|X_t - X_s|^p \leq c|t - s|^{1+\varepsilon}, \qquad s, t \in D.$$

Then with probability one, X_t is uniformly continuous on D.

Proof. Let $\lambda_n = 2^{-n\varepsilon/2p}$ and

$$A_n = \big\{|X_s - X_t| \geq \lambda_n \text{ for some } s, t \in D \text{ with } |t - s| \leq 2^{-n}\big\}.$$

We will show $\mathbb{P}(A_n) \leq c2^{-n\varepsilon/2}$. Then by the Borel-Cantelli lemma, we have $\mathbb{P}(A_n \text{ i.o.}) = 0$. This will show that, except for a null set, X_t is uniformly continuous on D.

Define $a(n, t) = k/2^n$ if $t \in [k/2^n, (k+1)/2^n)$. If $t \in D$ and $n \geq 0$, we can write

$$X_t = X_{a(n,t)} + [X_{a(n+1,t)} - X_{a(n,t)}] + [X_{a(n+2,t)} - X_{a(n+1,t)}] + \cdots,$$

where the sum is actually finite since $t \in D$, and hence $a(j, t) = t$ for j large enough. We have a similar expression for X_s. If $|t - s| \leq 2^{-n}$, then $|a(n, t) - a(n, s)| \leq 2^{-n}$.

Now if $|X_s - X_t| > \lambda_n$ for some $s, t \in D$, then either
(a) $|X_{a(n,t)} - X_{a(n,s)}| \geq \lambda_n/2$ for some $s, t \in D$ with $|s - t| \leq 2^{-n}$, or
(b) $|X_{a(n+i+1,t)} - X_{a(n+i,t)}| \geq \lambda_n/40(i+1)^2$ for some t and some i.

For possibility (a) to hold, we must have $|X_r - X_q| \geq \lambda_n/2$ for some $q, r \in D_n$ with $|q - r| \leq 2^{-n}$. There are at most 2^n pairs, so the probability of possibility (a) is bounded by

$$2^n \sup_{s \leq 1} \mathbb{P}(|X_{s+2^{-n}} - X_s| \geq \lambda_n/2) \leq \frac{2^n 2^p}{\lambda_n^p} \mathbb{E}\, |X_{s+2^{-n}} - X_s|^p$$

$$\leq \frac{c2^n}{\lambda_n^p}(2^{-n})^{1+\varepsilon} \leq \frac{c2^{-n\varepsilon}}{\lambda_n^p}.$$

For $|X_{a(n+i+1,t)} - X_{a(n+i,t)}|$ to be greater than $\lambda_n/40(i+1)^2$ for some t and some i, then for some i we must have $|X_r - X_q| \geq \lambda_n/40(i+1)^2$ for some $r \in D_{n+i}$, $q \in D_{n+i+1}$ with $|r - q| \leq 2^{-n-i}$. There are at most 2^{n+i+2} pairs, and so the probability of possibility (b) is bounded by

$$\sum_{i=0}^{\infty} 2^{n+i+2} \sup_s \mathbb{P}\left(|X_{s+2^{-(n+i)}} - X_s| \geq \frac{\lambda_n}{40(i+1)^2}\right)$$

$$\leq \sum_{i=0}^{\infty} \frac{2^{n+i+2}2^{-(n+i)(1+\varepsilon)}(40(i+1)^2)^p}{\lambda_n^p}$$

$$\leq \sum_{i=0}^{\infty} \frac{c2^{-n\varepsilon}2^{-i\varepsilon/2}}{\lambda_n^p} \leq c\frac{2^{-n\varepsilon}}{\lambda_n^p}.$$

Therefore the probability of A_n is bounded by $c2^{-n\varepsilon/2}$ as required. $\qquad\square$

The proof of Theorem 3.11 is an example of what is known as a metric entropy argument.

We use Theorems 3.10 and 3.11 to construct one-dimensional Brownian motion as follows. Let $t_1, t_2, \ldots$ be an enumeration of D. Let Ω be the set of functions from D to $\mathbb{R}$. Let $X_t(\omega) = \omega(t)$, $t \in D$. We define μ_n: fix n and let $s_1, s_2, \ldots, s_n$ be the permutation of $t_1, \ldots, t_n$ so that $s_1 < s_2 < \cdots < s_n$. Define

$$\mu_n(A_1 \times \cdots \times A_n)$$

$$= \int_{A_n} \cdots \int_{A_1} p(s_1, 0, x_1)p(s_2 - s_1, x_1, x_2) \times$$

$$\ldots \times p(s_n - s_{n-1}, x_{n-1}, x_n)dx_1 \cdots dx_n.$$

This says that μ_n is the law of a Gaussian random vector and so one can check that the conditions of Theorem 3.10 are satisfied. Let μ be the resulting measure on D.

If Z is $\mathcal{N}(0, 1)$, then $\mathbb{E}\, Z^4 < \infty$. (In fact it equals 3: expand e^{iuZ} in a power series, take expectations and compare to the power series of $e^{-|u|^2/2}$.) From the definition of μ_n, we see that $X_t - X_s$ is $\mathcal{N}(0, t-s)$ if $s, t \in D$, and so $\mathbb{E}\, |X_t - X_s|^4 = (t - s)^2 \mathbb{E}\, Z^4$. Then by Theorem 3.11, for almost every ω, X_t is uniformly continuous on D. Now define $X_u = \lim_{t \in D, t \downarrow u} X_t$ for

$u \notin D$. Exercise 1 asks the reader to fill in the proof that X_u is Brownian motion on $[0, 1]$.

4 Martingales

Definitions

In this section we consider martingales, both discrete-time ones and ones related to Brownian motion. Let $\mathcal{F}_n$ be an increasing sequence of σ-fields. A sequence of random variables M_n is adapted to $\mathcal{F}_n$ if for each n, M_n is $\mathcal{F}_n$ measurable. Similarly a collection of random variables M_t is adapted to $\mathcal{F}_t$ if each M_t is $\mathcal{F}_t$ measurable.

We say the filtration $\mathcal{F}_t$ satisfies the *usual conditions* if $\mathcal{F}_t$ is right continuous (i.e., $\mathcal{F}_t = \mathcal{F}_{t+}$ for all t, where $\mathcal{F}_{t+} = \cap_{\varepsilon>0}\mathcal{F}_{t+\varepsilon}$) and each $\mathcal{F}_t$ is complete (i.e., $\mathcal{F}_t$ contains all $\mathbb{P}$-null sets). For the most part, in this and the following sections there will only be one probability measure and we assume the filtration $\mathcal{F}_t$ satisfies the usual conditions. Where we do have a family $\{\mathbb{P}^x\}$ of probability measures, the reader can check that the $\mathcal{F}_t$ constructed in Sect. 3 contain all the null sets that are needed.

(4.1) Definition. *M_n is a martingale if M_n is adapted to $\mathcal{F}_n$, M_n is integrable for all n, and*

$$(4.1) \qquad \mathbb{E}[M_n|\mathcal{F}_{n-1}] = M_{n-1}, \qquad \text{a.s. ,} \qquad n = 2, 3, \ldots.$$

Similarly, M_t is a martingale if M_t is integrable for all t, $\mathcal{F}_t$ adapted, and

$$\mathbb{E}[M_t|\mathcal{F}_s] = M_s, \qquad \text{a.s. if } s \leq t.$$

If we have $\mathbb{E}[M_t|\mathcal{F}_s] \geq M_s$ a.s., for every $s \leq t$, then M_t is a submartingale. If we have $\mathbb{E}[M_t|\mathcal{F}_s] \leq M_s$, we have a supermartingale. Submartingales have a tendency to increase, which may make the name seem odd until one realizes that this is the same behavior one observes for subharmonic functions.

Let us take a moment to look at some examples. If M_t is Brownian motion, then it is a martingale, since $\mathbb{E}[M_t|\mathcal{F}_s] = M_s + \mathbb{E}[M_t - M_s|\mathcal{F}_s] = M_s + \mathbb{E}[M_t - M_s] = M_s$, using independent increments. If X_t is Brownian motion and $M_t = X_t^2 - t$, then

$$\mathbb{E}[X_t^2|\mathcal{F}_s] = \mathbb{E}[(X_t - X_s)^2|\mathcal{F}_s] + 2X_s\mathbb{E}[X_t|\mathcal{F}_s] - X_s^2 = t - s + X_s^2,$$

using independent increments and the fact that the distribution of $X_t - X_s$ is that of an $\mathcal{N}(0, t - s)$ distribution. It follows that M_t is a martingale. Similarly, $\exp(aX_t - a^2t/2)$ is a martingale for any complex number a.

If $X = (X^1, \ldots, X^d)$ is d-dimensional Brownian motion, then $X_t^i X_t^j$ is a martingale when $i \neq j$. Another example of a continuous-time martingale is $M_t = \mathbb{E}[Z | \mathcal{F}_t]$ when Z is an integrable random variable.

For an example of a discrete martingale, let $\Omega = [0, 1]$, $\mathbb{P}$ Lebesgue measure, and f an integrable function on $[0, 1]$. Let $\mathcal{F}_n$ be the σ–field generated by the sets $\{[k/2^n, (k+1)/2^n), k = 0, 1, \ldots, 2^n - 1\}$. Let $f_n = \mathbb{E}[f | \mathcal{F}_n]$. If I is an interval in $\mathcal{F}_n$, (1.8) shows that

$$(4.2) \qquad f_n(x) = \frac{1}{|I|} \int_I f(y) dy \qquad \text{if } x \in I.$$

f_n is a particular example of what is known as a dyadic martingale. Of course, $[0, 1]$ could be replaced by any interval as long as we normalize so that the total mass of the interval is 1. We could also divide cubes in $\mathbb{R}^d$ into 2^d subcubes at each step and define f_n analogously. Such martingales are called dyadic martingales. In fact, we could replace Lebesgue measure by any finite measure μ, and instead of decomposing into equal subcubes, we could use any nested partition of sets we like, provided none of these sets had μ measure 0.

Optional stopping

Note that if one takes expectations of (4.1), one has $\mathbb{E}\, M_n = \mathbb{E}\, M_{n-1}$, or by induction $\mathbb{E}\, M_n = \mathbb{E}\, M_0$. The theorem about martingales that lies at the basis of all other results is Doob's optional stopping theorem, which says that the same is true if we replace n by a stopping time N. There are various versions, depending on what conditions one puts on the stopping times. We will give a version for discrete martingales and one for continuous martingales that will suffice for our purposes.

(4.2) Theorem. *If N is a bounded stopping time with respect to $\mathcal{F}_n$ and M_n a martingale, then $\mathbb{E}\, M_N = \mathbb{E}\, M_0$.*

Proof. Since N is bounded, let K be the largest value N takes. We write

$$\mathbb{E}\, M_N = \sum_{k=0}^{K} \mathbb{E}[M_N; N = k] = \sum_{k=0}^{K} \mathbb{E}[M_k; N = k].$$

Note $(N = k)$ is $\mathcal{F}_j$ measurable if $j \geq k$, so

$$\mathbb{E}[M_k; N = k] = \mathbb{E}[M_{k+1}; N = k]$$
$$= \mathbb{E}[M_{k+2}; N = k] = \ldots = \mathbb{E}[M_K; N = k].$$

Hence

$$\mathbb{E}\, M_N = \sum_{k=0}^{K} \mathbb{E}[M_K; N = k] = \mathbb{E}\, M_K = \mathbb{E}\, M_0.$$

This completes the proof. □

The same proof gives the following corollary.

(4.3) Corollary. *If N is bounded by K and M_n is a submartingale, then $\mathbb{E}\, M_N \leq \mathbb{E}\, M_K$.*

The difficulties come in considering unbounded stopping times or in continuous time. Recall that a collection of random variables is uniformly integrable if $\mathbb{E}\,[|X_n|; |X_n| \geq a] \to 0$ as $a \to \infty$, uniformly in n. Take a moment to read over Exercise 21.

In Proposition 1.11 we gave a conditional expectation version of Jensen's inequality. If X_t is a martingale or nonnegative submartingale, φ is convex, and $\varphi(|X_t|)$ is integrable, then $\mathbb{E}\,[\varphi(|X_t|)|\mathcal{F}_s] \geq \varphi(|\mathbb{E}\,[X_t|\mathcal{F}_s]|) \geq \varphi(|X_s|)$ if $s \leq t$, or $\varphi(|X_t|)$ is a submartingale.

With this preliminary, we can now give the following.

(4.4) Theorem. *If M_t is a right continuous martingale and T is a stopping time bounded by K, then $\mathbb{E}\, M_T = \mathbb{E}\, M_K = \mathbb{E}\, M_0$. If M is a nonnegative submartingale, we have $\mathbb{E}\, M_T \leq \mathbb{E}\, M_K$.*

Proof. We do the martingale case; the other one is similar. Define the stopping times T_n by $T_n(\omega) = (k+1)K/2^n$ if $kK/2^n \leq T(\omega) < (k+1)K/2^n$. $M_{kK/2^n}$ is a discrete-time martingale with respect to $\mathcal{F}_{kK/2^n}$, so by Theorem 4.2 we have $\mathbb{E}\, M_{T_n} = \mathbb{E}\, M_K$ for each n. Since M_K is integrable, there exists a nonnegative increasing convex function φ with $\varphi(x)/x \to \infty$ as $x \to \infty$ such that $\mathbb{E}\, \varphi(|M_K|) < \infty$ (Exercise 21). Since $\varphi(|M_t|)$ is a submartingale, $\mathbb{E}\, \varphi(|M_{T_n}|) \leq \mathbb{E}\, \varphi(|M_K|) < \infty$, or by Exercise 21 again, the random variables $|M_{T_n}|$ are uniformly integrable. If we let $n \to \infty$, $T_n \downarrow T$, and by right continuity, we conclude $M_{T_n} \to M_T$. Using Exercise 21 one more time, $\mathbb{E}\, M_T = \mathbb{E}\, M_K$. □

(4.5) Corollary. *If $S \leq T$ are stopping times bounded by K and M is a right continuous martingale, then $\mathbb{E}\,[M_T|\mathcal{F}_S] = M_S$, a.s.*

Proof. Suppose $A \in \mathcal{F}_S$. We need to show $\mathbb{E}\,[M_S; A] = \mathbb{E}\,[M_T; A]$. Define a new stopping time U by

$$U(\omega) = \begin{cases} S(\omega) & \text{if } \omega \in A \\ T(\omega) & \text{if } \omega \notin A. \end{cases}$$

It is easy to check that U is a stopping time, so $\mathbb{E}\, M_U = \mathbb{E}\, M_K = \mathbb{E}\, M_T$ implies

$$\mathbb{E}\,[M_S; A] + \mathbb{E}\,[M_T; A^c] = \mathbb{E}\,[M_T].$$

Subtracting $\mathbb{E}\,[M_T; A^c]$ from each side completes the proof. □

Using this, we make the important observation that if S and T are bounded stopping times, then

(4.3) $\mathbb{E}\left[(M_T - M_S)^2 | \mathcal{F}_S\right] = \mathbb{E}\left[M_T^2 | \mathcal{F}_S\right] - 2M_S \mathbb{E}\left[M_T | \mathcal{F}_S\right] + M_S^2$
$$= \mathbb{E}\left[M_T^2 - M_S^2 | \mathcal{F}_S\right].$$

Taking expectations we obtain

(4.4) $$\mathbb{E}\left[(M_T - M_S)^2\right] = \mathbb{E}\, M_T^2 - \mathbb{E}\, M_S^2.$$

Doob's inequalities

The first interesting consequences of the optional stopping theorems are Doob's inequalities. If M_t or M_n are martingales, denote $M_t^* = \sup_{s \le t} |M_s|$, and similarly M_n^*.

(4.6) Theorem. *If M_n is a martingale,*

$$\mathbb{P}(M_n^* \ge a) \le \mathbb{E}\, |M_n| / a.$$

The same result holds for M_t if M_t is a martingale or a positive submartingale with right continuous paths.

Proof. Let $N = \min\{j : |M_j| \ge a\}$. Since $|\cdot|$ is convex, $|M_n|$ is a submartingale. Since $|M_N| \ge a$ on $(N < \infty)$,

$$\mathbb{P}(M_n^* \ge a) = \mathbb{P}(N \le n) \le \mathbb{E}\left[|M_N|/a; N \le n\right]$$
$$\le \mathbb{E}\, |M_{N \wedge n}|/a \le \mathbb{E}\, |M_n|/a.$$

$\square$

For $p > 1$, we have the following inequality.

(4.7) Theorem. *If $p > 1$, there exists c depending only on p such that*

$$\mathbb{E}\,(M_n^*)^p \le c \mathbb{E}\, |M_n|^p,$$

with the same being true if M_t is a martingale or positive submartingale with right continuous paths.

The proof that follows is slightly different from the standard one. The value c that we obtain is $2^p p/(p-1)$, while the optimal value is $p/(p-1)$ (see Durrett [2], pp. 216–217).

Proof. $|M_j|$ is a submartingale and hence $|M_j| \le \mathbb{E}\left[|M_n| \,|\, \mathcal{F}_j\right] \le \|M_n\|_\infty$. Therefore the proof is trivial if $p = \infty$. Theorem 4.6 is what is known as a weak (1-1) inequality, so the result follows immediately by the Marcinkiewicz interpolation theorem. This completes the proof.

For the reader not familiar with the Marcinkiewicz interpolation theorem, let us give the details in this particular case. (See Sect. IV.7 for a more general version.) Let $M_n^1 = M_n 1_{(|M_n|>a/2)}, M_n^2 = M_n - M_n^1$. Let $M_j^i = \mathbb{E}[M_n^i|\mathcal{F}_j]$, $i = 1, 2$. Since $\max_j |M_j| \leq \max_j |M_j^1| + a/2$, the weak (1-1) inequality applied to M^1 shows

$$\mathbb{P}(M_n^* > a) \leq \mathbb{P}((M_n^1)^* > a/2) \leq 2\mathbb{E}|M_n^1|/a$$
$$= 2\mathbb{E}[|M_n|; |M_n| > a/2]/a.$$

So by Proposition 1.5,

$$\mathbb{E}(M_n^*)^p = \int_0^\infty pa^{p-1}\mathbb{P}(M_n^* > a)\,da$$
$$\leq \int_0^\infty 2pa^{p-2}\mathbb{E}[|M_n|1_{(|M_n|>a/2)}]\,da.$$

By Fubini's theorem, the last integral is

$$\mathbb{E}\int_0^{2|M_n|} 2pa^{p-2}\,da|M_n| = \mathbb{E}\,\frac{2^p p}{p-1}|M_n|^p,$$

as desired.   □

If we apply Doob's inequalities to dyadic martingales, we obtain a version of the Hardy-Littlewood maximal theorem for dyadic martingales; see Exercise 16. We will do the full Hardy-Littlewood theorem in Chap. IV.

As another application of Doob's inequalities, we have the following useful estimate.

(4.8) Proposition. *Let X_t be a Brownian motion. Then if $a, t > 0$,*

$$\mathbb{P}(\sup_{s\leq t}|X_s| \geq a) \leq 2e^{-a^2/2t}.$$

Proof. Since e^{bx} is convex, e^{bX_t} is a nonnegative submartingale. Hence by Theorem 4.6

$$\mathbb{P}(\sup_{s\leq t} X_s \geq a) = \mathbb{P}(\sup_{s\leq t} e^{bX_s} \geq e^{ab}) \leq e^{-ab}\mathbb{E}\,e^{bX_t} \leq e^{-ab+b^2t/2}.$$

Now take $b = a/t$, repeat the argument for $-X_t$, and add the resulting two inequalities.   □

Let us compute the exit distribution of Brownian motion from the interval $[a, b]$.

(4.9) Proposition. *If $a < x < b$, then $\tau_{[a,b]} < \infty$, a.s., and*

$$\mathbb{P}^x(X(\tau_{[a,b]}) = a) = \frac{b-x}{b-a}, \qquad \mathbb{P}^x(X(\tau_{[a,b]}) = b) = \frac{x-a}{b-a}.$$

Proof. Let us write just τ for $\tau_{[a,b]}$. $X_t^2 - t$ is a martingale, so by Theorem 4.4,

$$\mathbb{E}^x X_{\tau \wedge t}^2 = \mathbb{E}^x \tau \wedge t.$$

For $t \le \tau$, $|X_t| \le |a| + |b|$, so using Fatou's lemma, $\mathbb{E}^x \tau \le (|a| + |b|)^2$, or $\tau < \infty$ a.s.

Since X_t is a martingale, $\mathbb{E}^x X_{\tau \wedge t} = x$. Letting $t \to \infty$ and using dominated convergence,

$$(4.5) \qquad x = \mathbb{E}^x X_\tau = a\mathbb{P}^x(X_\tau = a) + b\mathbb{P}^x(X_\tau = b).$$

Since $\tau < \infty$ a.s.,

$$(4.6) \qquad 1 = \mathbb{P}^x(X_\tau = a) + \mathbb{P}^x(X_\tau = b).$$

Solving the system of linear equations in (4.5) and (4.6) in the two unknowns $\mathbb{P}^x(X_\tau = a)$ and $\mathbb{P}^x(X_\tau = b)$ gives our result. □

If we let $b \to \infty$ and $x > a$, then

$$(4.7) \qquad \mathbb{P}^x(T_{\{a\}} < \infty) = 1.$$

The same proof as Proposition 4.9 shows the following.

(4.10) Corollary. *Suppose M_t is a continuous martingale that exits $[a, b]$ with probability one, and $M_0 = x$. Then $\mathbb{P}(M_\tau = a) = (b-x)/(b-a)$ and $\mathbb{P}(M_\tau = b) = (x-a)/(b-a)$.*

Martingale convergence theorems

The martingale convergence theorems are another set of important consequences of optional stopping. The main step is the upcrossing lemma. The number of upcrossings of an interval $[a, b]$ is the number of times a process crosses from below a to above b.

To be more exact, let

$$S_1 = \min\{k : X_k \le a\}, \qquad T_1 = \min\{k > S_1 : X_k \ge b\},$$

and

$$S_{i+1} = \min\{k > T_i : X_k \le a\}, \qquad T_{i+1} = \min\{k > S_{i+1} : X_k \ge b\}.$$

The number of upcrossings U_n before time n is $U_n = \max\{j : T_j \le n\}$.

(4.11) Theorem. (Upcrossing lemma) *If X_k is a submartingale,*

$$\mathbb{E}\, U_n \leq (b-a)^{-1}\mathbb{E}\left[(X_n - a)^+\right].$$

Proof. First assume that $a = 0$ and $X_k \geq 0$ for each k. Fix n and define $X_m = X_n$ for $m \geq n$. This will still be a submartingale. Define the S_i, T_i as above, and let $S_i' = S_i \wedge (n+1)$, $T_i' = T_i \wedge (n+1)$.

We write

$$\mathbb{E}\, X_{n+1} = \mathbb{E}\, X_{S_1'} + \sum_{i=0}^{\infty} \mathbb{E}\left[X_{T_i'} - X_{S_i'}\right] + \sum_{i=0}^{\infty} \mathbb{E}\left[X_{S_{i+1}'} - X_{T_i'}\right].$$

All the summands in the third term on the right are nonnegative since X_k is a submartingale, and

$$\sum_{i=0}^{\infty}(X_{T_i'} - X_{S_i'}) \geq (b-a)U_n.$$

So

(4.8) $$\mathbb{E}\, U_n \leq \mathbb{E}\, X_{n+1}/b.$$

Now let us remove the assumption that $a = 0$ and $X_k \geq 0$. The number of upcrossings of $[a, b]$ by X_k is the same as the number of upcrossings of $[0, b - a]$ by $Y_k = (X_k - a)^+$. So we merely apply (4.8) to the number of upcrossings of the interval $[0, b - a]$ by the process $(X_k - a)^+$.   □

This leads to the martingale convergence theorem.

(4.12) Theorem. *If X_n is a submartingale such that $\sup_n \mathbb{E}\, X_n^+ < \infty$, then X_n converges a.s. as $n \to \infty$.*

Proof. Let $U(a,b) = \lim_{n\to\infty} U_n$. For each a, b rational, by monotone convergence, $\mathbb{E}\, U(a,b) \leq c(b-a)^{-1} < \infty$. So $U(a,b) < \infty$, a.s. Taking the union over all pairs of rationals a, b, we see that a.s. the sequence $X_n(\omega)$ cannot have $\limsup X_n > \liminf X_n$. Therefore X_n converges a.s., although we still have to rule out the possibility of the limit being infinite. Since X_n is a submartingale, $\mathbb{E}\, X_n \geq \mathbb{E}\, X_0$, and thus

$$\mathbb{E}\,|X_n| = \mathbb{E}\, X_n^+ + \mathbb{E}\, X_n^- = 2\mathbb{E}\, X_n^+ - \mathbb{E}\, X_n \leq 2\mathbb{E}\, X_n^+ - \mathbb{E}\, X_0.$$

By Fatou's lemma, $\mathbb{E}\, \lim_n |X_n| \leq \sup_n \mathbb{E}\,|X_n| < \infty$, or X_n converges a.s. to a finite limit.   □

(4.13) Corollary. *If X_n is a positive supermartingale or a martingale bounded above or below, X_n converges a.s.*

Proof. If X_n is a positive supermartingale, $-X_n$ is a submartingale bounded above by 0. Now apply Theorem 4.12.

If X_n is a martingale bounded above, by considering $-X_n$, we may assume X_n is bounded below. Looking at $X_n + M$ for fixed M will not affect the convergence, so we may assume X_n is bounded below by 0. Now apply the first assertion of the corollary. $\qquad\square$

(4.14) Corollary. *The assertions of Theorem 4.12 and Corollary 4.13 remain true if we consider continuous-time martingales or supermartingales or submartingales with right continuous paths.*

Proof. The proof that $\limsup |X_t|$ is finite is the same as in the last line of the proof of Theorem 4.12. So the possibility we have to rule out is the possibility of oscillation, i.e., $\limsup X_t > \liminf X_t$. The same argument as in Theorem 4.12 shows that $\mathbb{E}\, U_t(a,b) \le \mathbb{E}\,(X_t - a)^+$, where $U_t(a,b)$ is the number of upcrossings of $[a,b]$ by X_t by time t, and we can proceed as in those proofs. $\qquad\square$

(4.15) Corollary. *If X_t is a right continuous martingale with $\sup_t \mathbb{E}\,|X_t|^p < \infty$ for some $p > 1$, then the convergence is in L^p as well as a.s. This is also true when X_t is a right continuous submartingale. If X_t is a uniformly integrable martingale with right continuous paths, then the convergence is in L^1. If $X_t \to X_\infty$ in L^1, then $X_t = \mathbb{E}\,[X_\infty | \mathcal{F}_t]$.*

X_t is a uniformly integrable martingale if the collection of random variables X_t is uniformly integrable.

Proof. The L^p convergence assertion follows by using Doob's inequality (Theorem 4.7) and dominated convergence. The L^1 convergence assertion follows since a.s. convergence together with uniform integrability implies L^1 convergence (see Exercise 21). Finally, if $t < n$, $X_t = \mathbb{E}\,[X_n | \mathcal{F}_t]$. If $A \in \mathcal{F}_t$,

$$\mathbb{E}\,[X_t; A] = \mathbb{E}\,[X_n; A] \to \mathbb{E}\,[X_\infty; A]$$

by the L^1 convergence of X_n to X_∞. Since this is true for all $A \in \mathcal{F}_t$, $X_t = \mathbb{E}\,[X_\infty | \mathcal{F}_t]$. $\qquad\square$

Let $D_n = \{k/2^n : k \le 2^n\}$ and $D = \cup_n D_n$. Another consequence of the upcrossing inequality is the following.

(4.16) Corollary. *Suppose X_t is a submartingale with $\sup_t \mathbb{E}\, X_t^+ < \infty$. Then $\{X_t : t \in D\}$ has left and right limits a.s.*

Proof. Let $U_n(a,b)$ be the number of upcrossings of $[a,b]$ by $X_t, t \in D_n$, and $U(a,b)$ the number of upcrossings of $[a,b]$ by $X_t, t \in D$. By monotone convergence,

$$\mathbb{E}\, U(a,b) = \lim_n \mathbb{E}\, U_n(a,b) \le \frac{\mathbb{E}\,(X_1 - a)^+}{b - a} < \infty.$$

This is true for every pair of rationals a, b. This, together with the argument of Theorem 4.12, proves the corollary. □

The martingale convergence theorem also provides the basis of the Calderón-Zygmund lemma. We give the standard proof, phrased in martingale language.

(4.17) Lemma. (Calderón-Zygmund) *Let $f \geq 0$ be integrable. Let $\alpha > 0$. There exists a closed set F and countably many pairwise disjoint open cubes Q_i such that $|F^c \Delta \cup_i Q_i| = 0$, $f \leq \alpha$ a.e. on F, and for each i,*

$$(4.9) \qquad \alpha \leq \frac{1}{|Q_i|} \int_{Q_i} f \leq 2^d \alpha.$$

Moreover, $|F^c| \leq \int f / \alpha$.

Proof. First of all, note that the last assertion is a consequence of (4.9), since (4.9) implies $|Q_i| \leq \int_{Q_i} f / \alpha$. Summing over i then gives $|\cup_i Q_i| \leq \int_{\cup_i Q_i} f / \alpha \leq \int f / \alpha$.

Suppose next that R is a cube with $|R|^{-1} \int_R f < \alpha/2$ and look at the dyadic martingale $f_n = \mathbb{E}[f | \mathcal{F}_n]$, where $\mathcal{F}_n$ is the partition of R into 2nd equal cubes. Let $T = \inf\{n : f_n > \alpha\}$. Note by our assumption on R that $T > 0$. For each n, $(T = n)$ is the union of cubes in $\mathcal{F}_n$, and the boundary of $(T = n)$ has measure 0. Let

$$F^c = \cup_n \big(\text{int} \, (T = n) \big).$$

Then F is closed.

If $n < T$, then $f_n \leq \alpha$. By the martingale convergence theorem, on $(T = \infty)$ we have $f = \lim_n f_n \leq \alpha$. Since F differs from the set $(T = \infty) = (T < \infty)^c = (\cup_n (T = n))^c$ by a set of measure 0, then $f \leq \alpha$ a.e. on F.

If Q is one of the cubes in $\mathcal{F}_n$ contained in $(T = n)$ and $x \in \text{int} \, (Q)$, then

$$\frac{1}{|Q|} \int_Q f = f_n(x) \geq \alpha.$$

By our definition of T, $f_{n-1}(x) \leq \alpha$. Let Q' be the element of $\mathcal{F}_{n-1}$ containing Q. Then since $|Q'| = 2^d |Q|$,

$$\frac{1}{|Q|} \int_Q f \leq \frac{2^d}{|Q'|} \int_{Q'} f \leq 2^d \alpha.$$

Finally, take n_0 large enough so that $\|f\|_1 / 2^{n_0 d} < \alpha/2$. We will take each cube R_j that has side length 2^{n_0} and vertices at integer multiples of 2^{n_0} and decompose R_j into F_j and Q_i^j as above. If we then let $F = \cup_j F_j$, we have our result. □

Supermartingale decompositions

The next set of results, the Doob-Meyer decomposition and the limit results for $\langle M \rangle_t$, are used to construct a continuous increasing process $\langle M \rangle_t$ such that $M_t^2 - \langle M \rangle_t$ is a martingale. For Brownian motion $\langle M \rangle_t$ is simply t, and in each of the martingales we come across we will explicitly display the corresponding bracket process. Readers may skip to the beginning of the next subsection.

The Doob-Meyer decomposition says that, under mild hypotheses, a supermartingale can be decomposed into a martingale minus an increasing process. We give a proof for the continuous-time continuous-path process case. It starts with a discrete-time approximation, and then shows L^2 convergence of the approximations.

We are limiting ourselves to continuous supermartingales, and the theorem we want to prove is the following.

(4.18) Theorem. *Suppose Z_t is a supermartingale with continuous paths. Then there exists a martingale M_t and an increasing process A_t, both with continuous paths and adapted to the filtration of Z_t, such that $Z_t = M_t - A_t$. Moreover, M_t and A_t are uniquely determined.*

Let us show uniqueness first. It is a consequence of the following proposition.

(4.19) Proposition. *Suppose M_t is a continuous martingale with paths of bounded variation and $M_0 \equiv 0$. Then M is identically 0.*

Proof. Let $T_N = \inf\{t > 0 : |M_t| > N \text{ or } \int_0^t |dM_s| > N\}$. ($\int_0^t |dM_s|$ is the total variation of M_s up until time t.) If we show $M_{t \wedge T_N}$ is identically 0 and let $N \to \infty$, we will have our result, so let us suppose M is bounded and has total variation bounded by N.

Let $\varepsilon > 0$, let $S_0 = 0$, and let $S_{i+1} = \inf\{t > S_i : |M_t - M_{S_i}| > \varepsilon\}$. Since M_t is continuous, $S_i \to \infty$. We have by (4.4) that

$$\mathbb{E} \, M_\infty^2 = \mathbb{E} \sum_{i=0}^\infty (M_{S_{i+1}} - M_{S_i})^2$$

$$\leq \varepsilon \mathbb{E} \sum |M_{S_{i+1}} - M_{S_i}| \leq \varepsilon \mathbb{E} \int_0^\infty |dM_s|.$$

Since ε is arbitrary, $\mathbb{E} \, M_\infty^2 = 0$. Then by Doob's inequality (Theorem 4.7),

$$\mathbb{E} \sup_{s < \infty} M_s^2 = 0.$$

$\square$

Proof of Theorem 4.18, uniqueness. If $Z_t = M_t - A_t = N_t - B_t$, where M_t and N_t are martingales and A_t and B_t are processes of bounded variation, then $M_t - N_t = A_t - B_t$ is a martingale with continuous paths that is also of bounded variation. Applying Proposition 4.19 to $(M - N)_t - (M - N)_0$ proves $M_t = N_t$, $A_t = B_t$, and hence uniqueness.   □

The following two lemmas are also very useful for applications other than the existence proof. Later we will see the continuous-time versions.

(4.20) Lemma. *Suppose A_k is an increasing process with $A_0 \equiv 0$. Suppose A_k is $\mathcal{F}_{k-1}$ measurable and suppose*

$$(4.10) \qquad\qquad \mathbb{E}\left[A_\infty - A_k | \mathcal{F}_k\right] \leq N,$$

$k = 0, 1, 2, \ldots.$ *Then $\mathbb{E} A_\infty^2 \leq 2N^2$.*

Proof. Let A_k be any process of bounded variation with $A_0 \equiv 0$. Let $a_k = A_{k+1} - A_k$. Some algebra shows that

$$(4.11) \qquad\qquad A_\infty^2 = 2 \sum_{k=0}^{\infty} (A_\infty - A_k) a_k - \sum_{k=0}^{\infty} a_k^2.$$

If A_k is also increasing, so that $a_k \geq 0$ for all k, then

$$(4.12) \qquad \mathbb{E} A_\infty^2 = 2\mathbb{E}\left[\sum_{k=0}^{\infty} \mathbb{E}\left[A_\infty - A_k | \mathcal{F}_k\right] a_k\right] - \mathbb{E} \sum_{k=0}^{\infty} (a_k)^2$$
$$\leq 2N\mathbb{E} \sum_{k=0}^{\infty} a_k = 2N\mathbb{E} A_\infty.$$

Since $\mathbb{E}\left[A_\infty - A_0 | \mathcal{F}_0\right] \leq N$ and $A_0 \equiv 0$, taking expectations shows $\mathbb{E} A_\infty \leq N$. Substituting in (4.12) gives the assertion.   □

(4.21) Lemma. *Suppose $A_k^{(1)}$ and $A_k^{(2)}$ are increasing processes satisfying the hypotheses of Lemma 4.20. Let $B_k = A_k^{(1)} - A_k^{(2)}$. Suppose there exists $W \geq 0$ with $\mathbb{E} W^2 < \infty$ such that for all k*

$$(4.13) \qquad\qquad \left|\mathbb{E}\left[B_\infty - B_k | \mathcal{F}_k\right]\right| \leq \mathbb{E}\left[W | \mathcal{F}_k\right].$$

Then there exists c such that

$$\mathbb{E}\left[\sup_k B_k^2\right] \leq c\mathbb{E} W^2 + cN(\mathbb{E} W^2)^{1/2}.$$

Proof. Let $b_k = B_{k+1} - B_k$. Let $a_k^{(i)} = A_{k+1}^{(i)} - A_k^{(i)}$, $i = 1, 2$. Since (4.11) was valid for any process of bounded variation,

$$\mathbb{E}\, B_\infty^2 = 2\mathbb{E} \sum_{k=0}^{\infty} \mathbb{E}\, [B_\infty - B_k | \mathcal{F}_k] b_k - \mathbb{E} \sum_{k=0}^{\infty} b_k^2$$

$$\leq 2\mathbb{E} \sum_{k=0}^{\infty} \mathbb{E}\, [W | \mathcal{F}_k] (a_k^{(1)} + a_k^{(2)})$$

$$\leq 2\mathbb{E}\, [W (A_\infty^{(1)} + A_\infty^{(2)})].$$

The Cauchy-Schwarz inequality and the bounds for $\mathbb{E}\, (A_\infty^{(i)})^2$ show $\mathbb{E}\, B_\infty^2 \leq cN(\mathbb{E}\, W^2)^{1/2}$.

Regarding the L^2 bound on the supremum of the B_ks, let $M_k = \mathbb{E}\, [B_\infty | \mathcal{F}_k]$, $N_k = \mathbb{E}\, [W | \mathcal{F}_k]$, and $X_k = M_k - B_k$. We have

$$|X_k| = |\mathbb{E}\, [B_\infty - B_k | \mathcal{F}_k]| \leq N_k,$$

so by Doob's inequality,

$$\mathbb{E} \sup_k X_k^2 \leq \mathbb{E} \sup_k N_k^2 \leq c\mathbb{E}\, N_\infty^2 = c\mathbb{E}\, W^2.$$

Again by Doob's inequality,

$$\mathbb{E} \sup_k M_k^2 \leq c\mathbb{E}\, M_\infty^2 = c\mathbb{E}\, B_\infty^2.$$

Since $\sup_k |B_k| \leq \sup_k |X_k| + \sup_k |M_k|$, we therefore have

$$\mathbb{E} \sup_k B_k^2 \leq c\mathbb{E}\, W^2 + cN(\mathbb{E}\, W^2)^{1/2}.$$

$\square$

Here is the Doob-Meyer decomposition in the discrete-time case.

(4.22) Proposition. *If Z_k is a discrete-time supermartingale, there exists a martingale M_k and an increasing process A_k such that A_{k+1} is $\mathcal{F}_k$ measurable and $Z_k = M_k - A_k$.*

Proof. Let $a_k = \mathbb{E}\, [Z_k - Z_{k+1} | \mathcal{F}_k]$. Since Z is a supermartingale, the a_k are nonnegative and clearly $\mathcal{F}_k$ measurable. Let $A_k = \sum_{j=1}^{k-1} a_j$. It is trivial to check that $Z_k + A_k$ is a martingale. $\square$

Proof of existence in Theorem 4.18. Since Z_t is continuous, it suffices to show that $Z_{t \wedge T_N}$ has the desired decomposition for each N, where $T_N = \inf\{t : |Z_t| > N\}$, because we then let $N \to \infty$ and use the uniqueness result. So we may assume Z is bounded. Also, if we have the decomposition up to any fixed time M, we can let M go to ∞ and get our existence result. So we may assume that Z_t is constant for $t \geq M$, and hence that almost surely the paths of Z_t are uniformly continuous.

Fix M and n. Let $\mathcal{F}_k^n = \mathcal{F}_{k/2^n}$. Construct A_k^n using Proposition 4.22. Let $\overline{A}_t^n = A_k^n$ and $\overline{\mathcal{F}}_t^n = \mathcal{F}_k^n$ if $(k-1)/2^n < t \leq k/2^n$.

Let $W(\delta) = \sup_{s \le M, s \le t \le s+\delta} |Z_t - Z_s|$. Since Z is bounded, so is $W(\delta)$. Since the paths of Z are uniformly continuous, $W(\delta) \to 0$ a.s. as $\delta \to 0$. Hence $W(\delta) \to 0$ in L^2.

The first thing we want to show is that $\overline{A}_t^n$ converges in L^2 as $n \to \infty$, uniformly over t. We will do that by showing $\mathbb{E} \sup_t |\overline{A}_t^m - \overline{A}_t^n|^2 \to 0$ as $n, m \to \infty$. Then since Cauchy sequences converge, the $\overline{A}_t^n$ converge. We will estimate the L^2 norm of the supremum of the difference using Lemma 4.21. Suppose $m \ge n$. Since $\overline{A}_t^n$ and $\overline{A}_t^m$ are constant over intervals $(k/2^m, (k+1)/2^m]$, the supremum of the difference will take place at some $k/2^m$. Fix $t = k/2^m$ for some k, and we will bound the difference of the conditional expectations with respect to $\mathcal{F}_k^m$. Let u be the smallest multiple of 2^{-n} bigger than or equal to t. We have by Proposition 4.22

$$\mathbb{E}\left[\overline{A}_\infty^m - \overline{A}_t^m | \mathcal{F}_t^m\right] = \mathbb{E}\left[A_\infty^m - A_k^m | \mathcal{F}_{k/2^m}\right] = \mathbb{E}\left[Z_t - Z_\infty | \mathcal{F}_t\right].$$

On the other hand

$$\begin{aligned}
\mathbb{E}\left[\overline{A}_\infty^n - \overline{A}_t^n | \mathcal{F}_t^m\right] &= \mathbb{E}\left[A_\infty^n - A_{u2^n}^n | \mathcal{F}_t\right] \\
&= \mathbb{E}\left[\mathbb{E}\left[A_\infty^n - A_{u2^n}^n | \mathcal{F}_u\right] | \mathcal{F}_t\right] \\
&= \mathbb{E}\left[\mathbb{E}\left[Z_u - Z_\infty | \mathcal{F}_u\right] | \mathcal{F}_t\right] \\
&= \mathbb{E}\left[Z_u - Z_\infty | \mathcal{F}_t\right],
\end{aligned}$$

using the definition of A_k^n. Then the difference of the conditional expectations is bounded:

$$\begin{aligned}
\left|\mathbb{E}\left[\overline{A}_\infty^m - \overline{A}_t^m | \mathcal{F}_t\right] - \mathbb{E}\left[\overline{A}_\infty^n - \overline{A}_t^n | \mathcal{F}_t\right]\right| & \\
\le \mathbb{E}\left[|Z_t - Z_u| \, | \mathcal{F}_t\right] &\le \mathbb{E}\left[W(2^{-n}) | \mathcal{F}_t\right].
\end{aligned}$$

Using Lemma 4.21 shows that $\mathbb{E} \sup_t |\overline{A}_t^m - \overline{A}_t^n|^2$ tends to 0 as $n \to \infty$.

Next we want to show the limit is continuous. The jumps of $\overline{A}_t^n$ are

$$\Delta \overline{A}_t^n = \mathbb{E}\left[Z_{(k-1)/2^n} - Z_{k/2^n} | \mathcal{F}_{(k-1)/2^n}\right], \qquad t = k/2^n,$$

which are bounded by $\mathbb{E}\left[W(2^{-n}) | \mathcal{F}_{(k-1)/2^n}\right]$. So

$$\begin{aligned}
\mathbb{E} \sup_t (\Delta \overline{A}_t^n)^2 &\le \mathbb{E}\left[\sup_k (\mathbb{E}\left[W(2^{-n}) | \mathcal{F}_{(k-1)/2^n}\right)^2\right] \\
&\le c\mathbb{E}\left[W(2^{-n})\right]^2 \to 0,
\end{aligned}$$

by Doob's inequality, since $\mathbb{E}\left[W(2^{-n}) | \mathcal{F}_{(k-1)/2^n}\right]$ is a martingale. By looking at a suitable subsequence n_j, $\sup_t |\Delta A_t^{n_j}| \to 0$, a.s., so the limit is continuous.

Finally, we show that if A_t is the limit of the $\overline{A}_t^n$, then $Z_t + A_t$ is a martingale. Since Z_t and A_t are both continuous and square integrable, it suffices to show that for $s, t \in D_n$, $s < t$, and $B \in \mathcal{F}_s$,

$$\mathbb{E}\left[Z_t + A_t; B\right] = \mathbb{E}\left[Z_s + A_s; B\right].$$

We know this for each $Z_t + \overline{A}_t^n$, and the result follows from a passage to the limit, using the Cauchy-Schwarz inequality.   □

A variation of this method can be used to prove the full Doob-Meyer decomposition (where Z_t could have jumps); see Bass [2].

Quadratic variation

If M_t is a continuous square integrable martingale (i.e., $\mathbb{E}\, M_\infty^2 < \infty$), then M_t^2 is a submartingale and $-M_t^2$ is a supermartingale. By Theorem 4.18 there exists a continuous increasing process, denoted $\langle M \rangle_t$, the quadratic variation of M, such that $M_t^2 - \langle M \rangle_t$ is a martingale. If we have two martingales M, N, we define $\langle M, N \rangle_t$ by polarization:

$$(4.14) \qquad \langle M, N \rangle_t = \frac{1}{2}\Big(\langle M + N \rangle_t - \langle M \rangle_t - \langle N \rangle_t \Big).$$

We want to show the following.

(4.23) Theorem. *For each ε suppose $S_i(\varepsilon)$ is a sequence of stopping times that increase to infinity as $i \to \infty$ such that $\sup_i |M_{S_{i+1}(\varepsilon)} - M_{S_i(\varepsilon)}| \to 0$ as $\varepsilon \to 0$. Then $\sum_{i=0}^{\infty}(M_{S_{i+1}(\varepsilon)} - M_{S_i(\varepsilon)})^2 \to \langle M \rangle_\infty$ in probability.*

One could prove this theorem by computing the second moment of the difference and showing it goes to 0. We use instead Lemma 4.21.

Proof. By stopping the first time M_t or $\langle M \rangle_t$ exceeds N, we may assume M and $\langle M \rangle$ are both bounded. We apply Lemma 4.21. Dropping the (ε) from the notation, let $a_i^{(1)} = (M_{S_{i+1}} - M_{S_i})^2$, $a_i^{(2)} = \langle M \rangle_{S_{i+1}} - \langle M \rangle_{S_i}$, $b_i = a_i^{(1)} - a_i^{(2)}$. Let $A^{(1)}, A^{(2)}$ denote the sums, and let $\mathcal{F}_k = \sigma(M_{S_{i+1}}, i \le k)$. Since M and $\langle M \rangle$ are bounded, it is clear the hypotheses of Lemma 4.21 are satisfied for $A^{(1)}$ and $A^{(2)}$. Let $B_k = A_k^{(1)} - A_k^{(2)}$. Let $W(\varepsilon) = \sup_i(|M_{S_{i+1}} - M_{S_i}|^2 + \langle M \rangle_{S_{i+1}} - \langle M \rangle_{S_i})$.

We have $B_\infty - B_k = \sum_{i=k}^{\infty} b_i$. Note that $\mathbb{E}\,[b_i | \mathcal{F}_k] = 0$ if $i > k$. So

$$|\mathbb{E}\,[B_\infty - B_k | \mathcal{F}_k]| \le a_k^{(1)} + a_k^{(2)}.$$

Observe also that $\sup_k a_k^{(1)} \le \sup_k \mathbb{E}\,[W(\varepsilon)|\mathcal{F}_k]$ and similarly for $\sup_k a_k^{(2)}$. Hence applying Lemma 4.21, we obtain the bound

$$\mathbb{E}\,[\sup_k B_k^2] \le c(\mathbb{E}\,W(\varepsilon)^2)^{1/2} + \mathbb{E}\,W(\varepsilon)^2),$$

which goes to 0 as $\varepsilon \to 0$.   □

Nothing precludes the stopping times in the above from being fixed times.

5 Stochastic integrals

Construction

By Exercise 19, the paths of Brownian motion are nowhere differentiable. This means one cannot expect to make sense of $\int f(s)\,dB_s$ by means of a Lebesgue-Stieltjes integral for general integrands f. Nevertheless, it is possible to define such a stochastic integral via L^2 means. Our goal in this section is to define $\int H_s\,dM_s$, where $H_s = H_s(\omega)$ is a suitably adapted random process and M_s is a continuous martingale, most often Brownian motion.

The class of integrands H_s is the following. Let $\mathcal{F}_t$ be a filtration satisfying the usual conditions. Let $\mathcal{P}$ be the σ–field on $\Omega \times [0, \infty)$ generated by all the left continuous processes Y_t that are adapted to $\mathcal{F}_t$. $\mathcal{P}$ is called the predictable σ-field. We require that $H(s, \omega)$ be $\mathcal{P}$ measurable. At first we also require $\mathbb{E} \int_0^\infty H_s^2 d\langle M\rangle_s < \infty$, and later we weaken that to $\int_0^\infty H_s^2 d\langle M\rangle_s < \infty$, a.s.

Let us sketch the definition of the stochastic integral in the case of Brownian motion first, since it is really very simple. Note that if H is $\mathcal{F}_a$ measurable and K is $\mathcal{F}_c$ measurable, W_t is a Brownian motion, and $a \leq b \leq c \leq d$, then

$$(5.1) \quad \mathbb{E}\left[H(W_b - W_a)K(W_d - W_c)\right]$$
$$= \mathbb{E}\left[H(W_b - W_a)K\mathbb{E}\left[W_d - W_c|\mathcal{F}_c\right]\right] = 0.$$

Also, since $W_t^2 - t$ is a martingale, then by (4.3),

$$\mathbb{E}\left[(W_b - W_a)^2|\mathcal{F}_a\right] = \mathbb{E}\left[W_b^2 - W_a^2|\mathcal{F}_a\right] = \mathbb{E}\left[(b-a)|\mathcal{F}_a\right] = b - a,$$

and hence

$$(5.2) \quad \mathbb{E}\left[H^2(W_b - W_a)^2\right] = \mathbb{E}\left[H^2\mathbb{E}\left[(W_b - W_a)^2|\mathcal{F}_a\right]\right] = \mathbb{E}\left[H^2(b-a)\right].$$

Now, let us construct $\int_0^1 H_s dW_s$. If H_s is elementary, that is, it equals $H(\omega)1_{[a,b]}(s)$ where H is $\mathcal{F}_a$ measurable, then define $\int_0^1 H_s dW_s$ to be $H(W_b - W_a)$. This is just what one would get if one could define a Lebesgue-Stieltjes integral. If now H_s is simple, that is, it is a finite linear combination of elementary integrands, define $\int_0^1 H_s dW_s$ by linearity. Finally, if H_s satisfies $\mathbb{E} \int_0^1 H_s^2 ds < \infty$, approximate H by simple integrands H^n and define the stochastic integral as the L^2 limits of the stochastic integrals of the approximating integrands. How does one know that the L^2 limit exists and is independent of the choice of the H^ns? If H is simple, we can write H as $\sum_{j=1}^N K_j 1_{[a_j, b_j]}(s)$, where the K_js are bounded, $\mathcal{F}_{a_j}$ measurable, and $a_1 \leq b_1 \leq a_2 \leq \ldots \leq b_N$. Then

$$\mathbb{E}\left[\left(\int_0^1 H_s dW_s\right)^2\right] = \mathbb{E}\left[\sum_{j=1}^N K_j^2 (W_{b_j} - W_{a_j})^2\right]$$
$$+ \mathbb{E}\left[2\sum_{i<j} K_i(W_{b_i} - W_{a_i})K_j(W_{b_j} - W_{a_j})\right].$$

By (5.1), the second summand is 0. So by (5.2),

$$\mathbb{E}\left[\left(\int_0^1 H_s dW_s\right)^2\right] = \mathbb{E}\left[\int_0^1 H_s^2 ds\right],$$

or there is an isometry between $\int_0^1 H_s dW_s$ with the $L^2(\mathbb{P})$ norm and H_s with the L^2 norm $\left(\mathbb{E}\int_0^1 (\cdot)^2 ds\right)^{1/2}$. This is used to show that the limit exists and is independent of the choice of approximating sequences.

We now turn to the general case of continuous martingales and give details. We suppose M_t is square integrable and continuous, so $\sup_t \mathbb{E}\,M_t^2 < \infty$. By Doob's inequality and the martingale convergence theorem, $M_\infty = \lim_{t\to\infty} M_t$ exists and $\mathbb{E}\,M_\infty^2 < \infty$. The integrands H we consider are predictable and $\mathbb{E}\int_0^\infty H_s^2 d\langle M\rangle_s < \infty$.

Let K be a bounded $\mathcal{F}_a$ measurable random variable, and define

$$N_t = K(M_{t\wedge b} - M_{t\wedge a}).$$

(5.1) Lemma. N_t *is a continuous martingale*, $\mathbb{E}\,N_\infty^2 = \mathbb{E}\left[K^2(\langle M\rangle_b - \langle M_a\rangle)\right]$, *and*

$$\langle N\rangle_t = \int_0^t K^2 1_{[a,b]}(s)d\langle M\rangle_s.$$

Proof. The continuity of N_t is clear. We need to show that if $s < t$, then $\mathbb{E}[N_t|\mathcal{F}_s] = N_s$. There are a number of cases to consider, depending on where s, t are located relative to a, b. We do the cases $a < s < t < b$ and $s < a < t < b$; the others are similar. Then

$$\mathbb{E}[K(M_t - M_a)|\mathcal{F}_s] = K\mathbb{E}[M_t - M_a|\mathcal{F}_s] = K(M_s - M_a),$$

in the first case, as desired. For the second case,

$$\mathbb{E}[K(M_t - M_a)|\mathcal{F}_s] = \mathbb{E}\left[K\mathbb{E}[M_t - M_a|\mathcal{F}_a]|\mathcal{F}_s\right] = 0,$$

again as required.

For the second assertion,

$$(5.3)\qquad \mathbb{E}\,N_\infty^2 = \mathbb{E}\left[K^2\mathbb{E}[(M_b - M_a)^2|\mathcal{F}_a]\right] = \mathbb{E}\left[K^2\mathbb{E}[M_b^2 - M_a^2|\mathcal{F}_a]\right]$$

by (4.3). Since $M_t^2 - \langle M\rangle_t$ is a martingale,

$$\mathbb{E}[M_b^2 - M_a^2|\mathcal{F}_a] = \mathbb{E}[\langle M\rangle_b - \langle M\rangle_a|\mathcal{F}_a].$$

Substituting in (5.3) gives the second assertion.

For the third assertion, we must show

$$\mathbb{E}\left[K^2(M_{t\wedge b} - M_{t\wedge a})^2 - K^2(\langle M\rangle_{t\wedge b} - \langle M\rangle_{t\wedge a})|\mathcal{F}_s\right]$$
$$= K^2(M_{s\wedge b} - M_{s\wedge a})^2 - K^2(\langle M\rangle_{s\wedge b} - \langle M\rangle_{s\wedge a}).$$

Again it is matter of checking several cases. We leave the details to the reader, as they are very similar to the proof of the first assertion. □

Let us say that a process H_s is simple if it can be written in the form $\sum_{j=1}^{J} H_j 1_{[a_j,b_j]}(s)$, where for each j, H_j is $\mathcal{F}_{s_j}$ measurable and bounded. For simple processes H_s, define

$$N_t = \int_0^t H_s dM_s = \sum_{j=1}^{J} H_j(M_{b_j\wedge t} - M_{a_j\wedge t}).$$

(5.2) Proposition. *If H_s^n is a sequence of simple processes such that*

$$\mathbb{E}\left[\int_0^\infty (H_s^n - H_s^m)^2 d\langle M\rangle_s\right] \to 0$$

as $n, m \to \infty$, then

$$\mathbb{E}\left[\sup_{s<\infty}(N_s^n - N_s^m)^2\right] \to 0$$

as $n, m \to \infty$.

Proof. If H is a simple process, it can be rewritten as $\sum_{j=1}^{J} H_j 1_{[a_j,b_j]}(s)$, where the H_j are bounded and $\mathcal{F}_{s_j}$ measurable and the intervals $[a_j, b_j]$ satisfy $a_1 \le b_1 \le a_2 \le b_2 \le a_3 \le \ldots \le b_J$. Then by Lemma 5.1, N_t is a martingale, and by Doob's inequality,

$$\mathbb{E}\left[\sup_{s<\infty} N_s^2\right] \le c\mathbb{E}\left[N_\infty^2\right]$$

$$= c\mathbb{E}\left[\sum_{j=1}^{J} H_j^2(M_{b_j} - M_{a_j})^2\right]$$

$$+ c\mathbb{E}\left[2\sum_{i<j} H_i H_j(M_{b_i} - M_{a_i})(M_{b_j} - M_{a_j})\right].$$

Conditioning on $\mathcal{F}_{a_j}$, each summand in the second sum on the right is

$$\mathbb{E}\left[H_i H_j(M_{b_i} - M_{a_i})\mathbb{E}\left[M_{b_j} - M_{a_j}|\mathcal{F}_{a_j}\right]\right] = 0.$$

Since

$$\mathbb{E}\left[H_j^2(M_{b_j} - M_{a_j})^2\right] = \mathbb{E}\left[H_j^2\mathbb{E}\left[(M_{b_j} - M_{a_j})^2|\mathcal{F}_{a_j}\right]\right]$$
$$= \mathbb{E}\left[H_j^2\mathbb{E}\left[M_{b_j}^2 - M_{a_j}^2|\mathcal{F}_{a_j}\right]\right]$$
$$= \mathbb{E}\left[H_j^2\mathbb{E}\left[\langle M\rangle_{b_j} - \langle M\rangle_{a_j}|\mathcal{F}_{a_j}\right]\right]$$
$$= \mathbb{E}\left[H_j^2(\langle M\rangle_{b_j} - \langle M\rangle_{a_j})\right],$$

then

$$\mathbb{E}\left[\sup_{s<\infty} N_s^2\right] \le c\mathbb{E}\left[\int_0^\infty H_s^2 d\langle M\rangle_s\right].$$

The difference of two simple processes is again a simple process, so applying this estimate to $N^n - N^m$ proves the proposition. □

(5.3) Theorem. *Suppose H_s^n are simple processes such that*

$$\mathbb{E}\left[\int_0^\infty (H_s^n - H_s)^2 d\langle M\rangle_s\right] \to 0$$

as $n \to \infty$. Then N_s^n converges in L^2, uniformly over $s \in [0, \infty)$, to a continuous martingale. The limit, which we denote $\int_0^t H_s\, dM_s$, is independent of the choice of which sequence H^n is used to approximate H_s.

Proof. If H_s^n is as in the hypothesis, $\mathbb{E}\left[\int_0^\infty (H_s^n - H_s^m)^2 d\langle M\rangle_s\right] \to 0$ as $n, m \to \infty$. By Proposition 5.2, we see that N_s^n forms a Cauchy sequence with respect to the L^2 norm of the supremum over s. So the sequence actually converges. If we call the limit N_s, we have the assertion about the L^2 convergence. By taking an appropriate subsequence, we can assert that with probability one, $N_s^{n_k}$ converges to N_s uniformly over $s \in [0, \infty)$. Since each N_s^n is continuous, this proves that N_s is also continuous.

Suppose $s \le t$. By the L^2 convergence and Proposition 1.11,

$$\mathbb{E}\left[\left(\mathbb{E}[N_t^n|\mathcal{F}_s] - \mathbb{E}[N_t|\mathcal{F}_s]\right)^2\right] \le \mathbb{E}\left[\mathbb{E}[(N_t^n - N_t)^2|\mathcal{F}_s]\right]$$
$$= \mathbb{E}\left[(N_t^n - N_t)^2\right] \to 0.$$

Since $\mathbb{E}[N_t^n|\mathcal{F}_s] = N_s^n$, passing to the limit as $n \to \infty$ shows N_t is a martingale.

Finally, if $H^n, H^{n'}$ are two simple processes converging to H, Proposition 5.2 shows that $\mathbb{E}[\sup_{s<\infty}(N_s^n - N_s^{n'})^2] \to 0$, or the limit is independent of which approximating sequence is used. □

(5.4) Corollary. $\langle N\rangle_t = \int_0^t H_s^2 d\langle M\rangle_s.$

Proof. The assertion is true when H is simple as in Lemma 5.1. Now approximate general H by simple ones. □

Extensions

We make a number of extensions to the definition that are fairly routine. We do this in several steps.

First, if $\int_0^\infty H_s^2 d\langle M\rangle_s < \infty$, a.s, but is not necessarily integrable, let

$$T_N = \inf\left\{t > 0 : \int_0^T H_s^2 d\langle M\rangle_s > N\right\},$$

and define $\int_0^t H_s dM_s$ to be $\int_0^t H_s dM_{s \wedge T_N}$ if $t \leq T_N$. Since $\langle M \rangle_{s \wedge T_N}$ does not increase on $[T_N, \infty)$, we can use the definition in the integrable case to define this stochastic integral. Since $T_N \to \infty$ as $N \to \infty$, we have thus defined N_t for all t, provided the definitions are consistent. That they are is the content of Exercise 26.

A process M_t is a local martingale if there exist stopping times $S_n \to \infty$ such that $M_{t \wedge S_n}$ is a square integrable martingale for each n. If M_t is a continuous process, the stopping times $S_n = \inf\{t : |M_t| > n\}$ will do the job, since $M_{t \wedge S_n}$ is bounded by n. (For an example of a local martingale that is not a square integrable martingale, look at Brownian motion.) To define $\int_0^t H_s dM_s$, define it to be $\int_0^t H_s dM_{s \wedge S_n}$ if $t \leq S_n$. Define $\langle M \rangle_t$ to be $\langle M \rangle_{t \wedge S_n}$ if $t \leq S_n$. Again there is consistency to check (see Exercise 26).

A process is locally of bounded variation if there exist stopping times R_n such that $A_{R_n \wedge t}$ is of bounded variation for each n. (Here the appropriate R_n to use in the continuous case is $R_n = \inf\{t : \int_0^t |dA_s| > n\}$.) A semimartingale X_t is a process that is the sum of a local martingale M_t and a process that is locally of bounded variation A_t. If $\int_0^t H_s^2 d\langle M \rangle_s + \int_0^t |H_s||dA_s| < \infty$ for all t, we define the stochastic integral $\int_0^t H_s dX_s = \int_0^t H_s dM_s + \int_0^t H_s dA_s$, where the first integral is the stochastic integral of the type we have discussed, and the second is a Lebesgue-Stieltjes integral. If X_t is a semimartingale, $\langle X \rangle_t$ is defined to be $\langle M \rangle_t$.

Itô's formula

The most important fact about stochastic integration is the change of variables formula or Itô's formula.

(5.5) Theorem. *Let X_t be a semimartingale with continuous paths. Suppose $f \in C^2$. Then with probability one we have*

$$f(X_t) = f(X_0) + \int_0^t f'(X_s) dX_s + \frac{1}{2} \int_0^t f''(X_s) d\langle X \rangle_s, \qquad t \geq 0.$$

Proof. Itô's formula is an assertion about paths. We know $X_t = M_t + A_t$, where M is a local martingale and A is a process that is locally of bounded variation. Define

$$T_N = \inf\left\{ t > 0 : |M_t| > N \text{ or } \langle M \rangle_t > N \text{ or } \int_0^t |dA_s| > N \right\}.$$

Since $T_N \to \infty$ as $N \to \infty$, if we prove our result for $X_{t \wedge T_N}$, then we have it for all t. So we may assume M and A are bounded, hence X is; we may then assume that f is C^2 and that f, f', and f'' are bounded.

Fix $t_0 > 0$ and let $\varepsilon > 0$, let $S_0(\varepsilon) = 0$, and define

$$S_{i+1}(\varepsilon) = \inf \left\{ t > S_i(\varepsilon) : |M_t - M_{S_i(\varepsilon)}| > \varepsilon \text{ or } \langle M \rangle_t - \langle M \rangle_{S_i(\varepsilon)} > \varepsilon \right.$$

$$\left. \text{or } \int_{S_i(\varepsilon)}^t |dA_s| > \varepsilon \text{ or } t - S_i(\varepsilon) > \varepsilon \right\} \wedge t_0.$$

We will drop the ε from the notation.

Since we are dealing with continuous processes, $S_i \to t_0$ as $i \to \infty$. The key idea is that

$$(5.4) \quad f(X_{t_0}) - f(X_0) = \sum_{i=0}^\infty \left[f(X_{S_{i+1}}) - f(X_{S_i}) \right]$$

$$= \sum_{i=0}^\infty f'(X_{S_i})(X_{S_{i+1}} - X_{S_i})$$

$$+ \sum_{i=0}^\infty \frac{1}{2} f''(X_{S_i})(X_{S_{i+1}} - X_{S_i})^2 + \sum_{i=0}^\infty R_i,$$

where we have used Taylor's theorem and the R_i denotes the remainder. We want to show that the first term on the right of (5.4) converges to the stochastic integral term in Itô's formula, the second term converges to the bounded variation term, and the remainder term goes to 0.

Let us first look at the stochastic integral term. Let $H_s^\varepsilon = f'(X_{S_i})$ if $S_i \leq s < S_{i+1}$. By the continuity of X_s and f', H_s^ε converges boundedly to $f'(X_s)$. Then $\sum_{i=0}^\infty f'(X_{S_i})(A_{S_{i+1}} - A_{S_i}) = \int_0^t H_s^\varepsilon dA_s$ converges to $\int_0^t f'(X_s)dA_s$ by ordinary dominated convergence. Note

$$\sum_{i=0}^\infty f'(X_{S_i})(M_{S_{i+1}} - M_{S_i}) = \int_0^t H_s^\varepsilon dM_s.$$

By Proposition 5.2, $\sum_{i=0}^\infty f'(X_{S_i})(M_{S_{i+1}} - M_{S_i})$ converges to $\int_0^t f'(X_s)dM_s$.

Let us write

$$f''(X_{S_i})(X_{S_{i+1}} - X_{S_i})^2 = f''(X_{S_i})(M_{S_{i+1}} - M_{S_i})^2$$

$$+ 2f''(X_{S_i})(M_{S_{i+1}} - M_{S_i})(A_{S_{i+1}} - A_{S_i})$$

$$+ f''(X_{S_i})(A_{S_{i+1}} - A_{S_i})^2.$$

By Theorem 4.23, for each t, $V_t^\varepsilon = \sum_{\{i: S_{i+1} \leq t\}} (M_{S_{i+1}} - M_{S_i})^2$ converges to $\langle M \rangle_t$ in probability. So if ε_k is a sequence tending to 0 fast enough, the convergence will be almost sure, and $\sup_{\varepsilon_k} V_t^{\varepsilon_k} < \infty$. Since $f''(X_s)$ is a continuous process, $\int_0^{t_0} f''(X_s)dV_s^{\varepsilon_k}$ converges to $\int_0^{t_0} f''(X_s)d\langle M \rangle_s$ as $\varepsilon_k \to 0$ (Exercise 27). On the other hand, since $f \in C^2$, given δ we can find ε so that $|f''(x) - f''(y)| < \delta$ if $|x - y| < \varepsilon$. Therefore

$$\left| \sum f''(X_{S_i})(M_{S_{i+1}} - M_{S_i})^2 - \int_0^t f''(X_s)dV_s^\varepsilon \right| \leq \delta V_{t_0}^\varepsilon.$$

Hence $\sum f''(X_s)(M_{S_{i+1}} - M_{S_i})^2 \to \int_0^t f''(X_s)d\langle M\rangle_s$.

The other terms are easier.

$$\left|\sum_{i=0}^{\infty} f''(X_{S_i})(M_{S_{i+1}} - M_{S_i})(A_{S_{i+1}} - A_{S_i})\right|$$

$$\le \|f''\|_{\infty}\varepsilon \sum_{i=0}^{\infty} |A_{S_{i+1}} - A_{S_i}| \le \|f''\|_{\infty}\varepsilon N,$$

since we are assuming $\int_0^t |dA_s|$ is bounded and $|M_{S_{i+1}} - M_{S_i}| \le \varepsilon$. Hence this term goes to 0 as $\varepsilon \to 0$.

The $\sum f''(X_{S_i})(A_{S_{i+1}} - A_{S_i})^2$ term is exactly similar. By Taylor's theorem, $R_i \le \eta(\varepsilon)(X_{S_{i+1}} - X_{S_i})^2$, where $\eta(\varepsilon) \to 0$ as $\varepsilon \to 0$. As above, $\mathbb{E}\sum(X_{S_{i+1}} - X_{S_i})^2$ stays bounded, which shows $\sum R_i \to 0$ in L^2.   □

Given two continuous semimartingales X and Y, we define $\langle X, Y\rangle$ by polarization:

(5.5)
$$\langle X, Y\rangle_t = \frac{1}{2}\Big[\langle X + Y\rangle_t - \langle X\rangle_t - \langle Y\rangle_t\Big].$$

By polarizing Theorem 4.23, if M and N are martingales, then

$$\sum(M_{S_{i+1}\wedge t} - M_{S_i\wedge t})(N_{S_{i+1}\wedge t} - N_{S_i\wedge t}) \to \langle M, N\rangle_t.$$

With this fact, the multivariate version of Itô's formula is proved exactly the same way as Theorem 5.5.

(5.6) Corollary. *Suppose X_t is a d-dimensional process, each of whose components is a continuous semimartingale. Let $f \in C^2(\mathbb{R}^d)$. Then with probability one*

$$f(X_t) - f(X_0) = \int_0^t \sum_{i=1}^{d} \frac{\partial f}{\partial x^i}(X_s)dX_s^i$$

$$+ \frac{1}{2}\int_0^t \sum_{i,j=1}^{d} \frac{\partial^2 f}{\partial x^i \partial x^j}(X_s)d\langle X^i, X^j\rangle_s, \qquad t > 0.$$

The following is the integration-by-parts formula for stochastic integrals.

(5.7) Corollary. *If X and Y are continuous semimartingales, then*

$$X_t Y_t = X_0 Y_0 + \int_0^t X_s dY_s + \int_0^t Y_s dX_s + \langle X, Y\rangle_t.$$

Proof. From Itô's formula with $f(x) = x^2$,

(5.6) $$X_t^2 = X_0^2 + 2\int_0^t X_s dX_s + \langle X\rangle_t.$$

The corollary follows from this by polarization.

Alternately, we could apply Corollary 5.6 with $f(x,y) = xy$. $\square$

Applications

Using martingale methods, we see another connection of Brownian motion with the Dirichlet problem. Let D be a domain and suppose u is C^2 inside the domain, $\Delta u = 0$ in D, and suppose u is continuous on the closure of D. If X is d-dimensional Brownian motion, then $\langle X^i, X^j\rangle_t = 0$ unless $i = j$, in which case it equals t. Itô's formula for Brownian motion is therefore

(5.7) $$f(X_t) = f(X_0) + \int_0^t \nabla f(X_s) \cdot dX_s + \frac{1}{2}\int_0^t \Delta f(X_s)ds,$$

where $\cdot$ is the inner product in $\mathbb{R}^d$. Replacing f by u, we see that $u(X_t)$ is a bounded martingale. By optional stopping, $u(x) = \mathbb{E}^x u(X_{t\wedge\tau_D})$. Letting $t \to \infty$ and using dominated convergence, $u(x) = \mathbb{E}^x u(X_{\tau_D})$. So if u is the solution to the Dirichlet problem with boundary function h and u is continuous on the closure of D, then u is simply $\mathbb{E}^x h(X_{\tau_D})$, or h evaluated according to the hitting distribution of X_{τ_D}.

We use Itô's formula to derive the hitting probabilities of annuli for d-dimensional Brownian motion.

(5.8) Proposition. *Let X_t be d-dimensional Brownian motion.*

(a) *If $d = 2$, $r < |x| < R$, then*
   (i) $\mathbb{P}^x(T_{B(0,r)} < \tau_{B(0,R)}) = (\log R - \log|x|)/(\log R - \log r)$;
   (ii) $\mathbb{P}(T_{\{0\}} < \infty) = 0$;
   (iii) $\mathbb{P}^x(T_{B(0,r)} < \infty) = 1$;
   (iv) $\mathbb{P}^x(X_t \in B(0,r) \text{ i.o.}) = 1$.

(b) *If $d \geq 3$, $r < |x| < R$, then*
   (i) $\mathbb{P}^x(T_{B(0,r)} < \tau_{B(0,R)}) = (|x|^{2-d} - R^{2-d})/(r^{2-d} - R^{2-d})$;
   (ii) $\mathbb{P}^x(T_{\{0\}} < \infty) = 0$;
   (iii) $\mathbb{P}^x(T_{B(0,r)} < \infty) = (|x|/r)^{2-d}$.

Proof. If $d = 2$, let $u(x) = -\log(|x|)$. If $d \geq 3$, let $u(x) = |x|^{2-d}$. In either case, we do a direct calculation and see that $\Delta u = 0$ on $B(0,R) - B(0,r)$. (Recall $\partial|x|/\partial x^i = x^i/|x|$ since $|x| = \left(\sum_{i=1}^d (x^i)^2\right)^{1/2}$.) This implies $M_t = u(X_{t\wedge T(B(0,r))})$ is a martingale. Since one-dimensional Brownian motion will exceed R in absolute value with probability one, then $|X_t| \geq |X_t^1|$ will also exceed R with probability one. So if $S = T_{B(0,r)} \wedge \tau_{B(0,R)}$, then $\mathbb{P}^x(S < \infty) = 1$. Then (a)(i) and (b)(i) follow by Corollary 4.10.

If in (a)(i) we let first $r \to 0$, we have $\mathbb{P}^x(T_{\{0\}} < \tau_{B(0,R)}) = 0$. Letting $R \to \infty$ gives (a)(ii). If instead in (a)(i) we first let $R \to \infty$, we get (a)(iii). If we let $r \to 0$ and then $R \to \infty$ in (b)(i), we see (b)(ii), while if we let $R \to \infty$, we obtain (b)(iii).

To see (a)(iv), note for any s,

$$\mathbb{P}^x(X_t \in B(0,r) \text{ for some } t \geq s) = \mathbb{E}^x \mathbb{P}^{X_s}(\tau_{B(0,r)} < \infty) = 1$$

by the Markov property and (a)(iii). This says that if $L = \sup\{t : X_t \in B(0,r)\}$, then $L \geq s$ with probability 1. Since s is arbitrary, $L = \infty$, a.s. $\square$

Here is another important application of Itô's formula. Corollary 5.10 is the key to Chap. V.

(5.9) Theorem. *Suppose M_t is a continuous local martingale with $\langle M \rangle_t \equiv t$ and $M_0 \equiv 0$. Then M_t is a Brownian motion.*

Proof. We need to show that $M_t - M_s$ is independent of $\mathcal{F}_s$ and has an $\mathcal{N}(0, t-s)$ distribution. If we consider $M'_u = M_{s+u} - M_s$, $\mathcal{F}'_u = \mathcal{F}_{s+u}$, we can reduce consideration to the case $s = 0$.

By Itô's formula,

$$(5.8) \qquad e^{iuM_t} - 1 = iu \int_0^t e^{iuM_s} dM_s - \frac{u^2}{2} \int_0^t e^{iuM_s} ds.$$

Suppose $A \in \mathcal{F}_0$. Let $T_N = \inf\{t : |M_t| > N\}$. Since the first term on the right of (5.8) is a mean zero martingale, we have by optional stopping that $\mathbb{E}\left[\int_0^{t \wedge T_N} e^{iuM_s} dM_s; A\right] = 0$. So replacing t by $t \wedge T_N$ in (5.8), multiplying by 1_A, taking expectations, and then letting $N \to \infty$, we obtain the equation

$$(5.9) \qquad J(t) = \mathbb{P}(A) - \frac{u^2}{2} \int_0^t J(s) ds,$$

where $J(s) = \mathbb{E}\left[e^{iuM_s}; A\right]$.

Since J is bounded, (5.9) shows us that J is continuous; since it is continuous, (5.9) shows us again that it is C^1. So $J'(t) = -u^2 J(t)/2$, or $(\log J(t))' = -u^2/2$, or $J(t) = \mathbb{P}(A)e^{-u^2t/2}$.

This equation can be rewritten as

$$\mathbb{E}\left[e^{iuM_t}; A\right] = \mathbb{P}(A)e^{-u^2t/2}.$$

Taking $A = \Omega$ shows that M_t has an $\mathcal{N}(0,t)$ distribution. Approximating $1_B(x)$ by linear combinations of e^{iux}s shows

$$\mathbb{E}\left[1_B(M_t); A\right] = \mathbb{P}(A)\mathbb{P}(M_t \in B),$$

which proves the independence assertion. $\square$

Let δ_{ij} equal 0 if $i \neq j$, 1 if $i = j$.

(5.10) Corollary. *If X_t is a d-dimensional process, each of whose coordinates is a continuous local martingale, $X_0 \equiv 0$, and $\langle X^i, X^j \rangle_t \equiv \delta_{ij} t$, then X_t is d-dimensional Brownian motion.*

Proof. Apply Theorem 5.9 to $\sum_{j=1}^{d} \lambda_j X_t^j$ with $\sum_{j=1}^{d} \lambda_j^2 = 1$. So

$$\mathbb{E}\left[e^{iu \sum \lambda_j X_t^j}; A\right] = \mathbb{P}(A) e^{-u^2 t/2}.$$

Hence letting $u = |v|$ and $\lambda = v/|v|$,

$$\mathbb{E}\left[e^{iv \cdot X_t}; A\right] = \mathbb{E}\left[e^{iu\lambda \cdot X_t}; A\right] = e^{-|v|^2 t/2} \mathbb{P}(A).$$

As in the last paragraph of the proof of Theorem 5.9, this is sufficient to prove the corollary. □

(5.11) Theorem. *Suppose M is a continuous martingale with $\langle M \rangle_t$ strictly increasing and $\langle M \rangle_\infty = \infty$. Then M is a time change of Brownian motion and there exists a one-dimensional Brownian motion X such that $M_t = X_{\langle M \rangle_t}$.*

M_t is a time change of X_t if for some increasing process $\tau(t)$ we have $M_t = X_{\tau(t)}$.

Proof. Define $\tau(u) = \inf\{t : \langle M \rangle_t > u\}$ and let $X_u = M_{\tau(u)}$. Since M has continuous paths, then so does $\langle M \rangle_t$. Hence $\tau(u)$ will be continuous in u, and therefore so will X. If $u_1 < u_2$, then $\tau(u_1) < \tau(u_2)$ are stopping times, and by optional stopping,

$$\mathbb{E}\left[X_{u_2} | \mathcal{F}_{\tau(u_1)}\right] = \mathbb{E}\left[M_{\tau(u_2)} | \mathcal{F}_{\tau(u_1)}\right] = M_{\tau(u_1)} = X_{u_1}.$$

Therefore X_u is a martingale with respect to the σ-fields $\mathcal{F}_{\tau(u)}$. Similarly, $X_t^2 - t$ is a martingale, hence $\langle X \rangle_t \equiv t$. By Theorem 5.9, X is a Brownian motion, and undoing the time change, $M_t = X_{\langle M \rangle_t}$. □

The assumption that $\langle M \rangle_t$ is strictly increasing is unnecessary. Provided we consider stopped Brownian motions, so is the assumption that $\langle M \rangle_\infty \equiv \infty$. This is Exercise 25.

Martingale representations

In this subsection we show that every martingale adapted to the filtration generated by a Brownian motion can be written as a stochastic integral with respect to the Brownian motion.

(5.12) Theorem. *If $Y \in L^2$ is $\mathcal{F}_t$ measurable, there exists H_s predictable with $\mathbb{E} \int_0^t H_s^2 ds < \infty$ such that*

(5.10) $$Y = \mathbb{E}\,Y + \int_0^t H_s dX_s.$$

Proof. There are three main steps. By Itô's formula and the product formula,

$$
\begin{aligned}
e^{u^2 t/2} e^{iuX_t} &= 1 + \int_0^t e^{u^2 r/2} d(e^{iuX_r}) + \int_0^t e^{iuX_r} d(e^{u^2 r/2}) \\
&= 1 + \int_0^t e^{u^2 r/2} iu e^{iuX_r} dX_r + \left(\frac{-u^2}{2}\right) \int_0^t e^{u^2 r/2} e^{iuX_r} dr \\
&\quad + \int_0^t e^{iuX_r} \frac{u^2}{2} e^{u^2 r/2} dr \\
&= 1 + \int_0^t iu e^{u^2 r/2} e^{iuX_r} dX_r.
\end{aligned}
$$

Multiplying by the deterministic function $e^{-u^2 t/2}$,

$$e^{iuX_t} = e^{-u^2 t/2} + \int_0^t iu e^{u^2 (r-t)/2} e^{iuX_r} dX_r.$$

If we apply this to $X_t' = X_{t+s} - X_s$,

(5.11) $$e^{iu(X_{t+s}-X_s)} = e^{-u^2 t/2} + \int_s^{t+s} iu e^{u^2 (r-t)/2} e^{iu(X_{s+r}-X_s)} dX_r,$$

or (5.10) is true when $Y = e^{iu(X_{t+s}-X_s)}$.

For the second step, suppose we have

$$Y_i = \mathbb{E}\,Y_i + \int_0^t H_i(r) dX_r, \qquad i = 1, \dots, n$$

with $H_i(r) H_j(r) = 0$ if $i \neq j$. We claim (5.10) is true for $Y = Y_1 \cdots Y_n$, the product. We will do the case where $n = 2$, and an induction will give the case of general n. Let $Y_i(s) = \mathbb{E}\,[Y_i | \mathcal{F}_s] = \mathbb{E}\,Y_i + \int_0^s H_i(r) dX_r$, $i = 1, 2$. By the product formula,

$$
\begin{aligned}
Y_1 Y_2 &= Y_1(t) Y_2(t) \\
&= Y_1(0) Y_2(0) + \int_0^t Y_1(r) dY_2(r) + \int_0^t Y_2(r) dY_1(r) + \langle Y_1, Y_2 \rangle_t \\
&= \mathbb{E}\,Y_1 \mathbb{E}\,Y_2 + \int_0^t [Y_1(r) H_2(r) + Y_2(r) H_1(r)] dX_r + \langle Y_1, Y_2 \rangle_t.
\end{aligned}
$$

Note, however, that $\langle Y_1, Y_2 \rangle_t = \int_0^t H_1(r) H_2(r) dr = 0$.

For the third step, suppose $Y_n = \mathbb{E}\,Y_n + \int_0^t H_n(r) dX_r$ and $Y_n \to Y$ in L^2. Then $\mathbb{E}\,Y_n \to \mathbb{E}\,Y$ and

$$\mathbb{E} \int_0^t (H_n(r) - H_m(r))^2 dr = \mathbb{E}\left[\left(\int_0^t (H_n(r) - H_m(r))dX_r\right)^2\right]$$

$$= \mathbb{E}\left[((Y_n - Y_m) - (\mathbb{E}\,Y_n - \mathbb{E}\,Y_m))^2\right]$$

$$\leq 2\mathbb{E}\left[(Y_n - Y_m)^2\right] + 2(\mathbb{E}\,Y_n - \mathbb{E}\,Y_m)^2 \to 0$$

as $m, n \to \infty$. So $H_n(r)$ forms a Cauchy sequence with respect to the norm $\left(\mathbb{E} \int_0^t (\cdot)^2 dr\right)^{1/2}$. Therefore there exists H such that $H_n \to H$ in this norm, and $\mathbb{E} \int_0^t H^2(r)dr < \infty$. We have

$$\mathbb{E}\left[\left(Y - \mathbb{E}\,Y - \int_0^t H(r)dX_r\right)^2\right]$$

$$= \lim_{n\to\infty} \mathbb{E}\left[\left(Y_n - \mathbb{E}\,Y_n - \int_0^t H_n(r)dX_r\right)^2\right] = 0,$$

which implies (5.10) holds for Y.

We now finish the proof. By the first and second steps, random variables of the form $\prod_{j=1}^m e^{iu_j(X_{s_{j+1}} - X_{s_j})}$, where $s_1 \leq \cdots \leq s_{m+1}$, satisfy (5.10). Linear combinations of such random variables are dense in L^2, so (5.10) holds for all $Y \in L^2$.   □

(5.13) Corollary. *If M_t is a martingale with $M_0 = 0$ and $\mathbb{E}\,M_t^2 < \infty$, then there exists H_r predictable such that for each s we have $M_s = \int_0^s H_r dX_r$, a.s.*

Proof. Apply the above theorem to $Y = M_t$. (5.10) is true at time t. Since both sides of (5.10) are martingales, taking conditional expectations with respect to $\mathcal{F}_s$ establishes the corollary.   □

In particular, because stochastic integrals of Brownian motion have continuous paths, if M_t is a martingale with respect to the filtration of a Brownian motion, then M_t has a version that has continuous paths. (Two processes are versions of each other if for each t they are equal almost surely.)

Suppose $X_t = (X_t^1, \ldots, X_t^d)$ is d-dimensional Brownian motion.

(5.14) Corollary. *If Y is in $L^2(\mathbb{P})$ and $\mathcal{F}_t$ measurable, there exists $H_r = (H_r^1, \ldots, H_r^d)$ predictable with $\mathbb{E} \int_0^t (H_r^j)^2 dr < \infty$ for each j such that*

(5.12)
$$Y = \mathbb{E}\,Y + \int_0^t H_r \cdot dX_r.$$

Proof. We show that (5.12) holds for $e^{iu \cdot X_t}$, where $u = (u^1, \ldots, u^d)$, and then follow the proof of Theorem 5.12. By Theorem 5.12,

$$Y_j = e^{iu_j X_t^j} = \mathbb{E}\, e^{iu_j X_t^j} + \int_0^t K_r^j dX_r^j.$$

Let $H_j(r) = (0, \ldots, K_r^j, 0, \ldots, 0)$. If $Y_j(s) = \mathbb{E}\,[Y_j|\mathcal{F}_s] = \mathbb{E}\,Y_j + \int_0^s H_j(r) \cdot dX_r$, then $\langle Y_j, Y_k \rangle_t = \int_0^t H_j(r) \cdot H_k(r)dr = 0$ if $j \neq k$. By the argument of the second step of Theorem 5.12, $\prod_{j=1}^d Y_j$ is a random variable that satisfies (5.12). Since $\prod_{j=1}^d Y_j = e^{iu \cdot X_t}$, we are done. □

6 Stochastic calculus

This section is concerned with some aspects of stochastic calculus, namely stochastic differential equations, change of measure, inequalities, and additive functionals.

Stochastic differential equations

Let W_t be a one-dimensional Brownian motion. We want to show that if the coefficients are smooth enough there exists a solution to the stochastic differential equation

$$(6.1) \qquad dX_t = \sigma(X_t)dW_t + b(X_t)dt, \qquad X_0 = x_0,$$

and that any two solutions agree path by path. The equation (6.1) means

$$(6.2) \qquad X_t = x_0 + \int_0^t \sigma(X_s)dW_s + \int_0^t b(X_s)ds.$$

Suppose σ and b are two bounded Lipschitz functions. That is, there exists c such that

$$(6.3)\ \ |\sigma(x) - \sigma(y)| \leq c(|x-y| \wedge 1) \qquad \text{and} \qquad |b(x) - b(y)| \leq c(|x-y| \wedge 1).$$

We have the following.

(6.1) Theorem. *Let W_t be a one-dimensional Brownian motion, $x_0 \in \mathbb{R}$. There exists a process X_t satisfying (6.2). If X_t' is another process satisfying (6.2), then $X_t' = X_t$ for all t, a.s.*

Proof. For the existence of X_t, define $X_0(t) \equiv x_0$ and set

$$(6.4) \qquad X_{i+1}(t) = x_0 + \int_0^t \sigma(X_i(s))dW_s + \int_0^t b(X_i(s))ds, \qquad i \geq 0.$$

Since σ and b are bounded, for any t,

(6.5) $\mathbb{E}\,[X_1(t)^2] \le c\Big[x_0^2 + \int_0^t \sigma(x_0)^2 ds + \Big[\Big(\int_0^t b(x_0)ds\Big)^2\Big]\Big] \le c(x_0^2 + t + t^2).$

By Doob's inequality and the Cauchy-Schwarz inequality,

(6.6) $\mathbb{E}\,\sup_{r\le t}(X_{i+1}(r) - X_i(r))^2$

$$\le c\mathbb{E}\,\Big[\sup_{r\le t}\Big(\int_0^r [\sigma(X_i(s)) - \sigma(X_{i-1}(s))]dW_s\Big)^2\Big]$$

$$+ c\mathbb{E}\,\Big[\sup_{r\le t}\Big(\int_0^r |b(X_i(s)) - b(X_{i-1}(s))|ds\Big)^2\Big]$$

$$\le c\mathbb{E}\,\Big[\Big(\int_0^t [\sigma(X_i(s)) - \sigma(X_{i-1}(s))]dW_s\Big)^2\Big]$$

$$+ c\mathbb{E}\,\Big[\Big(\int_0^t |b(X_i(s)) - b(X_{i-1}(s))|ds\Big)^2\Big]$$

$$\le c\mathbb{E}\,\int_0^t |X_i(s) - X_{i-1}(s)|^2 ds + c\mathbb{E}\,\Big[\Big(\int_0^t |X_i(s) - X_{i-1}(s)|ds\Big)^2\Big]$$

$$\le c\int_0^t \mathbb{E}\,|X_i(s) - X_{i-1}(s)|^2 ds + ct\int_0^t \mathbb{E}\,|X_i(s) - X_{i-1}(s)|^2 ds.$$

Let $g_i(t) = \mathbb{E}\,\sup_{r\le t} |X_i(r) - X_{i-1}(r)|^2$. What (6.6) says, then, is that

(6.7) $$g_{i+1}(t) \le a\int_0^t g_i(s)ds, \qquad t \le t_0$$

for some constant a depending on t_0. By (6.5) we can choose a so that $g_1(t)$ is bounded by a as long as $t \le t_0$. By (6.7),

$$g_2(t) \le a\int_0^t a\,ds \le a^2 t.$$

Then

$$g_3(t) \le a\int_0^t a^2 s\,ds \le a^3 t^2/2.$$

Continuing, $g_i(t) \le a^i t^{i-1}/i!$. Since $\sum a^i t^{i-1}/i! < \infty$, this implies that $\mathbb{E}\,\sup_{s\le t} |X_m(s) - X_n(s)|^2$ can be made small if m and n are large. In other words X_n is a Cauchy sequence with respect to the norm $(\mathbb{E}\,\sup_{s\le t_0}(\cdot)^2)^{1/2}$. This space is complete, the sequence X_n converges in this norm, and by taking a subsequence, we see that the limit X_t is almost surely the uniform limit of continuous processes and hence is continuous. Taking the limit in (6.4) shows that X_t satisfies (6.3).

Now we look at the uniqueness. If X_t and X_t' are two solutions, let

$$g(t) = \mathbb{E}\,\sup_{r\le t} |X_r - X_r'|^2.$$

Since σ and b are bounded,

$$\mathbb{E} \sup_{r \le t} X_r^2 \le c x_0^2 + c \int_0^t \|\sigma\|_\infty^2 ds + c \left(\int_0^t \|b\|_\infty ds \right)^2,$$

and similarly with X' in place of X. Just as in (6.6) we have

$$(6.8) \qquad g(t) \le c\mathbb{E} \left[\sup_{r \le t} \left(\int_0^r [\sigma(X_s) - \sigma(X'_s)] dW_s \right)^2 \right]$$

$$+ c\mathbb{E} \left[\sup_{r \le t} \left(\int_0^r [b(X_s) - b(X'_s)] ds \right)^2 \right]$$

$$\le c\mathbb{E} \left[\left(\int_0^t [\sigma(X_s) - \sigma(X'_s)] dW_s \right)^2 \right]$$

$$+ c\mathbb{E} \left[\left(\int_0^t |b(X_s) - b(X'_s)| ds \right)^2 \right]$$

$$\le c\mathbb{E} \int_0^t |X_s - X'_s|^2 ds + ct\mathbb{E} \left[\int_0^t |X_s - X'_s|^2 ds \right]$$

$$\le c \int_0^t g(s) ds.$$

We then have that there exists a constant a depending on t_0 such that $g(t) \le a$ if $t \le t_0$ and $g(t) \le a \int_0^t g(s) ds$ for all $t \le t_0$. Similarly to the existence argument, $g(t) \le a \int_0^t a\, ds = a^2 t$, and then $g(t) \le a \int_0^t a^2 s\, ds = a^3 t^2/2$, etc. Since $a^i t^{i-1}/i! \to 0$ as $i \to \infty$, then $g(t) = 0$, and the uniqueness assertion follows. $\qquad \square$

The above proof goes through with virtually no changes if W_t is a d-dimensional Brownian motion, b is vector-valued, and σ is matrix-valued, and each of the components of b and σ are bounded and Lipschitz continuous.

(6.2) Theorem. *Suppose W_t, b, and σ are as just described. Let $x_0 \in \mathbb{R}^d$. Then there exists a solution X_t to the stochastic differential equation*

$$(6.9) \qquad X_t^i = x_0^i + \int_0^t \sum_{j=1}^d \sigma_{ij}(X_s) dW_s^j + \int_0^t b_i(X_s) ds,$$

and the solution is pathwise unique.

Let σ^t denote the transpose of σ, and let a be the matrix $\sigma\sigma^t$. Let L be the operator on C^2 defined by

$$(6.10) \qquad Lf(x) = \sum_{i,j=1}^d a_{ij}(x) \frac{\partial^2 f}{\partial x^i \partial x^j}(x) + \sum_{i=1}^d b_i(x) \frac{\partial f}{\partial x^i}(x).$$

L is an example of an elliptic operator.

The connection between probability theory and elliptic operators is given by the following simple theorem.

(6.3) Theorem. *Suppose X_t is a solution to (6.9), where σ and b are bounded and measurable, but no other assumptions are placed on them. If $f \in C^2(\mathbb{R}^d)$, then*

$$(6.11) \qquad f(X_t) = f(X_0) + M_t + \int_0^t Lf(X_s)ds,$$

where

$$(6.12) \qquad M_t = \int_0^t \sum_{i,j=1}^d \frac{\partial f}{\partial x^i}(X_s)\sigma_{ij}(X_s)dW_s^j$$

is a local martingale.

Proof. This is just Itô's formula. Note $\langle W^k, W^l \rangle_s = \delta_{k\ell}s$, so

$$d\langle X^i, X^j \rangle_s = d\Big\langle \int_0^{\cdot} \sum_{k=1}^d \sigma_{ik}(X_r)dW_r^k, \int_0^{\cdot} \sum_{\ell=1}^d \sigma_{j\ell}(X_r)dW_r^\ell \Big\rangle_s$$

$$= \sum_{k,\ell=1}^d \sigma_{ik}(X_s)\sigma_{j\ell}(X_s)d\langle W^k, W^\ell \rangle_s = a_{ij}(X_s)ds.$$

Then by Itô's formula,

$$f(X_t) = f(X_0) + \int_0^t \nabla f(X_s) \cdot dX_s$$

$$+ \frac{1}{2}\int_0^t \sum_{i,j=1}^d \frac{\partial^2 f}{\partial x^i \partial x^j}(X_s)d\langle X^i, X^j \rangle_s$$

$$= f(X_0) + \sum_{i,j=1}^d \int_0^t \frac{\partial f}{\partial x^i}(X_s)\sigma_{ij}(X_s)dW_s^j$$

$$+ \sum_{i=1}^d \int_0^t \frac{\partial f}{\partial x^i}(X_s)b_i(X_s)ds + \frac{1}{2}\int_0^t \sum_{i,j=1}^d a_{ij}(X_s)\frac{\partial^2 f}{\partial x^i \partial x^j}(X_s)ds$$

$$= f(X_0) + M_t + \int_0^t Lf(X_s)ds.$$

$\square$

See Stroock and Varadhan [1] and Ikeda and Watanabe [1] for some of the implications of this connection.

Change of measure

If Y_t is a continuous local martingale, let us apply Itô's formula to $X_t = Y_t - (1/2)\langle Y \rangle_t$ with $f(x) = e^x$. Then

$$Z_t = \exp(Y_t - \langle Y \rangle_t/2) = 1 + \int_0^t e^{X_s} dX_s + \frac{1}{2} \int_0^t e^{X_s} d\langle X \rangle_s.$$

Since $\langle X \rangle_s = \langle Y \rangle_s$ (recall that the quadratic variation of a semimartingale is defined to be the quadratic variation of the martingale part), then

$$(6.13) \qquad\qquad Z_t = 1 + \int_0^t Z_s dY_s.$$

The martingale Z_t is called an exponential martingale and is sometimes written $\mathcal{E}(Y)_t$. (6.13) can be written $dZ_t = Z_t dY_t$.

Conversely, if Z_t is a strictly positive continuous martingale with $Z_0 \equiv 1$, let $Y_t = \int_0^t (Z_s)^{-1} dZ_s$. The stochastic integral can be defined because we assumed that $Z_t > 0$ for all t; by continuity, with probability one, $(Z_t)^{-1}$ is bounded on any finite time interval. Then using Exercise 23, $Z_t = 1 + \int_0^t Z_s dY_s$.

Now let M_t be a positive continuous martingale with $M_0 = 1$. Let us define a new probability measure $\mathbb{Q}$ by $\mathbb{Q}(A) = \mathbb{E}[M_t; A]$ if $A \in \mathcal{F}_t$, i.e., the Radon-Nikodym derivative of $\mathbb{Q}$ with respect to $\mathbb{P}$ on $\mathcal{F}_t$ is M_t. If $A \in \mathcal{F}_s \subseteq \mathcal{F}_t$, then $\mathbb{E}[M_t; A] = \mathbb{E}[M_s; A]$, since M is a martingale, and so this definition is consistent.

Let X_t be a continuous martingale with respect to $\mathbb{P}$. Then it turns out that X_t is a semimartingale with respect to $\mathbb{Q}$, and we can explicitly write its decomposition into martingale and bounded variation parts.

(6.4) Theorem. *If X_t is a continuous martingale with respect to $\mathbb{P}$, then $X_t - D_t$ is a martingale with respect to $\mathbb{Q}$, where $D_t = \int_0^t (M_s)^{-1} d\langle X, M \rangle_s$. The quadratic variation of $X_t - D_t$ under the probability $\mathbb{Q}$ is the same as that of X_t under $\mathbb{P}$.*

This theorem is known both as the Girsanov transformation and the Cameron-Martin formula. If $M_t = \mathcal{E}(Y)_t$, then D_t is equal to $\langle X, Y \rangle_t$.

Proof. Assume without loss of generality that $X_0 = 0$. Denote expectations by $\mathbb{E}_{\mathbb{Q}}$ or $\mathbb{E}_{\mathbb{P}}$. If $A \in \mathcal{F}_s$,

$$\mathbb{E}_{\mathbb{Q}}[X_t; A] = \mathbb{E}_{\mathbb{P}}[M_t X_t; A]$$

$$= \mathbb{E}_{\mathbb{P}}\Big[\int_0^t M_r dX_r; A\Big] + \mathbb{E}_{\mathbb{P}}\Big[\int_0^t X_r dM_r; A\Big] + \mathbb{E}_{\mathbb{P}}[\langle X, M \rangle_t; A]$$

$$= \mathbb{E}_{\mathbb{P}}\Big[\int_0^s M_r dX_r; A\Big] + \mathbb{E}_{\mathbb{P}}\Big[\int_0^s X_r dM_r; A\Big] + \mathbb{E}_{\mathbb{P}}[\langle X, M \rangle_t; A]$$

$$= \mathbb{E}_{\mathbb{Q}}[X_s; A] + \mathbb{E}_{\mathbb{P}}[\langle X, M \rangle_t - \langle X, M \rangle_s; A],$$

using the integration-by-parts formula (Corollary 5.7).

On the other hand,

$$\mathbb{E}_{\mathbb{Q}}[D_t - D_s; A] = \mathbb{E}_{\mathbb{P}}[(D_t - D_s)M_t; A] = \mathbb{E}_{\mathbb{P}}\Big[\int_s^t M_t dD_r; A\Big]$$

$$= \mathbb{E}_{\mathbb{P}}\Big[\int_s^t \mathbb{E}_{\mathbb{P}}[M_t|\mathcal{F}_r]dD_r; A\Big] = \mathbb{E}_{\mathbb{P}}\Big[\int_s^t M_r dD_r; A\Big]$$

$$= \mathbb{E}_{\mathbb{P}}\Big[\int_s^t d\langle X, M\rangle_r; A\Big] = \mathbb{E}_{\mathbb{P}}[\langle X, M\rangle_t - \langle X, M\rangle_s; A].$$

We used Exercise 28 in the third equality. This shows that $X_t - D_t$ is a martingale with respect to $\mathbb{Q}$.

The assertion regarding the quadratic variation can be shown the same way, but it is easier to use Theorem 4.23, which says that $\langle X\rangle_t$ is the $\mathbb{P}$-a.s. limit along some subsequence of $\sum_{j=1}^n (X_{jt/n} - X_{(j-1)t/n})^2$. Since, under $\mathbb{Q}$, $\langle X\rangle_t = \langle X - D\rangle_t$, and $\mathbb{Q}$ and $\mathbb{P}$ are equivalent, the limit must be the same. $\qquad\square$

We will use the Girsanov theorem to give a proof of the support theorem for Brownian motion. We start with the following.

(6.5) Proposition. *If X_t is a d-dimensional Brownian motion and $\varepsilon, t > 0$, then*

$$\mathbb{P}^0(\sup_{s\leq t} |X_s| < \varepsilon) > 0.$$

In Sect. II.4 we will give an exact expression for this quantity when $d = 1$.

Proof. We show this result first for one-dimensional Brownian motion. Then the general case follows from

$$\mathbb{P}^0(\sup_{s\leq t} |X_s| < \varepsilon) \geq \mathbb{P}^0(\sup_{s\leq t} |X_s^i| < \varepsilon/\sqrt{d}, i = 1, \ldots, d)$$

$$\geq \Big(P^0(\sup_{s\leq t} |X_s^1| < \varepsilon/\sqrt{d})\Big)^d > 0.$$

If we show, in the case $d = 1$, that

(6.14) $$\mathbb{P}^0(\sup_{s\leq u} |X_s| < 1) > 0,$$

we will achieve our desired result by replacing u by $t/\sqrt{\varepsilon}$ and using scaling.

Note that if $|x| \leq 1/4$ and n is sufficiently large, then $\mathbb{P}^x(|X_{1/n}| \leq 1/4) \geq 1/4$. This is because $\mathbb{P}^x(|X_{1/n} - X_0| < 1/4$ and $X_{1/n} < X_0) \to 1/2$ as $n \to \infty$ by the symmetry of the normal distribution. By Proposition 1.9,

$$\sup_{|x|\leq 1/4} \mathbb{P}^x(\sup_{s\leq 1/n} |X_s - X_0| > 1/4) \leq 2e^{-n/32}.$$

Therefore

$$\inf_{|x|\leq 1/4} \mathbb{P}^x\left(\sup_{s\leq 1/n} |X_s| < 1/2, |X_{1/n}| \leq 1/4\right) \geq (1/4) - 2e^{-n/32}.$$

If we take n sufficiently large, this will be greater than $1/8$.

Let $I_m = (\sup_{s\leq 1/n} |X_{s+m/n} - X_{m/n}| < 1/2, |X_{(m+1)/n}| \leq 1/4)$. By the Markov property,

$$\mathbb{P}(I_1 \cap I_2 \cap \cdots \cap I_m) \geq (1/8)^m.$$

Note $(\sup_{s\leq m/n} |X_s| < 1) \supseteq I_1 \cap \cdots \cap I_m$, hence taking $m > nu$ proves (6.14) and hence the proposition.   □

(6.6) Theorem. *If* $\psi : [0,t] \to \mathbb{R}^d$ *is continuous,* $\varepsilon > 0$, *and* $x = \psi(0)$, *then*

$$\mathbb{P}^x(\sup_{s\leq t} |X_s - \psi(s)| < \varepsilon) > c,$$

where c *can be taken to depend only on* t, ε, *and the modulus of continuity of* ψ.

Proof. Let us take $\widetilde{\psi}$ such that $\widetilde{\psi}$ has a bounded derivative and such that $\sup_{s\leq t} |\psi(s) - \widetilde{\psi}(s)| < \varepsilon/2$. We can choose $\widetilde{\psi}$ so that $\|\widetilde{\psi}'\|_\infty$ depends only on t, ε, and the modulus of continuity of ψ. We will show

$$(6.15) \qquad \mathbb{P}^x(\sup_{s\leq t} |X_s - \widetilde{\psi}(s)| < \varepsilon/2) > c,$$

where c depends only on t, ε, and $\|\widetilde{\psi}'\|_\infty$; this will give us our result.

Define $\mathbb{Q}$ by $d\mathbb{Q}/d\mathbb{P} = M_t$, where

$$M_t = \exp\left(\int_0^t \widetilde{\psi}'(s) \cdot dX_s - (1/2)\int_0^t (|\widetilde{\psi}'(s)|^2 ds\right).$$

Note

$$\left\langle X, \int_0^\cdot \widetilde{\psi}'(s) \cdot dX_s \right\rangle_t = \int_0^t \widetilde{\psi}'(s)\, ds = \widetilde{\psi}(t) - \widetilde{\psi}(0).$$

So $X_t - \widetilde{\psi}(t)$ is a process, each of whose components under $\mathbb{Q}$ is a continuous martingale, and the quadratic variation is the same as that of X under $\mathbb{P}$. Hence, by Corollary 5.10, under $\mathbb{Q}$, $X(t) - \widetilde{\psi}(t)$ is a Brownian motion started at 0. By Proposition 6.5, $\mathbb{Q}(A) \geq c$, where $A = (\sup_{s\leq t} |X(s) - \widetilde{\psi}(s)| < \varepsilon/2)$, and c depends only on ε and t. Since $|\widetilde{\psi}'|$ is bounded, $\mathbb{E}\, M_t^2 < \infty$ and

$$c \leq \mathbb{Q}(A) = \int_A M_t\, d\mathbb{P} \leq \left(\mathbb{E}\, M_t^2\right)^{1/2}\left(\mathbb{P}(A)\right)^{1/2},$$

which proves (6.15).   □

As another example of the use of change of measure, we have Doob's h-path transforms. Let D be a domain and let X_t be Brownian motion killed on exiting the domain. One would like to give a precise meaning to the intuitive notion of Brownian motion conditioned to exit the domain at a certain point. Let h be a positive harmonic function (i.e., h is C^2 in D, and $\Delta h = 0$ (see Sect. II.1)) that, say, is 0 everywhere on the boundary of D except at one point z. We will see later that the Poisson kernel for the ball or half-space gives an example of such a harmonic function. Then, heuristically, we have by the Markov property at time t,

$$
\begin{aligned}
\mathbb{P}^x(X_t \in dy | X_{\tau_D} = z) &= \frac{\mathbb{P}^x(X_t \in dy, X_{\tau_D} = z)}{\mathbb{P}^x(X_{\tau_D} = z)} \\
&= \frac{\mathbb{P}^x(X_t \in dy)\mathbb{P}^y(X_{\tau_D} = z)}{\mathbb{P}^x(X_{\tau_D} = z)}.
\end{aligned}
$$

If $p^0(t, x, dy)$ represents the probability that Brownian motion started at x and killed on leaving D is in dy at time t, we then expect that the analogous probability for Brownian motion conditioned to exit D at z ought to be $h(y)p^0(t, x, dy)/h(x)$.

With this in mind, let D be a domain, h a positive harmonic function on D. Let $\widehat{X}$ be Brownian motion killed on exiting D. Since $\Delta h = 0$ in D, (5.7) tells us that $h(\widehat{X}_{t \wedge \tau_D})$ is a martingale. So if we let $M_t = h(\widehat{X}_{t \wedge \tau_D})/h(\widehat{X}_0)$, M_t is a positive continuous martingale with $M_0 = 1$, $\mathbb{P}^x$-a.s.

We define the h-path transform of Brownian motion by setting

$$
(6.16) \qquad \mathbb{P}^x_h(A) = \mathbb{E}^x[M_t; A], \qquad A \in \mathcal{F}_t.
$$

Let us examine some properties of the $\mathbb{P}^x_h$s.

First of all, since $M_0 = 1$, $\mathbb{P}^x_h(\Omega) = 1$. By Itô's formula,

$$
(6.17) \qquad M_t = 1 + \int_0^{t \wedge \tau_D} \nabla h(\widehat{X}_s) dX_s / h(X_0),
$$

so by Theorem 6.4 $\widehat{X}_t - D_t$ is a martingale with

$$
(6.18) \qquad D_t = \int_0^t (M_s)^{-1} d\langle X, M \rangle_s = \int_0^{t \wedge \tau_D} \frac{\nabla h(X_s)}{h(X_s)} ds.
$$

By Theorems 6.4 and 5.9, under $\mathbb{P}^x_h$ the process $W_t = \widehat{X}_t - D_t$ is a Brownian motion up until the first exit of D. $\widehat{X}_t$ satisfies the stochastic differential equation

$$
(6.19) \qquad \widehat{X}_t = \widehat{X}_0 + W_{t \wedge \tau_D} + \int_0^{t \wedge \tau_D} \frac{\nabla h(\widehat{X}_s)}{h(\widehat{X}_s)} ds.
$$

We have the following.

(6.7) Proposition. $(\mathbb{P}_h^x, \widehat{X}_t)$ *forms a strong Markov process.*

Proof. Suppose $A \in \mathcal{F}_s$. Then

$$\mathbb{E}_h^x[f(\widehat{X}_{t+s}); A] = \frac{\mathbb{E}^x[f(\widehat{X}_{t+s})h(\widehat{X}_{t+s}); A]}{h(x)}$$

$$= \frac{\mathbb{E}^x[\mathbb{E}^{\widehat{X}_s}[f(\widehat{X}_t)h(\widehat{X}_t)]; A]}{h(x)}$$

by the Markov property for $\widehat{X}$. This is equal to

$$\mathbb{E}^x\big[\mathbb{E}_h^{\widehat{X}_s}[f(\widehat{X}_t)]h(\widehat{X}_s)1_A\big]/h(x) = E_h^x[\mathbb{E}_h^{\widehat{X}_s}f(\widehat{X}_t); A].$$

As in Proposition 3.1, the Markov property follows from this. The strong Markov property is proved in almost the identical fashion.   □

Further information on h-path transforms can be found in Sect. III.2.

Inequalities

Next we turn to three sets of basic inequalities. The first set consists of the inequalities of Burkholder, Davis, and Gundy.

(6.8) Theorem. *Let M_t be a continuous martingale with $M_0 \equiv 0, 0 < p < \infty$. There exist constants c_1 and c_2 such that for any stopping time T,*

$$\mathbb{E}\,(M_T^*)^p \le c_1 \mathbb{E}\,T^{p/2} \qquad and \qquad \mathbb{E}\,T^{p/2} \le c_2 \mathbb{E}\,(M_T^*)^p.$$

Recall $M_t^* = \sup_{s \le t} |M_s|$. The proof of Theorem 6.8 relies on the following lemma; (6.20) is known as the "good-λ" inequality.

(6.9) Lemma. *Suppose $0 < p < \infty$. Suppose X, Y are positive random variables, $\beta > 1$, $\delta \in (0,1)$, and $\varepsilon < \beta^{-p}/2$ such that for all $\lambda > 0$*

$$(6.20) \qquad \mathbb{P}(X > \beta\lambda, Y < \delta\lambda) \le \varepsilon\mathbb{P}(X \ge \lambda).$$

Then there exists c depending only on p, β, δ, and ε such that $\mathbb{E}\,X^p \le c\mathbb{E}\,Y^p$.

Proof. First suppose X is bounded. We have

$$\mathbb{P}(X > \beta\lambda) \le \mathbb{P}(X > \beta\lambda, Y < \delta\lambda) + \mathbb{P}(Y \ge \delta\lambda)$$
$$\le \varepsilon\mathbb{P}(X \ge \lambda) + \mathbb{P}(Y \ge \delta\lambda).$$

So

$$\mathbb{P}(X/\beta > \lambda) \le \varepsilon\mathbb{P}(X \ge \lambda) + \mathbb{P}(Y/\delta \ge \lambda).$$

Multiply by $p\lambda^{p-1}$, integrate over λ from 0 to ∞, and use Proposition 1.5. Then

$$\mathbb{E}\left[(X/\beta)^p\right] \le \varepsilon \mathbb{E}\, X^p + \mathbb{E}\left[(Y/\delta)^p\right].$$

Subtracting $\varepsilon \mathbb{E}\, X^p$ from both sides, we have

$$\frac{\mathbb{E}\, X^p}{2\beta^p} \le \frac{\mathbb{E}\, Y^p}{\delta^p},$$

which gives our result when X is bounded.

If X is not bounded, notice that if inequality (6.20) holds for X, it also holds for $X \wedge N$, so we get

$$\mathbb{E}\left[(X \wedge N)^p\right] \le \frac{2\beta^p}{\delta^p}\mathbb{E}\, Y^p.$$

Let $N \to \infty$ and use monotone convergence. $\qquad\square$

We now prove Theorem 6.8.

Proof of Theorem 6.8. Since any continuous martingale is a time change of a Brownian motion, it suffices to prove the result for Brownian motion. Let $U = \inf\{t : X_t^* > \lambda\}$. We have

$$
\begin{aligned}
\mathbb{P}(X_T^* &> \beta\lambda, T^{1/2} < \delta\lambda) \\
&= \mathbb{P}(X_T^* > \beta\lambda, T < \delta^2\lambda^2, U < \infty) \\
&\le \mathbb{P}\big(\sup_{U \le t \le T} |X_t - X_U| > (\beta - 1)\lambda, T < \delta^2\lambda^2, U < \infty\big) \\
&\le \mathbb{P}\big(\sup_{U \le t \le U + \delta^2\lambda^2} |X_t - X_U| > (\beta - 1)\lambda, U < \infty\big) \\
&= \mathbb{E}\Big[\mathbb{P}^{X_U}\big(\sup_{t \le \delta^2\lambda^2} |X_t - X_0| > (\beta - 1)\lambda\big); U < \infty\Big],
\end{aligned}
$$

using the strong Markov property at the last step. By scaling,

$$\sup_y \mathbb{P}^y\big(\sup_{t \le \delta^2\lambda^2} |X_t - X_0| > (\beta - 1)\lambda\big) = \mathbb{P}^0\big(X_{\delta^2}^* > (\beta - 1)\big),$$

which can be made less than ε if we let $\beta = 2$ and take δ sufficiently small. So we see

$$\mathbb{P}(X_T^* > \beta\lambda, T^{1/2} < \delta\lambda) \le \varepsilon \mathbb{P}(U < \infty) = \varepsilon \mathbb{P}(X_T^* > \lambda).$$

Now use Lemma 6.9 to obtain the first assertion.

The second is along the same lines. We have

$$
\begin{aligned}
\mathbb{P}(T^{1/2} &> \beta\lambda, X_T^* < \delta\lambda) \\
&\le \mathbb{P}\big(T > \beta^2\lambda^2, \sup_{\lambda^2 \le t \le \beta^2\lambda^2} |X_t - X_{\lambda^2}| < 2\delta\lambda\big) \\
&\le \mathbb{P}\big(T > \lambda^2, \sup_{\lambda^2 \le t \le \beta^2\lambda^2} |X_t - X_{\lambda^2}| < 2\delta\lambda\big) \\
&= \mathbb{E}\Big[\mathbb{P}^{X_{\lambda^2}}\big(\sup_{t \le (\beta^2 - 1)\lambda^2} |X_t - X_0| < 2\delta\lambda\big); T > \lambda^2\Big].
\end{aligned}
$$

With $\beta = 2$, if we take δ sufficiently small,

$$\sup_y \mathbb{P}^y \Big(\sup_{t \leq (\beta^2 - 1)\lambda^2} |X_t - X_0| < 2\delta\lambda \Big) = \mathbb{P}^0(X^*_{\beta^2 - 1} < 2\delta) \leq \varepsilon$$

by scaling. Hence

$$\mathbb{P}(T^{1/2} > \beta\lambda, X^*_T < \delta\lambda) \leq \varepsilon \mathbb{P}(T > \lambda^2).$$

Now use Lemma 6.9 again.   □

The next type of inequality is for increasing processes.

(6.10) Theorem. *Suppose A_t is a continuous increasing process with $A_0 = 0$ and suppose there exists a random variable B such that*

$$(6.21) \qquad\qquad \mathbb{E}\left[A_\infty - A_t | \mathcal{F}_t\right] \leq \mathbb{E}\left[B | \mathcal{F}_t\right], \qquad \text{a.s.}$$

for each t. Then if $p \geq 1$, there exists c such that $\mathbb{E}\, A^p_\infty \leq c\mathbb{E}\, B^p$; c depends only on p and not A or B.

Proof. The case $p = 1$ follows simply by setting $t = 0$ and taking expectations. So let us suppose $p > 1$. First assume A is bounded. Since $A^p_\infty = A_\infty \int_0^\infty d(A^{p-1}_t)$ and $\int_0^\infty A_t d(A^{p-1}_t) = A^p_\infty - \int_0^\infty A^{p-1}_t dA_t = A^p_\infty - A^p_\infty/p$, we have

$$\mathbb{E}\, A^p_\infty = p\mathbb{E} \int_0^\infty (A_\infty - A_t) dA^{p-1}_t$$

$$= p\mathbb{E} \int_0^\infty \mathbb{E}\left[A_\infty - A_t | \mathcal{F}_t\right] dA^{p-1}_t$$

$$\leq p\mathbb{E} \int_0^\infty \mathbb{E}\left[B | \mathcal{F}_t\right] dA^{p-1}_t$$

$$= p\mathbb{E} \int_0^\infty B\, dA^{p-1}_t$$

$$\leq p(\mathbb{E}\, B^p)^{1/p} (\mathbb{E}\, A^p_\infty)^{(p-1)/p}.$$

Here we used Exercise 28 twice and Hölder's inequality. Dividing through by $(\mathbb{E}\, A^p_\infty)^{(p-1)/p}$ gives our result.

If A_t is not bounded, note that $A_{t \wedge T_N}$ satisfies (6.21) with $T_N = \inf\{t : A_t > N\}$. Apply the above argument to $A_{t \wedge T_N}$ to deduce $\mathbb{E}\, A^p_{t \wedge T_N} \leq c\mathbb{E}\, B^p$, let $N \to \infty$, and apply monotone convergence.   □

The next inequality is extremely useful, and will, for example, be used in Chap. IV to give us the John-Nirenberg inequality for BMO functions as a consequence.

If X_t is a process with left and right limits at each t, let the jump at time t be given by

$$\Delta X_t = \lim_{u \to t, u > t} X_u - \lim_{s \to t, s < t} X_s.$$

(6.11) Theorem. *Suppose X_t has paths that are left continuous with right limits and for every stopping time T,*

$$\mathbb{E}\left[|X_\infty - X_T|\,|\mathcal{F}_T\right] \leq 1, \qquad \text{a.s.}$$

Suppose also that

$$\sup_t |\Delta X_t| \leq 1, \qquad \text{a.s.}$$

Then there exists a constant α, not depending on X, such that

$$\mathbb{E}\,\exp(\alpha X_\infty^*) \leq 2.$$

Proof. If A is $\mathcal{F}_T$ measurable, the hypotheses tell us that

$$(6.22) \qquad \mathbb{E}\left[|X_\infty - X_T|; A\right] \leq \mathbb{P}(A).$$

Then

$$(6.23) \quad \mathbb{P}(|X_\infty - X_T| > b; A) \leq \mathbb{E}\left[\frac{|X_\infty - X_T|}{b}; A \cap (|X_\infty - X_T| > b)\right]$$
$$\leq b^{-1}\mathbb{E}\left[|X_\infty - X_T|; A\right] \leq \mathbb{P}(A)/b.$$

Note

$$(6.24) \qquad \mathbb{P}(X_\infty^* > \lambda + 20) \leq \mathbb{P}(X_\infty^* > \lambda + 20, |X_\infty| \leq \lambda + 10)$$
$$+ \mathbb{P}(|X_\infty| > \lambda + 10).$$

To bound this, let $T_1 = \inf\{t : |X_t| \geq \lambda + 15\}$. By the assumption on the jumps of X_t,

$$|X_{T_1}| \geq \lambda + 14.$$

So by (6.23) the first term on the right is less than or equal to

$$\mathbb{P}(|X_\infty| < \lambda + 10, T_1 < \infty) \leq \mathbb{P}(|X_\infty - X_{T_1}| > 4; T_1 < \infty)$$
$$\leq \mathbb{P}(T_1 < \infty)/4 \leq \mathbb{P}(X_\infty^* > \lambda)/4.$$

Let $T_2 = \inf\{t : |X_t| \geq \lambda + 5\}$. Then $|X_{T_2}| \leq \lambda + 5$ and the second term on the right of (6.24) is less than or equal to

$$\mathbb{P}(|X_\infty| > \lambda + 10, T_2 < \infty) \leq \mathbb{P}(|X_\infty - X_{T_2}| > 5; T_2 < \infty)$$
$$\leq \mathbb{P}(T_2 < \infty)/5 \leq \mathbb{P}(X_\infty^* > \lambda)/5.$$

Therefore

$$\mathbb{P}(X_\infty^* > \lambda + 20) \leq \mathbb{P}(X_\infty^* > \lambda)/2.$$

Taking $\lambda = 0, 20, 40, \ldots$, induction gives $\mathbb{P}(X_\infty^* > 40m) \leq 2^{-m}$. The result follows easily. $\qquad\square$

(6.12) Corollary. *There exists* $\alpha > 0$ *such that if* X_n *is a sequence with* $\mathbb{E}\left[|X_\infty - X_n| \,|\mathcal{F}_n\right] \leq 1$, *a.s., and* $|X_{n+1} - X_n| \leq 1$, *a.s., for all* n, *then* $\mathbb{E}\exp(\alpha X_\infty^*) \leq 2$.

Proof. Define $Y_t = X_k$, $\mathcal{F}_t = \mathcal{F}_k$ if $k < t \leq k+1$. Then Y_t is left continuous with jumps bounded by 1. Looking at the proof of Theorem 6.11, we see that minor modifications will give us our result provided we can replace (6.22) by

$$(6.25) \qquad \mathbb{E}\left[|Y_\infty - Y_S|; S < \infty\right] \leq 2\mathbb{P}(S < \infty)$$

for $S = \inf\{t : |Y_t| > b\}$. Since our Y_t is piecewise constant, S takes only integer values. Also if $A \in \mathcal{F}_k$, our hypothesis implies that

$$\mathbb{E}\left[|Y_\infty - Y_k|; A\right] \leq \mathbb{E}\left[|X_\infty - X_k|; A\right] + \mathbb{E}\left[|X_k - X_{k+1}|; A\right] \leq 2\mathbb{P}(A),$$

since $Y_k = X_{k+1}$. Then

$$\begin{aligned}
\mathbb{E}\left[|Y_\infty - Y_S|; S < \infty\right] &= \sum_{k=0}^\infty \mathbb{E}\left[|Y_\infty - Y_S|; S = k\right] \\
&= \sum_{k=0}^\infty \mathbb{E}\left[|Y_\infty - Y_k|; S = k\right] \\
&\leq \sum_{k=0}^\infty 2\mathbb{P}(S = k) = 2\mathbb{P}(S < \infty),
\end{aligned}$$

which is (6.25). $\qquad\qquad\square$

An interesting application of Corollary 6.12 is Khintchine's inequalities. Let X_i be a sequence of independent random variables with $\mathbb{P}(X_i = 1) = \mathbb{P}(X_i = -1) = 1/2$, $i = 0, 1, 2, \ldots$.

(6.13) Theorem. (Khintchine's inequalities) *If* $p > 1$, *there exist constants* c_1, c_2 *such that*

$$\left(\mathbb{E}\left|\sum_{i=0}^\infty a_i X_i\right|^p\right)^{1/p} \leq c_1 \left(\sum_{i=0}^\infty a_i^2\right)^{1/2} \leq c_2 \left(\mathbb{E}\left|\sum_{i=0}^\infty a_i X_i\right|^p\right)^{1/p}.$$

Proof. Let us normalize so that $\sum_{i=0}^\infty a_i^2 = 1$. Let $M_n = \sum_{i=0}^n a_i X_i$ and let $\mathcal{F}_n$ be the σ-field generated by $X_1, \ldots, X_n$. M_n is a martingale, and since $\mathbb{E}\, X_i X_j = \mathbb{E}\, X_i \mathbb{E}\, X_j = 0$ if $i \neq j$, then $\mathbb{E}\, M_\infty^2 = \sum_i a_i^2 = 1$. The inequalities $\|M_\infty\|_p \leq \|M_\infty\|_2$ for $p \leq 2$ and $\|M_\infty\|_2 \leq \|M_\infty\|_p$ for $p \geq 2$ are immediate consequences of Jensen's inequality.

Let us now prove

$$(6.26) \qquad \|M_\infty\|_p \leq c\|M_\infty\|_2, \qquad p > 2.$$

Using Proposition 1.11 and the independence of the X_is,

$$\left(\mathbb{E}\left[|M_\infty - M_n| \,|\mathcal{F}_n \right] \right)^2 \le \mathbb{E}\left[(M_\infty - M_n)^2 |\mathcal{F}_n \right]$$

$$= \mathbb{E}\left[\left(\sum_{i=n+1}^\infty a_i X_i \right)^2 |\mathcal{F}_n \right] = \mathbb{E}\left[\left(\sum_{i=n+1}^\infty a_i X_i \right)^2 \right]$$

$$= \sum_{i=n+1}^\infty a_i^2 \le 1.$$

Note that for each k, $|M_{k+1} - M_k| \le |a_{k+1}| \le (\sum_{i=1}^\infty a_i^2)^{1/2} = 1$. So by Corollary 6.12, there exists α independent of M such that $\mathbb{E}\, e^{\alpha M_\infty^*} \le 2$, hence $\mathbb{E}\, M_\infty^p \le c$. This proves (6.26).

The other inequality,

(6.27) $$\|M_\infty\|_2 \le c\|M_\infty\|_p, \qquad 1 < p < 2,$$

follows from this by duality. Choose q so that $p^{-1} + q^{-1} = 1$, hence $q > 2$. Let $L_n = \sum_{i=0}^n b_i X_i$, and suppose $\sum_i b_i^2 \le 1$. By what we have just proved,

$$\sum_{k=0}^\infty a_k b_k = \mathbb{E} \sum (M_{k+1} - M_k)(L_{k+1} - L_k)$$

$$= \mathbb{E}\, M_\infty L_\infty$$
$$\le \|M_\infty\|_p \|L_\infty\|_q$$
$$\le c\|M_\infty\|_p \|L_\infty\|_2 \le c\|M_\infty\|_p.$$

The second equality follows from (4.4) by polarization. Now take the supremum over bs with $\sum_i b_i^2 \le 1$ and use the duality of ℓ^2 with itself to obtain (6.27). (ℓ^2 is the Hilbert space of sequences $a = (a_0, a_1, \ldots)$ with inner product $\langle a, b \rangle = \sum_{i=0}^\infty a_i \bar{b}_i$.) $\square$

Khintchine's inequalities are usually stated in terms of the Rademacher functions $r_i(t)$ defined by $r_i(t) = [k/2^i, (k+1)/2^i) = +1$ if k is even, $= -1$ if k is odd. It is quite easy to see that the the r_i, viewed as random variables on $[0, 1)$ with $\mathbb{P}$ equal to Lebesgue measure, are mean 0, are independent, and $\mathbb{P}(r_i = 1) = \mathbb{P}(r_i = -1) = 1/2$.

As another application of Theorem 6.11, we have the following.

(6.14) Proposition. *Suppose $A_t^{(1)}$, $A_t^{(2)}$ are increasing continuous processes with $A_0^{(i)} \equiv 0$, $i = 1, 2$, and let $B_t = A_t^{(1)} - A_t^{(2)}$. Suppose there exist N and δ such that for all stopping times T*

(6.28) $$\mathbb{E}\,[A_\infty^{(i)} - A_T^{(i)} |\mathcal{F}_T] \le N, \qquad \text{a.s.,} \qquad i = 1, 2,$$

and

(6.29) $|\mathbb{E}\left[B_\infty - B_T | \mathcal{F}_T\right]| \leq \delta,$ a.s.

Then if $p \geq 1$, there exists c such that

$$\mathbb{E} \sup_t |B_t|^p \leq c\delta^{p/2} N^{p/2}.$$

Proof. We have

$$(B_\infty - B_T)^2 = 2 \int_T^\infty (B_\infty - B_t)dB_t,$$

and

$$\mathbb{E}\left[(B_\infty - B_T)^2 | \mathcal{F}_T\right] = 2\mathbb{E}\left[\int_T^\infty \mathbb{E}\left[B_\infty - B_t | \mathcal{F}_t\right]dB_t | \mathcal{F}_T\right]$$

$$\leq 2\delta\mathbb{E}\left[\int_T^\infty |dB_t| \, | \mathcal{F}_T\right]$$

$$\leq 2\delta\mathbb{E}\left[(A_\infty^{(1)} - A_T^{(1)}) + (A_\infty^{(2)} - A_T^{(2)}) | \mathcal{F}_T\right] \leq 2\delta N.$$

So

$$\mathbb{E}\left[|B_\infty - B_T| \, | \mathcal{F}_T\right] \leq \left(\mathbb{E}\left[(B_\infty - B_T)^2 | \mathcal{F}_T\right]\right)^{1/2} \leq (2\delta N)^{1/2}.$$

Then by Theorem 6.11, $\mathbb{E} \exp(\alpha B_\infty^* / (2\delta N)^{1/2}) \leq 2$. The result follows.   □

Local times

We finish this section with some results on local times. These will not be used in the remainder of this text but are useful in studying reflecting Brownian motion and the Neumann problem.

Let X_t be one-dimensional Brownian motion. By Jensen's inequality, $|X_t|$ is a submartingale, and by the Doob-Meyer decomposition, it can be written as a martingale plus an increasing process. Since X_t is itself a martingale, the increasing process grows only at times when the Brownian motion is at 0. This increasing process is known as local time at 0. We want to derive its simplest properties and use it to construct reflecting Brownian motion.

Rather than appealing to the Doob-Meyer decomposition, we give the explicit decomposition. First of all, since for each t, $\mathbb{P}(X_t = 0) = \int_{\{0\}} p(t, 0, y)dy = 0$, then $\mathbb{E} \int_0^t 1_{\{0\}}(X_t)dt = 0$. So $\int_0^t 1_{\{0\}}(X_t)dt = 0$, a.s.

Suppose f_n is a sequence of C^2 functions converging to f boundedly and with f_n', f_n'' converging to f', f'' boundedly. Suppose f'' is continuous at all but finitely many points. Then it is easy to see by taking a limit that Itô's formula still holds for f.

We apply it, then, to f_ε, where $f_\varepsilon(0) = f_\varepsilon'(0) = 0$ and $f_\varepsilon''(x) = (2\varepsilon)^{-1} 1_{[-\varepsilon,\varepsilon]}(x)$. As we let $\varepsilon \to 0$, we see that $f_\varepsilon(x) \to |x|$ uniformly, and $f_\varepsilon'(x) \to \operatorname{sgn}(x)$ boundedly. By Doob's inequality,

$$\mathbb{E} \sup_{t \leq 1} \left| \int_0^t f_\varepsilon'(X_s)dX_s - \int_0^t \operatorname{sgn}(X_s)dX_s \right|^2 \to 0.$$

Therefore $\int_0^t (2\varepsilon)^{-1} 1_{[-\varepsilon,\varepsilon]}(X_s)ds$ converges in L^2, uniformly over $t \in [0, 1]$, to some increasing process L_t^0. By taking a suitable subsequence, we see that for almost every ω there is pathwise convergence, uniformly over t, hence L_t^0 is continuous in t. It is also clear from the limit that L_t^0 increases only on the set when $X_t = 0$.

We have the equation, called the Tanaka formula:

$$(6.30) \qquad |X_t| - |X_0| = \int_0^t \operatorname{sgn}(X_s)dX_s + L_t^0.$$

L is called local time at 0.

In the Tanaka formula, the stochastic integral term is a martingale, of course, say N_t. Note $\langle N \rangle_t = t$ by Corollary 5.4, since $\operatorname{sgn}(x)^2 = 1$ unless $x = 0$, and we already observed that Brownian motion spends 0 time at 0. So we have exhibited reflecting Brownian motion, namely $|X_t|$, as the sum of another Brownian motion, N_t, and a continuous process that increases only when X is at the boundary. Tanaka's formula is a special case of what is known as the Skorokhod equation.

Let M_t denote $\sup_{s \leq t} X_s$. (This is not the same as X^* because we do not have an absolute value in this case.)

The following, due to Lévy, is often useful.

(6.15) Theorem. *Under* $\mathbb{P}^0$, *the processes* $(|X_t|, L_t^0)$ *and* $(M_t - X_t, M_t)$ *have the same law.*

Proof. Let $W_t = -N_t$ in Tanaka's formula, so that

$$(6.31) \qquad |X_t| = -W_t + L_t^0.$$

Let $S_t = \sup_{s \leq t} W_s$. We will show $S_t = L_t^0$. This proves the result, since $(M_t - X_t, M_t)$ is equal in law to $(S_t - W_t, S_t)$, which in turn is equal pointwise to $(|X_t|, L_t^0)$.

From (6.31) $W_t = L_t^0 - |X_t|$, or $W_t \leq L_t^0$ for all t, hence $S_t \leq L_t^0$, since L is increasing. L_t^0 increases only when $X_t = 0$ and at those times $L_t^0 = W_t \leq S_t$. Hence $L_t^0 = S_t$ for all t. $\qquad \square$

Just as L_t^0 came from Tanaka's formula, we can construct local time at a by the formula

$$(6.32) \qquad |X_t - a| - |X_0 - a| = \int_0^t \operatorname{sgn}(X_s - a)dX_s + L_t^a,$$

and the same proof as above shows that L_t^a is the limit (uniformly over t) of

$$(2\varepsilon)^{-1} \int_0^t 1_{[a-\varepsilon,a+\varepsilon]}(X_s)ds.$$

As an illustration of the inequalities we had above, we prove the following.

(6.16) Proposition. *With probability 1, there exist versions of the processes L_t^a that are jointly continuous in a and t.*

Two processes are versions of each other if for each t they are equal almost surely.

Proof. It suffices to show this for times less than or equal to M for every $M > 0$. Let $N_t^a = \int_0^{M \wedge t} \mathrm{sgn}\,(X_s - a)dX_s$. Since $|X_t - a|$ is uniformly continuous in t and a, it suffices to establish the same fact for N_t^a. Let T be a stopping time bounded by M.

$$\mathbb{E}\left[((N_M^a - N_M^b)-(N_T^a - N_T^b))^2|\mathcal{F}_T\right]$$

$$= \mathbb{E}\left[\int_T^M (\mathrm{sgn}\,(X_s - a) - \mathrm{sgn}\,(X_s - b))^2 ds|\mathcal{F}_T\right]$$

$$= 4\mathbb{E}\left[\int_T^M 1_{[a,b]}(X_s)ds|\mathcal{F}_T\right]$$

$$\leq 4\mathbb{E}^{X_T}\left[\int_0^M 1_{[a,b]}(X_s)ds\right]$$

$$\leq 4\int_0^M \sup_y \mathbb{E}^y[1_{[a,b]}(X_s)]ds$$

$$\leq c|b - a|.$$

So

$$\mathbb{E}\left[|(N_M^a - N_M^b) - (N_T^a - N_T^b)|\,|\mathcal{F}_T\right] \leq c|a - b|^{1/2}$$

by Jensen's inequality. By Theorem 6.11 there exists α independent of a and b such that

$$\mathbb{E}\exp\left(\alpha \sup_{t \leq M} |N_t^a - N_t^b|/c|a - b|^{1/2}\right) \leq 2,$$

or

(6.33) $$\mathbb{E}\left[\sup_{t \leq M} |N_t^a - N_t^b|^4\right] \leq c|a - b|^2.$$

By Theorem 3.11, N_t^a is jointly continuous on the dyadic rationals, and so L_a^t is jointly continuous on the dyadic rationals. Also, (6.30) and (6.32) imply

(6.34) $$\mathbb{E}\left[\sup_{t\leq M}|L_t^a - L_t^b|^4\right] \leq c\big(|a - b| \wedge 1\big)^2.$$

We note that if we form a jointly continuous process by letting $\widetilde{L}_t^a = \lim L_t^{b_n}$ where the limit is as $b_n \to a$ and b_n is in the dyadic rationals, then (6.34) implies that $\widetilde{L}_t^a = L_t^a$, a.s. $\square$

If we integrate local times over a set, we obtain occupation times. More precisely, we have the following.

(6.17) Proposition. *If f is nonnegative,*

(6.35) $$\int f(y)L_t^y dy = \int_0^t f(X_s)ds, \qquad \text{a.s.}$$

with the null set independent of f and t.

Proof. Suppose we prove the above equality for each C^2 function f with compact support and denote the null set by N_f. Taking a countable collection $\{f_i\}$ of C^2 functions with compact support that are dense in the set of continuous functions on $\mathbb{R}$ with compact support and letting $N = \cup_i N_{f_i}$, then if $\omega \notin N$ we have the above equality for all f_i, and by taking limits, for all bounded and continuous f. A further limiting procedure implies our result.

Suppose f is bounded and C^2 with compact support. Integrating Tanaka's formula, we see that if $g(x) = \int f(y)|x - y|dy$, then $g(X_t) - g(X_0)$ is equal to a martingale plus the left-hand side of (6.35). Now it is not hard (Exercise 2) to see that $g'' = 2f$, so using Itô's formula on g, $g(X_t) - g(X_0)$ is a martingale plus the right-hand side of (6.35). The difference of the left and right sides of (6.35) is a continuous martingale of bounded variation, hence by Proposition 4.19, it is identically 0. $\square$

7 Weak convergence

Definitions

Let Y_i be a sequence of independent identically distributed random variables with $\mathbb{P}(Y_i = 1) = 1/2$ and $\mathbb{P}(Y_i = -1) = 1/2$. Let $S_n = \sum_{i=0}^n Y_i$. S_n is called a simple symmetric random walk, and is a model of one's fortune if one is betting with a fair coin and win \$1 whenever it turns up heads and lose \$1 whenever it turns up tails. Since the Y_is are mean 0 and independent, it is easy to see that S_n is a discrete-time martingale.

Now define a process Z_t^n by $Z_t^n = S_{t2^n}/\sqrt{2^n}$ if t is a multiple of 2^{-n} and define Z_t^n for other ts by linear interpolation between the values of $Z_{j/2^n}$.

For each n, Z_t^n is a continuous process, and it turns out they converge to Brownian motion in a certain sense as $n \to \infty$. We will now make this precise.

(7.1) Definition. *A sequence of probabilities $\mathbb{P}^n$ on some metric space S is said to converge weakly to $\mathbb{P}$ if $\int f d\mathbb{P}_n \to \int f d\mathbb{P}$ for every bounded and continuous function f on the metric space.*

Weak convergence is what is known in functional analysis as weak-$*$ convergence. Other characterizations of weak convergence are given by the following theorem.

(7.2) Theorem. *Suppose $\{\mathbb{P}_n, n = 1, 2, \ldots\}$ and $\mathbb{P}$ are probabilities on a metric space. The following are equivalent.*
(a) $\mathbb{P}_n$ converges weakly to $\mathbb{P}$.
(b) $\limsup_n \mathbb{P}_n(F) \leq \mathbb{P}(F)$ for all closed sets F.
(c) $\liminf_n \mathbb{P}_n(G) \geq \mathbb{P}(G)$ for all open sets G.
(d) $\lim_n \mathbb{P}_n(A) = \mathbb{P}(A)$ for all Borel sets A such that $\mathbb{P}(\partial A) = 0$.

Proof. We begin by showing (a) implies (b). Suppose F is closed. Let $\varepsilon > 0$. Take $\delta > 0$ small enough so that $\mathbb{P}(\{x : d(x, F) \leq \delta\}) - \mathbb{P}(F) < \varepsilon$. Then take f continuous so that it is 1 on F, has support in $\{x : d(x, F) \leq \delta\}$, and is bounded between 0 and 1. For example, $f = 1 - (1 \wedge \delta^{-1} d(x, F))$ would do. Then

$$\limsup_n \mathbb{P}_n(F) \leq \limsup_n \int f d\mathbb{P}_n = \int f d\mathbb{P}$$
$$\leq \mathbb{P}(\{x : d(x, F) \leq \delta\}) \leq \mathbb{P}(F) + \varepsilon.$$

Since this is true for all ε, (b) follows.

The equivalence of (b) and (c) is easy because if F is closed, then $G = F^c$ is open and $\mathbb{P}_n(G) = 1 - \mathbb{P}_n(F)$.

To see that (b) and (c) imply (d), suppose $\mathbb{P}(\partial A) = 0$. Let $\text{int}\,(A)$ denote the interior of A. Then

$$\limsup_n \mathbb{P}_n(A) \leq \limsup_n \mathbb{P}_n(\overline{A}) \leq \mathbb{P}(\overline{A}) = \mathbb{P}(A)$$
$$= \mathbb{P}(\text{int}\,(A)) \leq \liminf_n \mathbb{P}_n(\text{int}\,(A)) \leq \liminf_n \mathbb{P}_n(A).$$

Next, let us show (d) implies (b). Let F be closed. Let $F_\delta = \{x : d(x, F) \leq \delta\}$. If $y \in \partial F_\delta$, then $d(y, F) = \delta$. So the sets ∂F_δ are disjoint for different δ. At most countably many of them can have positive $\mathbb{P}$-measure, hence there exists a sequence $\delta_k \downarrow 0$ such that $\mathbb{P}(\partial F_{\delta_k}) = 0$. Then

$$\limsup_n \mathbb{P}_n(F) \leq \limsup_n \mathbb{P}_n(F_{\delta_k}) = \mathbb{P}(F_{\delta_k})$$

for each k. Since $\mathbb{P}(F_{\delta_k}) \downarrow \mathbb{P}(F)$ as $\delta_k \to 0$, this gives (b).

Finally, let us show (b) implies (a). Let f be bounded and continuous. If we show

$$(7.1) \qquad \limsup_n \int f d\mathbb{P}_n \le \int f d\mathbb{P},$$

for every such f, then applying this inequality to both f and $-f$ will give (a). By adding a sufficiently large positive constant to f, we may assume f is bounded below by 1. By dividing by $2\|f\|_\infty$, f is bounded by numbers between 0 and 1. Let $F_i = \{x : f(x) \ge i/k\}$, which is closed.

$$\int f d\mathbb{P}_n \le \sum_{i=0}^{k} \frac{i}{k} \mathbb{P}_n\left(\frac{i-1}{k} \le f(x) < \frac{i}{k}\right)$$

$$= \sum_{i=0}^{k} \frac{i}{k}[\mathbb{P}_n(F_{i-1}) - \mathbb{P}_n(F_i)] \le \frac{1}{k} + \frac{1}{k}\sum_i \mathbb{P}_n(F_i).$$

Similarly, $\int f d\mathbb{P} \ge (1/k)\sum \mathbb{P}(F_i)$. Then

$$\limsup_n \int f d\mathbb{P}_n \le \frac{1}{k} + \frac{1}{k}\sum \limsup_n \mathbb{P}_n(F_i)$$

$$\le \frac{1}{k} + \frac{1}{k}\sum \mathbb{P}(F_i) \le \frac{1}{k} + \int f d\mathbb{P}.$$

Since k is arbitrary, this gives (7.1). $\square$

If $x_n \to x$, $\mathbb{P}_n = \delta_{x_n}$, and $\mathbb{P} = \delta_x$, it is easy to see $\mathbb{P}_n$ converges weakly to $\mathbb{P}$. Letting $A = \{x\}$ shows that one cannot, in general, have $\lim_n \mathbb{P}_n(F) = \mathbb{P}(F)$ for all closed sets F.

Prohorov's theorem

It turns out there is a simple condition that ensures that a sequence of probability measures has a weakly convergent subsequence.

(7.3) Definition. *A sequence of probabilities $\mathbb{P}_n$ on a metric space S is tight if for every ε there exists a compact set K such that $\sup_n \mathbb{P}_n(K^c) \le \varepsilon$.*

The most important result here is Prohorov's theorem.

(7.4) Theorem. *If a sequence of probability measures on a metric space S is tight, it is relatively compact in the topology of weak convergence.*

Proof. Suppose first that the metric space $S = [0,1]^\infty$. (For this particular S, the theorem is an immediate consequence of the Banach-Alaoglu theorem, but we will ignore that and proceed.) Note $|a-b| \wedge 1$ is a metric on $\mathbb{R}$ giving rise to the ordinary topology on $\mathbb{R}$, and then

$$(7.2) \quad d(a, b) = \sum_{i=1}^{\infty} 2^{-i}(|a^i - b^i| \wedge 1), \qquad a = (a^1, a^2, \ldots), b = (b^1, b^2, \ldots)$$

is a metric on $[0, 1]^{\infty}$. S is compact and Hausdorff. Let $C(S)$ be the bounded and continuous functions on S. Let f_i be a countable collection of positive elements of $C(S)$ whose span is dense in $C(S)$. For each i, $\int f_i d\mathbb{P}_n$ is a bounded sequence, so we have a convergent subsequence. By a diagonalization procedure, we can find a subsequence n' such that $\int f_i d\mathbb{P}_{n'}$ converges for all i. (Readers not familiar with the diagonalization procedure should look at the proof of the Ascoli–Arzelà theorem in any basic undergraduate analysis book, e.g., Rudin [2].) Call the limit Lf_i. Clearly $0 \leq Lf_i \leq \|f_i\|_{\infty}$, L is linear, and so we can extend L to a bounded linear functional on S. By the Riesz representation theorem (see Rudin [1]), there exists a measure $\mathbb{P}$ such that $Lf = \int f d\mathbb{P}$. Since $\int f_i d\mathbb{P}_{n'} \to \int f_i d\mathbb{P}$ for all f_i, $\int f d\mathbb{P}_{n'} \to \int f d\mathbb{P}$ for all $f \in C(S)$, or $\mathbb{P}_{n'}$ converges weakly to $\mathbb{P}$. Since $Lf \geq 0$ if $f \geq 0$, then $\mathbb{P}$ is a positive measure. 1 is bounded and continuous, so $1 = \mathbb{P}_{n'}(S) = \int 1 d\mathbb{P}_{n'} \to \int 1 d\mathbb{P}$, or $\mathbb{P}(S) = 1$.

Now let S be an arbitrary metric space. For each m there exists a compact set K_m such that $\sup_n \mathbb{P}_n(K_m^c) \leq 1/m$. Since all the $\mathbb{P}_n$s are supported on $\cup_m K_m$, we can replace S by $\cup_m K_m$, or we may as well assume that S is σ-compact, and hence separable. If ρ is the metric on S, $\rho \wedge 1$ will also be a compatible metric, so we may assume ρ is bounded by 1. Now S can be embedded in $[0, 1]^{\infty}$. In fact, if d_j is a countable dense subset of S, let $I : S \to [0, 1]^{\infty}$ be defined by $I(x) = (\rho(x, d_1), \rho(x, d_2), \ldots)$.

If $x_n \to x$, then each component of $I(x_n)$ converges to the corresponding component of $I(x)$. Using (7.2), this implies $I(x_n) \to I(x)$. Conversely, if x_n is a sequence such that $\rho(x_n, x) > \varepsilon$ for all n, we can find d_j such that $\rho(d_j, x) < \varepsilon/2$, and hence $\rho(x_n, d_j) > \varepsilon/2$ for all n. In this case $I(x_n)$ cannot converge to $I(x)$. Therefore I is an open mapping as well as a continuous one. Taking $x_n = y$ for all n, if $I(y) = I(x)$, then $y = x$, or I is one to one. Since S is σ-compact, and the continuous image of compact sets is compact, then $I(S)$ is a Borel set.

We may define $\mathbb{P}_n$ on $[0, 1]^{\infty}$ by letting $\mathbb{P}_n(A) = \mathbb{P}_n(I^{-1}(A \cap I(S)))$. If we were to identify S and $I(S)$, all we would be doing is extending $\mathbb{P}_n$ to all of $[0, 1]^{\infty}$ by setting it to be 0 on S^c. By the first paragraph of the proof, there exists a subsequence $\mathbb{P}_{n'}$ that converges to a probability $\mathbb{P}$ on $[0, 1]^{\infty}$ (relative to the topology of $[0, 1]^{\infty}$). Since $I(K_m)$ is compact in $[0, 1]^{\infty}$, by Theorem 7.2

$$\mathbb{P}(I(K_m)) \geq \limsup_{n'} \mathbb{P}_{n'}(I(K_m)) \geq 1 - 1/m,$$

hence $\mathbb{P}(I(S)) = 1$.

Finally, if G is open in S, then $I(G) = H \cap I(S)$ for some H open in $[0, 1]^{\infty}$. Also

$$\liminf_{n'} \mathbb{P}_{n'}(G) = \liminf_{n'} \mathbb{P}_{n'}(H) \geq \mathbb{P}(H) = \mathbb{P}(H \cap I(S)) = \mathbb{P}(G).$$

So by Theorem 7.2, $\mathbb{P}_{n'}$ converges weakly to $\mathbb{P}$ relative to the topology on S. □

Clearly, Prohorov's theorem is easily modified to handle the case of finite measures on S.

Another construction of Brownian motion

We will now use Prohorov's theorem to give a third construction of Brownian motion. Given that we already have two constructions of Brownian motion, this is still not totally redundant, because weak convergence can be used to construct reflecting Brownian motion and solutions to certain elliptic partial differential equations.

Let $Z_t^n = S_{t2^n}/\sqrt{2^n}$ for t a multiple of $1/2^n$ and define Z_t^n by linear interpolation for other t. So if $k/2^n \le t \le (k+1)/2^n$, then

$$(7.3) \qquad Z_t^n = \frac{(k+1) - 2^n t}{\sqrt{2^n}} S_k + \frac{2^n t - k}{\sqrt{2^n}} S_{k+1}.$$

Let $\mathbb{P}_n$ be the law of Z_t^n. Recall from Sect. 1 that this means: if Ω is the set of continuous paths from $[0, 1]$ to $\mathbb{R}$ and A is a measurable subset of $C[0, 1]$, then $\mathbb{P}_n(A) = \mathbb{P}(Z^n \in A)$.

(7.5) Theorem. *The sequence $\mathbb{P}_n$ converges to a probability measure $\mathbb{P}_\infty$ on $C[0, 1]$, and under $\mathbb{P}_\infty$ the coordinate maps $X_t(\omega) = \omega(t)$ form a Brownian motion.*

Proof. We first prove that the $\mathbb{P}_n$ are tight. We then show that any subsequential limit point is a Brownian motion.

A simple computation shows that

$$(7.4) \qquad \mathbb{E}\, S_n^4 = \sum_{i=1}^n \mathbb{E}\, Y_i^4 + \sum_{i \ne j} \mathbb{E}\, Y_i^2 \mathbb{E}\, Y_j^2 \le cn^2,$$

since $\mathbb{E}\, Y_i$ and $\mathbb{E}\, Y_i^3$ are both 0 and the Y_is are independent.

Let

$$K_M = \{\omega : \omega(0) = 0, |\omega(t) - \omega(s)| \le M|t - s|^{1/8}, \, s, t \le 1\}.$$

K_M is compact by the Ascoli-Arzelà theorem, for each M. We need to show that, given $\varepsilon > 0$,

$$(7.5) \qquad \mathbb{P}_n(K_M^c) = \mathbb{P}(Z_t^n \notin K_M) \le \varepsilon$$

uniformly in n if M is large enough.

If $|Z_t^n - Z_s^n| \ge M|t - s|^{1/8}$ for some $s, t \in [0, 1]$, $s < t$, then this will also be true for some s, t that are both multiples of 2^{-n}. Let i be the smallest integer such that $|t - s| \leq 2^i/2^n$ and let m be the largest integer such that $m2^i/2^n \leq s$. Then

$$\sup_{m2^{i-n} \leq t \leq (m+2)2^{i-n}} |Z_t^n - Z_{m/2^n}^n| \geq M(2^{i-n})^{1/8}.$$

Therefore using Theorem 4.7 with $p = 2$, (7.4), and the fact that $S_{k+j} - S_j$ has the same distribution as S_k,

$$\mathbb{P}(Z_t^n \notin K_M)$$

$$\leq \mathbb{P}\Big(\sup_{m2^i \leq k \leq (m+2)2^i} |\frac{S_k}{\sqrt{2^n}} - \frac{S_{m2^i}}{\sqrt{2^n}}| \geq M(2^{i-n})^{1/8}$$

$$\text{for some } i \leq n \text{ and } m \leq 2^{n-i}\Big)$$

$$\leq \sum_{i=1}^{n} 2^{n-i} \mathbb{P}\Big(\sup_{k \leq 2 \cdot 2^i} \frac{|S_k|}{\sqrt{2^n}} \geq M(2^{i-n})^{1/8}\Big)$$

$$\leq \sum_{i=1}^{n} 2^{n-i} \frac{\mathbb{E}|S_{2^{i+1}}|^4}{M^4 2^{2n} 2^{(i-n)/2}}$$

$$\leq c \sum_{i=0}^{n} 2^{n-i} \frac{2^{2i+2}}{M^4 2^{2n} 2^{i/2} 2^{-n/2}} \leq cM^{-4}.$$

Taking M large enough shows (7.5) uniformly in n. Thus the $\mathbb{P}_n$ are tight.

Any subsequential limit point is a probability on $C[0, 1]$, so to show that the limit is Brownian motion, it is enough to show that the finite-dimensional distributions (i.e., the law of $(Z_{t_1}, \ldots, Z_{t_m})$ under the limit law $\mathbb{P}_\infty$) agree with those of Brownian motion. For this, see Exercise 14.

□

8 Exercises and further results

Exercise 1. Let D be the dyadic rationals in $[0, 1]$. Suppose $\{X_t, t \in D\}$ is a mean zero Gaussian process with paths that are uniformly continuous on D and with $\text{Var}(X_t - X_s) = t - s$, $s < t$. Let Y_t be defined by $Y_t = \lim_{u \downarrow t, u \in D} X_u$ for $t \in [0, 1]$. Show Y_t is a Brownian motion.

Exercise 2. If $f \in C^2$ with compact support and $g(x) = \int f(y)|x - y|\, dy$, show $g \in C^2$ and $g'' = 2f$.

Exercise 3. If Z is a random variable with an $\mathcal{N}(0, 1)$ law, show there exists c such that $\mathbb{P}(Z > a) \geq ca^{-1}e^{-a^2/2}$ if $a \geq 1$.

Exercise 4. Prove the law of the iterated logarithm for Brownian motion:

$$\limsup_{t \to \infty} \frac{|X_t|}{\sqrt{2t \log \log t}} = 1, \qquad \text{a.s.}$$

Use time inversion to obtain the law of the iterated logarithm near 0:

$$\limsup_{t \to 0} \frac{|X_t|}{\sqrt{2t \log \log(1/t)}} = 1.$$

Hint: To get the upper bound, it suffices to show that for each $\varepsilon > 0$,

$$\mathbb{P}(X_t > (1 + \varepsilon)\sqrt{2t \log \log t} \ \text{i.o.}) = 0.$$

To do this, let q be greater than 1 but very close to 1, then use Proposition 4.8 to bound

$$\mathbb{P}(\sup_{s \le q^n} |X_s| > (1 + \varepsilon)\sqrt{2q^n \log n})$$

above by terms in a convergent series. Finish by using the first half of the Borel-Cantelli lemma.

To get the lower bound, let q be very large, use Exercise 3 to bound

$$\mathbb{P}(X_{q^{n+1}} - X_{q^n} > (1 - \varepsilon)\sqrt{2q^{n+1} \log n})$$

below by terms in a divergent series. Use the second half of the Borel-Cantelli lemma to see that with probability one,

$$X_{q^{n+1}} - X_{q^n} > (1 - \varepsilon)\sqrt{2q^{n+1} \log n} \qquad \text{i.o.}$$

Then use the upper bound to complete the proof of the lower bound.

Exercise 5. If $a < 0 < b$, find $\mathbb{E}\left[e^{-\lambda T_{[a,b]}}; X_{T_{[a,b]}} = a\right]$. Let $b \to \infty$ to show

$$\mathbb{E}\left[\exp(-\lambda T_{\{a\}})\right] = \exp\left(-\sqrt{2\lambda}|a|\right).$$

Hint: Imitate the proof of Proposition 4.9, and use the fact that $\exp(\pm \lambda X_t - \lambda^2 t/2)$ is a martingale.

Exercise 6. Show the infinitesimal generator of Brownian motion is $(1/2)\Delta$ on C^2 functions.

Hint: Calculate $\mathbb{E}\, f(X_t) - f(x)$ by using a Taylor series expansion.

Exercise 7. Let X_t be d-dimensional Brownian motion. Let $D_A = \inf\{t \ge 0 : X_t \in A\}$. D_A is called the first entry time into A. Show that if A is either open or closed, then D_A is a stopping time with respect to any filtration that is right continuous and is such that X_t is $\mathcal{F}_t$ adapted.

Exercise 8. Prove Proposition 2.6.

Exercise 9. Suppose T is a stopping time for a Brownian motion X_t. Show $\mathcal{F}_T$ as defined by (3.7) is a σ-field. Show T and X_T are both $\mathcal{F}_T$ measurable.

Exercise 10. Suppose Y_i are mean 0 independent random variables with $\sup_i \mathbb{E} Y_i^2 \leq c$. Show the strong law of large numbers in this case: $\sum_{i=1}^n Y_i/n \to 0$ a.s.

Hint: The second moment of $|\sum_{i=1}^n Y_i|$ can be bounded by cn since the sum is a martingale. By Doob's inequality, we have that

$$\mathbb{P}(\sup_{n \leq 2^k} |\sum_{i=1}^n Y_i| > \varepsilon 2^k) \leq c/\varepsilon^2 2^k.$$

Now use the Borel-Cantelli lemma.

Exercise 11. Show that for every $\varepsilon > 0$, with probability one, the paths of Brownian motion are Hölder continuous of order $1/2 - \varepsilon$.

Hint: Imitate the proof of Theorem 2.4.

Exercise 12. Suppose Y is $\mathcal{F}_\infty$ measurable and bounded. If $\varphi(x) = \mathbb{E}^x Y$, show φ is Lebesgue measurable.

Exercise 13. If M_t is a continuous martingale, show

$$\mathbb{P}(\sup_{s \leq t} |M_s| > a, \langle M \rangle_t < b) \leq 2e^{-a^2/2b}.$$

Exercise 14. Suppose Y_i is a sequence of independent random variables with

$$\mathbb{P}(Y = -1) = \mathbb{P}(Y = +1) = 1/2.$$

Let $S_n = \sum_{i=1}^n Y_i$.
(a) Show the central limit theorem in this case: the law of $S_n/\sqrt{n}$ converges weakly to the law of a $\mathcal{N}(0,1)$ random variable.

Hint: Since the second moments of $S_n/\sqrt{n}$ are uniformly bounded, Chebyshev's inequality shows the sequence of laws is tight. If $\mathbb{P}_\infty$ is any subsequential limit point, show

$$\int e^{iux} d\mathbb{P}_\infty = \lim \mathbb{E} e^{iuS_n/\sqrt{n}} = \lim(\mathbb{E} e^{iuY_1/\sqrt{n}})^n$$

$$= \lim(\cos(u/\sqrt{n}))^n = e^{-u^2/2}.$$

Then use the uniqueness of the Fourier transform.
(b) Show that if $s_1, \ldots, s_k$ is an increasing sequence of dyadic rationals, then the law of $(S_{s_1 2^n}/\sqrt{2^n}, \ldots, S_{s_m 2^n}/\sqrt{2^n})$ converges weakly to the law of a Gaussian sequence, and find the covariance matrix of the limit.

Exercise 15. Suppose $\{f_m\}$ is a uniformly bounded equicontinuous family of functions on a metric space and $\mathbb{P}_n$ converges weakly to $\mathbb{P}$. Show $\int f_m d\mathbb{P}_n$ converges to $\int f_m d\mathbb{P}$, uniformly over m.

Exercise 16. Suppose $f : [0,1] \to \mathbb{R}$. Let $\mathcal{F}_n$ be the collection of intervals of length 2^{-n} that have endpoints at integer multiples of 2^{-n} and let I_{xn} be the leftmost interval in $\mathcal{F}_n$ containing x. Show

(a) if $f \in L^1[0,1]$, then

$$\lim_{n \to \infty} \frac{1}{|I_{xn}|} \int_{I_{xn}} f \, dx = f(x), \qquad \text{a.e.}$$

(b) there is a constant c depending only on p such that if $f \in L^p$, $1 < p \le \infty$, then

$$\left\| \sup_n \frac{1}{|I_{xn}|} \int_{I_{xn}} f \, dx \right\|_p \le c \|f\|_p.$$

Hint: Compute $\mathbb{E}\,[f|\mathcal{F}_n]$.

Exercise 17. Suppose Y_i is a sequence of independent random variables with mean 0 and identical law and such that $\mathbb{E}\, e^{\lambda Y_1} < \infty$ for all λ in some neighborhood of 0. Let $S_n = \sum_{i=1}^n Y_i$. Show that there exists c such that $\mathbb{P}(|S_n/n| > a) \le e^{-ca^2 n}$.

Hint: Compute $\mathbb{E}\, \exp(\lambda S_n)$ and use Chebyshev's inequality.

Exercise 18. Suppose $p > 1/2$. Let Y_i be a sequence of independent identically distributed random variables with $\mathbb{P}(Y_1 = 1) = p$, $\mathbb{P}(Y_1 = -1) = 1-p$. Let $S_n = \sum_{i=1}^n Y_i$. Show $S_n \to +\infty$, a.s.

Hint: $S_n - an$ is a martingale for some $a > 0$. Imitate Exercise 10 to show $S_n/n \to a$, a.s.

Exercise 19. Show the paths of Brownian motion are nowhere differentiable. That is, with probability 1, the function $t \to X_t(\omega)$ has a derivative at no t.

Hint: It suffices to let $M > 0$ and show that $\mathbb{P}(A_n) \to 0$ as $n \to \infty$, where

$$A_n = \{ \text{ there exists } s \in [0,1] \text{ with } |X_t - X_s| \le M|t - s|$$
$$\text{if } |t - s| \le 2/n \}.$$

The set A_n is contained in B_n, where

$$B_n = \{ \text{ for some } k \le n, |X_{k/n} - X_{(k-1)/n}| \le 4M/n,$$
$$|X_{(k+1)/n} - X_{k/n}| \le 4M/n, \text{ and } |X_{(k+2)/n} - X_{(k+1)/n}| \le 4M/n \}.$$

However

$$\mathbb{P}(B_n) \le n(\mathbb{P}(|X_{1/n}| \le 4M/n))^3 \to 0.$$

Exercise 20. If X_t is a Brownian motion, $a > 0$, and $T = \inf\{t : X_t > a+bt\}$, then show $\mathbb{E}^0 e^{-\lambda T} = \exp(-a(b + (b^2 + 2\lambda)^{1/2})$. Take $b = 0$ to obtain $\mathbb{E}\, e^{-\lambda T_{\{a\}}} = e^{-a\sqrt{2\lambda}}$.

Hint: Use the fact that $\exp(rX_t - r^2 t/2)$ and $\exp(-rX_t - r^2 t/2)$ are martingales for suitable r and optional stopping.

Exercise 21. Suppose X_n is a family of random variables. Show X_n is uniformly integrable if and only if there exists $\varphi : [0, \infty) \to [0, \infty)$ continuous, convex, and increasing with $\varphi(0) = 0$ and $\varphi(x)/x \to \infty$ as $x \to \infty$ such that $\sup_n \mathbb{E}\,\varphi(|X_n|) < \infty$. Show that if X_n is uniformly integrable and $X_n \to X$, a.s., then $\mathbb{E}\,|X_n - X| \to 0$ as $n \to \infty$.

Exercise 22. If $\int_0^t H_s^2 d\langle M\rangle_s$ is integrable, and $N_t = \int_0^t H_s dM_s$, show $\langle N\rangle_t = \int_0^t H_s^2 d\langle M\rangle_s$.

Exercise 23. Suppose $\int_0^t H_s^2 K_s^2 d\langle M\rangle_s$ is integrable and $N_t = \int_0^t K_s dM_s$ is square integrable. Show $\int_0^t H_s dN_s = \int_0^t H_s K_s dM_s$.

Exercise 24. Prove the Kunita-Watanabe inequality:

$$\int_0^\infty |H_s K_s| d\langle M, N\rangle_s \le \left(\int_0^\infty H_s^2 d\langle M\rangle_s \right)^{1/2} \left(\int_0^\infty K_s^2 d\langle N\rangle_s \right)^{1/2}$$

where M, N are martingales.

   Hint: By replacing H by $|H|$ and K by $|K|$, suppose $H, K \ge 0$. Approximate H and K by processes that are piecewise constant. Use the fact that $0 \le \langle M + \lambda N\rangle_t - \langle M + \lambda N\rangle_s$ is quadratic in λ, hence the discriminant is nonnegative.

Exercise 25. If M is a continuous martingale, show that M is the time change of a Brownian motion, possibly stopped at a stopping time.

Exercise 26. Let X_t be a semimartingale with a decomposition as $M_t + A_t$, where M_t is a continuous local martingale and A_t is a continuous process of locally bounded variation. Suppose R_n is any sequence of stopping times such that $\int_0^{R_n} H_s^2 d\langle M\rangle_s + \int_0^{R_n} |H_s| d|A_s|$ is bounded for each n. Show that if we define $\int_0^t H_s dX_s$ by $\int_0^{t \wedge R_n} H_s dX_s$ if $t < R_n$, then the definition is independent of the choice of the sequence R_n.

Exercise 27. Suppose A_t^n is a sequence of nondecreasing right continuous processes that converge for each t to A_t, a.s. Show that if H_s is a process that is continuous in s, then $\int_0^t H_s dA_s^n$ converges to $\int_0^t H_s dA_s$ for each t, a.s.

Exercise 28. Suppose H_s is a continuous process that for each s is adapted to $\mathcal{F}_\infty$ (but not necessarily to $\mathcal{F}_s$). Show that if A_t is a continuous nonincreasing adapted process, then

$$\mathbb{E} \int_0^t H_s dA_s = \mathbb{E} \int_0^t \mathbb{E}\,[H_s|\mathcal{F}_s] dA_s.$$

   Hint: Use a Riemann sum approximation.

Exercise 29. Show there exists c such that the c_1 and c_2 in Theorem 6.8 can both be bounded by $(c\sqrt{p})^p$, $p \ge 2$.

Exercise 30. If X and Y are semimartingales, the Stratonovich stochastic integral

$$\int_0^t X_s \circ dY_s$$

is defined to be $\int_0^t X_s \, dY_s + (1/2)\langle X, Y \rangle_t$. Show that if $f \in C^3$, then

$$f(Y_t) = f(Y_0) + \int_0^t f'(Y_s) \circ dY_s.$$

Exercise 31. Show that if X_t is d-dimensional Brownian motion, then $R_t = |X_t|$ satisfies the stochastic differential equation

$$dR_t = dW_t + \frac{d-1}{2R_t} dt, \qquad R_0 = |X_0|,$$

where W_t is one-dimensional Brownian motion. A process solving this equation is said to have the law of a Bessel process of order d.

Exercise 32. Suppose for $i = 1, 2$, the process $X_t^{(i)}$ satisfies the equation

$$dX_t^{(i)} = \sigma_i(X_t^{(i)})dW_t + b_i(X_t^{(i)})dt, \qquad X_0^{(i)} = x_0.$$

Suppose $\sigma_1 = \sigma_2$ and $b_1 = b_2$ in a neighborhood V of x_0 and both functions are Lipschitz in V. Show $X_t^{(1)} = X_t^{(2)}$, a.s., for $t < \tau_V$.

Exercise 33. Suppose σ and b are functions that are Lipschitz in every bounded interval and suppose there exists c such that

$$|\sigma(x)| + |b(x)| \leq c(1 \vee |x|).$$

Show there exists a unique solution to (6.2) in this case.

   Hint: If X_t is a solution, show there exists c such that

$$\mathbb{E} \sup_{s \leq t} X_s^2 \leq c\left(1 + \int_0^t \mathbb{E} \sup_{r \leq s} X_r^2 \, ds\right).$$

Exercise 34. Show that if $d \geq 3$ and X_t is d-dimensional Brownian motion, then $|X_t| \to \infty$, a.s.

   Hint: Cf. Proposition 5.8.

Notes

Good references for basic probability theory include Billingsley [2], Breiman [1], Chung [1], Dudley [1], and Durrett [2]. For material on Brownian motion and stochastic integrals, consult Karatzas and Shreve [1], Revuz and Yor

[1], Ikeda and Watanabe [1], Protter [1], or Dellacherie and Meyer [2]. A good reference for Markov processes and Markov properties is Blumenthal and Getoor [1] or Sharpe [1]; see also Durrett [1]. See Billingsley [1] for more information on weak convergence. The Calderón-Zygmund lemma and Khintchine's inequalities can also be found in Stein [1].

Most of the material in Chap. I is standard and was taken, with some modifications, from the above references. The Doob-Meyer decomposition was proved in the discrete case by Doob and in the considerably more difficult case continuous case by Meyer. Rao [1] showed how one could obtain the decomposition by starting with discrete approximations. The proof of Theorem 4.18 given here is new and uses L^2 convergence instead of $\sigma(L^1, L^\infty)$ convergence. The proof of Theorem 6.11 is also new.

II
POTENTIAL THEORY

1 The Dirichlet problem

Harmonic functions

Given a domain D and a continuous function f on the boundary of D, the Dirichlet problem is to find a function u that is harmonic in D, continuous on $\overline{D}$, and agrees with f on ∂D, the boundary of D. We begin with a discussion of harmonic functions.

(1.1) Definition. *Let D be a domain. h is harmonic in D if h is locally integrable and for all $x \in D$ and all $r < \operatorname{dist}(x, \partial D)$,*

$$(1.1) \qquad h(x) = \frac{1}{|B(0,r)|} \int_{B(x,r)} h(y)\,dy.$$

If h is locally integrable, this averaging property implies h is bounded on compact subsets of D. Then by dominated convergence, h is continuous on compact subsets of D.

(1.2) Proposition. *If h is locally bounded, then h is harmonic if and only if for all $x \in D$ and all $r < \operatorname{dist}(x, \partial D)$*

$$(1.2) \qquad h(x) = \int_{\partial B(0,r)} h(x + y)\sigma_r(dy),$$

where σ_r is surface measure on $\partial B(0, r)$, normalized to have total mass 1.

Proof. Without loss of generality, we can take $x = 0$. Then, changing to polar coordinates,

$$
(1.3) \quad \frac{1}{|B(0,r)|} \int_{B(0,r)} h(y) dy
$$

$$
= \frac{1}{r^d |B(0,1)|} \int_0^r \int_{\partial B(0,s)} h(y) \sigma_s(dy) |\partial B(0,s)| \, ds
$$

$$
= d r^{-d} \int_0^r \int_{\partial B(0,s)} h(y) \sigma_s(dy) s^{d-1} ds.
$$

If (1.2) holds, then

$$
\frac{1}{|B(0,r)|} \int_{B(0,r)} h(y) dy = d r^{-d} \int_0^r h(0) s^{d-1} ds = h(0),
$$

or (1.1) holds.

Conversely, if h is harmonic, then (1.3) implies

$$
r^d h(0) = d \int_0^r \int_{\partial B(0,s)} h(y) \sigma_s(dy) s^{d-1} ds.
$$

The inside integral is a continuous function of s since h is continuous. Differentiating both sides with respect to r,

$$
d r^{d-1} h(0) = d \int_{\partial B(0,r)} h(y) \sigma_r(dy) r^{d-1},
$$

which implies (1.2).   □

(1.3) Proposition. *If h is harmonic in D, then h is C^∞ in D.*

In general, the derivatives of h at x blow up as x approaches the boundary of D. As an example, let D be the half-plane in $\mathbb{R}^2$ and consider $h(x^1, x^2) = x^2 / ((x^1)^2 + (x^2)^2)$ if $x = (x^1, x^2)$.

Proof. If $r < \text{dist}\,(x, \partial D)/4$,

$$
|h(y) - h(x)| = \frac{1}{|B(0,r)|} \left| \int_{B(y,r)} h(z) dz - \int_{B(x,r)} h(z) dz \right|
$$

$$
\leq \frac{1}{|B(0,r)|} \int_{B(y,r) \triangle B(x,r)} h(z) dz.
$$

If x and y are close together, then $B(x,r) \subseteq B(y, r + |x - y|)$ and $B(y,r) \subseteq B(x, r + |x - y|)$, and so

$$
|B(y,r) \triangle B(x,r)| \leq c r^{d-1} |x - y|.
$$

Therefore

(1.4) $$|h(y) - h(x)| \le c\frac{|x-y|}{r} \sup_{z \in B(x,5r/4)} |h(z)|.$$

This shows that h is a Lipschitz function. (h Lipschitz means there exists a constant c such that $|h(x) - h(y)| \le c|x-y|$ for all x and y.) Hence the partials derivatives of h exist a.e. (Exercise 8) and are uniformly bounded.

Let e_i be the unit vector in the ith direction. Since

(1.5) $$\frac{h(x + \varepsilon e_i) - h(x)}{\varepsilon} = \frac{1}{|B(0,r)|} \int_{B(x,r)} \frac{h(z + \varepsilon e_i) - h(z)}{\varepsilon} dz,$$

just as we derived (1.4) we obtain

(1.6) $$\left| \frac{h(x + \varepsilon e_i) - h(x)}{\varepsilon} - \frac{h(y + \varepsilon e_i) - h(y)}{\varepsilon} \right|$$
$$\le c\frac{|x-y|}{r} \sup_{z \in B(x,5r/4)} \left| \frac{h(z + \varepsilon e_i) - h(z)}{\varepsilon} \right|.$$

Using (1.4), the right-hand side of (1.6) is bounded by

$$c\frac{|x-y|}{r^2} \sup_{z \in B(x,7r/4)} |h(z)|.$$

This tells us the functions $k_\varepsilon(x) = (h(x + \varepsilon e_i) - h(x))/\varepsilon$, $\varepsilon > 0$, are equicontinuous. Since $\lim_{\varepsilon \to 0} k_\varepsilon(x) = (\partial h/\partial x^i)(x)$ for a.e. x, then the limit exists for every x, the limit $\partial h/\partial x^i$ is continuous in x, and the convergence is uniform. Letting $\varepsilon \to 0$ in (1.5) shows that

(1.7) $$\frac{\partial h}{\partial x^i}(x) = \frac{1}{|B(0,r)|} \int_{B(x,r)} \frac{\partial h}{\partial x^i}(z) dz$$

for each $x \in D$ if $r < \text{dist}(x, \partial D)/4$. By (1.7) the first partial derivatives of h are themselves harmonic in $B(x, \text{dist}(x, \partial D)/4)$ for each $x \in D$. Repeating the argument with h replaced by $\partial h/\partial x^i$ shows the second partials are continuous and harmonic, and so on. $\square$

(1.4) Corollary. *If $|h|$ is bounded by M in $B(x, r)$, then the kth-order partial derivatives of h at x are bounded in absolute value by cM/r^k.*

Proof. The case $k = 1$ comes from (1.4). For all $y \in B(x, r/2)$,

$$\left| \frac{\partial h}{\partial x^i}(y) \right| \le \frac{cM}{r}, \qquad i = 1, 2, \ldots, d.$$

Repeating with r replaced by $r/2$ and h by $\partial h/\partial x^i$,

$$\left| \frac{\partial^2 h}{\partial x^i \partial x^j}(y) \right| \le \frac{cM}{r^2}$$

if $y \in B(x, r/4)$. Repeating the procedure gives the result for all k. $\qquad\square$

(1.5) Proposition. *If h is harmonic in D, then $\Delta h = 0$ in D.*

Proof. Suppose $\Delta h(x_0) > 0$. Without loss of generality, we may suppose $x_0 = 0$. Since h is C^∞, there exists $r < \text{dist}(0, \partial D)/2$ such that $\Delta h > 0$ on $B(0, r)$. By Itô's formula,

$$(1.8) \quad h(X_{t \wedge \tau(B(0,r))}) - h(X_0) = \text{martingale} + \frac{1}{2} \int_0^{t \wedge \tau(B(0,r))} \Delta h(X_s) ds.$$

Taking the expectation with respect to $\mathbb{P}^0$ and letting $t \to \infty$, we see

$$(1.9) \qquad \mathbb{E}^0 h(X_{\tau(B(0,r))}) - h(0) = \frac{1}{2} \mathbb{E}^0 \int_0^{\tau(B(0,r))} \Delta h(X_s) ds > 0.$$

By Proposition I.2.8, the left-hand side of (1.9) is $\int_{\partial B(0,r)} h(y) \sigma_r(dy) - h(0)$, and since h is harmonic, this is 0, a contradiction. Therefore $\Delta h(0) \leq 0$. The same argument shows that $\Delta h \geq 0$, hence $\Delta h = 0$. $\qquad\square$

Conversely, we have the following.

(1.6) Proposition. *If h is C^2 and $\Delta h = 0$ in D, then h is harmonic.*

Proof. We will show that for $r < \text{dist}(x, \partial D)$,

$$(1.10) \qquad\qquad h(x) = \int_{B(0,r)} h(x + y) \sigma_r(dy).$$

This with Proposition 1.2 will show h is harmonic. By Itô's formula, $h(X_{t \wedge \tau(B(0,r))}) - h(X_0)$ is a martingale. Take the expectation with respect to $\mathbb{P}^x$, let $t \to \infty$, and we obtain $\mathbb{E}^x h(X_{\tau(B(0,r))}) - h(x) = 0$. By Proposition I.2.8, this is (1.10). $\qquad\square$

We state the following as a corollary

(1.7) Corollary. *If h is harmonic, then $h(X_{t \wedge \tau_D})$ is a martingale.*

If h is harmonic in a neighborhood of x for each $x \in D$, then by Propositions 1.3 and 1.5, h is C^∞ and $\Delta h = 0$ in D. Then by Proposition 1.6, h is harmonic in D. Hence to show a function h is harmonic in D it suffices that each point in D has a neighborhood in which (1.1) [or (1.2)] holds for each x in the neighborhood and for all r less than some r_0, where r_0 depends on the neighborhood.

One of the most important properties of harmonic functions is the maximum principle.

(1.8) Theorem. *If D is a bounded domain and h is harmonic on D and continuous on $\overline{D}$, then*

$$\sup_{\overline{D}} h = \sup_{\partial D} h.$$

Proof. Suppose $x \in D$. Since h is continuous on $\overline{D}$, it is bounded on $\overline{D}$. By Itô's formula, $h(X_{t \wedge \tau_D}) - h(X_0)$ is a martingale. Taking expectations with respect to $\mathbb{P}^x$, letting $t \to \infty$ and using dominated convergence,

$$(1.11) \qquad\qquad h(x) = \mathbb{E}^x h(X_{\tau_D}).$$

So $h(x) \leq \sup_{\partial D} h$. The other direction, $\sup_{\partial D} h \leq \sup_{\overline{D}} h$, is trivial.   □

This need not be true if D is unbounded. For example, if D is the upper half-plane in $\mathbb{R}^2$ and $h(x^1, x^2) = x^2$, then h is harmonic, since $\Delta h = 0$ and h is 0 on the boundary of D but certainly not bounded by 0.

Regular points

The Dirichlet problem is not solvable on every domain. An important consideration is whether Brownian motion started on the boundary of D hits D^c immediately. Recall T_A is the hitting time of the set A.

(1.9) Definition. *A point y is regular for a set A if $\mathbb{P}^y(T_A = 0) = 1$.*

A point y is regular for a set A provided, starting at y, the process does not go a positive length of time before hitting A. In the next section we will see that T_A is measurable for all Borel sets A; for the current section we may restrict attention to the cases where A is open or closed.

The right condition for the existence of a solution to the Dirichlet problem turns out to be that every point y in ∂D is regular for D^c. Starting at any point on the boundary, one hits D^c immediately. We say a domain D is regular for the Dirichlet problem if every boundary point is regular for D^c.

The corresponding analytic definition is a bit less intuitive. A function w is superharmonic at a point x if $\int_{\partial B(0,r)} w(x + y) \sigma_r(dy) \leq w(x)$ for all r sufficiently small (see Sect. 6). Let D be open. The analytic definition is that $y \in \partial D$ is regular for D^c if there exists a barrier there; a barrier is a continuous function w on $\overline{D}$ such that w is superharmonic at each point of D, $w > 0$ in $\overline{D} - \{y\}$, and $w(y) = 0$.

A function f is lower semicontinuous if $\{x : f(x) > a\}$ is open for all a. A function f is upper semicontinuous if $-f$ is lower semicontinuous. A key example for us of a lower semicontinuous function is the following.

(1.10) Proposition. $x \to \mathbb{P}^x(\tau_D \leq t)$ *is lower semicontinuous.*

Proof. Let $0 < s < t$. Let $\varphi(y) = \mathbb{P}^y(\tau_D \leq t - s)$. By the Markov property,

$$w_s(x) = \mathbb{P}^x(X_u \in D^c \text{ for some } u \in [s, t]) = \mathbb{E}^x[\mathbb{P}^{X_s}(\tau_D \leq t - s)]$$
$$= \mathbb{E}^x\varphi(X_s) = P_s\varphi(x),$$

where $p(s, x, y)$ and P_s are defined in (I.3.1) and (I.3.2). Since φ is bounded, by the continuity of $p(s, x, y)$ in x and dominated convergence, w_s is continuous in x. Also $w_s(x) \uparrow \mathbb{P}^x(\tau_D \leq t)$. By Exercise 3, the latter function is lower semicontinuous. $\qquad\square$

(1.11) Corollary. *Suppose $y \in \partial D$ is regular for D^c. If $x_n \in D$ converges to y, then for all $t > 0$*

$$\lim_{n \to \infty} \mathbb{P}^{x_n}(\tau_D \leq t) = 1.$$

Proof. By the preceding proposition, $\{x : \mathbb{P}^x(\tau_D \leq t) > 1 - \varepsilon\}$ is open. $\mathbb{P}^y(\tau_D \leq t) = 1$, so y is in that open set. Thus for n sufficiently large, $\mathbb{P}^{x_n}(\tau_D \leq t) \geq 1 - \varepsilon$. Since probabilities are bounded above by 1, that proves the corollary. $\qquad\square$

The whole point of regularity is the following.

(1.12) Proposition. *Suppose f is bounded and continuous on ∂D. Suppose $y \in \partial D$ is regular for D^c. If $x_n \in D$ and $x_n \to y$, then $\mathbb{E}^{x_n}f(X_{\tau_D}) \to f(y)$.*

Proof. We first want to show that if a Brownian motion starts near $y \in \partial D$, then with high probability it exits D near y. More precisely, we want to show that if $\delta > 0$,

$$(1.12) \qquad\qquad \lim_{n \to \infty} \mathbb{P}^{x_n}(X_{\tau_D} \in B(y, \delta)) = 1.$$

Let $\varepsilon > 0$. Pick $t > 0$ small so that $\mathbb{P}^0(\sup_{s \leq t} |X_s| \geq \delta/2) < \varepsilon$. Then if n is large enough, we have $|x_n - y| < \delta/2$, and by Corollary 1.11, we see that $\mathbb{P}^{x_n}(\tau_D \leq t) \geq 1 - \varepsilon$. Hence

$$\mathbb{P}^{x_n}(X_{\tau_D} \in B(y, \delta)) \geq \mathbb{P}^{x_n}(\tau_D \leq t, \sup_{s \leq t} |X_s - x_n| \leq \delta/2)$$

$$\geq \mathbb{P}^{x_n}(\tau_D \leq t) - \mathbb{P}^0(\sup_{s \leq t} |X_s| > \delta/2)$$

$$\geq (1 - \varepsilon) - \varepsilon.$$

This implies (1.12).

Now we use the continuity and boundedness of f to complete the proof. Pick δ small so that $|f(z) - f(y)| < \varepsilon$ if $|z - y| < \delta$, $z \in \partial D$.

$$(1.13) \qquad \mathbb{E}^{x_n}f(X_{\tau_D}) = \mathbb{E}^{x_n}\big[f(X_{\tau_D}); X_{\tau_D} \in B(y, \delta)\big]$$
$$+ \mathbb{E}^{x_n}\big[f(X_{\tau_D}); X_{\tau_D} \notin B(y, \delta)\big].$$

The second term on the right is bounded by $\|f\|_\infty \mathbb{P}^{x_n}(X_{\tau_D} \notin B(y,\delta))$, which tends to 0 as $n \to \infty$ by (1.12). On the other hand,

$$(1.14) \quad \left| \mathbb{E}^{x_n}\left[f(X_{\tau_D}); X_{\tau_D} \in B(y,\delta)\right] - f(y)\mathbb{P}^{x_n}(X_{\tau_D} \in B(y,\delta)) \right|$$
$$\leq \varepsilon \mathbb{P}^{x_n}(X_{\tau_D} \in B(y,\delta)).$$

By (1.12), (1.13), and (1.14) we conclude $\mathbb{E}^{x_n} f(X_{\tau_D}) \to f(y)$. $\square$

We will put the pieces together for the solution to the Dirichlet problem in a moment. First, however, let us give some conditions and examples concerning regularity.

Let $\widetilde{V}_a = \{(x^1, \ldots, x^d) : x^1 > 0, |(x^2, \ldots, x^d)| < ax^1\}$. The vertex of $\widetilde{V}_a$ is the origin. A cone V in $\mathbb{R}^d$ is a translation and rotation of $\widetilde{V}_a$ for some a.

(1.13) Proposition. *(Poincaré cone condition) Suppose there exists a cone V with vertex $y \in \partial D$ such that $V \cap B(y,r) \subseteq D^c$ for some $r > 0$. Then y is regular for D^c.*

Proof. By translation and rotation of the coordinates, we may suppose $y = 0$ and $V = \widetilde{V}_a$ for some a. Then for each t,

$$\mathbb{P}^0(\tau_D \leq t) \geq \mathbb{P}^0(X_t \in D^c) \geq \mathbb{P}^0(X_t \in V \cap B(0,r))$$
$$\geq \mathbb{P}^0(X_t \in V) - \mathbb{P}^0(X_t \notin B(0,r)).$$

By scaling, the last term is $\mathbb{P}^0(X_1 \in V) - \mathbb{P}^0(X_1 \notin B(0, r/\sqrt{t}))$, which converges to $\mathbb{P}^0(X_1 \in V) > 0$ as $t \to 0$. The left hand side converges to $\mathbb{P}^0(\tau_D = 0)$. So by the zero-one law, $\mathbb{P}^0(\tau_D = 0) = 1$. $\square$

For $d = 2$, if $y \in \partial D$ and there exists a line segment contained in D^c with one end at y, then y is regular for D^c. This is Exercise 6 and is proved similarly to the following.

(1.14) Proposition. *Suppose $d = 2$ and D is simply connected. If $y \in \partial D$, then y is regular for D^c.*

Proof. Without loss of generality, take $y = 0$. Let $\varphi : [0, 3 + 4\pi] \to \mathbb{R}^2$ be the curve that moves at constant unit speed, starts at 0, moves horizontally to the right to the point $(2, 0)$, then left to the point $(1, 0)$, and then goes counterclockwise twice around the unit circle. Let $\varepsilon = 1/16$. By the support theorem (Theorem I.6.6),

$$c_1 = \mathbb{P}^0\left(\sup_{s \leq 3+4\pi} |X_s - \varphi(s)| < \varepsilon \right) > 0.$$

Note c_1 does not depend on D. Since $y \in \partial D$ but D is simply connected, it follows that X_t hits D^c before time $3 + 4\pi$ with probability at least c_1; that is, $\mathbb{P}^0(\tau_D \leq 3 + 4\pi) \geq c_1$. By scaling, $\mathbb{P}^0(\tau_D \leq t) \geq c_1$ for each t, c_1

independent of t, and then $\mathbb{P}^0(\tau_D = 0) \geq c_1$. The zero-one law shows that $\mathbb{P}^0(\tau_D = 0) = 1$.   □

Now let us see an example of a domain with a point on the boundary that is not regular for the complement. Recall from Proposition I.5.8 that two-dimensional Brownian motion never hits 0 if it starts from any point other than 0. Even starting from 0, $X_{1/n} \neq 0$, a.s. and hence by the Markov property X_t is never 0 for any $t > 0$.

Let $D \subseteq \mathbb{R}^3$ be the domain $B(0,1) - \{(x^1, 0, 0) : x^1 \geq 0\}$, the ball minus the positive x^1-axis. Let y be the origin. Then

$$\mathbb{P}^y(\tau_D \leq \varepsilon) \leq \mathbb{P}^0\big(\sup_{s \leq \varepsilon} |X_s| \geq 1\big)$$

$$+ \mathbb{P}^0\big((X_t^2, X_t^3) = (0,0) \text{ for some } t > 0\big);$$

the second term on the right is 0 and the first tends to 0 as $\varepsilon \to 0$. So the origin is not regular for the complement of D.

The above example is not all that interesting, but since 0 is regular for D^c when $D = B(0,1) - \widetilde{V}_a$ for any $a > 0$ and not for the example we just did, it is not unreasonable to expect that 0 will not be regular for D^c if D is the ball minus some sufficiently sharp cusp. This turns out to be correct and we will even see exactly how sharp the cusp must be (see Proposition 5.17).

Later in this chapter we will make frequent use of the notation

(1.15) $A^r = \{x : x \text{ is regular for } A\}.$

Dirichlet problem

We now put the pieces together to solve the Dirichlet problem.

(1.15) Theorem. *Suppose D is a bounded domain such that every point on the boundary is regular for D^c. Suppose f is continuous on ∂D. Then there exists one and only one function u that is harmonic in D, continuous on $\overline{D}$, and agrees with f on ∂D. Furthermore u is given by the formula*

(1.16) $u(x) = \mathbb{E}^x f(X(\tau_D)).$

Proof. First the uniqueness: if u_1 and u_2 are two solutions, then $u_1 - u_2$ is harmonic on D and equals 0 on ∂D. By the maximum principle, $u_1 - u_2 \leq 0$ in D, or $u_1 \leq u_2$. Reversing the roles of u_1 and u_2, $u_2 \leq u_1$ on D. This gives uniqueness.

Define $u(x) = \mathbb{E}^x f(X(\tau_D))$. By (I.3.8), $u(x) = \mathbb{E}^x u(X_{\tau(B(x,r))})$ for $r < \text{dist}(x, \partial D)$. By Propositions 1.2 and I.2.8, u is harmonic in D. If $x_n \in D$, $y \in \partial D$ and $x_n \to y$, then $u(x_n) \to f(y)$ by Theorem 1.12. So u is the desired function that solves the Dirichlet problem.   □

Poisson kernel

Since $\mathbb{E}^x f(X(\tau_D)) = \int f(y)\mathbb{P}^x(X_{\tau_D} \in dy)$, it would be nice to find an explicit formula for $\mathbb{P}^x(X_{\tau_D} \in dy)$. In Chap. V we will see how conformal mapping can be used when $d = 2$. Here we will confine ourselves to finding $\mathbb{P}^x(X_{\tau_D} \in dy)$ when D is either the half-space or a ball, and $d \geq 2$.

The case of a half-space is a little easier, so we will do that first. Suppose

$$D = H = \{(x^1, \ldots, x^d) : x^d > 0\}.$$

To avoid confusion with dth powers, we will temporarily write x_d for x^d.

(1.16) Theorem. $\mathbb{P}^x(X_{\tau_D} \in dy) = P_H(x, y)\,dy$, *where*

$$(1.17) \qquad P_H(x, y) = c_d \frac{x_d}{(|\widetilde{x} - \widetilde{y}|^2 + x_d^2)^{d/2}}, \qquad c_d = \frac{\Gamma(d/2)}{\pi^{d/2}},$$

$\widetilde{x} = (x^1, \ldots, x^{d-1})$ *and* $\widetilde{y} = (y^1, \ldots, y^{d-1})$. *Moreover, if* $u = (u^1, \ldots, u^{d-1})$, *then*

$$\int e^{iu \cdot y} P_H(x, y)\, dy = e^{-|u|x_d}.$$

The constant c_d is such that $\int_{\partial D} P_H(x, y)dy = 1$. $P_H(x, y)$ is called the Poisson kernel for D. When $d = 2$, $P_H(x, y)$ is the density of the Cauchy distribution.

There are a number of proofs. One of the quickest is the following.

Proof. By the translation invariance of Brownian motion, we may assume $\widetilde{x} = 0$. Since $\mathbb{P}^x(X_{\tau_D} \in dy)$ is a probability measure on ∂D, we will find its characteristic function (i.e., Fourier transform). Thus we calculate

$$\mathbb{E}^x[e^{iu \cdot X_{\tau_D}}] = \mathbb{E}^x\left[\int_0^\infty e^{iu \cdot X_t}; \tau_D \in dt\right],$$

where $u \in \mathbb{R}^{d-1}$. τ_D depends only on X_t^d, which is independent of $\widetilde{X}_t = (X_t^1, \ldots, X_t^{d-1})$, and $\mathbb{E}^0 e^{iu \cdot \widetilde{X}_t} = e^{-|u|^2 t/2}$ (I.1.6). So

$$\mathbb{E}^x[e^{iu \cdot X_{\tau_D}}] = \int_0^\infty e^{-|u|^2 t/2}\mathbb{P}^{x_d}(X_{\tau_D}^d \in dt) = \mathbb{E}^{x_d} e^{-\lambda \tau_D},$$

where $\lambda = |u|^2/2$.

If $\beta \geq 0$ and $M_t = \exp(-\beta X_t^d - \beta^2 t/2)$, we saw in Sect. I.4 that M_t is a martingale. By optional stopping, $e^{-\beta x_d} = \mathbb{E}^{x_d} M_{t \wedge \tau_D}$. Since $M_{t \wedge \tau_D}$ is bounded below by 0 and above by 1, we may let $t \to \infty$ and use dominated convergence to obtain

$$e^{-\beta x_d} = \mathbb{E}^{x_d} M_{\tau_D} = \mathbb{E}^{x_d} \exp(-\beta X_{\tau_D}^d - \beta^2 \tau_D/2).$$

Note $X^d_{\tau_D} = 0$. Hence, setting $\beta = \sqrt{2\lambda}$,

$$\mathbb{E}^{x_d} \exp(-\lambda\tau_D) = \exp(-\sqrt{2\lambda}x_d).$$

(This was also proved in Exercises I.8.5 and I.8.20.) Therefore

$$\int e^{iu\cdot y} P_H(x,y)\, dy = \mathbb{E}^x e^{iu\cdot X_{\tau_D}} = e^{-|u|x_d}.$$

We now look in a table of Fourier transforms to invert this (see Erdélyi [1]), and obtain (1.17).   □

Durrett [1] gives a proof that is similar in spirit: the density of $\widetilde{X}_t$ is known for each t, the density of $T_{\{0\}}$ under $\mathbb{P}^{x^d}$ is known, and by doing some messy but routine integrals, one arrives at (1.17).

Another way to deduce (1.17) is to use the fact that the hitting distribution of the boundary of a ball starting at the center is the uniform distribution on the surface. Using the Kelvin transform, one inverts the ball through a certain sphere to obtain (1.17). We will use a similar procedure in reverse to get the hitting distribution for a ball started at any point; see Theorem 1.17.

For fixed y, $P_H(x,y)$ is harmonic as a simple calculation shows, and one can show that $\varphi_{x^d}(\widetilde{x} - \widetilde{y}) = P_H(x,y)$ is an approximation to the identity (see Chap. IV). Hence if f is continuous and bounded on ∂D, then $\int f(z)P_H(x,z)dz \to f(y)$ as $x \in D$ tends to $y \in \partial D$. So if one were told that (1.17) should be the correct formula, one could use the above facts to verify that it was indeed the Poisson kernel.

Finally, as we mentioned, when $d = 2$, one has what is known as the Cauchy density. The probabilistic reasoning is the following. If $S_r = \inf\{t : X^d_t = r\}$ and $Y_r = \widetilde{X}_{S_r}$, the strong Markov property shows that Y_r has independent increments. The translation invariance of Brownian motion shows that Y_r has stationary increments. Scaling shows that Y_r has the same law as rY_1. By symmetry, Y_r must be symmetric. The only process that satisfies all these properties is the symmetric Cauchy process (with some parameter). Provided one can identify the parameter, one has (1.17). This argument also works for $d \geq 3$.

We now want the Poisson kernel for a ball, i.e., $\mathbb{P}^x(X_{\tau_D} \in dy)$ when D is a ball in $\mathbb{R}^d$, $d \geq 2$.

(1.17) Theorem. *Let* $D = B(0,r)$. *Then* $\mathbb{P}^x(X_{\tau_D} \in dy) = P_B(x,y)\sigma_r(dy)$, *where*

$$P_B(x,y) = r^{d-2}\frac{(r^2 - |x|^2)}{|y - x|^d}, \qquad x \in D, y \in \partial D.$$

To prove this we proceed as follows.

(1.18) Lemma. *Define $I(x) = x/|x|^2$. If u is harmonic in a domain D, then $|x|^{2-d} u(I(x))$ is harmonic in $I(D)$.*

The mapping $x \to I(x)$ is called inversion through the unit sphere.

Proof. The proof is a tedious but routine calculation and is left to the reader. A few hints: check that

$$\Delta(fg) = (\Delta f)g + 2\nabla f \cdot \nabla g + f(\Delta g),$$

and that if $f = |x|^{2-d}$, then $\Delta f = 0$ if $x \neq 0$. Also, recall that $\partial(|x|)/\partial x^i = x^i/|x|$. □

Note that if $D = B(e_d, 1)$, where $e_d = (0, 0, \ldots, 0, 1)$ is the unit vector in the x^d direction, then $y \in \partial D$ implies that $(y^1)^2 + \cdots + (y^{d-1})^2 + (y^d - 1)^2 = 1$, or $y^d = |y|^2/2$. The dth component of $I(x)$ is $x^d/|x|^2$, so the mapping I takes D onto $H = \{x : x^d > 1/2\}$.

We claim that

(1.18)
$$|I(x) - I(y)| = \frac{|y - x|}{|y| \, |x|}.$$

To see this, note that the square of the left-hand side of (1.18) is

$$(I(x) - I(y)) \cdot (I(x) - I(y)) = |I(y)|^2 - 2I(x) \cdot I(y) + |I(y)|^2$$
$$= \frac{1}{|y|^2} - \frac{2x \cdot y}{|x|^2 |y|^2} + \frac{1}{|x|^2}.$$

Since the square of the right-hand side is

$$\frac{(x - y) \cdot (x - y)}{|x|^2 |y|^2} = \frac{|x|^2 - 2x \cdot y + |y|^2}{|x|^2 |y|^2},$$

which is the same thing, (1.18) is proved.

If in (1.18) we let $x = y + \varepsilon e_i$, where e_i is the unit vector in the ith direction, divide both sides by ε, and let $\varepsilon \to 0$, we see that $I(\cdot)$ dilates each direction by a factor $1/|y|^2$.

Proof of Theorem 1.17. First of all, if we show

$$P_{B(0,1)}(x, y) = c \frac{1 - |x|^2}{|x - y|^d}, \qquad y \in \partial B(0, 1), x \in B(0, 1),$$

for some constant c, taking $x = 0$ shows that the constant c must be 1. Secondly, we want to work with $\widehat{B} = B(e_d, 1)$ instead of $B(0, 1)$. To avoid confusion with dth powers, let us temporarily write x_d for the dth coordinate of x.

$$1 - |x - e_d|^2 = 1 - (|x|^2 - 2x \cdot e_d + |e_d|^2) = 2(x_d - |x|^2/2).$$

So it is enough to show

(1.19)
$$P_{\widehat{B}}(x,y) = c\frac{x_d - |x|^2/2}{|y-x|^d}$$

Let $H_{1/2} = \{(x^1, \ldots, x^{d-1}, x_d) : x_d > 1/2\}$. Since the Poisson kernel for the upper half-space in $\mathbb{R}^d$ is $cx_d/|y-x|^d$, the Poisson kernel for $H_{1/2}$ is

(1.20)
$$P_{H_{1/2}}(x,y) = c\frac{x_d - 1/2}{|x-y|^d}, \qquad x \in H_{1/2}, y \in \partial H_{1/2}.$$

We can now derive (1.19). Let f be a smooth function on $\partial \widehat{B}$ and let h be the solution to the Dirichlet problem in $\widehat{B}$ with boundary values f. Then $|x|^{2-d}h(x/|x|^2)$ is harmonic in $H_{1/2}$ with boundary values $|y|^{2-d}f(y/|y|^2)$. By the definition of $P_{H_{1/2}}$,

$$|x|^{2-d}h(x/|x|^2) = \int_{\partial H_{1/2}} P_{H_{1/2}}(x,y)|y|^{2-d}f(y/|y|^2)\,dy.$$

Let $\bar{x} = x/|x|^2$ so that $|x| = 1/|\bar{x}|$. Make the substitution $\bar{y} = y/|y|^2$. I maps the boundary of $H_{1/2}$ onto the boundary of $\widehat{B}$, both of which are $(d-1)$-dimensional. As we saw following (1.18), $I(x)$ dilates each direction by a factor of $|y|^{-2}$. Hence $dy = (|\bar{y}|^{-2})^{d-1}d\bar{y} = |\bar{y}|^{2-2d}d\bar{y}$, where $d\bar{y}$ is surface measure on $\partial \widehat{B}$. Then

$$|\bar{x}|^{d-2}h(\bar{x}) = \int_{\partial \widehat{B}} P_{H_{1/2}}\left(\frac{\bar{x}}{|\bar{x}|^2}, \frac{\bar{y}}{|\bar{y}|^2}\right)|\bar{y}|^{d-2}f(\bar{y})|\bar{y}|^{2-2d}d\bar{y}$$

$$= c\int_{\partial \widehat{B}} \frac{\bar{x}_d/|\bar{x}|^2 - 1/2}{\left|\bar{x}/|\bar{x}|^2 - \bar{y}/|\bar{y}|^2\right|^d}|\bar{y}|^{-d}f(\bar{y})d\bar{y}.$$

Using (1.18),

$$h(\bar{x}) = c\int_{\partial \widehat{B}} \frac{\bar{x}_d/|\bar{x}|^2 - 1/2}{|\bar{x}-\bar{y}|^d/(|\bar{x}||\bar{y}|)^d}|\bar{x}|^{2-d}|\bar{y}|^{-d}f(\bar{y})d\bar{y}$$

(1.21)
$$= c\int_{\partial \widehat{B}} \frac{\bar{x}_d - |\bar{x}|^2/2}{|\bar{x}-\bar{y}|^d}f(\bar{y})d\bar{y}.$$

h also satisfies

(1.22)
$$h(\bar{x}) = \int_{\partial \widehat{B}} P_{\widehat{B}}(\bar{x},\bar{y})f(\bar{y})d\bar{y}.$$

Since the right-hand sides of (1.21) and (1.22) are equal for all smooth f,

$$P_{\widehat{B}}(\bar{x},\bar{y}) = c\frac{\bar{x}_d - |\bar{x}|^2/2}{|\bar{x}-\bar{y}|^d},$$

which is (1.19).   □

One can also find $\mathbb{P}^x(X_{T_B} \in dy)$ when $|x| > 1$, the exterior Poisson kernel; see Port and Stone [1].

Harnack inequality

One important use of the Poisson kernel for a ball is to prove the Harnack inequality.

(1.19) Theorem. *Suppose $r < R$. There exists c such that if u is nonnegative and harmonic in $B(0, R)$ and $x, y \in B(0, r)$, then $u(x) \le cu(y)$. In fact, c can be taken to be*

$$c_1(r, R) = \frac{R^2}{R^2 - r^2} \left(\frac{r + R}{R - r} \right)^d.$$

Proof. For any $x \in B(0, r)$, if $|z| = R$, then

$$P_{B(0,R)}(x, z) = \frac{R^{d-2}(R^2 - |x|^2)}{|z - x|^d}$$

by Theorem 1.17 and scaling. $P_{B(0,R)}$ is bounded above by $R^d/(R - r)^d$ and below by $R^{d-2}(R^2 - |x|^2)/(R + r)^d$.
Suppose that $u \in C(\overline{B(0, R)})$. Then if

$$c_2 = \sup_{w \in B(0,r)} P_B(w, z) \quad \text{and} \quad c_3 = \inf_{w \in B(0,r)} P_B(w, z),$$

we have

$$u(x) = \int_{\partial B(0,R)} P_B(x, z)u(z)\sigma_R(dz) \le c_2 \int_{\partial B(0,R)} u(z)\sigma_R(dz)$$

$$\le \frac{c_2}{c_3} \int_{\partial B(0,R)} P_B(y, z)u(z)\sigma_R(dz) = \frac{c_2}{c_3} u(y).$$

Note $c_1 = c_2/c_3$.
If u is not in $C(\overline{B(0, R)})$, it will still be in $C(\overline{B(0, R - \varepsilon)})$ for every ε, hence $u(x) \le c_1(r, R - \varepsilon)u(y)$. Now let $\varepsilon \to 0$.   □

It is often convenient to use the Harnack inequality repeatedly on a chain of balls.

(1.20) Theorem. *Suppose $x, y \in D$ and x and y can be connected by a curve γ in D such that $\inf_{z \in \gamma} \text{dist}(z, \partial D) \ge R$. If u is nonnegative and harmonic on D, then $u(x) \le c_1 u(y)$, where c_1 depends only on R and the length of γ.*

Proof. Let $x_0 = x$ and choose $x_1, \ldots, x_n \in \gamma$ such that $|x_{i+1} - x_i| \leq R/2$, for $i = 0, 1, \ldots, n - 1$, and $x_n = y$. The minimum number n needed depends only on R and the length of γ. By the Harnack inequality in $B(x_i, R)$, $u(x_{i+1}) \geq c_1(R/2, R)u(x_i)$. By induction,

$$u(y) = u(x_n) \geq c_1^n u(x_0) = c_1^n u(x).$$

□

One can prove Harnack's inequality without having an explicit formula for the Poisson kernel. A nice probabilistic argument using coupling can be obtained by the methods of Lindvall and Rogers [1] and Cranston [3].

Representation of harmonic functions

Recall that $\sigma_r(dy)$ is normalized surface measure on $\partial B(0, r)$. If h is the harmonic function in $B(0, 1)$ with boundary values f, then we saw $h(x) = \int P_{B(0,1)}(x, y)f(y)\sigma_1(dy)$. We now want to show that the class of all positive harmonic functions has the same form provided we replace $f(y)\sigma_1(dy)$ by positive finite measures. That is the content of the next two propositions.

(1.21) Proposition. *If μ is a finite measure on $\partial B(0, 1)$, then $h(x) = \int P_{B(0,1)}(x, y)\mu(dy)$ is harmonic inside $B(0, 1)$.*

Proof. Let us write B for $B(0, 1)$. Let f_n be continuous functions on ∂B such that $f_n(y)\sigma_1(dy)$ converges weakly to μ (see Exercise 9 for the construction of such f_n). If $x_0 \in B$ and $r < (1 - |x_0|)/2$, then $\{P_B(x, \cdot), x \in B(x_0, r)\}$ is an equicontinuous family of functions. By Exercise I.8.15, $h_n(x) = \int P_B(x, y)f_n(y)\sigma_1(dy)$ converges uniformly to $h(x)$ for $x \in B(x_0, r)$. Note h_n is harmonic, so

$$(1.23) \qquad h_n(x_0) = \frac{1}{|B(x_0, r)|} \int_{B(x_0, r)} h_n(x)\, dx.$$

Letting $n \to \infty$ and using the uniform convergence of h_n to h shows that (1.23) holds with h_n replaced by h. This says h is harmonic.   □

(1.22) Proposition. *Suppose h is a positive harmonic function in $B(0, 1)$. Then there exists a finite measure μ on $\partial B(0, 1)$ such that*

$$h(x) = \int_{\partial B(0,1)} P_{B(0,1)}(x, y)\mu(dy).$$

Proof. Fix $x \in B(0, 1)$ and let $1 > r > |x|$. Since h is harmonic, hence C^∞ on $\overline{B}(0, r)$,

$$h(x) = \mathbb{E}^x h(X(\tau_{B(0,r)})) = \int_{\partial B(0,r)} P_{B(0,r)}(x, y)h(y)\sigma_r(dy).$$

Let μ_r be the measure on $\partial B(0,1)$ defined by $\mu_r(dy) = h(ry)\sigma_1(dy)$. Note μ_r does not depend on x as long as $r > |x|$. The total mass of the μ_r is uniformly bounded by $h(0)$ since

$$\mu_r(\partial B(0,1)) = \int h(ry)\sigma_1(dy) = \int P_{B(0,r)}(0,y)h(ry)\sigma_1(dy)$$
$$= \mathbb{E}^0 h(X(\tau_{B(0,r)})) = h(0).$$

By Theorem I.7.4 and the remark following the proof, there exists a subsequence r_j such that μ_{r_j} converges weakly to a positive measure, say μ. We have

(1.24)
$$h(x) = \int_{\partial B(0,1)} P_{B(0,r)}(x,ry)\mu_r(dy).$$

Since $P_{B(0,r)}(x,\cdot)$ converges uniformly to $P_{B(0,1)}(x,\cdot)$, $P_{B(0,1)}(x,y)$ is continuous in y, and μ_{r_j} converges weakly to μ, taking the limit in (1.24) along the subsequence r_j gives

$$h(x) = \int_{\partial B(0,1)} P_{B(0,1)}(x,y)\mu(dy).$$

This is precisely what we wanted.   □

2 Choquet capacities

Analytic sets

This section is devoted to proving Choquet's capacity theorem, and using it to show that hitting times of Borel sets are measurable and can be approximated by hitting times of compact sets. The reader willing to believe this can skip this section. These results will be used in the remainder of Chap. II, but not in Chapters III through V.

It turns out the problem is actually simpler if we try to do more. We will prove some results for a class of sets called analytic sets, which include the Borel sets.

A collection of sets $\mathcal{M}$ is called a monotone class if (a) $A_n \in \mathcal{M}$ and A_n increasing implies $\cup_n A_n \in \mathcal{M}$, and (b) $A_n \in \mathcal{M}$ and A_n decreasing implies $\cap_n A_n \in \mathcal{M}$.

A very useful result is the following.

(2.1) Theorem. (Monotone class theorem) *Let $\mathcal{M}$ be a monotone class containing the open sets of a metric space. Then $\mathcal{M}$ contains all the Borel sets.*

Proof. Let $\mathcal{G}$ be the collection of open sets and let $\mathcal{M}'$ be the smallest monotone class containing $\mathcal{G}$, that is, $\mathcal{M}'$ is the intersection of all monotone classes containing $\mathcal{G}$. It is routine to show that the intersection of monotone classes is a monotone class. We will show $\mathcal{M}'$ contains all the Borel sets. Since the collection of Borel sets is itself a monotone class, this will show that in fact $\mathcal{M}'$ is equal to the Borel σ-field.

We first want to show that $\mathcal{M}'$ is closed under the operation of taking finite intersections. Fix $A \in \mathcal{G}$ and let $\mathcal{N}_1 = \{B \in \mathcal{M}' : A \cap B \in \mathcal{M}'\}$. Clearly $\mathcal{G} \subseteq \mathcal{N}_1$. It is easy to see that $\mathcal{N}_1$ is a monotone class itself, and therefore $\mathcal{N}_1 = \mathcal{M}'$. We have shown that if $A \in \mathcal{G}$ and $B \in \mathcal{M}'$, then $A \cap B \in \mathcal{M}'$.

Next fix $A \in \mathcal{M}'$ and let $\mathcal{N}_2 = \{B \in \mathcal{M}' : A \cap B \in \mathcal{M}'\}$. By what we have just shown, $\mathcal{G} \subseteq \mathcal{N}_2$. Again, $\mathcal{N}_2$ is a monotone class, hence $\mathcal{N}_2 = \mathcal{M}'$. This implies that the intersection of two sets in $\mathcal{M}'$ is again in $\mathcal{M}'$.

We now show $\mathcal{M}'$ is closed under the operation of taking complements. Let $\mathcal{N}_3 = \{A \in \mathcal{M}' : A^c \in \mathcal{M}'\}$. If A is open, then A^c is closed and can be written as the intersection of a countable decreasing sequence of open sets. Hence $A^c \in \mathcal{M}'$. This shows $\mathcal{G} \subseteq \mathcal{N}_3$. However, $\mathcal{N}_3$ is a monotone class contained in $\mathcal{M}'$ and containing $\mathcal{G}$, so $\mathcal{N}_3 = \mathcal{M}'$, or $\mathcal{M}'$ is closed under the operation of taking complements.

A monotone class that is closed under taking finite intersections and complements is a σ-field. Since $\mathcal{M}'$ contains the open sets, it must contain all the Borel sets. $\qquad\square$

Let $\mathcal{K} = \mathcal{K}(F)$ denote the collection of compact subsets of F, $\mathcal{K}_\sigma = \mathcal{K}_\sigma(F)$ the collection of subsets that are the union of a countable increasing sequence of compact subsets of F, and $\mathcal{K}_{\sigma\delta} = \mathcal{K}_{\sigma\delta}(F)$ the collection of subsets that are the intersection of a countable decreasing sequence of $\mathcal{K}_\sigma$ sets. The δ and σ are supposed to suggest "d" and "s," which are the first letters of the German words "Durchschnitt" and "Summe" for intersection and union, respectively.

(2.2) Definition. *A set $A \subseteq F$ is analytic if there exists a compact Hausdorff space X and a $\mathcal{K}_{\sigma\delta}(X \times F)$ set B such that A is the projection onto F of B. That is, if $\pi(x, f) = f$ for $(x, f) \in X \times F$, then $\pi(B) = A$.*

(2.3) Proposition. *Borel sets are analytic.*

Proof. Since any open set A is a $\mathcal{K}_\sigma$ set and hence a $\mathcal{K}_{\sigma\delta}$ set, the case when A is open is easy; just let $X = [0, 1]$ and $B = [0, 1] \times A$. If we show that the class of analytic sets is a monotone class, our result will follow by Theorem 2.1.

Suppose there exist compact Hausdorff spaces X_n and $\mathcal{K}_{\sigma\delta}(X_n \times F)$ sets B_n whose projections are A_n. Let $X = \prod_{n=1}^{\infty} X_n$, the product space with the product topology. Then X is compact and Hausdorff. Let C_n be the largest set in $X \times F$ whose projection onto $X_n \times F$ is B_n. More precisely, if π_n is the projection of $X \times F$ onto $X_n \times F$, let $C_n = \pi_n^{-1}(B_n)$. Then each $C_n \in \mathcal{K}_{\sigma\delta}(X \times F)$, and so $\cap_n C_n \in \mathcal{K}_{\sigma\delta}(X \times F)$. If π is the projection of $X \times F$ onto F, then noting that $\cap_n A_n = \cap_n \pi(C_n) = \pi(\cap_n C_n)$ shows that $\cap_n A_n$ is analytic.

Now, for the union, let X_n, B_n be as before. The space $\sum_{n=1}^{\infty} X_n$ is defined to be the space $\cup_n (X_n \times \{n\})$ with the topology generated by the sets $\{G_n \times \{n\} : G_n \text{ open in } X_n\}$. Let X be the one point compactification of $\sum_{n=1}^{\infty} X_n$. Each $B_n = \cap_m B_{nm}$, where $B_{nm} \in \mathcal{K}_\sigma(X_n \times F)$. Let

$$\widetilde{B}_{nm} = \{((x,n),f) \in X \times F : x \in X_n, (x,f) \in B_{nm}\}.$$

It is easy to see that each $\widetilde{B}_{nm} \in \mathcal{K}_\sigma(X \times F)$. If $\widetilde{B}_n = \cap_m \widetilde{B}_{nm}$, then $\pi(\widetilde{B}_n) = A_n$. Let $\widetilde{B} = \cup_n \widetilde{B}_n$. Since $\cup_n \widetilde{B}_{nm} \in \mathcal{K}_\sigma(X \times F)$, then

$$\widetilde{B} = \cup_n(\cap_m(\widetilde{B}_{nm} \times \{n\})) = \cap_m(\cup_n \widetilde{B}_{nm} \times \{n\}) \in \mathcal{K}_{\sigma\delta}(X \times F).$$

Finally, $\pi(\widetilde{B}) = \cup_n \pi(\widetilde{B}_n) = \cup_n A_n$, or $\cup_n A_n$ is analytic.   □

Capacities

We now give the definition of capacity (or Choquet capacity). Newtonian capacity will turn out to be a special case of this (see Sect. 5).

(2.4) Definition. C^* is an outer capacity if it is a nonnegative function on all subsets of F taking values in $[0, \infty]$ with the properties
(a) if $A \subseteq B$, then $C^*(A) \leq C^*(B)$;
(b) if A_n is an increasing sequence of sets, then $C^*(\cup_n A_n) = \sup_n C^*(A_n)$;
(c) if A_n is a decreasing subsequence of compact sets, then $C^*(\cap_n A_n) = \inf_n C^*(A_n)$.

Given C^*, we define the inner capacity C_* by

$$C_*(A) = \sup\{C^*(B) : B \subseteq A, B \text{ compact}\}.$$

A set A is capacitable if $C^*(A) = C_*(A)$, in which case we write $C(A)$ and call it the capacity of A.

(2.5) Theorem. *(Choquet) Every analytic set is capacitable.*

This theorem will be an easy consequence of the following two lemmas.

(2.6) Lemma. $\mathcal{K}_{\sigma\delta}$ *sets are capacitable.*

Proof. Let $A \in \mathcal{K}_{\sigma\delta}$, so that $A = \cap_n A_n$ with each $A_n = \cup_m A_{nm}$ and for each n the sets A_{nm} form an increasing sequence of compacts. Let $a < C^*(A)$.

We will construct a compact set B such that $B \subseteq A$ and $C^*(B) \geq a$; this will prove A is capacitable.

We will begin by constructing by induction a sequence of compact sets B_n contained in A_n such that if $D_n = A \cap B_1 \cap \cdots \cap B_n$, then $C^*(D_n) > a$. We start with $n = 1$. By Definition 2.4(b),

$$C^*(A) = C^*(A \cap A_1) = \sup_m C^*(A \cap A_{1m}).$$

Choose m large enough so that $C^*(A \cap A_{1m}) > a$, let $B_1 = A_{1m}$, and let $D_1 = A \cap B_1$.

Suppose now that we have constructed $D_{n-1}, B_1, \ldots, B_{n-1}$. Since $A \subseteq A_n$,

$$C^*(D_{n-1}) = C^*(D_{n-1} \cap A_n),$$

and so

$$C^*(D_{n-1}) = \sup_m C^*(D_{n-1} \cap A_{nm})$$

by Definition 2.4(b). Take m large enough so that $C^*(D_{n-1} \cap A_{nm}) > a$, let $B_n = A_{nm}$, and $D_n = D_{n-1} \cap B_n$. Thus $D_n \subseteq A_n$ and $C^*(D_n) > a$.

Now let $E_n = B_1 \cap \ldots \cap B_n$ and $B = \cap_n B_n = \cap_n E_n$. The E_ns are compact and decrease, so B is compact. $D_n \subseteq E_n$, so $C^*(E_n) \geq C^*(D_n) > a$, and by Definition 2.4(c), $C^*(B) \geq a$. Also $B_n \subseteq A_n$, so $B \subseteq A$. This is the desired B. $\qquad \square$

(2.7) Lemma. *Suppose A is analytic and $B \in \mathcal{K}_{\sigma\delta}(X \times F)$ with $A = \pi(B)$, where π is the projection of $X \times F$ onto F. Define a set function D^* on subsets of $X \times F$ by $D^*(H) = C^*(\pi(H))$. Then D^* is an outer capacity on $X \times F$.*

Proof. Parts (a) and (b) of Definition 2.4 are immediate. For part (c),

$$D^*(\cap_n H_n) = C^*(\pi(\cap_n H_n)) = C^*(\cap_n \pi(H_n)).$$

If the H_n are compact in $X \times F$, then since projections are continuous functions, the $\pi(H_n)$ are compact in F, hence

$$C^*(\cap_n \pi(H_n)) = \inf_n C^*(\pi(H_n)) = \inf_n D^*(H_n).$$

$\qquad \square$

Proof of Theorem 2.5. Let $\varepsilon > 0$. If A is analytic, there exists $B \in \mathcal{K}_{\sigma\delta}(X \times F)$ such that $\pi(B) = A$. By Lemmas 2.6 and 2.7, B is capacitable relative to D^*, so there exists a compact set $G \subset X \times F$ such that $G \subseteq B$ and $D^*(G) \geq D^*(B) - \varepsilon$. Let $J = \pi(G)$. Then J is compact, contained in A, and $C^*(J) = D^*(G) \geq D^*(B) - \varepsilon = C^*(A) - \varepsilon$. Since ε is arbitrary, this proves A is capacitable, provided $C^*(A) < \infty$. The case $C^*(A) = \infty$ is very similar. $\qquad \square$

Sometimes one does not start with C^* defined on all subsets. Suppose there exists a set function $\overline{C}$ defined on $\mathcal{K}(F)$ such that

(a) if $A, B \in \mathcal{K}$ and $A \subseteq B$, then $\overline{C}(A) \leq \overline{C}(B)$;

(b) if $A \in \mathcal{K}$ and $\varepsilon > 0$, there exists an open set G containing A such that whenever $A \subseteq B \subseteq G$ and $B \in \mathcal{K}$, then $\overline{C}(B) - \overline{C}(A) < \varepsilon$;

(c) $\overline{C}(A \cup B) + \overline{C}(A \cap B) \leq \overline{C}(A) + \overline{C}(B)$ if $A, B \in \mathcal{K}$.

One can then extend $\overline{C}$ to open sets by defining $\overline{C}(G) = \sup\{\overline{C}(K) : K \subseteq A, K \in \mathcal{K}\}$, and define $C^*(A) = \inf\{\overline{C}(G) : G \text{ open}, A \subseteq G\}$. One can then prove that C^* is an outer capacity. See Blumenthal and Getoor [1] for a proof. In what follows we do not need this result but only Theorem 2.5.

Hitting times

Let us first assume that the $\mathcal{F}_t$ are complete and right continuous with respect to a single probability $\mathbb{P}$. Recall that the first entry time $D_A = \inf\{t \geq 0 : X_t \in A\}$ is a stopping time if A is open or closed (Exercise I.8.7). As we will see from (2.1), it is easier to first prove measurability for D_A.

(2.8) Theorem. *For all A Borel, T_A is a stopping time.*

Proof. Fix t and define

$$(2.1) \qquad R_t(A) = \{\omega : X_s(\omega) \in A \text{ for some } s \in [0, t]\} = (D_A \leq t).$$

Let $\mathbb{P}^*$ be the outer measure associated to the probability $\mathbb{P}$ (that is, $\mathbb{P}^*(A) = \inf\{\mathbb{P}(B) : A \subseteq B, B \in \mathcal{F}\}$.) If A is compact, $R_t(A)$ is $\mathcal{F}_t$ measurable and $\mathbb{P}^*(R_t(A)) = \mathbb{P}(R_t(A))$. For any Borel set A, let $C^*(A) = \mathbb{P}^*(R_t(A))$. We first show that C^* is an outer capacity.

Parts (a) and (b) of Definition 2.4 are easy. As for part (c), suppose A_n are compact sets decreasing to a compact set A. Then T_{A_n} increases to a stopping time, say S. We claim $S = T_A$. Since $T_A \geq T_{A_n}$ for all n, then $T_A \geq S$. If $m \leq n$, then since the A_n are closed and X_t has continuous paths, $X(T_{A_n}) \in A_n \subseteq A_m$, hence $X_S \in A_m$ for all m, so $X_S \in A$. Therefore $S \geq T_A$, and so $S = T_A$.

It follows that $D_{A_n} \uparrow D_A$. For if $X_0(\omega) \in A_n$ for all n, then $X_0(\omega) \in A$ and $D_{A_n}(\omega) = D_A(\omega) = 0$; if $X_0(\omega) \notin A_n$ for some n, then $X_0(\omega) \notin A_m$ for all $m \geq n$, and then $D_{A_m}(\omega) = T_{A_m}(\omega)$ for m large and $D_A(\omega) = T_A(\omega)$.

We then have

$$C^*(A_n) = \mathbb{P}(R_t(A_n)) = \mathbb{P}(D_{A_n} \leq t) \downarrow \mathbb{P}(D_A \leq t)$$
$$= \mathbb{P}(R_t(A)) = C^*(A).$$

Thus C^* is an outer capacity.

By Theorem 2.5, A is capacitable, or there exist increasing compact sets K_n contained in A with $\mathbb{P}(D_{K_n} \leq t) = C^*(K_n) \uparrow C^*(A) = \mathbb{P}^*(D_A \leq t)$.

Hence $(D_A \leq t) - \cup_n(D_{K_n} \leq t)$ has $\mathbb{P}^*$ outer measure 0. Since $\mathcal{F}_t$ is complete, $(D_A \leq t) \in \mathcal{F}_t$.

If we repeat the whole procedure for $X_t \circ \theta_u$, we see that D_A^u is $\mathcal{F}_{t+u}$ measurable, where $D_A^u = \inf\{t \geq u : X_t \in A\}$. Since $1/m + D_A^{1/m} \circ \theta_{1/m} \downarrow T_A$ as $m \to \infty$, this proves T_A is a stopping time. $\square$

We actually want to be able to work with the situation where we have a family of probabilities $\mathbb{P}^x$, and $\mathcal{F}_t$ is defined as in Sect. I.3. To assert T_A is a stopping time relative to $\mathcal{F}_t$, we need to show that $(T_A \leq t)$ differs from some $\mathcal{F}_{t+}^{00}$ set by a null set that does not depend on x. This is Exercise 10.

Besides knowing T_A is measurable, the approximation of A from below is important.

(2.9) Proposition. *For each x, there exists an increasing sequence of compact sets K_n contained in B such that $D_{K_n} \downarrow D_B$ on $(D_B < \infty)$, $\mathbb{P}^x$-a.s.*

Proof. For each t we can find an increasing sequence of compacts L_n^t contained in B with $\mathbb{P}^x(R_t(L_n^t)) \uparrow \mathbb{P}^x(R_t(B))$. Let q_j be an enumeration of the rationals. Let $K_n = L_n^{q_1} \cup \cdots \cup L_n^{q_n}$. Then the K_n are compact, form an increasing sequence, and are all contained in B. So $D_{K_n} \downarrow$, say to S, and since $D_{K_n} \geq D_B$ for all n, then $S \geq D_B$. If we prove $S \leq D_B$, $\mathbb{P}^x$-a.s., then $S = D_B$, and we have our result.

If $D_B < S$, there exists a rational q_j with $D_B < q_j < S$. So it suffices to prove $\mathbb{P}^x(D_B < q_j < S) = 0$ for all j. If $D_B < q_j$, then $\omega \in R_{q_j}(B)$. Since $R_{q_j}(L_n^{q_j}) \uparrow R_{q_j}(B)$, a.s., then except for a null set, ω will be in $R_{q_j}(L_n^{q_j})$ for all n large enough, hence in $R_{q_j}(K_n)$ if n is large enough. Then $D_{K_n}(\omega) \leq q_j < D_B$ or $S \leq q_j$. So $\mathbb{P}^x(D_b < q_j < S) = 0$. $\square$

(2.10) Theorem. *There exists an increasing sequence of compacts K_n contained in B such that $T_{K_n} \downarrow T_B$.*

Proof. Applying the above proposition to $X_t \circ \theta_{1/m}$, for each m there exists L_n^m compact, increasing in n, and contained in B such that $D(L_n^m) \circ \theta_{1/m} \downarrow D_B \circ \theta_{1/m}$. Let $K_n = L_n^1 \cup \cdots \cup L_n^n$. Then K_n is an increasing sequence of compacts contained in B, and $D_{K_n} \circ \theta_{1/m} \downarrow D_B \circ \theta_{1/m}$. Also, for each n, $1/m + D_{K_n} \circ \theta_{1/m} \downarrow T_{K_n}$ and $1/m + D_B \circ \theta_{1/m} \downarrow T_B$. We write

$$T_B = \lim_m(1/m + D_B \circ \theta_{1/m}) = \lim_m \lim_n(1/m + D_{K_n} \circ \theta_{1/m})$$
$$= \lim_n \lim_m(1/m + D_{K_n} \circ \theta_{1/m}) = \lim_n T_{K_n}.$$

Since $1/m + D_{K_n} \circ \theta_{1/m}$ is decreasing in both m and n, the change in the order of taking limits is justified. Since T_{K_n} is decreasing, this completes the proof. $\square$

3 Newtonian potentials and Green functions

Newtonian potentials

Suppose we put an electron at the origin. Another electron at the point x will feel a repulsive force away from the origin with a magnitude proportional to $1/|x|^2$, or $F = cx/|x|^3$ if F denotes the force vector at x. Since electrostatic fields are conservative force fields, F is supposed to be the gradient of a potential u. Since u is symmetric, it is a function of $|x|$, and u must be of the form $c/|x|$. A quick calculation (recall $\partial(|x|)/\partial x^i = x^i/|x|$) shows that $\nabla u = F$.

For simplicity we will take $d \geq 3$ (see below for some comments on the $d = 2$ case). Let us define

$$(3.1) \qquad u(x, y) = \frac{c_d}{|x - y|^{d-2}}, \qquad c_d = \frac{\Gamma(d/2 - 1)}{(2\pi)^{d/2}}.$$

The reason for the choice of c_d will become apparent later.

If we have n electrons at $x_1, \ldots, x_n$, potentials should add and the potential at the origin should be $\sum_{i=1}^n c_d |x_i|^{2-d}$. Looking at Riemann sum approximations suggests that it makes sense to define the Newtonian potential of f by

$$(3.2) \qquad Uf(x) = c_d \int \frac{f(y)}{|x - y|^{d-2}}\, dy$$

if f is nonnegative or bounded, and the potential of a finite measure μ by

$$(3.3) \qquad U\mu(x) = c_d \int \frac{\mu(dy)}{|x - y|^{d-2}}.$$

$U\mu$ need not be finite at every point (e.g., take $\mu = \delta_x, y = x$). However, changing to polar coordinates shows that

$$Uf(x) = c \int_0^\infty \int_{\partial B(0,r)} f(x + y)\, \sigma_r(dy)\, r^{d-1} r^{2-d} dr,$$

and so if f is bounded with compact support, then $Uf(x)$ will be bounded.

One might wonder why the power is $d - 2$ in (3.1) and not -1. u must be translation invariant, and if we take $x = 0$, then $u(0, y)$ should be a function of $|y|$ by symmetry. What functions of $|y|$ are harmonic? Write the Laplacian in polar coordinates:

$$\Delta f(r, \theta) = f_{rr} + \frac{d-1}{r} f_r + \frac{1}{r^2} \Delta_{\theta\theta},$$

for $r \in [0, \infty), \theta \in \partial B(0,1)$, and where $f_r = \partial f/\partial r$, $f_{rr} = \partial^2 f/\partial r^2$, and $\Delta_{\theta\theta}$ is the spherical Laplacian on the unit ball. If f depends only on r and not θ and $\Delta f = 0$ for $r > 0$, we have

(3.4)
$$f_{rr} + \frac{d-1}{r}f_r = 0$$

and either f is a constant or $f = cr^{2-d}$.

The same reasoning shows that for $d = 2$ we should let the Newtonian potential density be $u(x, y) = c\log|x - y|$. Note that for $d = 2$, no matter what choice of c we make, u will take both positive and negative values.

Let us now return to the case $d \geq 3$.

(3.1) Proposition. *If* $f \geq 0$, $d \geq 3$, *then*

$$Uf(x) = \mathbb{E}^x \int_0^\infty f(X_s)ds.$$

Proof. The right-hand side is

$$\int_0^\infty P_s f(x)ds = \int f(y)\left[\int_0^\infty (2\pi s)^{-d/2}e^{-|y-x|^2/2s}ds\right]dy,$$

where P_s is defined in (1.3.2).

If we make the substitution $t = |y - x|^2/2s$, the integral inside the brackets is

$$(2\pi)^{-d/2}|y - x|^{2-d}2^{d/2-1}\int_0^\infty t^{d/2-2}e^{-t}dt = c_d|y - x|^{2-d} = u(x, y).$$

Hence $\mathbb{E}^x\int_0^\infty f(X_s)ds = \int f(y)u(x, y)dy$.   □

Note that $u(\cdot, y)$ is harmonic in $\mathbb{R}^d - \{y\}$. To see this, use translation invariance to suppose $y = 0$ and then calculate:

(3.5) $$\frac{\partial u}{\partial x^i} = (2 - d)c_d\frac{x^i}{|x|^d}, \qquad \frac{\partial^2 u}{\partial(x^i)^2} = (2 - d)c_d\frac{|x|^2 - d(x^i)^2}{|x|^{d+2}},$$

and hence $\Delta u = 0$.

A corollary of this calculation is the following.

(3.2) Proposition. *Suppose μ is supported on a closed set B, i.e., $\mu(B^c) = 0$. Then $U\mu$ is harmonic on B^c.*

Proof. Let $x \in B^c$ and let $r < \text{dist}(x, \partial B)$. Let $S_r = \inf\{t : |X_t - x| > r\}$. By Proposition I.2.8, $\mathbb{E}^x u(X_{S_r}, y) = u(x, y)$. Then

$$\mathbb{E}^x U\mu(X_{S_r}) = \mathbb{E}^x \int u(X_{S_r}, y)\mu(dy) \int_B \mathbb{E}^x u(X_{S_r}, y)\mu(dy)$$

$$= \int_B u(x, y)\mu(dy) = U\mu(x).$$

By Proposition 1.2, this proves $U\mu$ is harmonic at x.   □

Let us give an example of how the preceding can be used to do some calculations. Let us fix R and calculate $U1_{B(0,R)}(x)$. Let $\sigma_s(dy)$ be normalized surface measure on $\partial B(0,s)$. By rotational invariance, $U\sigma_s$ must be constant for $x \in \partial B(0,s)$. Let the value be c_s. Since $U\sigma_s$ is harmonic in $B(0,s)$, then if $|x| < s$,

$$U\sigma_s(x) = \mathbb{E}^x U\sigma_s(X(\tau_{B(0,s)})) = c_s.$$

Hence $U\sigma_s$ is constant with value c_s in the interior of $B(0,s)$. To find c_s,

$$c_s = U\sigma_s(0) = \int \frac{c_d}{|y|^{d-2}}\sigma_s(dy) = c_d s^{2-d}.$$

If $|x| > s$, then by Proposition 3.2

$$U\sigma_s(x) = \mathbb{E}^x\left[U\sigma_s(X(T_{B(0,s)})); T_{B(0,s)} < \infty\right]$$
$$= c_s \mathbb{P}^x(T_{B(0,s)} < \infty) = c_s\left(\frac{s}{|x|}\right)^{d-2} = c_d|x|^{2-d},$$

using Proposition I.5.8.

We see then that

(3.6) $$\qquad\qquad U\sigma_s(x) = c_d(s^{2-d} \wedge |x|^{2-d}).$$

So if $|x| \leq R$,

(3.7) $\quad U1_{B(0,R)}(x) - |\partial B(0,1)| \displaystyle\int_0^R U\sigma_s(x)s^{d-1}ds$

$$= c_d|\partial B(0,1)|\int_0^R s^{d-1}(s^{2-d} \wedge |x|^{2-d})ds$$
$$= c_d|\partial B(0,1)|\left(\frac{R^2}{2} - \frac{d-2}{2d}|x|^2\right) = \frac{R^2}{d-2} - \frac{|x|^2}{d}.$$

If $|x| > R$,

(3.8) $$\qquad\qquad U1_{B(0,R)}(x) = \frac{2R^d|x|^{2-d}}{d(d-2)}.$$

Newtonian potentials can be used to solve Poisson's equation in $\mathbb{R}^d$, that is, to find g such that $\Delta g = f$.

(3.3) Proposition. *If $f \in C^2$ with compact support, then*

$$-\frac{1}{2}\Delta Uf = f.$$

Proof. By translation invariance, $\partial^2 (Uf)/\partial x^i \partial x^j = U(\partial^2 f/\partial x^i \partial x^j)$ is finite and continuous. So Uf is in C^2 (although not with compact support). By Itô's formula,

$$(3.9) \qquad Uf(X_t) - Uf(X_0) - \frac{1}{2}\int_0^t \Delta Uf(X_s)ds$$

is a local martingale, zero at 0. On the other hand,

$$(3.10) \quad Uf(X_t) = \mathbb{E}^{X_t}\int_0^\infty f(X_s)ds = \mathbb{E}^x\Big[\int_0^\infty f(X_{s+t})ds|\mathcal{F}_t\Big]$$

$$= \mathbb{E}^x\Big[\int_t^\infty f(X_s)ds|\mathcal{F}_t\Big]$$

$$= \mathbb{E}^x\Big[\int_0^\infty f(X_s)ds|\mathcal{F}_t\Big] - \int_0^t f(X_s)ds.$$

(Note that since f has compact support and $|X_t| \to \infty$ as $t \to \infty$ (Exercise I.8.34), the integral $\int_0^\infty f(X_s)ds$ is finite for almost every path.) Hence $Uf(X_t) - Uf(X_0) + \int_0^t f(X_s)ds$ is also a local martingale, zero at 0.

Combining (3.9) and (3.10), $\int_0^t[(1/2)\Delta Uf(X_s) + f(X_s)]ds$ is a local martingale that is zero at 0. It is also continuous and of bounded variation, hence it is identically 0 (Proposition I.4.19). Therefore for almost every s, $-(1/2)\Delta Uf(X_s) = f(X_s)$, a.s. Since X_s, ΔUf, and f are continuous, with probability 1 we have

$$(3.11) \qquad (-1/2)\Delta Uf(X_s) = f(X_s), \qquad s \geq 0.$$

If $-(1/2)\Delta Uf(y) > f(y)$ for some y, then by continuity, we have the inequality for a neighborhood V of y. With positive probability, however, X_t enters V (see Proposition I.5.8), contradicting (3.11). Therefore $-(1/2)\Delta Uf(y) \leq f(y)$. The same argument with the inequalities reversed shows we have equality for all y. $\square$

This proposition is true under much weaker assumptions on f. All we need is that f be continuous and Uf be in C^2. This will hold, for example, if f is in C^α with compact support (see Theorem 3.14).

(3.4) Proposition. *If $f = U\mu(x)$, then $P_h f \leq f$ for all h and $P_h f(x) \uparrow f(x)$ as $h \downarrow 0$.*

A function f satisfying the conclusion of this proposition is called excessive. We will investigate excessive functions further in Sect. 6.

Proof. By the Markov property (I.3.4),

$$(3.12) \quad P_h f(x) = P_h \int_0^\infty P_s\mu(x)ds = \int_0^\infty P_{h+s}\mu(x)ds = \int_h^\infty P_s\mu(x)ds.$$

Clearly this is bounded above by $\int_0^\infty P_s\mu(x)ds = f(x)$ and increases to $f(x)$ as $h \downarrow 0$ by monotone convergence. $\qquad\square$

More interesting is the following uniqueness result.

(3.5) Proposition. *If $U\mu = U\nu$ a.e., then $\mu = \nu$.*

Proof. Since $P_h g$ is the integral of g against the density $p(h, x, y)$, we have

$$P_h U\mu(x) = \int p(h, x, y) U\mu(y)dy = \int p(h, x, y) U\nu(y)dy = P_h U\nu(x),$$

or

$$\int_h^\infty P_s\mu(x)ds = \int_h^\infty P_s\nu(x)ds$$

for all h and all x. It follows that for each x, $P_s\mu(x) = P_s\nu(x)$ for almost every s (the null set depends on x). Since the x-section of the set of $\{(x, s)\}$ where $P_s\mu(x)$ and $P_s\nu(x)$ differ has Lebesgue measure 0 for every x, Fubini's theorem says that it has measure 0 for every s-section, i.e., for almost every s,

$$P_s\mu(x) = P_s\nu(x), \qquad \text{a.e. } (x).$$

(The null set of xs depends on s).

As above, except for s in an exceptional set N, $P_{s+h}\mu(x) = P_h P_s\mu(x) = P_h P_s\nu(x) = P_{s+h}\nu(x)$ for all h and all x. Since for any t we can find h such that $s = t - h$ is not in the exceptional set N, we conclude that for all t, $P_t\mu(x) = P_t\nu(x)$ for all x.

Now we take f continuous with compact support. Then

$$(3.13) \qquad \int f(x)P_t\mu(x)dx = \int\int f(x)p(t, x, y)dx\mu(dy) = \int P_t f(y)\mu(dy)$$

and similarly for μ replaced by ν. Since it is easy to see that $P_t f(y) \to f(y)$ uniformly as $t \to 0$, we have $\int f(y)\mu(dy) = \int f(y)\nu(dy)$ for all f continuous with compact support, which implies the proposition. $\qquad\square$

Green potentials

If f is continuous with compact support and B is a Borel measurable set, define

$$(3.14) \qquad\qquad G_B f(x) = \mathbb{E}^x \int_0^{T_B} f(X_s)ds.$$

We will generally use the letter D for domains and B for arbitrary Borel sets.

Clearly,

$$G_B f(x) \leq \mathbb{E}^x \int_0^\infty f(X_s)ds = Uf(x),$$

or $G_B f(x)$ is finite. By the strong Markov property,

$$(3.15) \qquad G_B f(x) = \mathbb{E}^x \int_0^\infty f(X_s)ds - \mathbb{E}^x \int_{\tau_B}^\infty f(X_s)ds$$

$$= Uf(x) - \mathbb{E}^x \int_0^\infty f(X_{\tau_B+s})ds$$

$$= Uf(x) - \mathbb{E}^x \mathbb{E}^{X_{\tau_B}} \int_0^\infty f(X_s)ds$$

$$= Uf(x) - \mathbb{E}^x Uf(X_{\tau_B}).$$

If we define a measure $G_B(x, dy)$ by $G_B(x, A) = G_B 1_A(x)$ then

$$G_B(x, A) \leq U 1_A(x) = \int_A u(x, y)dy,$$

or $G_B(x, \cdot) \ll dy$. A suitable version of the density will be the Green function with pole at x.

Let us define

$$(3.16) \qquad g_B(x, y) = u(x, y) - \mathbb{E}^x u(X_{\tau_B}, y).$$

Then for f smooth,

$$\int f(y)g_B(x, y)dy = \int f(y)\big[u(x, y) - \mathbb{E}^x u(X_{\tau_B}, y)\big] dy$$

$$= \int f(y)u(x, y)dy - \mathbb{E}^x \int f(y)u(X_{\tau_B}, y)dy$$

$$- Uf(x) - \mathbb{E}^x Uf(X_{\tau_B}) = G_B f(x) = \int f(y)G_B(x, dy).$$

Hence $g_B(x, \cdot)$ is a density for the measure $G_B(x, dy)$. In particular, $0 \leq g_B(x, y) \leq u(x, y)$ for almost every y. $g_B(x, y)$ is called the Green function for B.

For $x = y \in B$, we let $g_B(x, y) = \infty$. If $x \in B$, then $g_B(x, y)$ is continuous in y for $y \in B - \{x\}$ since the right-hand side of (3.16) is. If $y \in B$, then $\mathbb{E}^x u(X_{\tau_B}, y)$ is harmonic in x for $x \in \text{int}\,(B - \{y\})$, so $g_B(x, y)$ is continuous for such x. From (3.16) it is immediate that if x is regular for B^c, then $g_B(x, y) = 0$. We will come back to this in Sect. 4 when we prove that g_B is symmetric in x and y.

For B open, $\mathbb{E}^x u(X_{\tau_B}, y)$ is harmonic in B, hence $g_B(x, y)$ equals $u(x, y)$ minus a harmonic function. Thus our definition agrees with the analytic one, namely, a function $\tilde{g}$ is the Green function with pole at y if $\tilde{g}$ is equal to $u(x, y)$ minus a harmonic function and $\tilde{g}$ is zero on the points of ∂B that are regular for B^c.

Suppose we have a region $D \subseteq \mathbb{R}^3$ that is a vacuum and ∂D is made of a conducting metal that is grounded. Let ∂D be initially uncharged. Now place an electron at $y \in D$. This causes electrons on ∂D to move about and sets up an induced charge distribution μ on ∂D. The potential inside at a point x is now $c_d/|x - y| - c_d \int |x - y|^{-1} \mu(dy)$, which is the Newtonian potential u minus a harmonic function. The potential should be 0 on the boundary, since it is grounded, hence the potential inside is $g_D(x, y)$, the Green function.

One interesting application of Green functions and Green potentials is to solve Poisson's equation: $\Delta h = f$ in D, $h = 0$ on ∂D.

(3.6) Proposition. *Suppose D is an open domain. Suppose every point of ∂D is regular for D^c and suppose $f \in C^2$ with compact support. Then $h = -G_D f/2$ solves Poisson's equation.*

Proof. That h is 0 on ∂D is clear. By the proof of Proposition 3.3, $Uf(x)$ is C^2. Since

$$G_D f(x) = Uf(x) - \mathbb{E}^x Uf(X_{\tau_D})$$

and $\mathbb{E}^x Uf(X_{\tau_D})$ is harmonic on D,

$$\Delta h = -(1/2)\Delta G_D f = -(1/2)\Delta U f = f$$

by Proposition 3.3. $\qquad\square$

Feynman-Kac formula

The Feynman-Kac formula is a simple idea, but one that turns out to be extremely useful.

Let D be a bounded domain such that every point of ∂D is regular for D^c, let V be a smooth function on D, and let f be a continuous function on ∂D. Suppose u is a function that is continuous on $\overline{D}$, C^2 in D, agrees with f on ∂D, and satisfies

$$(3.17) \qquad \frac{1}{2}\Delta u(x) - V(x)u(x) = 0, \qquad x \in D.$$

(3.7) Theorem. *Let D, f, V, and u be as above. If $x \in D$,*

$$(3.18) \qquad u(x) = \mathbb{E}^x\left[f(X_{\tau_D})e^{-\int_0^{\tau_D} V(X_s)ds}\right].$$

The equation (3.17) is sometimes called the probabilistic Schrödinger equation or just the Schrödinger equation.

Proof. Let $U_t = u(X_{t \wedge \tau_D})$, $B_t = \int_0^{t \wedge \tau_D} V(X_s)\, ds$. B_t is of bounded variation and has no martingale part. By the product formula (Corollary I.5.7) and Itô's formula,

$$U_t e^{-B_t} = U_0 e^{-B_0} - \int_0^t U_s e^{-B_s}\, dB_s + \int_0^t e^{-B_s}\, dU_s$$

$$= u(X_0) - \int_0^{t \wedge \tau_D} u(X_s) V(X_s) e^{-B_s}\, ds + \text{martingale}$$

$$+ \int_0^{t \wedge \tau_D} e^{-B_s} \frac{1}{2} \Delta u(X_s)\, ds.$$

Take expectations with respect to $\mathbb{P}^x$ and let $t \to \infty$. Then

$$\mathbb{E}^x U_{\tau_D} e^{-B_{\tau_D}} = u(x) + \mathbb{E}^x \int_0^{\tau_D} e^{-B_s} \left(\frac{1}{2}\Delta u - Vu \right)(X_s)\, ds = u(x).$$

Substituting in the definitions of U_{τ_D} and B_{τ_D}, this is (3.18). □

There are variations of this formula that can be obtained by replacing τ_D by $t_0 \wedge \tau_D$ for some fixed time t_0 or by $S \wedge \tau_D$, where S is a random variable that has an exponential distribution and is independent of X_t.

The Feynman-Kac formula can be used either to derive information about solutions of (3.17) from the representation (3.18) or to derive information about certain functionals of Brownian motion from (3.17). We indicate briefly how the latter works. Let h be a function on D, let $V_\lambda = \lambda h$, and let u_λ be the solution of (3.17) with V replaced by V_λ. Let $f = 1$ on ∂D. Then (3.18) tells us that

$$\mathbb{E}^x e^{-\lambda \int_0^{\tau_D} h(X_s)\, ds} = u_\lambda(x),$$

and we thus have a formula for the Laplace transform of $\int_0^{\tau_D} h(X_s)\, ds$ in terms of the solutions u_λ to the differential equations $(1/2)\Delta u - \lambda h u = 0$, $\lambda > 0$.

Occupation time densities

Let us now find $g_D(x, y)$ for D the half-space and the unit ball. For the half-space we could proceed using the Kelvin transform, but it is easier to think as follows. If $y \in D$, let $\widetilde{y} = (y^1, \ldots, y^{d-1}, -y^d)$, the reflection of y across the boundary of D. Think of starting a particle at x that follows a Brownian motion path. Think of its reflection across ∂D starting at $\widetilde{x}$ and imagine that this second particle is made of antimatter. The first particle is a Brownian motion until it hits the boundary; at that point it meets its evil twin and is annihilated. So the first particle is Brownian motion killed on hitting ∂D. With this intuition one might expect the transition density of Brownian motion killed on hitting ∂D to be $p(t, x, y) - p(t, \widetilde{x}, y)$ (see

Exercise 43) and hence $g_D(x, y) = u(x, y) - u(\tilde{x}, y)$. This turns out to be true.

(3.8) Proposition. $g_H(x, y) = u(x, y) - u(\tilde{x}, y)$ is the Green function for H.

Proof. $g(x, y) = u(x, y) - u(\tilde{x}, y)$ is equal to u minus a harmonic function, equals 0 on ∂D, and tends to 0 uniformly as $|x| \to \infty$. The Green function for H, $g_H(x, y)$, also equals u minus a harmonic function, equals 0 on ∂D, and since by (3.16), $g_H(x, y) \leq u(x, y)$, $g_H(x, y) \to 0$ uniformly as $|x| \to \infty$. Hence $g(x, y) - g_H(x, y)$ is bounded in D, harmonic, 0 on ∂D, and tends to 0 uniformly as $|x| \to \infty$. By the maximum principle applied to $g - g_H$ and $g_H - g$,

$$\sup_{x \in D \cap B(0,M)} |g(x, y) - g_H(x, y)| \leq \sup_{x \in \partial D \cap B(0,M)} |g(x, y) - g_H(x, y)| \to 0$$

as $M \to \infty$, or $g \equiv g_H$. $\qquad\square$

For the ball, there are several approaches to deriving $g_D(x, y)$. One can use the Kelvin transform as in finding the exit distribution, one can let $g_D(x, y) = u(x, y) - \mathbb{E}^x u(X_{\tau_D}, y) = u(x, y) - \int u(z, y) P_B(x, z) dz$ and do the integral, or one can guess the answer and verify that g_D equals u minus a harmonic function and vanishes on the boundary. It can also be done using stochastic differential equations.

Guessing the answer is not as hard as it might first appear. Let us work with the unit ball, and imitating the half-space case, think of a Brownian motion made of antimatter started at $x/|x|^2$. A first guess for the Green function would be $u(x, y) - u(x/|x|^2, y)$. This is wrong because $u(x/|x|^2, y)$ is not harmonic. On the other hand, by Lemma 1.18 $|x|^{2-d} u(x/|x|^2, y)$ is harmonic, and it will have the same boundary values as $u(x/|x|^2, y)$. So a better guess is $u(x, y) - |x|^{2-d} u(x/|x|^2, y)$, and this turns out to be correct.

(3.9) Proposition. *Let $d \geq 3$ and for $y \in B(0, 1)$, let*

$$(3.19) \qquad g_{B(0,1)}(x, y) = \begin{cases} u(x, y) - |x|^{2-d} u(x/|x|^2, y), & x \neq 0, \\ u(0, y) - c_d, & x = 0. \end{cases}$$

Then $g_{B(0,1)}$ is the Green function for $B(0, 1)$.

Proof. By Lemma 1.18, $g(x, y) = u(x, y) - |x|^{2-d} u(x/|x|^2, y)$ is equal to u minus a harmonic function (a separate calculation is needed for the case $x = 0$). When $|x| = 1$, $g(x, y) = 0$, hence it is 0 on ∂B. Therefore it must be the Green function. $\qquad\square$

The same intuition leads to the formula for $g_{B(0,1)^c}$, and we have

$$(3.20) \quad g_{B(0,1)^c}(x, y) = u(x, y) - |x|^{2-d} u(x/|x|^2, y), \qquad x, y \in B(0, 1)^c.$$

To prove this, we proceed similarly to the proofs of Propositions 3.8 and 3.9.

Two-dimensional Brownian motion

We have not talked much about the case $d = 2$; the difficulty is that two-dimensional Brownian motion is recurrent. This leads to problems: if $f \geq 0$ is smooth, pick a point y such that $f(y) > 0$. Then f will be greater than some δ in a neighborhood $B(y, \varepsilon)$. Two-dimensional Brownian motion will hit $B(y, \varepsilon/2)$ infinitely often and one can show that the expected time spent in $B(y, \varepsilon)$ starting at any $z \in \partial B(y, \varepsilon/2)$ is greater than $c\varepsilon^2$ (see Exercise 17). Hence $\mathbb{E}^x \int_0^\infty f(X_s)ds \geq \delta \mathbb{E}^x \int_0^\infty 1_{B(y,\varepsilon)}(X_s)ds = \infty$ for all x.

What one has to do is somehow make two-dimensional Brownian motion transient. The most common way is to look at X_t killed on exiting $B(0, R)$ or killed on exiting the upper plane (see Proposition V.1.13).

Another way of making the process transient, which can be used in all dimensions, is to kill the process at a time S, where S has the distribution of an exponential random variable with parameter λ and S is independent of the Brownian motion. (S exponential means $\mathbb{P}(S > t) = e^{-\lambda t}$ if $t > 0$, $\mathbb{P}(S \leq 0) = 0$.) We have

$$\mathbb{E}^x \int_0^S f(X_s)ds = \int_0^\infty \mathbb{E}^x f(X_s)1_{(S>s)}ds.$$

By the independence of S and X_s, this is

$$\int_0^\infty P_s f(x)\mathbb{P}(S > s)ds = \int_0^\infty e^{-\lambda s} P_s f(x)ds.$$

This is what we called $U^\lambda f(x)$ in Sect. I.3. We define

$$(3.21) \qquad u^\lambda(x, y) = \int_0^\infty e^{-\lambda s} p(s, x, y)ds,$$

and

$$(3.22) \quad U^\lambda f(x) = \int u^\lambda(x, y)f(y)dy, \qquad U^\lambda \mu(x) = \int u^\lambda(x, y)\mu(dy).$$

We will return to these λ–potentials or λ–resolvents later.

By the discussion following (3.4) one might wonder if one could develop a potential theory around the kernel $c \log |x - y|$ for some c. If e_1 is the unit vector in the x^1 direction, note

$$(3.23) \quad \int_0^\infty \left[p(t,0,x) - p(t,0,e_1) \right] dt = \int_0^\infty \frac{1}{2\pi t} \left(e^{-|x|^2/2t} - e^{-1/2t} \right) dt$$

$$= \frac{1}{2\pi} \int_0^\infty \frac{1}{t} \int_{|x|^2/2t}^{1/2t} e^{-s} ds \, dt$$

$$= \frac{1}{2\pi} \int_0^\infty e^{-s} \left[\int_{|x|^2/2s}^{1/2s} \frac{1}{t} dt \right] ds$$

$$= \frac{1}{2\pi} \int_0^\infty e^{-s} ds \left(-\log(|x|^2) \right)$$

$$= -\frac{1}{\pi} \log |x|.$$

This might suggest that the constant c should be $-1/\pi$. It turns out one can discuss logarithmic potentials, which are potentials defined in terms of the kernel

$$(3.24) \qquad k(x,y) = \frac{1}{\pi} \log(1/|x-y|).$$

See Port and Stone [1] for details; see also Exercises 44 and 45.

Normal derivatives and Green's identities

If one has the Green function for a smooth domain, one can then derive the Poisson kernel. This is because the Poisson kernel turns out to be the normal derivative of the Green function. To prove this, we recall the divergence theorem. Let n denote the outward normal vector of a domain D, and given $F = (f^1, \ldots, f^d)$, $\operatorname{div} F = \sum_{i=1}^d \partial f^i / \partial x^i$. The divergence theorem (see Marsden and Tromba [1]) holds for a domain D if for all $F \in C^1$,

$$(3.25) \qquad \int_D \operatorname{div} F(x) dx = \int_{\partial D} F \cdot n(y) \sigma(dy),$$

where σ denotes surface measure. Lipschitz domains (see Chap. III) are examples of domains for which the divergence theorem holds.

(3.10) Theorem. (Green's first and second identities) *Let D be a domain for which the divergence theorem holds, and suppose $u, v \in C^2(D) \cap C^1(\overline{D})$. Then*

$$(3.26) \qquad \int_D u \Delta v + \int_D \nabla u \cdot \nabla v = \int_{\partial D} u \frac{\partial v}{\partial n} \sigma(dy)$$

and

$$(3.27) \qquad \int_D (u \Delta v - v \Delta u) = \int_{\partial D} \left(u \frac{\partial v}{\partial n} - v \frac{\partial u}{\partial n} \right) \sigma(dy).$$

Proof. For the first identity, let $F = u \nabla v$. Then $f^i = u(\partial v / \partial x^i)$ and

$$\frac{\partial f^i}{\partial x^i} = \frac{\partial u}{\partial x^i}\frac{\partial v}{\partial x^i} + u\frac{\partial^2 v}{\partial(x^i)^2},$$

or $\operatorname{div} F = u\Delta v + \nabla u \cdot \nabla v$. Also, $F \cdot n = u(\partial v/\partial n)$. Applying the divergence theorem gives (3.26).

Reversing the roles of u and v, we have

(3.28)
$$\int_D (v\Delta u + \nabla u \cdot \nabla v) = \int_{\partial D} v\frac{\partial u}{\partial n}\sigma(dy).$$

Subtracting (3.28) from (3.26) gives (3.27). $\qquad\qquad\square$

(3.11) Proposition. *Let $D \subseteq \mathbb{R}^3$, $d \geq 3$, be a domain for which the divergence theorem holds and which has the properties that*
(a) whenever f is C^2 on ∂D, then the harmonic function with boundary values f is in $C^2(\overline{D})$, and
(b) if $x_0 \in D$, then $g_D(x_0, y)$ is in $C^1(\overline{D} - \{x_0\})$.
Then for each x_0,

$$\frac{\partial g_D(x_0, y)}{\partial n} = (2 - d)c_d\frac{\mathbb{P}^{x_0}(X_{\tau_D} \in dy)}{\sigma(dy)}.$$

Proof. Let f be smooth on ∂D, $v(y) = \mathbb{E}^y f(X_{\tau_D})$, and $w(y) = g_D(x_0, y)$. We want to show

(3.29)
$$\int_{\partial D} f\frac{\partial w}{\partial n}\sigma(dy) = (2 - d)c_d v(x_0).$$

Let us apply Green's second identity to $D - B(x_0, \varepsilon)$, where we take $\varepsilon <$ dist $(x_0, \partial D)/2$. Both v and w are harmonic in $D - B(x_0, \varepsilon)$, so the left-hand side of (3.27) is 0. Thus

$$0 = \int_{\partial D}\left(w\frac{\partial v}{\partial n} - v\frac{\partial w}{\partial n}\right) - \int_{\partial B(x_0,\varepsilon)}\left(w\frac{\partial v}{\partial n} - v\frac{\partial w}{\partial n}\right).$$

Since v is harmonic on D, by Corollary 1.4 we have that $|\nabla v(x)| \leq c\|v\|_\infty$ for $x \in B(x_0, \text{dist}(x_0, \partial D)/2)$. Then

$$\int_{\partial B(x_0,\varepsilon)}\left|w\frac{\partial v}{\partial n}\right| \leq c\int_{\partial B(x_0,\varepsilon)} w \leq c\int_{\partial B(x_0,\varepsilon)} u(x_0, y)\sigma(dy)$$

$$\leq c\varepsilon^{2-d}\int_{\partial B(x_0,\varepsilon)}\sigma(dy) = c\varepsilon \to 0$$

as $\varepsilon \to 0$. Similarly, since $w = u(x, y)$ minus a harmonic function h,

$$\int_{\partial B(x_0,\varepsilon)}\frac{\partial h}{\partial n}v \to 0.$$

We have

$$\int_{\partial B(x_0,\varepsilon)} \frac{\partial u(x_0,y)}{\partial n} v(y)\sigma(dy) = (2-d)c_d\varepsilon^{1-d}\int_{\partial B(x_0,\varepsilon)} v(y)\sigma(dy)$$

$$\to (2-d)c_d v(x_0).$$

Now $w = 0$ on ∂D and $v = f$ on ∂D, so we obtain

$$(2-d)c_d\mathbb{E}^{x_0}f(X_{\tau_D}) = (2-d)c_d v(x_0) = \int_{\partial D} f(y)\frac{\partial w(y)}{\partial n}\sigma(dy),$$

which is (3.29). □

Smoothness of potentials

In this subsection we will show that if $d \geq 3$ and if $f \in C^\alpha$, the space of Hölder continuous functions [see (3.33)], then $Uf \in C^{2+\alpha}$, where Uf is the Newtonian potential of f. We will give another proof of this in Chap. IV using Riesz transforms. We say $f \in C^\alpha$, $\alpha \in (0,1)$, if there exists c such that

$$|f(x+t) - f(x)| \leq c|t|^\alpha \qquad \text{and} \qquad |f(x)| \leq c, \qquad x,t \in \mathbb{R}^d.$$

$f \in C^{k+\alpha}$ means that all the kth order partial derivatives of f are in C^α.

For the remainder of this section we will use the notation $\partial_i f$ for $\partial f/\partial x^i$. We will use the following fact (see Rudin [2]):

(3.30) If $v_m \to v$ and $w_m \to w$, both uniformly in a neighborhood of x, and $w_m = \partial_i v_m$ is continuous, then $\partial_i v$ exists and equals w.

We will also use the fact that if $M > 0$ and $f \in C^1$, then

$$(3.31) \qquad \int_{B(0,3M)} \partial_i f(x)dx = \int_{\partial B(0,3M)} f(y)n^i(y)\sigma(dy),$$

where n is the outward normal derivative, n^i is the ith component of n, and σ is surface measure on $\partial B(0,3M)$. This follows immediately from the divergence theorem with $F = (0,\ldots,0,f,0,\ldots,0)$.

Let $\varphi : [0,\infty) \to [0,1]$ be a C^2 function such that $\varphi(x) = 1$ if $x \geq 1$ and $\varphi(x) = 0$ if $x \leq 1/2$. Let $u_m(x) = u(x)\varphi(m|x|)$. Note $u_m \in C^2$, $u_m \uparrow u$, as $m \to \infty$, and $|u - u_m| \leq u 1_{B(0,1/m)}$.

(3.12) Proposition. *If f is bounded and integrable, then $Uf \in C^1$ and*

$$(3.32) \qquad \partial_i Uf(x) = \int \partial_i u(x-y)f(y)dy.$$

Proof. Let $v_m = U_m f$ and $w_m = \partial_i U_m f$. Since f is bounded, $v_m \to Uf$ uniformly on compacts. Let $w(x)$ be the right side of (3.32). Then

$$|w(x) - w_m(x)| \leq c \int_{|x-y| \leq 1/m} |x - y|^{1-d} f(y) dy \to 0$$

since f is bounded. The convergence is uniform, hence by the observation (3.30) above, $\partial_i v = w$. □

For $\alpha \in (0, 1)$, let

(3.33) $$\|f\|_{C^\alpha} = \|f\|_\infty + \sup_{t \neq 0, x} |f(x+t) - f(x)|/|t|^\alpha$$

and let C^α be the set of continuous functions such that $\|f\|_{C^\alpha}$ is finite. Suppose $M > 1$.

(3.13) Proposition. *Suppose $f \in C^\alpha$ and f has support in $B(0, M)$. Then for $x \in B(0, M)$*

(3.34) $$\partial_i \partial_j U f(x) = \int_{B(0,3M)} [\partial_i \partial_j u(x - y)][f(y) - f(x)] dy$$
$$+ f(x) \int_{\partial B(0,3M)} \partial_j u(x - y) n^i(y) \sigma(dy).$$

Proof. Let $v_m = \partial_j U_m f$, $w_m = \partial_i \partial_j U_m f$, and w the right-hand side of (3.34). We just saw that $v_m \to v$ uniformly on compacts. By (3.31),

(3.35) $$w_m(x) = \partial_i v_m(x) = \int_{B(0,3M)} \partial_i \partial_j u_m(x - y) f(y) dy$$
$$= \int_{B(0,3M)} [\partial_i \partial_j u_m(x - y)][f(y) - f(x)] dy$$
$$+ f(x) \int_{B(0,3M)} \partial_i \partial_j u_m(x - y) dy$$
$$= \int_{B(0,3M)} [\partial_i \partial_j u_m(x - y)][f(y) - f(x)] dy$$
$$+ f(x) \int_{\partial B(0,3M)} \partial_j u_m(x - y) n^i(y) \sigma(dy).$$

The second term on the right-hand side of (3.35) converges to

$$f(x) \int_{B(0,3M)} \partial_j u(x - y) n^i(y) \sigma(dy)$$

by Proposition 3.12. Since

$$\int_{B(0,3M)} |\partial_i \partial_j u_m(x - y) - \partial_i \partial_j u(x - y)| \, |f(y) - f(x)| dy$$
$$\leq c \int_{B(0,1/m)} |x - y|^{-d} |x - y|^\alpha dy \to 0$$

as $m \to \infty$, the first term on the right in (3.35) converges uniformly to the first term on the right in (3.34). $\qquad\square$

(3.14) Theorem. *Suppose $M > 1$, $f \in C^\alpha$, and f has support in $B(0, M)$. If $w = \partial_i \partial_j U f(x)$, then*

$$\sup_{B(0,M)} |w(x)| + \sup_{x \in B(0,M)} \sup_{0 < |t| < M} |w(x + t) - w(x)|/|t|^\alpha \le c\|f\|_{C^\alpha}.$$

Proof. We use the expression for w given by (3.34). The second integral, the one on the boundary of $B(0, 3M)$, causes no trouble. Let

$$r(x) = \int_{\partial B(0,3M)} \partial_j u(x - y) n^i(y) \sigma(dy)$$

$$= (2 - d) c_d \int_{\partial B(0,3M)} \frac{x^j - y^j}{|x - y|^d} n^i(y) \sigma(dy).$$

$r(x)$ is bounded and it is easy to see that for $x, x' \in B(0, 2M)$,

$$|f(x)r(x) - f(x')r(x')| \le |f(x)| \, |r(x) - r(x')| + |r(x')| \, |f(x) - f(x')|$$
$$\le c\|f\|_\infty |x - x'|^\alpha + c\|f\|_{C^\alpha} |x - x'|^\alpha.$$

Now let us look at the integral over $B(0, 3M)$. This is bounded, since

$$\int_{B(0,3M)} |\partial_i \partial_j u(x - y)| \, |f(y) - f(x)| dy$$

$$\le c\|f\|_{C^\alpha} \int_{B(0,3M)} |x - y|^{-d} |x - y|^\alpha dy \le c\|f\|_{C^\alpha}.$$

Let $\delta = |x - x'|$ and suppose $\delta < 1$. Let $s(z) = \partial_i \partial_j u(z)$.

$$(3.36) \quad \int_{B(0,3M)} s(x - y)[f(y) - f(x)]dy - \int_{B(0,3M)} s(x' - y)[f(y) - f(x')]dy$$

$$= \int_{B(0,3M) - B(x,2\delta)} [f(y) - f(x')][s(x' - y) - s(x - y)]dy$$

$$+ [f(x') - f(x)] \int_{B(0,3M) - B(x,2\delta)} s(x - y)dy$$

$$+ \int_{B(x,2\delta)} s(x - y)[f(y) - f(x)]dy$$

$$+ \int_{B(x,2\delta)} s(x' - y)[f(y) - f(x')]dy.$$

The third integral on the right in (3.36) is bounded in absolute value by

$$c\|f\|_{C^\alpha} \int_{B(x,2\delta)} |x - y|^{-d} |y - x|^\alpha dy \le c\delta^\alpha \|f\|_{C^\alpha},$$

and the fourth integral is similar.

By the mean value theorem

$$|s(x-y) - s(x'-y)| \le |\nabla s(\bar{x}-y)| \, |x-x'|$$

for some point $\bar{x}$ on the line segment between x and x'. If $y \notin B(x,2\delta)$, then $|\bar{x}-y| \ge c|x-y|$ and

$$|\nabla s(\bar{x}-y)| \le \frac{c}{|\bar{x}-y|^{d+1}} \le c|x-y|^{-d-1}.$$

Therefore the first integral on the right in (3.36) is bounded in absolute value by

$$c\delta\|f\|_{C^\alpha} \int_{B(0,3M)-B(x,2\delta)} |y-x|^\alpha |x-y|^{-d-1} dy \le c\|f\|_{C^\alpha}\delta^\alpha.$$

Finally the second integral on the right-hand side of (3.36) is bounded by

$$c\|f\|_{C^\alpha} |x-x'|^\alpha \left| \int_{B(0,3M)-B(x,2\delta)} s(x-y)dy \right|.$$

Observe

$$(3.37) \quad \int_{B(0,3M)-B(x,2\delta)} s(x-y)dy = \int_{\partial B(0,3M)} \partial_j u(x-y)n^i(dy)\sigma(dy)$$

$$- \int_{\partial B(x,2\delta)} \partial_j u(x-y)n^i(y)\sigma(dy).$$

The first integral on the right in (3.37) is bounded by a constant, and the second integral is also, since

$$\left| \int_{B(x,2\delta)} \partial_j u(x-y)n^i(y)\sigma(dy) \right| \le c \int_{\partial B(x,2\delta)} |x-y|^{1-d}\sigma(dy)$$

$$= c(2\delta)^{1-d}\sigma(\partial B(x,2\delta)) = c.$$

$\square$

Since U is translation invariant, then $\partial_i Uf = U(\partial_i f)$ if f is C^1. An immediate consequence, then, is that if $f \in C^{k+\alpha}$, $k \ge 0$, $\alpha \in (0,1)$, and f has support in $B(0,M)$, then $Uf \in C^{k+2+\alpha}$.

(3.15) Corollary. *Suppose $f \in C^{k+\alpha}$ in a domain D. Then $G_D f \in C^{k+2+\alpha}$ in D.*

Proof. Fix $x_0 \in D$. We will show that $G_D f$ is $C^{k+2+\alpha}$ in a neighborhood of x_0, which will prove the corollary. Let $\delta < \text{dist}(x_0, \partial D)/2$. Let φ be a C^∞ function such that φ is 1 in $B(x_0, \delta/2)$ and 0 on $B(x_0, \delta)^c$. By scaling and Theorem 3.14 applied to $f\varphi$, $U(f\varphi)$ is $C^{k+2+\alpha}$ on $B(x_0, \delta)$. $U(f(1-\varphi))$

is harmonic in $B(x_0, \delta/2)$ and therefore C^∞ there. Hence Uf is $C^{k+2+\alpha}$ in $B(x_0, \delta/2)$. Recall $G_D f$ differs from Uf in D by a harmonic function. Since harmonic functions are C^∞, the corollary follows. $\qquad\square$

The same proof shows that if f is bounded in a bounded domain D, then $G_D f$ is continuous.

Additive functionals

Let $A_t = A_t^f = \int_0^t f(X_s)ds$. It follows from the definition of θ_s [see (I.3.3)] that

$$(3.38) \qquad\qquad A_t = A_s + A_{t-s} \circ \theta_s.$$

We call any nonnegative increasing process satisfying (3.38) an additive functional.

Suppose $d \geq 3$. We say that an additive functional is associated with a measure μ if

$$(3.39) \qquad\qquad \mathbb{E}^x A_\infty = \int u(x, y)\mu(dy), \qquad x \in \mathbb{R}^d.$$

If D is a bounded domain and we look at Brownian motion killed on exiting D, we replace u in (3.39) by g_D; this works in any dimension. From the above, we see that the additive functional A_t^f is associated to the measure $f(x)dx$. In one dimension, where we look at Brownian motion killed at first reaching 1 or negative 1, say, the additive functional associated to δ_a is L_t^a, the local time at a.

(3.16) Theorem. *Suppose μ is a measure with compact support with the property that there exists c such that*

$$(3.40) \qquad\qquad \mu(B(z, s)) \leq c(s^{d-1} \wedge 1), \qquad s > 0, z \in \mathbb{R}^d.$$

Then there exists a continuous additive functional A^μ associated to μ.

If μ is the surface measure of some nice domain, it is easy to see that (3.40) is satisfied.

Proof. Let φ be a nonnegative radially symmetric C^∞ function with integral 1 and compact support and let $\varphi_\varepsilon(x) = \varepsilon^{-d}\varphi(x/\varepsilon)$. Let $f_\varepsilon = \mu * \varphi_\varepsilon$. We want to show that A^{f_ε} converges (uniformly in t) in probability to some process, A^μ. We will first show $\int u(x, y)f_\varepsilon(y)dy \to \int u(x, y)\mu(dy)$ uniformly.

Note

$$\mu * \varphi_\varepsilon(B(z, s)) = \int \varphi_\varepsilon(x)\mu(x + B(z, s))dx = \int \varphi_\varepsilon(x)\mu(B(z + x, s))dx$$

$$\leq c(s^{d-1} \wedge 1) \int \varphi_\varepsilon(x)dx = c(s^{d-1} \wedge 1).$$

Let us check the boundedness of $\int u(x,y)f_\varepsilon(y)dy$.

$$(3.41) \quad \int u(x,y)f_\varepsilon(y)dy \le c \int_{|y-x|\ge 1} f_\varepsilon(y)dy$$

$$+ c \sum_{j=0}^{\infty} \int_{B(x,2^{-j})-B(x,2^{-j-1})} |x-y|^{2-d} f_\varepsilon(y)dy$$

$$\le c \int_{\mathbb{R}^d} f_\varepsilon(y)dy + c \sum_{j=0}^{\infty} 2^{j(d-2)} \int_{B(x,2^{-j})} f_\varepsilon(y)dy$$

$$\le c\mu * \varphi_\varepsilon(\mathbb{R}^d) + c \sum_{j=0}^{\infty} 2^{j(d-2)} \mu * \varphi_\varepsilon(B(x,2^{-j}))$$

$$\le c + c \sum_{j=0}^{\infty} 2^{j(d-2)}(2^{-j})^{d-1} \le c.$$

The same argument shows $\int_{|y-x|\le\delta} u(x,y)f_\varepsilon(y)dy \to 0$ uniformly over ε and x as $\delta \to 0$, and similarly, $\int_{|y-x|\le\delta} u(x,y)\mu(dy) \to 0$ as $\delta \to 0$.

Given η, take δ small so that $\int_{|y-x|\le\delta} u(x,y)f_\varepsilon(y)dy \le \eta$ for all x and ε and $\int_{|y-x|\le\delta} u(x,y)\mu(dy) \le \eta$ for all x. Take M large so that if $u(x,y) \ge M$, then $|x-y| \le \delta$. The functions $u(x,\cdot)\wedge M$ (indexed by x) are equicontinuous by our assumptions on u. Since $f_\varepsilon(y)\,dy$ converges weakly to $\mu(dy)$, Exercise I.8.15 shows that $\int[u(x,y) \wedge M]f_\varepsilon(y)\,dy$ converges to $\int[u(x,y) \wedge M]\mu(dy)$, uniformly over x, as $\varepsilon \to 0$.

$$\left| \int [u(x,y) - u(x,y) \wedge M]f_\varepsilon(y)dy \right| \le \int_{|y-x|\le\delta} u(x,y)f_\varepsilon(y)\,dy \le \eta,$$

and similarly, $\int[u(x,y) - u(x,y) \wedge M]\mu(dy) \le \eta$. Therefore

$$\limsup_{\varepsilon\to 0} \left| \int u(x,y)f_\varepsilon(y)dy - \int u(x,y)\mu(dy) \right| \le 2\eta,$$

which shows the uniform convergence.

Let $\varepsilon, \varepsilon' > 0$ and let $B_t = A_t^{f_\varepsilon} - A_t^{f_{\varepsilon'}}$. If T is a stopping time,

$$\mathbb{E}^x[A_\infty^{f_\varepsilon} - A_T^{f_\varepsilon}|\mathcal{F}_T] = \mathbb{E}^x[A_\infty^{f_\varepsilon} \circ \theta_T|\mathcal{F}_T] = \mathbb{E}^{X_T}A_\infty^{f_\varepsilon}.$$

For any z, by (3.41)

$$\mathbb{E}^z A_\infty^{f_\varepsilon} = \int u(z,y)f_\varepsilon(y)\,dy \le c.$$

Similarly, given $\gamma > 0$

$$|\mathbb{E}^z B_\infty| = \left| \int u(x,y)f_\varepsilon(y)dy - \int u(x,y)f_{\varepsilon'}(y)dy \right| < \gamma$$

for all $z \in \mathbb{R}^d$ if ε and ε' are small enough. Hence

$$\left| \mathbb{E}^x[B_\infty - B_T | \mathcal{F}_T] \right| = \left| \mathbb{E}^{X_T} B_\infty \right| \le \gamma$$

if ε and ε' are small enough. By Proposition I.6.14, $\mathbb{E}^x \sup_t |A_t^{f_\varepsilon} - A_t^{f_{\varepsilon'}}|^2 \to 0$ as $\varepsilon, \varepsilon' \to 0$. So with respect to the norm $\left(\mathbb{E}^x \sup_t |\cdot|^2 \right)^{1/2}$ the processes $A_t^{f_\varepsilon}$ form a Cauchy sequence. Let A_t^μ be the limit in this norm. If we take $\varepsilon \to 0$ fast enough, the convergence is almost sure and uniform over t, hence A_t^μ is continuous, nondecreasing, and satisfies (3.38). $\qquad\square$

The other result on additive functionals that we will prove is the following.

(3.17) Proposition. *Suppose μ satisfies (3.40). If h is nonnegative, then the continuous additive functional associated to $h(x)\mu(dx)$ is $\int_0^t h(X_s) dA_s^\mu$.*

Proof. We want to show $\mathbb{E}^x \int_0^\infty h(X_s) \, dA_s^\mu = \int u(x,y)h(y)\mu(dy)$ for all x. By monotone convergence, it suffices to suppose h is bounded and continuous. Define f_ε as above. Since $A_t^{f_\varepsilon}$ converges to A_t^μ uniformly over t, then approximating $h(X_s)$ (as a function of s) by a step function shows that $\int_0^\infty h(X_s) dA_s^{f_\varepsilon} \to \int_0^\infty h(X_s) dA_s^\mu$.

On the other hand, $\int u(x,y)h(y)f_\varepsilon(y)dy \to \int u(x,y)h(y)\mu(dy)$ by an argument very similar to the proof of Theorem 3.16: just use the fact that $h(y)f_\varepsilon(y) \, dy$ converges weakly to $h(y)\mu(dy)$. By the proof of Theorem 3.16, $\int_0^t h(X_s)f_\varepsilon(X_s)ds$ converges to the additive functional associated to the measure $h(y)\mu(dy)$. This implies the result. $\qquad\square$

4 Transition densities

Symmetry

We want to show the important theorem that $p_B(t,x,y)$ is symmetric in x and y, where p_B is the transition density for Brownian motion killed on exiting B. We will derive from this the symmetry of the Green function $g_B(x,y)$ in x and y.

Here is a heuristic calculation that will motivate what follows.

$$
\begin{aligned}
(4.1) \quad p(t,x,y)dy &= \mathbb{P}^x(X_t \in dy) \\
&= \mathbb{P}^x(X_t \in dy, \tau_B \ge t) + \mathbb{P}^x(X_t \in dy, \tau_B < t) \\
&= p_B(t,x,y) + \mathbb{E}^x[\mathbb{P}^{X_{\tau_B}}(X_{t-\tau_B} \in dy); \tau_B < t)] \\
&= p_B(t,x,y) + \mathbb{E}^x[p(t - \tau_B, X_{\tau_B}, y); \tau_B < t].
\end{aligned}
$$

We will use this equation to define $p_B(t, x, y)$. Let us set

$$(4.2) \qquad r(t, x, y) = \mathbb{E}^x[p(t - \tau_B, X_{\tau_B}, y); \tau_B < t]$$

and define

$$(4.3) \qquad p_B(t, x, y) = p(t, x, y) - r(t, x, y).$$

Integrating (4.3) against $1_A(y) dy$ we have

$$\int 1_A(y) p_B(t, x, y) dy = \mathbb{P}^x(X_t \in A) - \mathbb{E}^x\big[\mathbb{E}^{X(\tau_B)}[1_A(X_{t-\tau_B})]; \tau_B < t\big].$$

By the strong Markov property, the right-hand side is

$$\mathbb{E}^x 1_A(X_t) - \mathbb{E}^x[1_A(X_t); \tau_B < t] = \mathbb{E}^x[1_A(X_t); \tau_B \geq t]$$
$$= \mathbb{P}^x(X_t \in A; \tau_B \geq t).$$

It follows that $p_B(t, x, y)$ is a density for the measure $\mathbb{P}^x(X_t \in A; \tau_B \geq t)$.

We first prove the following.

(4.1) Proposition. $p_B(t, x, y) = p_B(t, y, x)$ *for almost every pair* $(x, y) \in \mathbb{R}^d \times \mathbb{R}^d$.

Proof. The idea is to write $p_B(t, x, y)$ in terms of $p(t, x, y)$ and then use the symmetry of $p(t, x, y)$ in x and y. Let $t_{nj} = jt/n$. If A and C are Borel sets and B is open,

$$\int_C \int_A p_B(t, x, y)\, dy\, dx = \int_C \mathbb{P}^x(X_t \in A, \tau_B \geq t) dx$$
$$= \lim_{n \to \infty} \int_C \mathbb{P}^{x_0}(X(t_{nn}) \in A, X(t_{n1}) \in B, \ldots, X(t_{n,n-1}) \in B) dx_0,$$

using the continuity of the paths of X_t.

By the Markov property, this is

$$\lim_n \int \cdots \int p(t/n, x_0, x_1) \ldots p(t/n, x_{n-1}, x_n)$$
$$\times 1_C(x_0) 1_B(x_1) \ldots 1_B(x_{n-1}) 1_A(x_n) dx_0 \ldots dx_n.$$

Writing y_i for x_{n-i} and using the symmetry of $p(t, x, y)$, we obtain the expression

$$\lim_n \int \cdots \int p(t/n, y_0, y_1) \cdots p(t/n, y_{n-1}, y_n)$$
$$\times 1_A(y_0) 1_B(y_1) \cdots 1_B(y_{n-1}) 1_C(y_n) dy_0 \cdots dy_n.$$

As above, this is equal to

(4.4)
$$\int_C \int_A p_B(t, y, x) dy\, dx.$$

So we have

(4.5)
$$\int_C \mathbb{P}^x(X_t \in A, \tau_B \geq t) dx = \int_A \mathbb{P}^y(X_t \in C, \tau_B \geq t) dy.$$

If we take open sets B_m decreasing to a compact set B, taking the limit in (4.5) gives (4.5) for B compact (note Exercise 11). Taking B_m compact in (4.5), increasing, and contained in B so that $T_{B_m} \downarrow T_B$, we have (4.5) for arbitrary Borel sets B.

Rewriting (4.5), we have

$$\int_C \int_A p_B(t, x, y)\, dy\, dx = \int_A \int_C p_B(t, x, y)\, dx\, dy$$

for all Borel sets B. The assertion now follows. □

(4.2) Lemma. $p_B(t, x, y) \geq 0$ for all x and y.

Proof. The functions $\mathbb{E}^x[p(t - \tau_B, X_{\tau_B}, y); \tau_B < t - \varepsilon]$, $\varepsilon > 0$, are continuous in y, since $p(s, z, y)$ is bounded and jointly continuous in z, y, and s for $s \geq \varepsilon$. These functions increase up to $r(t, x, y)$ as $\varepsilon \to 0$, hence $r(t, x, y)$ is lower semicontinuous. So $p_B(t, x, y)$ is upper semicontinuous. Since $p_B(t, x, \cdot) \geq 0$ a.e. on $\mathbb{R}^d$ (it is a density), this implies that it is nonnegative everywhere. □

We will shortly prove the technical lemma.

(4.3) Lemma. *For all x and y,*

(4.6)
$$\int p_B(t - a, x, u) p(a, u, y) du \downarrow p_B(t, x, y) \text{ as } a \to 0$$

and

(4.7)
$$\int p(a, x, u) p_B(t - a, u, y) du \downarrow p_B(t, x, y) \text{ as } a \to 0.$$

Once we prove this lemma we then have the following.

(4.4) Theorem. *For all x, y, and all $t > 0$, $p_B(t, x, y) = p_B(t, y, x)$.*

Proof of Theorem 4.4. Since $p_B(t, x, y) = p_B(t, y, x)$ for almost every pair (x, y), then

$$\int \int p(a, x, u) p_B(t - a - b, u, v) p(b, v, y) du\, dv$$

$$= \int \int p(b, y, v) p_B(t - a - b, v, u) p(a, u, x) du\, dv.$$

Now let $b \downarrow 0$, then $a \downarrow 0$, and use Lemma 4.3. $\qquad\square$

Proof of Lemma 4.3. (4.6) is easy. We have

$$\int \mathbb{E}^x \left[p(t - a - \tau_B, X_{\tau_B}, u); \tau_B < t - a \right] p(a, u, y) du$$
$$= \mathbb{E}^x \left[p(t - \tau_B, X_{\tau_B}, y); \tau_B < t - a \right],$$

which increases to

$$\mathbb{E}^x \left[p(t - \tau_B, X_{\tau_B}, y); \tau_B < t \right].$$

Using (4.3), this proves (4.6).

(4.7) is a bit harder. We first show that for all $b < t$

(4.8) $$r(t, x, y) \geq \int p(b, x, z) r(t - b, z, y) \, dz.$$

To see this, write

$$r(t, x, y) = \mathbb{E}^x \left[p(t - \tau_B, X_{\tau_B}, y); \tau_B < b \right]$$
$$+ \mathbb{E}^x \left[p(t - \tau_B, X_{\tau_B}, y); b \leq \tau_B < t \right].$$

By the Markov property, the first term on the right is

$$\int \mathbb{E}^x \left[p(b - \tau_B, X_{\tau_B}, z); \tau_B < b \right] p(t - b, z, y) dz,$$

and the second is

$$\mathbb{E}^x \left[\mathbb{E}^{X_b} \left[p(t - b - \tau_B, X_{\tau_B}, y); \tau_B < t - b \right] \right]$$
$$= \int p_B(b, x, z) \mathbb{E}^z \left[p(t - b - \tau_B, X_{\tau_B}, y); \tau_B < t - b \right] dz.$$

Adding,

$$r(t, x, y) = \int r(b, x, z) p(t - b, z, y) dz$$
$$+ \int p_B(b, x, z) r(t - b, z, y) dz.$$

By Lemma 4.2, $p(t - b, z, y) \geq r(t - b, z, y)$. So

$$r(t, x, y) \geq \int \left[r(b, x, z) + p_B(b, x, z) \right] r(t - b, z, y) dz,$$

which is (4.8).

We next show that (4.8) implies that $\int p(b, x, z) r(t - b, z, y) dz$ increases as b decreases. For if $b' < b$, (4.8) shows

$$\int p(b, x, z) r(t - b, z, y) dz$$

$$= \int \int p(b', x, w) p(b - b', w, z) r(t - b, z, y) dz dw$$

$$\leq \int p(b', x, w) r(t - b', w, y) dw.$$

We now look at two cases. Suppose $x \in (B^c)^r$. Note that we have $\int_{|z-x|>\delta} p(b, x, z) dz \to 0$ as $b \to 0$ for each $\delta > 0$. Since

$$r(t - b, z, y) = \mathbb{E}^z[p(t - b - \tau_B, X_{\tau_B}, y); \tau_B < t - b],$$

by Theorem 1.12, Corollary 1.11, and Fatou's lemma,

$$(4.9) \qquad \liminf_{b \to 0} \int p(b, x, z) r(t - b, z, y) dz \geq r(t, x, y).$$

With (4.8), (4.9) implies (4.7) for such x.

The other case is if x is not regular for B^c. It still suffices to show (4.9). We are looking at

$$(4.10) \quad \int p(b, x, u) \mathbb{E}^z[p(t - b - \tau_B, X_{\tau_B}, y); \tau_B < t - b] dz$$

$$= \mathbb{E}^x \mathbb{E}^{X_b}[p(t - b - \tau_B, X_{\tau_B}, y); \tau_B < t - b]$$

$$= \mathbb{E}^x \left[\mathbb{E}^{X_b}[p(t - b - \tau_B, X_{\tau_B}, y); \tau_B < t - b]; \tau_B > b \right]$$

$$+ \mathbb{E}^x \left[\mathbb{E}^{X_b}[p(t - b - \tau_B, X_{\tau_B}, y); \tau_B < t - b]; \tau_B \leq b \right].$$

On the set $(\tau_B > b)$ we have $\tau_B = b + \tau_B \circ \theta_b$ and $X_{\tau_B} \circ \theta_b = X_{\tau_B}$, and so the first term on the right-hand side of (4.10) is

$$\mathbb{E}^x[p(t - b - \tau_B, X_{\tau_B}, y) \circ \theta_b; \tau_B \circ \theta_b < t - b, \tau_B > b]$$

$$= \mathbb{E}^x[p(t - \tau_B, X_{\tau_B}, y); b < \tau_B < t].$$

If $x \notin (B^c)^r$, then $\tau_B > 0$, a.s., and so this increases to the quantity $\mathbb{E}^x[p(t - \tau_B, X_{\tau_B}, y); \tau_B < t] = r(t, x, y)$.

For the second term on the right-hand side of (4.10), note that

$$\mathbb{E}^{X_b}[p(t - b - \tau_B, X_{\tau_B}, y); \tau_B < t - b] = r(t - b, X_b, y)$$

$$\leq p(t - b, X_b, y) \leq \frac{c}{(t - b)^{d/2}}.$$

So if $x \notin (B^c)^r$, then the second term on the right of (4.10) is bounded by the quantity $c(t - b)^{-d/2} \mathbb{P}^x(\tau_B \leq b)$, which tends to 0 as $b \to 0$. Adding the two bounds for the right-hand side of (4.10) gives (4.9). $\qquad \square$

If we integrate $r(t, x, y)$ over t from 0 to ∞, we get

$$\mathbb{E}^x \int_{\tau_B}^{\infty} p(t - \tau_B, X_{\tau_B}, y)dt = \mathbb{E}^x \int_0^{\infty} p(t, X_{\tau_B}, y)dt = \mathbb{E}^x u(X_{\tau_B}, y).$$

So by (3.16) and (4.3),

(4.11) $$g_B(x, y) = \int_0^{\infty} p_B(t, x, y)dt.$$

Hence, as a corollary to Theorem 4.4, we have the following.

(4.5) Corollary. $g_B(x,y)$ *is symmetric in* x *and* y. $g_B(x,y)$ *is jointly continuous in* x *and* y *if* $x, y \in$ int (B) *and* $x \neq y$ *and equals* 0 *if either* x *or* y *is regular for* B^c.

Proof. The symmetry follows by (4.11) and Theorem 4.4. If $x \in (B^c)^r$, $g_B(x,y) = 0$ by its definition, and by symmetry, the same holds if $y \in (B^c)^r$. Let $\varepsilon > 0$. $g_B(\cdot, y)$ is continuous on int $(B - \{y\})$ since it is harmonic on int $(B - B(y, \varepsilon))$. By symmetry the same is true for $g_B(x, \cdot)$ on int $(B - \{x\})$. By Corollary 1.4, $g_B(\cdot, y)$ and $g_B(x, \cdot)$ are equicontinuous in int $(B - B(y, 2\varepsilon))$ and int $(B - B(x, 2\varepsilon))$, respectively. Hence $g_B(x, y)$ is jointly continuous on $[$int $(B - B(y, \varepsilon))] \times [$int $(B - B(x, \varepsilon))]$. Since ε is arbitrary, the joint continuity follows. □

Note, also, that if one defines

$$g_B^{\lambda}(x, y) = \int_0^{\infty} e^{-\lambda t} p_B(t, x, y)dt,$$

then g_B^{λ} will also be symmetric in x and y.

Continuity of transition densities

We have seen that $p_B(t, x, y) = 0$ if x or y is in $(B^c)^r$. Let us now show that $p_B(t, x, y)$ is continuous in the interior of B.

(4.6) Proposition. $p_B(t, x, y)$ *is jointly continuous in* x *and* y *for* $x, y \in$ int (B).

Proof. Since $p(t, x, y)$ is jointly continuous, it suffices to consider $r(t, x, y)$. Suppose $y_0 \in$ int (B) and $\delta < $ dist $(y_0, \partial B)$.

Let $\varepsilon > 0$. Since

$$\sup_t \sup_{|z-w|>\delta/2} p(t, z, w) < \infty \quad \text{and} \quad \limsup_{t \to 0} \sup_{|z-w|>\delta/2} p(t, z, w) = 0,$$

then

$$\sup_{|y-y_0|<\delta/2} \mathbb{E}^x[p(t - \tau_B, X_{\tau_B}, y); t - \eta < \tau_B < t]$$

can be made less than ε by taking η small. $\mathbb{E}^x[p(t - \tau_B, X_{\tau_B}, y); \tau_B \leq t - \eta]$ is continuous in $y \in B(y_0, \delta/2)$ by the continuity of $p(t, \cdot, \cdot)$. Adding, we see that $|r(t, x, y) - r(t, x, y_0)|$ can be made small by first taking η small, then y close to y_0. This proves continuity of $p_B(t, x, y)$ in y. Continuity in x follows by symmetry.

Let us now prove the joint continuity in x and y. By the semigroup property,

$$(4.12) \qquad p_B(t, x, y) = \int_B p_B(t/2, x, z) p_B(t/2, z, y) dz = \int_{B - B_\eta} + \int_{B_\eta},$$

where $B_\eta = \{z : \text{dist}\,(z, \partial B) < \eta\}$. We have joint continuity over $(x, y) \in \text{int}\,(B) \times \text{int}\,(B)$ of the first integral on the right of (4.12) by the continuity of $p_B(t/2, x, z)$ in x for each z, the continuity of $p_B(t/2, z, y)$ in y for each z, and dominated convergence. Since

$$\int_{B_\eta} p_B(t/2, x, z) p_B(t/2, z, y) dz \leq ct^{-d/2} t^{-d/2} |B_\eta|,$$

and $B_\eta \downarrow \emptyset$ as $\eta \to 0$, we can make the second integral on the right of (4.12) uniformly small by taking η small. Joint continuity follows. $\qquad \square$

A minor modification of the above gives us the joint continuity of $p_B(t, x, y)$ over (t, x, y) in $(0, \infty) \times \text{int}\,(B) \times \text{int}\,(B)$.

Now let D be a bounded domain such that every point of ∂D is regular for D^c. Define

$$(4.13) \qquad P_t^D f(x) = \int f(y) p_D(t, x, y) \, dy.$$

(4.7) Proposition. $p_D(t, x, y)$ is jointly continuous in $(x, y) \in \overline{D} \times \overline{D}$.

Proof. Fix y and let $h(z) = p_D(t/2, z, y)$. By the semigroup property, $p_D(t, x, y) = P_{t/2}^D h(x)$. h is bounded by $ct^{-d/2}$. Let $\varepsilon > 0$. By taking η small enough,

$$(4.14) \quad P_{t/2}^D (h 1_{D_\eta})(x) = \int_{D_\eta} h(z) p_D(t/2, x, z) dz \leq c \|h\|_\infty t^{-d/2} |D_\eta| < \varepsilon,$$

where $D_\eta = \{z : \text{dist}\,(z, \partial D) < \eta\}$.

If $w \in \partial D$, $\mathbb{P}^w(\tau_D \leq t/2) = 1$. By Proposition 1.10, $\mathbb{P}^x(\tau_D \leq t/2) \geq 1 - \varepsilon$ for x in a neighborhood of w. By a compactness argument, there exists δ such that

$$\mathbb{E}^x[h 1_{D_\eta^c}(X_{t/2}); \tau_D \geq t/2] \leq \|h\|_\infty \mathbb{P}^x(\tau_D \geq t/2) \leq \varepsilon \|h\|_\infty$$

if $\text{dist}\,(x, \partial D) < \delta$. Combining with (4.14), $p_D(t, x, y) = P_{t/2}^D h(x) \to 0$ as $x \to w \in \partial D$, uniformly over w. By symmetry the same is true for y. This, with Proposition 4.6, proves our result. $\qquad \square$

(4.8) Corollary. $\{P_t^D f : \|f\|_\infty \le 1\}$ *is an equicontinuous family.*

Proof. If $\varepsilon > 0$ and δ is chosen so that $|p_D(t,x,y) - p_D(t,x',y')| < \varepsilon$ if $|(x,y) - (x',y')| < \delta$, then

$$|P_t^D f(x) - P_t^D f(x')| \le \int_D f(y)|p_D(t,x,y) - p_D(t,x',y)|dy$$

$$\le \varepsilon|D|\,\|f\|_\infty.$$

$\square$

Define $\langle f, g \rangle = \int_D f\bar{g}$. Note that if $f \in L^2$,

$$|P_{t/2}^D f(x)| \le \int_D |f(y)|p_D(t/2,x,y)dy \le c\|f\|_2$$

by the Cauchy-Schwarz inequality, or $\|P_{t/2}^D f\|_\infty \le c\|f\|_2$. Since $P_t^D f = P_{t/2}^D(P_{t/2}^D f)$, we see that

(4.15) $\{P_t^D f : \|f\|_2 \le 1\}$ *is an equicontinuous family.*

Hilbert-Schmidt expansion theorem

Suppose $Q(x,y)$ is a symmetric real jointly continuous kernel on $K \times K$, where K is a compact subset of $\mathbb{R}^d$ with nonempty interior. We want to derive the Hilbert-Schmidt expansion of Q in terms of its eigenvalues and eigenfunctions. (The assumption that $K \subseteq \mathbb{R}^d$ with $\text{int}\,(K) \ne \emptyset$ is made only to ensure that $L^2(K)$ is infinite dimensional.)

We start by showing that every such Q has at least one eigenvalue. Recall $\|Q\|_2 = \sup\{\|Qf\|_2 : \|f\|_2 \le 1\}$.

(4.9) Proposition. *There exists $g \ne 0$ such that $Qg = \lambda g$ with $|\lambda| = \|Q\|_2$.*

Proof. Let $\beta = \sup\{|\langle Qf, f\rangle| : \|f\|_2 = 1\}$. We claim

(4.16) $$\beta = \|Q\|_2.$$

By the Cauchy-Schwarz inequality, $|\langle Qf, f\rangle| \le \|Qf\|_2 \|f\|_2$, so certainly $\beta \le \|Q\|_2$. On the other hand, if $\|f\|_2 = \|g\|_2 = 1$, then since Q is symmetric,

$$\langle Qf, f\rangle + \langle Qg, g\rangle + 2\langle Qf, g\rangle = \langle Q(f+g), f+g\rangle \le \beta\|f+g\|_2^2$$

and

$$\langle Qf, f\rangle + \langle Qg, g\rangle - 2\langle Qf, g\rangle = \langle Q(f-g), f-g\rangle \ge -\beta\|f-g\|_2^2.$$

Taking the difference,

$$4\langle Qf, g\rangle \le \beta(\|f+g\|_2^2 + \|f-g\|_2^2) = 2\beta(\|f\|_2^2 + \|g\|_2^2) = 4\beta.$$

If $\|Qf\|_2 = 0$, then $\beta \geq \|Qf\|_2$. Otherwise, let $g = Qf/\|Qf\|_2$. We then obtain

$$4\langle Qf, Qf \rangle / \|Qf\|_2 \leq 4\beta.$$

In either case, $\|Qf\|_2 \leq \beta$, and taking the supremum over fs with $\|f\|_2 = 1$ proves (4.16).

There then exists a sequence f_n such that $\|f_n\|_2 = 1$ and either $\langle Qf_n, f_n \rangle \to \beta$ or $\langle Qf_n, f_n \rangle \to -\beta$. We will do the first case; for the second, look at $-Q$. Since Q is bounded, $\{Qf_n\}$ is uniformly bounded by the Cauchy-Schwarz inequality. By the Cauchy-Schwarz inequality again,

$$|Qf_n(x) - Qf_n(x')| \leq \int |f_n(y)| \, |Q(x, y) - Q(x', y)|$$
$$\leq c \|f_n\|_2 \sup_{y \in K} |Q(x, y) - Q(x', y)|.$$

This and the joint continuity of Q shows that $\{Qf_n\}$ is an equicontinuous family. By looking at a subsequence, which we will also label f_n, $Qf_n \to h$ for some continuous h.

Let $g = Qh$. Since Qf_n converges to h uniformly, $Q^2 f_n \to Qh$.

$$\|Qg - \beta g\|_2^2 = \lim \|Q(Q^2 f_n) - \beta Q^2 f_n\|_2^2$$
$$\leq \lim \|Q^2\|_2^2 \|Qf_n - \beta f_n\|_2^2.$$

On the other hand

$$\|Qf_n - \beta f_n\|_2^2 = \langle Qf_n, Qf_n \rangle + \beta^2 \langle f_n, f_n \rangle - 2\beta \langle Qf_n, f_n \rangle$$
$$\leq \|Q\|_2^2 + \beta^2 - 2\beta \langle Qf_n, f_n \rangle$$
$$\to \|Q\|_2^2 - \beta^2 = 0.$$

So $\|Qg - \beta g\|_2 = 0$, and since both Qg and g are continuous, $Qg = \beta g$. $\square$

Let φ_n be the eigenfunctions of Q and λ_n the corresponding eigenvalues. The λ_n are real since

$$\lambda \langle \varphi_n, \varphi_n \rangle = \langle Q\varphi_n, \varphi_n \rangle = \overline{\langle \varphi_n, Q\varphi_n \rangle} = \overline{\lambda_n} \langle \varphi_n, \varphi_n \rangle.$$

If $\lambda_n \neq \lambda_m$, then φ_n and φ_m are orthogonal since

$$\lambda_n \langle \varphi_n, \varphi_m \rangle = \langle Q\varphi_n, \varphi_m \rangle = \overline{\langle \varphi_n, Q\varphi_m \rangle} = \lambda_m \langle \varphi_n, \varphi_m \rangle.$$

If there is more than one eigenfunction corresponding to a given eigenvalue λ, we may use the Gram-Schmidt orthogonalization procedure to select an orthonormal basis for the linear space spanned by the eigenfunctions corresponding to λ. So we may assume that the φs are orthonormal.

(4.10) Proposition. $\{\lambda_n\}$ *has no cluster point other than 0.*

Proof. First we show there do not exist infinitely many φ_n with the same eigenvalue λ. Suppose $\varphi_{n_1}, \varphi_{n_2}, \ldots$ all correspond to the same eigenvalue λ. Then

$$(4.17) \qquad \|\varphi_{n_i} - \varphi_{n_j}\|_2^2 = \langle \varphi_{n_i} - \varphi_{n_j}, \varphi_{n_i} - \varphi_{n_j} \rangle$$
$$= \langle \varphi_{n_i}, \varphi_{n_i} \rangle + \langle \varphi_{n_j}, \varphi_{n_j} \rangle = 2.$$

Since $\varphi_{n_i} = \lambda^{-1} Q \varphi_{n_i}$, then $\{\varphi_{n_i}\}$ forms an equicontinuous family, and some subsequence must converge, a contradiction.

Next we show there do not exist eigenvalues $\lambda_{n_1}, \lambda_{n_2}, \ldots$ converging to $\lambda \neq 0$. The proof is essentially the same: since $\varphi_{n_i} = \lambda_{n_i}^{-1} Q \varphi_{n_i}$, then $\{\varphi_{n_i}\}$ is an equicontinuous family, which is impossible because of (4.17). □

We will need the Bessel inequality.

(4.11) Proposition. $\|f\|_2^2 \geq \sum_{j=1}^{\infty} |\langle f, \varphi_j \rangle|^2$.

Proof. Let $g_n = f - \sum_{j=1}^n \langle f, \varphi_j \rangle \varphi_j$. Then

$$0 \leq \|g_n\|_2^2 = \|f\|_2^2 + \left\| \sum_{j=1}^n \langle f, \varphi_j \rangle \varphi_j \right\|_2^2 - 2 \left\langle f, \sum_{j=1}^n \langle f, \varphi_j \rangle \varphi_j \right\rangle$$
$$= \|f\|_2^2 + \sum_{j=1}^n |\langle f, \varphi_j \rangle|^2 - 2 \sum_{j=1}^n |\langle f, \varphi_j \rangle|^2.$$

Now let $n \to \infty$. □

(4.12) Theorem. (Hilbert-Schmidt expansion theorem) *Suppose Q is jointly continuous, real, and symmetric, and suppose no eigenvalue of Q is 0. Let $\lambda_1, \lambda_2, \ldots$ be the eigenvalues of Q arranged so that $|\lambda_1| \geq |\lambda_2| \geq \cdots$. Let the φ_j be the corresponding eigenfunctions and suppose they are orthonormal. Then*
(a) The φ_j form a complete orthonormal system.
(b) If f is continuous on K, then

$$Qf(x) = \sum_{i=1}^{\infty} \lambda_i \langle f, \varphi_i \rangle \varphi_i(x),$$

and the convergence is absolute and uniform on K.

Proof. First let us show there are infinitely many eigenvalues. Let Q_n be defined by

$$(4.18) \qquad Q_n f(x) = Qf(x) - \sum_{i=1}^n \lambda_i \langle f, \varphi_i \rangle \varphi_i(x).$$

Suppose $Q_n \neq 0$. Q_n has kernel $Q(x,y) - \sum_{i=1}^{n} \lambda_i \varphi_i(x)\varphi_i(y)$, and by Proposition 4.9, Q_n has an eigenfunction, say ψ, with eigenvalue μ and $|\mu| = \|Q_n\|_2 \neq 0$.

$$\mu\langle\psi,\varphi_i\rangle = \langle Q_n\psi,\varphi_i\rangle = \langle Q\psi,\varphi_i\rangle - \sum_{j=1}^{n}\lambda_j\langle\psi,\varphi_i\rangle\langle\varphi_i,\varphi_j\rangle$$

$$= \langle Q\psi,\varphi_i\rangle - \lambda_i\langle\psi,\varphi_i\rangle = \langle Q\psi,\varphi_i\rangle - \langle\psi,Q\varphi_i\rangle = 0$$

by the symmetry of Q. Hence ψ is orthogonal to φ_i for $i = 1,\ldots,n$.

Then

$$Q\psi = Q_n\psi + \sum_{i=1}^{n}\lambda_i\langle\psi,\varphi_i\rangle\varphi_i = Q_n\psi = \mu\psi,$$

or ψ is an eigenfunction for Q. If Q had only finitely many eigenfunctions, $\varphi_1,\ldots,\varphi_n$, then since $L^2(K)$ is infinite dimensional, there exists a nonzero $g \in L^2$ orthogonal to $\varphi_1,\ldots,\varphi_n$. By (4.18), $Q_n g = Qg$, and by assumption $Qg \neq 0$. So $\|Q_n\|_2 > 0$. This implies that there exists an eigenfunction ψ for Q orthogonal to all the φs, a contradiction.

Next, by Bessel's inequality applied to $f(y) = Q(x,y)$,

$$(4.19) \quad \|Q(x,\cdot)\|_2^2 \geq \sum_{j=1}^{\infty}|\langle Q(x,\cdot),\varphi_j\rangle|^2 = \sum_{j=1}^{\infty}\left|\int_K Q(x,y)\varphi_j(y)dy\right|^2$$

$$= \sum_{j=1}^{\infty}\lambda_j^2|\varphi_j(x)|^2.$$

Since Q is bounded,

$$(4.20) \quad \sum_{j=1}^{\infty}\lambda_j^2|\varphi_j(x)|^2 < \infty.$$

Let f be continuous. By the Cauchy-Schwarz inequality,

$$\sum_{i=m}^{n}|\lambda_i\langle f,\varphi_i\rangle\varphi_i(x)| \leq \left(\sum_{i=m}^{n}|\langle f,\varphi_i\rangle|^2\right)^{1/2}\left(\sum_{i=m}^{n}\lambda_i^2|\varphi_i(x)|^2\right)^{1/2}.$$

By Bessel's inequality and (4.20), this goes to 0 as $m,n \to \infty$. With (4.19) this shows that the sum in (b) converges absolutely and uniformly.

We want to show that

$$(4.21) \quad \sum_{i=1}^{\infty}\lambda_i\langle f,\varphi_i\rangle\varphi_i(x) = Qf(x), \qquad \text{a.e.}$$

Together with the uniform convergence of the sum in (b) and the fact that Qf and $\varphi_i = \lambda_i^{-1}Q\varphi_i$ are continuous, this will prove (b). As in the first paragraph of the proof, define Q_n by (4.18). What we proved there was that

Q_n had an eigenfunction ψ with corresponding eigenvalue μ, $|\mu| = \|Q_n\|_2$, and ψ was orthogonal to $\varphi_1, \ldots, \varphi_n$. Therefore $\mu = \lambda_k$ for some $k \geq n$, and $\|Q_n\|_2 = |\mu| = |\lambda_k| \leq |\lambda_n|$. There are infinitely many eigenvalues λ_n and the only cluster point is 0, hence $\lambda_n \to 0$ as $n \to \infty$. It follows that $\|Q_n\|_2 \to 0$ as $n \to \infty$, and (4.21) follows. (b) is proved.

Since the continuous functions are dense in L^2, (b) holds a.e. for $f \in L^2$. If $f \in L^2$, $f \neq 0$, and f is orthogonal to all of the φ_i, then by (b), $Qf = 0$, a contradiction to the fact that Q has no eigenvalues with the value 0. This proves (a). □

Eigenvalue expansions for transition densities

Let us see how the Hilbert-Schmidt expansion theorem applies to P_t^D.

(4.13) Theorem. *Suppose D is bounded and every point of ∂D is regular for D^c. There exist reals $0 < \mu_1 \leq \mu_2 \leq \cdots$ and functions φ_i such that φ_i is C^∞ in D, φ_i vanishes continuously on ∂D, $P_t^D \varphi_i = e^{-\mu_i t} \varphi_i$, $\Delta \varphi_i = -2\mu_i \varphi_i$, and for each t,*

$$(4.22) \qquad p_D(t, x, y) = \sum_{i=1}^{\infty} e^{-\mu_i t} \varphi_i(x) \varphi_i(y).$$

Moreover in (4.22) the convergence is absolute and uniform over x, y.

Proof. Suppose φ is an eigenfunction for P_t^D with corresponding eigenvalue λ.

$$\lambda \langle \varphi, \varphi \rangle = \langle \varphi, P_t^D \varphi \rangle = \langle P_{t/2}^D \varphi, P_{t/2}^D \varphi \rangle \geq 0,$$

since $P_t^D = P_{t/2}^D P_{t/2}^D$ and $P_{t/2}^D$ is symmetric. Hence all the eigenvalues are nonnegative.

Let us show that they are in fact strictly positive. Suppose $\varphi \neq 0$, a.e., and $\lambda = 0$. Then $\langle P_{t/2}^D \varphi, P_{t/2}^D \varphi \rangle = \langle P_t^D \varphi, \varphi \rangle = 0$, or $P_{t/2}^D \varphi = 0$. By induction, $P_{t/2^n}^D \varphi = 0$. Now $P_t^D f \to f$ if f is continuous with compact support. Such functions are dense in L^2 and P_t^D is a bounded operator for all t with a bound independent of t (see Exercise 46). Consequently $P_{t/2^n}^D f \to f$ in L^2 for all $f \in L^2$. Then $\varphi = \lim P_{t/2^n}^D \varphi = 0$, a contradiction. We conclude $\lambda > 0$.

We can also assert that all the eigenvalues are strictly less than 1. We have $P_{2t}^D \varphi = P_t^D P_t^D \varphi = \lambda P_t^D \varphi = \lambda^2 \varphi$, and by induction $P_{t2^n}^D \varphi = \lambda^{2^n} \varphi$. However, $\varphi = \lambda^{-1} P_t^D \varphi$ is continuous on $\overline{D}$, hence bounded, and

$$|P_{t2^n}^D \varphi(x)| \leq \mathbb{E}^x \left[\varphi(X_{t2^n}) |; t2^n \leq \tau_D \right] \leq \|\varphi\|_\infty \mathbb{P}^x (t2^n \leq \tau_D) \to 0$$

as $n \to \infty$ since D is bounded. Therefore λ must be less than 1.

We next show that if φ is an eigenfunction for P_1^D with eigenvalue λ, then φ is an eigenfunction for $P_{1/2}^D$ with eigenvalue $\lambda^{1/2}$. Let $\psi = P_{1/2}^D\varphi - \sqrt{\lambda}\varphi$. Then $(P_{1/2}^D + \sqrt{\lambda})\psi = P_1^D\varphi - \lambda\varphi = 0$. So

$$0 = \|P_{1/2}^D\psi + \sqrt{\lambda}\psi\|_2^2 = \|P_{1/2}^D\psi\|_2^2 + 2\sqrt{\lambda}\langle P_{1/2}^D\psi, \psi\rangle + \lambda\|\psi\|_2^2$$
$$= \|P_{1/2}^D\psi\|_2^2 + 2\sqrt{\lambda}\langle P_{1/4}^D\psi, P_{1/4}^D\psi\rangle + \lambda\|\psi\|_2^2.$$

Each term on the right is nonnegative. Since $\lambda \neq 0$, this implies $\psi = 0$. By induction, $P_{1/2^n}^D\varphi = \lambda^{2^{-n}}\varphi$. Then

$$P_{m/2^n}^D\varphi = (P_{1/2^n}^D)^m\varphi = \lambda^{m/2^n}\varphi,$$

and by continuity, $P_t^D\varphi = \lambda^t\varphi$. If we set $\mu = -\log\lambda$, then $0 < \mu$ and $P_t^D\varphi = e^{-\mu t}\varphi$ for all t.

Let $Q = P_{t/2}^D$ and $f(x) = p_D(t/2, x, y)$. Let φ_i be the normalized eigenfunctions for Q and let $e^{-\mu_i t/2}$ be the corresponding eigenvalues. By Theorem 4.12(b) with $K = \overline{D}$,

$$(4.23) \qquad Qf(x) = \sum_{i=1}^{\infty} e^{-\mu_i t/2}\langle f, \varphi_i\rangle\varphi_i(x),$$

and the convergence is absolute and uniform (in x). We also have

$$(4.24) \qquad \langle f, \varphi_i\rangle = \int p_D(t/2, z, y)\varphi_i(z)dz = P_{t/2}^D\varphi_i = e^{-\mu_i t/2}\varphi_i(y).$$

On the other hand,

$$(4.25) \qquad Qf(x) = \int p_D(t/2, x, z)p_D(t/2, z, y)dz = p_D(t, x, y).$$

Substituting (4.24) and (4.25) in (4.23) shows (4.22).

Let us show that the convergence in (4.22) is absolute and uniform in both x and y. If we set $y = x$ in (4.22) and integrate over $\overline{D}$,

$$(4.26) \qquad ct^{-d/2} \geq \int_{\overline{D}} p_D(t/4, x, x)\,dx = \sum_{i=1}^{\infty} e^{-\mu_i t/4}.$$

We have

$$|P_t^D\varphi_i(x)| \leq \int_{\overline{D}} p_D(t, x, y)\varphi_i(y)\,dy \leq ct^{-d/2}$$

by the Cauchy-Schwarz inequality. So

$$(4.27) \qquad |\varphi_i(x)| \leq e^{\mu_i t/4}|P_{t/4}^D\varphi_i(x)| \leq ce^{\mu_i t/4}t^{-d/2}.$$

Hence

$$(4.28) \qquad e^{-\mu_i t}|\varphi_i(x)|\,|\varphi_i(y)| \leq ce^{-\mu_i t/2}t^{-d}.$$

This and (4.26) show that the convergence is absolute and uniform in x and y.

Finally, we show that the φ_is are smooth and are eigenfunctions of the Laplacian. Note

$$G_D \varphi_i(x) = \int_0^\infty P_t^D \varphi_i(x)\, dx = \left(\int_0^\infty e^{-\mu_i t}\, dt \right) \varphi_i(x) = \mu_i^{-1} \varphi_i(x).$$

As we have already seen, $\varphi_i = e^{\mu_i t} P_t^D \varphi_i$ is continuous and vanishes on ∂D. By Proposition 3.12 and the remark following Corollary 3.15, $\varphi_i = \mu_i G_D \varphi_i$ is C^1 on D. By Corollary 3.15, $\varphi_i = \mu_i G_D \varphi_i$ is $C^{2+\alpha}$ on D for each $\alpha \in (0,1)$. Repeating, φ_i is C^∞ on D. Since $\varphi_i = \mu_i G_D \varphi_i$, then

$$\frac{1}{2} \Delta \varphi_i = \mu_i \left(\frac{1}{2} \Delta G_D \varphi_i \right) = -\mu_i \varphi_i.$$

This completes the proof. □

As an example, let $d = 1$ and $D = (0,1)$. The eigenfunctions for $(1/2)f''$ are $\sqrt{2}\sin(n\pi x)$ with corresponding eigenvalues $n^2 \pi^2 / 2$. Thus

$$(4.29) \qquad p_D(t, x, y) = \sum_{n=1}^\infty 2e^{-n^2 \pi^2 t / 2} \sin(n\pi x) \sin(n\pi y).$$

Integrating over y in $(0,1)$,

$$(4.30) \qquad \mathbb{P}^x(t < \tau_D) = \int_D p_D(t, x, y)\, dy = \sum_{n\,\mathrm{odd}} \frac{4}{n\pi} e^{-n^2 \pi^2 t / 2} \sin(n\pi x).$$

Heat equation

Suppose D is a bounded domain with every point of ∂D regular for D^c. The heat equation is the equation $\partial_t u(x, t) = (1/2)\Delta u(x, t)$, where ∂_t denotes $\partial u / \partial t$.

(4.14) Theorem. *Suppose f is continuous. Consider the heat equation with boundary values $u(t, x) = 0$ if $x \in \partial D$ and $u(t, 0) = f(x)$. The solution to the heat equation is given by*

$$u(x, t) = \mathbb{E}^x[f(X_t); t < \tau_D].$$

Proof. If φ_n are the eigenvalues of P_t^D, we saw in Theorem 4.13 that the φ_n are C^∞ and $(1/2)\Delta \varphi_n = -\mu_n \varphi_n$. Note

$$\mu_i e^{-\mu_i t} \leq \left(\sup_{\mu > 0} \mu e^{-\mu t / 2} \right) e^{-\mu_i t / 2} \leq c e^{-\mu_i t / 2}.$$

By the estimate (4.26), it follows that

$$\sum_{i=m}^{n} \mu_i e^{-\mu_i t} |\varphi_i(x)| \, |\varphi_i(y)| \to 0$$

as $n, m \to \infty$. Then we can differentiate the eigenvalue expansion for $p_D(t, x, y)$ term by term to deduce $(1/2)\Delta p_D(t, x, \cdot) = \partial_t p_D(t, x, \cdot)$. Hence by dominated convergence, $u(x, t) = \int_D f(y) p_D(t, x, y) dy$ is in C^2 and satisfies the heat equation.

By the continuity of f and X_t, $u(x, t) \to f(x)$ as $t \to 0$. The other boundary condition is obvious. □

5 Newtonian capacity

Maximum principle

We now look at capacity again, but this time Newtonian capacity. In all that follows, life will be a lot easier if we first have the result that Brownian motion never hits $B - B^r$ for any Borel set B, where B^r is the set of points regular for B.

In the last section we defined $g_B^\lambda(x, y) = \int_0^\infty c^{-\lambda t} p_B(t, x, y) dt$.

(5.1) Proposition.

(5.1) $$u^\lambda(x, y) = g_B^\lambda(x, y) + \mathbb{E}^x \left[e^{-\lambda \tau_B} u^\lambda(X_{\tau_B}, y) \right].$$

Proof. We multiply (4.3) by $e^{-\lambda t}$ and integrate over t from 0 to ∞. By the definition of $r(t, x, y)$ in (4.2),

$$\int_0^\infty e^{-\lambda t} r(t, x, y) dt$$

$$= \mathbb{E}^x \left[e^{-\lambda \tau_B} \int_0^\infty e^{-\lambda(t - \tau_B)} p(t - \tau_B, X_{\tau_B}, y) 1_{(\tau_B < t)} dt \right]$$

$$= \mathbb{E}^x \left[e^{-\lambda \tau_B} \int_{\tau_B}^\infty e^{-\lambda(t - \tau_B)} p_B(t - \tau_B, X_{\tau_B}, y) dt \right]$$

$$= \mathbb{E}^x \left[e^{-\lambda \tau_B} u^\lambda(X_{\tau_B}, y) \right].$$

□

The support of a measure is, of course, the smallest closed set whose complement is not charged by the measure. In order to be able to talk about more general sets than closed ones, we say a measure μ is concentrated on a Borel set B (not necessarily closed) if $\mu(B^c) = 0$.

The following is called the maximum principle for λ-potentials.

(5.2) Theorem. *Suppose $\lambda \geq 0$. Suppose μ is a measure concentrated on B. Then*

$$\sup_{\mathbb{R}^d} U^\lambda \mu(x) = \sup_B U^\lambda \mu(x).$$

Proof. Let $M = \sup_B U^\lambda \mu(x)$ and let us suppose without loss of generality that $M < \infty$. Let $\varepsilon > 0$ and let

$$A = \{x : U^\lambda \mu(x) \leq M + \varepsilon\}.$$

Since $\int_{1/n}^n e^{-\lambda s} P_s \mu(x) ds$ is continuous in x for each n and increases up to $U^\lambda \mu(x)$, $U^\lambda \mu$ is lower semicontinuous, and hence A is closed.

We want to show $B \subseteq A^r$. We have

$$U^\lambda \mu(x) \geq \int_t^\infty e^{-\lambda s} P_s \mu(x) ds = e^{-\lambda t} \int_0^\infty e^{-\lambda s} P_{s+t} \mu(x) ds$$

$$= e^{-\lambda t} P_t U^\lambda \mu(x) = e^{-\lambda t} \mathbb{E}^x U^\lambda \mu(X_t).$$

On the set $(T_A > t)$, we have $U^\lambda \mu(X_t) > M + \varepsilon$, hence

$$U^\lambda \mu(x) \geq e^{-\lambda t} \mathbb{E}^x U^\lambda \mu(X_t) \geq (M + \varepsilon) e^{-\lambda t} \mathbb{P}^x(T_A > t).$$

Letting $t \to 0$,

$$U^\lambda \mu(x) \geq (M + \varepsilon) \mathbb{P}^x(T_A > 0).$$

If $x \in B$, then $U^\lambda \mu(x) \leq M$, hence $\mathbb{P}^x(T_A > 0) < 1$. By the zero-one law, $\mathbb{P}^x(T_A = 0) = 1$, or $x \in A^r$. Therefore $B \subseteq A^r$.

We now complete the proof. Since A is closed, $X_{T_A} \in A$, and therefore

$$\mathbb{E}^x[e^{-\lambda T_A} U^\lambda \mu(X_{T_A})] \leq M + \varepsilon.$$

Also, by (5.1), $g_{A^c}^\lambda(x, y) = 0$ for $y \in B \subseteq A^r$, so $g_{A^c}^\lambda \mu(x) = 0$. Integrating (5.1) and recalling $\tau_{A^c} = T_A$, for any x,

$$U^\lambda \mu(x) = \mathbb{E}^x[e^{-\lambda T_A} U^\lambda \mu(X_{T_A})] \leq M + \varepsilon.$$

Since ε is arbitrary, we have our result. $\qquad \square$

We digress a moment to say a few more words about the maximum principle. The above result includes the case $\lambda = 0$, so we have $\sup_{\mathbb{R}^d} U\mu(x) = \sup_B U\mu(x)$. This is not an obvious result. $U\mu(x)$ is harmonic in B^c, hence $U\mu$ does not take its maximum in B^c. However, $U\mu(x)$ is not necessarily continuous, so it is conceivable that as $x_n \to x \in \partial B$, $\liminf U\mu(x_n) > \sup_B U\mu(x)$.

A corollary to the maximum principle is the continuity principle.

(5.3) Proposition. *Suppose μ is supported on a closed set B. If the restriction of $U\mu$ to B is continuous on B, then $U\mu$ is continuous on $\mathbb{R}^d$.*

Proof. $U\mu$ is continuous on B^c since it is harmonic there. Suppose $y_0 \in \partial B$. We must show $U\mu$ is continuous at y_0. Without loss of generality, take $y_0 = 0$.

Let $\varphi_\delta(x) = 1_{B(0,\delta)}(x)$. Let $\varepsilon > 0$.

$$U\mu(x) = \int u(x,y)\varphi_\delta(y)\mu(dy) + \int u(x,y)[1 - \varphi_\delta(y)]\mu(dy).$$

Write $\mu_\delta(dy)$ for $\varphi_\delta(y)\mu(dy)$. Since $U\mu(0) < \infty$, by dominated convergence we can take δ small enough so that $U\mu_\delta(0) < \varepsilon$.

$U(\mu - \mu_\delta)$ is harmonic off the support of $\mu - \mu_\delta$, hence it is continuous in a neighborhood of 0. $U\mu|_B$ is continuous on B, so taking the difference, $U\mu_\delta|_B$ is continuous on $B(0, \delta/2) \cap B$. Take $\delta' < \delta$ so that if $y \in \overline{B(0, 2\delta')} \cap B$, then $U\mu_{\delta'}(y) \leq U\mu_\delta(y) \leq U\mu_\delta(0) + \varepsilon \leq 2\varepsilon$.

Now $U(\mu - \mu_{\delta'})$ is continuous at 0 since it is harmonic in $B(0, \delta')$. $U\mu_{\delta'} \leq 2\varepsilon$ on the support of $\mu_{\delta'}$, hence by the maximum principle $U\mu_{\delta'} \leq 2\varepsilon$ on $\mathbb{R}^d$. Therefore

$$\limsup_{y \to 0} |U\mu(y) - U\mu(0)| \leq 2\varepsilon.$$

Since ε is arbitrary, this proves the proposition. $\qquad\square$

Irregular points

We now return to our main topic.

By (5.1), since u^λ and g_A^λ are both symmetric in x and y, then

$$\mathbb{E}^x\big[e^{-\lambda \tau_A} u^\lambda(X_{\tau_A}, y)\big] = \mathbb{E}^y\big[e^{-\lambda \tau_A} u^\lambda(X_{\tau_A}, x)\big].$$

Letting $A = B^c$,

(5.2) $$\mathbb{E}^x\big[e^{-\lambda T_B} u^\lambda(X_{T_B}, y)\big] = \mathbb{E}^y\big[e^{-\lambda T_B} u^\lambda(X_{T_B}, y)\big].$$

Note also that since $\int p(t,x,y)dy = 1$ for all t, then

(5.3) $$\int \lambda u^\lambda(x,y)dy = 1.$$

Define the measure μ_B^λ by

(5.4) $$\mu_B^\lambda(dz) = \int \lambda \mathbb{E}^y[e^{-\lambda T_B}; X_{T_B} \in dz]\,dy$$

and let

(5.5) $$C^\lambda(B) = \mu_B^\lambda(\mathbb{R}^d) = \int \lambda \mathbb{E}^y e^{-\lambda T_B}\,dy.$$

Using (5.2) and the symmetry of u^λ, we calculate

$$U^\lambda \mu_B^\lambda(x) = \int u^\lambda(x,z)\mu_B^\lambda(dz) = \int \lambda \mathbb{E}^y\left[e^{-\lambda T_B}u^\lambda(x,X_{T_B})\right]dy$$

$$= \int \lambda \mathbb{E}^x\left[e^{-\lambda T_B}u^\lambda(X_{T_B},y)\right]dy.$$

So by (5.3),

$$(5.6) \qquad\qquad U^\lambda \mu_B^\lambda(x) = \mathbb{E}^x e^{-\lambda T_B}.$$

μ_B^λ is sometimes called the λ-equilibrium measure, $U^\lambda \mu_B^\lambda$ the λ-equilibrium potential, and C^λ the λ-capacity, but we will not use these terms further.

(5.4) Proposition. *If* $\sup_B U^\lambda \mu_B^\lambda < 1$, *then* $\mathbb{P}^y(T_B < \infty) = 0$ *for all* y.

Proof. First assume B is compact. By the maximum principle,

$$\sup_{\mathbb{R}^d} U^\lambda \mu_B^\lambda(x) = M < 1.$$

Let $A_m = \{x : \text{dist}(x,B) \le 1/m\}$. Since B is in the interior of A_m, which is contained in A_m^r, then $U^\lambda \mu_{A_m}^\lambda = 1$ on B. Therefore, because μ_B^λ is supported in B,

$$C^\lambda(B) = \int U^\lambda \mu_{A_m}^\lambda(x)\mu_B^\lambda(dx) = \int\int u^\lambda(x,y)\mu_{A_m}^\lambda(dy)\mu_B^\lambda(dx)$$

$$= \int U^\lambda \mu_B^\lambda(y)\mu_{A_m}^\lambda(dy) \le M\mu_{A_m}^\lambda(\mathbb{R}^d).$$

As $m \to \infty$, $A_m \downarrow B$ and $\mathbb{E}^y e^{-\lambda T_{A_m}} \downarrow \mathbb{E}^y e^{-\lambda T_B}$. So by (5.5), $C^\lambda(B) \le MC^\lambda(B)$. Since $M < 1$, this implies that $C^\lambda(B) = 0$. Therefore $\mu_B^\lambda = 0$, hence $U^\lambda \mu_B^\lambda(x) = 0$ for all x, or by (5.6), $T_B = \infty$, $\mathbb{P}^x$-a.s. for all x.

If B is not compact, take B_n compact contained in B with $T_{B_n} \downarrow T_B$, $\mathbb{P}^x$-a.s. on $(T_B < \infty)$. Since $B_n \subseteq B$, $T_{B_n} \ge T_B$, hence by (5.6), $\sup_{B_n} U^\lambda \mu_{B_n}^\lambda < 1$. By what we just showed, $\mathbb{P}^x(T_{B_n} < \infty) = 0$. Letting $n \to \infty$ shows $\mathbb{P}^x(T_B < \infty) = 0$, and since x is arbitrary, this proves the proposition. $\qquad\square$

With this preliminary, we have one of the main theorems of this section.

(5.5) Theorem. *If* B *is Borel,* $\mathbb{P}^x(X_t$ *ever hits* $B - B^r) = 0$ *for all* x.

Sets that are never hit by Brownian motion from any starting point are called polar (see Definition 5.12), and this theorem says that the set of irregular points for B is polar for any set B.

Proof. Let $B_n = \{x \in B : U^\lambda \mu_B^\lambda(x) \le 1 - 1/n\}$. Since $B_n \subseteq B$, by (5.6),

$$U^\lambda \mu^\lambda_{B_n}(x) \leq U^\lambda \mu^\lambda_B(x) \leq 1 - 1/n$$

on B_n. Hence $\mathbb{P}^x(T_{B_n} < \infty) = 0$ for all x. If $y \in B - B^r$, then $\mathbb{P}^y(T_B > 0) = 1$ or $\mathbb{E}^y e^{-\lambda T_B} < 1$. Hence $B - B^r \subseteq \cup_n B_n$.   □

(5.6) Corollary. $\mathbb{P}^x(X_{T_B} \in dy)$ *is concentrated on* B^r.

Proof. We have $\mathbb{P}^x(X_{T_B} \in B - B^r) = 0$. By the definition of T_B, if $X_{T_B} \in B^c$, the process must hit B immediately afterwards, or $T_B \circ \theta_{T_B} = 0$ on $(X_{T_B} \in B^c)$. So

$$\mathbb{P}^x(X_{T_B} \in B^c) = \mathbb{P}^x(T_B < \infty, X_{T_B} \in B^c, T_B \circ \theta_{T_B} = 0)$$
$$= \mathbb{E}^x\left[\mathbb{P}^{X_{T_B}}(T_B = 0); T_B < \infty, X_{T_B} \in B^c\right].$$

This equals $\mathbb{P}^x(X_{T_B} \in B^c \cap B^r, T_B < \infty)$ since $\mathbb{P}^{X_{T_B}}(T_B = 0) = 0$ unless $X_{T_B} \in B^r$. Therefore $\mathbb{P}^x(X_{T_B} \in B^c - B^r) = 0$.   □

Newtonian capacity

Now that we have that $X_{T_B} \in B^r$, we can obtain many of the other results we want fairly quickly.

(5.7) Definition. *An equilibrium measure (or capacitary measure) is a measure* μ_B *concentrated on* B^r *such that* $U\mu_B = 1$ *on* B^r. $U\mu_B$ *is called the equilibrium potential. The capacity of* B *is* $C(B) = \mu_B(\mathbb{R}^d)$.

The electrostatic interpretation is the following. Suppose B is a conducting metal solid in the interior of a metal ball of radius R that is grounded. Since B conducts, if we introduce some charge on B, the electrons will distribute themselves so that the potential on B is constant (otherwise current would flow). The charge distribution is the equilibrium measure. Capacity is the amount of charge that can be stored given a fixed potential difference between B and $\partial B(0, R)$. This description is actually for the capacity with respect to the Green potential in the ball. If we let $R \to \infty$, we arrive at Newtonian capacity.

If B is bounded, we can give a nice probabilistic interpretation of capacity. Let

$$(5.7) \qquad \gamma_r(dy) = \frac{(2\pi)^{d/2} r^{d-2}}{\Gamma(d/2 - 1)} \sigma_r(dy),$$

where σ_r is normalized surface measure on $\partial B(0, r)$.

(5.8) Proposition. *If* $B \subseteq B(0, r)$ *and* μ_B *is defined by*

$$(5.8) \qquad \mu_B(A) = \int \mathbb{P}^y(X_{T_B} \in A; T_B < \infty)\gamma_r(dy),$$

then μ_B is an equilibrium measure,

(5.9) $\qquad U\mu_B(x) = \mathbb{P}^x(T_B < \infty), \qquad C(B) = \int \mathbb{P}^x(T_B < \infty)\gamma_r(dx).$

We will shortly show that there is only one equilibrium measure.

Proof. First of all, if $s > r$ and $x \in \partial B(0, s)$, then $\mathbb{P}^x(T_{B(0,r)} < \infty) = r^{d-2}/s^{d-2}$. Thus by the strong Markov property, our definition of μ_B and our assertions about $C(B)$ do not depend on the choice of r, provided $B \subseteq B(0, r)$.

By Corollary 5.6, μ_B is concentrated on B^r. Using (5.2) with $\lambda = 0$ and the symmetry of u, we have

(5.10) $\qquad U\mu_B(x) = \int u(x, z)\mu_B(dz)$

$$= \int \mathbb{E}^y\big[u(x, X_{T_B}); T_B < \infty\big]\gamma_r(dy)$$

$$= \int \mathbb{E}^x\big[u(X_{T_B}, y); T_B < \infty\big]\gamma_r(dy)$$

$$= \mathbb{E}^x[U\gamma_r(X_{T_B}); T_B < \infty].$$

By our calculation in (3.6), we saw that $U\sigma_r$ was equal to $c_d r^{2-d}$ inside $B(0, r)$, hence $U\gamma_r$ is identically 1 there. Substituting in (5.10), $U\mu_B(x) = \mathbb{P}^x(T_B < \infty)$. This is, of course, 1 on B^r, so μ_B is an equilibrium measure. Finally, $C(B) = \mu_B(\mathbb{R}^d) = \int \mathbb{P}^y(T_B < \infty)\gamma_r(dy).$ $\qquad \square$

As an immediate corollary, we have the following.

(5.9) Corollary. *μ_B is concentrated on ∂B.*

Let us next show that there is only one equilibrium measure.

(5.10) Proposition. *If μ and ν are concentrated on B^r and $U\mu$ and $U\nu$ are finite and equal on B^r, then $\mu = \nu$.*

Proof. If $y \in B^r$, $g_{B^c}(x, y) = 0$ and so $u(x, y) = \mathbb{E}^x u(X_{T_B}, y)$. Then for any x, by Theorem 5.5

$$U\mu(x) = \mathbb{E}^x U\mu(X_{T_B}) = \mathbb{E}^x U\nu(X_{T_B}) = U\nu(x).$$

The result now follows by Proposition 3.5. $\qquad \square$

It is immediate from Proposition 5.8 that

(5.11) $\qquad\qquad C(B(0, s)) = c_d^{-1} s^{d-2}.$

We have the following properties, whose proof is Exercise 21.

(5.11) Proposition.
a) $C(b + B) = C(B)$;
(b) $C(aB) = a^{d-2}C(B)$;
(c) $C(A \cup B) + C(A \cap B) \leq C(A) + C(B)$;
(d) $C(A) \leq C(B)$ if $A \subseteq B$;
(e) C is a Choquet capacity.

(5.12) Definition. *A set B is polar if $\mathbb{P}^x(T_B < \infty) = 0$ for all x.*

(5.13) Proposition. *B is polar if and only if $C(B) = 0$.*

Proof. Suppose first that B is bounded. If $C(B) = 0$, then $\mu_B = 0$, hence $U\mu_B \equiv 0$, which by Proposition 5.8 says that B is polar. On the other hand, if B is polar, by Proposition 5.8, $C(B) = 0$.
 For the unbounded case, look at $B \cap B(0, M)$ and let $M \to \infty$. $\qquad \square$

 The analytical definition of B polar is that there exists f superharmonic which is infinite on B but not identically infinite.

Wiener's test

A nice application of capacity is Wiener's test. In view of Theorem 1.15, it is important to know when a point is regular for a set. Wiener's test gives a criterion in terms of capacity.
 First we need a version of the Borel-Cantelli lemma for dependent events.

(5.14) Proposition. *Suppose there exists M such that $\mathbb{P}(A_m \cap A_n) \leq M\mathbb{P}(A_m)\mathbb{P}(A_n)$ for all $m \neq n$. Then $\mathbb{P}(A_n \text{ i.o.}) > 0$ if and only if $\sum_n \mathbb{P}(A_n) = \infty$.*

Proof. If $\sum \mathbb{P}(A_n) < \infty$, then $\mathbb{P}(A_n \text{ i.o.}) = 0$ by Proposition I.1.1. So let us suppose $\sum \mathbb{P}(A_n) = \infty$. Let $N_n = \sum_{i=1}^{n} 1_{A_i}$. Then $\mathbb{E}\, N_n = \sum_{i=1}^{n} \mathbb{P}(A_i) \to \infty$.
 We have the estimate

$$
\mathbb{E}\, N_n^2 = \sum_{i=1}^{n} \mathbb{P}(A_i) + \sum_{i \neq j} \mathbb{P}(A_i \cap A_j)
$$
$$
\leq \mathbb{E}\, N_n + M \sum_{i \neq j} \mathbb{P}(A_i)\mathbb{P}(A_j)
$$
$$
\leq \mathbb{E}\, N_n + M(\mathbb{E}\, N_n)^2.
$$

Since $\mathbb{E}\, N_n > 1$ for n large,

$$
\mathbb{E}\, N_n^2 \leq (M + 1)(\mathbb{E}\, N_n)^2
$$

for n large. Since

$$\mathbb{E}\,N_n = \mathbb{E}\left[N_n; N_n \geq \frac{\mathbb{E}\,N_n}{2}\right] + \mathbb{E}\left[N_n; N_n < \frac{\mathbb{E}\,N_n}{2}\right]$$

$$\leq \mathbb{E}\left[N_n; N_n \geq \frac{\mathbb{E}\,N_n}{2}\right] + \frac{\mathbb{E}\,N_n}{2},$$

then

$$\mathbb{E}\,N_n \leq 2\mathbb{E}\left[N_n; N_n \geq \frac{\mathbb{E}\,N_n}{2}\right].$$

By the Cauchy-Schwarz inequality,

$$\frac{1}{4}(\mathbb{E}\,N_n)^2 \leq \mathbb{E}\,N_n^2\,\mathbb{P}\left(N_n \geq \frac{\mathbb{E}\,N_n}{2}\right)$$

$$\leq (M+1)(\mathbb{E}\,N_n)^2\mathbb{P}\left(N_n \geq \frac{\mathbb{E}\,N_n}{2}\right).$$

Hence,

$$\mathbb{P}\left(N_\infty \geq \frac{\mathbb{E}\,N_n}{2}\right) \geq \mathbb{P}\left(N_n \geq \frac{\mathbb{E}\,N_n}{2}\right) \geq \frac{1}{4(M+1)} > 0.$$

Letting $n \to \infty$,

$$\mathbb{P}(N_\infty = \infty) \geq (4(M+1))^{-1} > 0.$$

$\square$

Note that if $\sum \mathbb{P}(A_n) = \infty$, then either $\sum \mathbb{P}(A_{2n}) = \infty$ or $\sum \mathbb{P}(A_{2n+1}) = \infty$. It suffices, then, in the above proposition to suppose only that we have $\mathbb{P}(A_n \cap A_m) \leq M\mathbb{P}(A_n)\mathbb{P}(A_m)$ whenever $|m - n| > 1$.

(5.15) Theorem. (Wiener's test) *Suppose $d \geq 3$, $\lambda \in (0,1)$. Let*

$$B_n = \{y \in B; \lambda^{n+1} \leq |y - x| < \lambda^n\}, \qquad n = 1, 2, \ldots.$$

Then $x \in B^r$ if and only if $\sum_n \lambda^{n(2-d)}C(B_n) = \infty$.

Proof. By Exercise 42,

(5.12) *$x \in B^r$ if and only if $\mathbb{P}^x(T_{B_n} < \infty$ i.o.$) > 0$.*

Suppose the hypotheses of the Borel-Cantelli lemma hold. Then we have

$$\mathbb{P}^x(T_{B_n} < \infty \text{ i.o.}) > 0 \text{ if and only if } \sum_n \mathbb{P}^x(T_{B_n} < \infty) = \infty.$$

Also

$$\mathbb{P}^x(T_{B_n} < \infty) = U\mu_{B_n}(x) = \int_{\overline{B}_n} \frac{c}{|y - x|^{d-2}}\mu_{B_n}(dy).$$

If $y \in B_n$, then $\lambda^{n+1} \leq |y - x| \leq \lambda^n$, so

$$\mathbb{P}^x(T_{B_n} < \infty) \le c\lambda^{n(2-d)} \int_{B_n} \mu_{B_n}(dy) = c\lambda^{n(2-d)}C(B_n),$$

and with a different constant, the reverse inequality also holds. The theorem then follows from (5.12) and Proposition 5.14.

Therefore it suffices to verify the hypotheses of Proposition 5.14. We use the remark following the proof of that proposition. If $z \in B_m$ and $|m - n| > 1$,

$$\mathbb{P}^z(T_{B_n} < \infty) = \int_{B_n} \frac{c}{|z - y|^{d-2}} \mu_{B_n}(dy)$$

$$\le M \int_{B_n} \frac{c}{|x - y|^{d-2}} \mu_{B_n}(dy) = M\mathbb{P}^x(T_{B_n} < \infty),$$

with $M = (\lambda - \lambda^2)^{2-d}$, since if $y \in B_n$, $|z - y| \ge \lambda^n(\lambda - \lambda^2) \ge (\lambda - \lambda^2)|x - y|$.

Then

$$\mathbb{P}^x(T_{B_m} \le T_{B_n} < \infty) = \mathbb{P}^x(T_{B_m} < \infty, T_{B_n} \circ \theta_{T_{B_m}} < \infty)$$

$$= \mathbb{E}^x\left[\mathbb{P}^{X(T_{B_m})}(T_{B_n} < \infty); T_{B_m} < \infty\right]$$

$$\le M\mathbb{P}^x(T_{B_n} < \infty)\mathbb{P}^x(T_{B_m} < \infty).$$

We have the same bound with m and n reversed, and since

$$\mathbb{P}^x(T_{B_m} < \infty, T_{B_n} < \infty) \le \mathbb{P}^x(T_{B_n} < \infty, T_{B_m} \circ \theta_{T_{B_n}} < \infty)$$

$$+ \mathbb{P}^x(T_{B_m} < \infty, T_{B_n} \circ \theta_{T_{B_m}} < \infty),$$

the hypotheses of Proposition 5.14 hold. $\qquad\square$

In order to apply Wiener's test to a concrete situation, we need the capacity of a cylinder. We will restrict attention to $d > 3$, the result being a little easier in this case. Let V_L denote the cylinder $\{(x^1, \ldots, x^d) : 0 \le x^1 \le L, (x^2)^2 + \cdots (x^d)^2 \le 1\}$.

(5.16) Proposition. If $d > 3$ and $L \ge 1$, there exist c_1, c_2 independent of L such that $c_1 L \le C(V_L) \le c_2 L$.

Proof. Since $V_{L_1+L_2}$ can be written as the union of V_{L_1} and $(L_1, 0, \ldots, 0) + V_{L_2}$, using Proposition 5.11 gives

$$C(V_{L_1+L_2}) \le C(V_{L_1}) + C\big((L_1, 0, \ldots, 0) + V_{L_2}\big) = C(V_{L_1}) + C(V_{L_2}).$$

Since $C(V_1) \le C(B(0, d)) < \infty$, this implies the upper bound.

For the lower bound, it suffices to assume L is a positive integer. Let $S_i = V_1 + (i, 0, \ldots, 0)$ and let μ_i be the equilibrium measure for S_i. We have $\mu_i(V_L) = \mu_i(S_i)$, which is a constant independent of i. Note $V_L = \cup_{i=0}^{L-1} S_i$. Let $\nu = \mu_1 + \cdots + \mu_L$. Since $U\mu_j(x) = \mathbb{P}^x(T_{S_j} < \infty)$ and S_j can be embedded in a ball of radius $\sqrt{2}$, then by Proposition I.5.8

$$U\mu_j(x) \le c|x - (j, 0, \ldots, 0)|^{2-d} \wedge 1.$$

If $x \in S_i$,

$$U\nu(x) = \sum_{j=0}^{L-1} U\mu_j(x) \le \sum_{j=0}^{L-1} c(|i - j|^{2-d} \wedge 1).$$

Since $d \ge 4$, the sum on the right can be bounded by a constant c_3 independent of L.

Hence $c_3^{-1}\nu$ is a measure supported on V_L whose potential is bounded by 1 on V_L. By Exercise 23,

$$C(V_L) \ge c_3^{-1}\nu(V_L) = c_3^{-1} \sum_{j=0}^{L-1} \mu_j(V_L) = c_1 L.$$

This is the lower bound. $\qquad\qquad\qquad\qquad\qquad\qquad\qquad\qquad\qquad\square$

Now consider the Lebesgue thorn. Suppose h is strictly increasing, $h(0) = 0$, and $h(r)/r$ is increasing in r for r small. Let $T_h = \{x : x^1 \ge 0, ((x^2)^2 + \cdots + (x^d)^2)^{1/2} \le h(x^1)\}$.

(5.17) Proposition. *Let $d > 3$. 0 is regular for T_h if and only if*

$$(5.13) \qquad\qquad \int_0^1 \left(\frac{h(r)}{r}\right)^{d-3} \frac{dr}{r} = \infty.$$

There is a corresponding integral for $d = 3$: $\int_0^1 |\log(h(r)/r)|^{-1} r^{-1} dr$. For the proof of this case, see Port and Stone [1].

Proof. If $\liminf h(r)/r > 0$, the Poincaré cone condition holds (Proposition 1.13), so we suppose the $\liminf$ is 0. Let $\lambda = 1/2$. Let

$$S_n = \{(x^1, \ldots, x^d) : \left(\frac{4}{3}\right)2^{-n-1} \le x^1 \le \left(\frac{3}{4}\right)2^{-n},$$

$$((x^2)^2 + \cdots + (x^d)^2)^{1/2} \le h(2^{-n-1})\}.$$

Then $S_n \subseteq B_n$ for n large. By scaling (Proposition 5.11), $C(S_n) \ge c2^{-n}(h(2^{-n}))^{d-3}$. If the integral (5.13) is infinite, then

$$\sum 2^{n(d-2)} C(B_n) \ge \sum 2^{n(d-2)} C(S_n) = \infty.$$

Now apply Theorem 5.15. The other direction is similar. $\qquad\qquad\qquad\square$

By varying h and letting $D = B(0, 1) - T_h$, we can find nontrivial examples of domains for which the Dirichlet problem cannot be solved.

Energy

Since electrostatic fields are conservative, the work to move a unit charge from a point x to a point y is the difference in the potentials of y and x. If there are no charges present, the work to move a charge q_1 from infinity to x_1 is zero. If $k - 1$ charges are in place with the ith charge being q_i at location x_i, the work to move a kth charge of size q_k from infinity to x_k is then $cq_k \sum_{i<k} q_i |x_i - x_k|^{-1}$. Therefore the total work to assemble k charges is $c \sum_{i \neq j} q_i q_j / |x_i - x_j|$.

With this in mind, we define

$$(5.14) \qquad E(\mu) = \int \int \mu(dx) u(x,y) \mu(dy) = \int U\mu(y)\mu(dy),$$

the energy of μ. The mutual energy of two measures μ and ν is

$$(5.15) \qquad \langle \mu, \nu \rangle = \int \mu(dx) u(x,y) \nu(dy).$$

The next proposition is essentially the Cauchy-Schwarz inequality for $\langle \cdot, \cdot \rangle$. Since we do not yet know that $E(\mu) \geq 0$ for signed measures μ, the standard proof for Hilbert spaces does not work.

(5.18) Proposition. *If μ and ν are positive measures, then $\langle \mu, \nu \rangle^2 \leq E(\mu)E(\nu)$.*

Proof. Since $p(t, x, y) = \int p(t/2, x, z)\, p(t/2, z, y)\, dz$, then

$$(5.16) \qquad \int P_t \mu(x)\nu(dx) = \int \int \mu(dx) p(t, x, y)\nu(dy)$$

$$= \int P_{t/2}\mu(z) P_{t/2}\nu(z) dz.$$

Hence

$$\langle \mu, \nu \rangle = \int U\mu \, d\nu = \int \int_0^\infty P_t\mu(x)\nu(dx)\, dt$$

$$= \int_0^\infty \int P_{t/2}\mu(z) P_{t/2}\nu(z)\, dz\, dt$$

$$\leq \left(\int_0^\infty \int P_{t/2}\mu(z)^2 dz\, dt \right)^{1/2} \left(\int_0^\infty \int P_{t/2}\nu(z)^2 dz\, dt \right)^{1/2},$$

using the Cauchy-Schwarz inequality. As in the derivation of (5.16), the right-hand side is $E(\mu)^{1/2} E(\nu)^{1/2}$. $\qquad \square$

Using Proposition 5.18, it is immediate that $\langle \cdot, \cdot \rangle$ becomes an inner product on the space of signed finite measures. A natural question, then, is whether the space of signed measures is a Hilbert space. It turns out that the collection of positive measures under $\langle \cdot, \cdot \rangle$ is in fact complete, but the class of finite signed measures is not. (The completion of the latter set is the class of BLD (Beppo-Levi-Deny) distributions.) See Port and Stone [1] and Cartan [1] for proofs.

If μ_B is the equilibrium measure on B, then

$$(5.17) \qquad E(\mu_B) = \int U\mu_B(x)\mu_B(dx) = \int_{B^r} U\mu_B(x)\mu_B(dx)$$
$$= \mu_B(B^r) = C(B).$$

(5.19) Proposition. μ_B *minimizes* $E(\nu)$ *over measures* ν *concentrated on* B^r *such that* $U\nu(x) \geq 1$ *on* B^r.

Proof. If $U\nu \geq 1$ on B^r, then by (3.16),

$$U\nu(x) \geq \mathbb{E}^x[U\nu(X_{T_B}); T_B < \infty] \geq \mathbb{P}^x(T_B < \infty) = U\mu_B(x)$$

for all x. Then

$$E(\nu) = \int U\nu(x)\nu(dx) \geq \int U\mu_B(x)\nu(dx)$$
$$= \int U\nu(x)\mu_B(dx) \geq \int U\mu_B(x)\mu_B(dx) = E(\mu_B).$$

$\square$

There is also a notion of the energy of a function, given by the Dirichlet form. If f is in the domain of $\mathcal{E}(\cdot, \cdot)$ given by (I.3.6), then the energy of f is defined to be

$$\frac{1}{2}\int |\nabla f(x)|^2 dx = \mathcal{E}(f, f).$$

If f is smooth with compact support, Green's first identity (on $\mathbb{R}^d$) shows that the energy of f is the same as the energy of the measure $f(y)\,dy$. Under mild conditions on B,

$$(5.18) \qquad C(B) = \inf\{\mathcal{E}(h, h) : h = 1 \text{ on } B, h \text{ is continuous,}$$
$$h \to 0 \text{ as } |x| \to \infty\},$$

and the minimizing function is $h = U\mu_B$. See Fukushima [1] for further information and proofs.

6 Excessive functions

Basic properties

We start by defining excessive functions and showing their relationship with superharmonic functions. Throughout this section P_t will be defined by (I.3.2).

(6.1) Definition. *A Borel measurable function $f \geq 0$ is excessive on $\mathbb{R}^d$, $d \geq 3$, if for all x, $P_t f(x) \leq f(x)$ and $P_t f(x) \to f(x)$ as $t \downarrow 0$.*

If f is excessive, then in fact $P_t f(x) \uparrow f(x)$ for all x. To see this, observe that if $t > s$, then

$$P_t f(x) = P_s(P_{t-s}f)(x) \leq P_s f(x).$$

(6.2) Proposition. *If f is excessive, there exist $g_n \geq 0$ such that $Ug_n \uparrow f$.*

Proof. Let $g_n = n(f - P_{1/n}f)$. Since f is excessive, $g_n \geq 0$. Then

$$
\begin{aligned}
Ug_n &= n\left(\int_0^\infty P_s f\,ds - \int_0^\infty P_{1/n}P_s f\,ds \right) \\
&= n\left(\int_0^\infty P_s f\,ds - \int_{1/n}^\infty P_s f\,ds \right) = n\int_0^{1/n} P_s f\,ds \uparrow f.
\end{aligned}
$$

$\square$

(6.3) Proposition. *If f is excessive, then f is lower semicontinuous.*

Proof. $p(t, x, y)$ is continuous for $t > 0$, so by Fatou's lemma, $P_t f$ is lower semicontinuous. Also $P_t f \uparrow f$ as $t \downarrow 0$. $\square$

(6.4) Proposition. *If f is excessive, then so is $1 - e^{-f}$.*

Proof. This follows immediately from Jensen's inequality. $\square$

(6.5) Proposition. *If f is excessive and T is a finite stopping time, then $\mathbb{E}^x f(X_T) \leq f(x)$.*

Although (as we shall see) $f(X_t)$ is a supermartingale, we do not yet know that it is right continuous, and we cannot use optional stopping.

Proof.

$$
\mathbb{E}^x Ug_n(X_T) = \mathbb{E}^x \mathbb{E}^{X_T} \int_0^\infty g_n(X_s)ds
$$

$$
= \mathbb{E}^x \int_0^\infty g_n(X_{s+T})ds = \mathbb{E}^x \int_T^\infty g_n(X_s)ds \leq Ug_n(x).
$$

Now let $n \to \infty$ and use monotone convergence. $\square$

(6.6) Definition. *A function f on $\mathbb{R}^d$ is superharmonic if it is locally integrable and*

$$(6.1) \qquad \int_{\partial B(0,r)} f(x+y)\sigma_r(dy) \le f(x)$$

for all x and r, where σ_r is normalized surface measure on $\partial B(0,r)$.

A function f is subharmonic if $-f$ is superharmonic. Consequently, results for superharmonic functions lead to analogs for subharmonic functions.

(6.7) Proposition. *f is excessive if and only if it is superharmonic and lower semicontinuous.*

Proof. Suppose f is excessive. If $S_r = \inf\{t : |X_t - x| \ge r\}$, then by Proposition 6.5 we have $\mathbb{E}^x f(X_{S_r}) \le f(x)$. This is just (6.1). By Proposition 6.3, f is lower semicontinuous.

Now suppose f is superharmonic and lower semicontinuous. We have

$$P_t f(x) = (2\pi t)^{-d/2} \int f(y) e^{-|x-y|^2/2t} dy$$

$$= (2\pi t)^{-d/2} |\partial B(0,1)| \int_0^\infty r^{d-1} e^{-r^2/2t} \int_{\partial B(0,r)} f(x+y)\sigma_r(dy) dr$$

$$\le (2\pi t)^{-d/2} |\partial B(0,1)| \int_0^\infty r^{d-1} e^{-r^2/2t} f(x) dr = f(x).$$

If $f(x) < \infty$, then the fact that f is lower semicontinuous implies that $A = \{y : f(y) > f(x) - \varepsilon\}$ is open, or there exists δ such that $B(x,\delta) \subseteq A$.

$$P_t f(x) = \mathbb{E}^x f(X_t) \ge \mathbb{E}^x[f(X_t); X_t \in B(x,\delta)]$$
$$\ge (f(x) - \varepsilon)\mathbb{P}^x(|X_t - x| \le \delta)$$
$$\to f(x) - \varepsilon$$

as $t \to 0$. Hence $\liminf_{t\to 0} P_t f(x) \ge f(x) - \varepsilon$. Since ε is arbitrary, $P_t f(x) \to f(x)$. If $f(x) = \infty$, a similar argument with $A = \{y : f(y) \ge N\}$ shows $\liminf P_t f(x) \ge N$ for all N, or again $P_t f(x) \to f(x)$. Hence f is excessive. $\square$

(6.8) Proposition. *Suppose $f \in C^2$. Then f is superharmonic if and only if $\Delta f \le 0$.*

Proof. Suppose f is superharmonic. Since f is then excessive, for all $t > s > 0$,

$$(6.2) \qquad \mathbb{E}^x[f(X_t)|\mathcal{F}_s] = \mathbb{E}^{X_s} f(X_{t-s}) = P_{t-s}f(X_s) \le f(X_s),$$

or $f(X_t)$ is a supermartingale. Since f is continuous, $f(X_t)$ is also. If $\Delta f(x) > 0$, then by continuity there exists $r > 0$ such that $\Delta f > 0$ in $B(0, 2r)$. Let $S_r = \inf\{t : |X_t - x| \geq r\}$. By Itô's formula, $f(X_{t \wedge S_r}) - (1/2) \int_0^{t \wedge S_r} \Delta f(X_s) ds$ is a martingale. Then $Y_t = (1/2) \int_0^{t \wedge S_r} \Delta f(X_s) ds$ is a supermartingale, or

$$0 = \mathbb{E}^x Y_0 \geq \mathbb{E}^x Y_{t \wedge S_r}.$$

Let $t \to \infty$. By dominated convergence, $0 \geq \mathbb{E}^x \int_0^{S_r} \Delta f(X_s) ds$, a contradiction. Therefore $\Delta f(x) \leq 0$.

On the other hand, suppose $\Delta f \leq 0$. By Itô's formula, $f(X_t) - (1/2) \int_0^t \Delta f(X_s) ds$ is a martingale, or $f(X_t)$ is a supermartingale. Hence $\mathbb{E}^x f(X_{t \wedge S_r}) \leq f(x)$. Let $t \to \infty$; by dominated convergence we see that f is superharmonic. $\qquad\square$

Path properties

We now want to show that if f is excessive, then $f(X_t)$ has continuous paths.

(6.9) Proposition. *Suppose f is bounded and excessive. If $S = \inf\{t : |f(X_t) - f(x)| > \varepsilon\}$, then $|f(X_S) - f(x)| \geq \varepsilon$, $\mathbb{P}^x$-a.s. on $(S < \infty)$.*

This is not immediate since f is not necessarily continuous.

Proof. Let $A = \{y : f(y) < f(x) - \varepsilon\}$, $B = \{y : f(y) > f(x) + \varepsilon\}$. Let $y \in A^r$ and let J_n be increasing compact sets contained in A so that $T_{J_n} \downarrow T_A = 0$, $\mathbb{P}^y$-a.s. If $g_m \geq 0$ such that $U g_m \leq f$ and $U g_m \uparrow f$,

$$U g_m(y) = \mathbb{E}^y \int_0^{T_{J_n}} g_m(X_s) ds + \mathbb{E}^y [U g_m(X(T_{J_n})); T_{J_n} < \infty]$$

$$\leq \mathbb{E}^y \int_0^{T_{J_n}} g_m(X_s) ds + \mathbb{E}^y [f(X(T_{J_n})); T_{J_n} < \infty]$$

$$\leq \mathbb{E}^y \int_0^{T_{J_n}} g_m(X_s) ds + (f(x) - \varepsilon) \mathbb{P}^y(T_{J_n} < \infty).$$

Letting $n \to \infty$, $\mathbb{P}^y(T_{J_n} < \infty) \to \mathbb{P}^y(T_A < \infty) = 1$ and $T_{J_n} \downarrow T_A = 0$, so $U g_m(y) \leq f(x) - \varepsilon$. Now let $m \to \infty$. Therefore $f(y) \leq f(x) - \varepsilon$ if $y \in A^r$. By Corollary 5.6, $X_{T_A} \in A^r$, $\mathbb{P}^x$-a.s., on $(T_A < \infty)$. Hence $f(X_{T_A}) \leq f(x) - \varepsilon$ on $(T_A < \infty)$.

If $y \in B^r$, take K_n increasing compact sets contained in B so that $T_{K_n} \downarrow T_B = 0$, $\mathbb{P}^y$-a.s. Then by Proposition 6.5,

$$f(y) \geq \mathbb{E}^y [f(X(T_{K_n})); T_{K_n} < \infty] \geq (f(x) + \varepsilon) \mathbb{P}^y(T_{K_n} < \infty)$$
$$\to (f(x) + \varepsilon) \mathbb{P}^y(T_B < \infty) = f(x) + \varepsilon.$$

By Corollary 5.6, $X_{T_B} \in B^r$, $\mathbb{P}^x$-a.s., on $(T_B < \infty)$, so $f(X_{T_B}) \geq f(x) + \varepsilon$ on $(T_B < \infty)$.

Since $S = T_A \wedge T_B$, the proposition follows. $\qquad\square$

(6.10) Theorem. *If f is excessive, $f(X_t)$ is a continuous supermartingale.*

Proof. By Proposition 6.4, we may assume f is bounded. We saw by (6.2) that $f(X_t)$ is a nonnegative supermartingale. By Corollary I.4.16, it has left and right limits on the dyadic rationals. Let $S_0 = 0$, $U_1 = \inf\{t : |f(X_t) - f(X_0)| > \varepsilon\}$, $U_{i+1} = U_i + U_1 \circ \theta_{U_i}$. Since $Y_i = f(X_{U_i})$ is a supermartingale, $f(X_{U_i})$ converges as $i \to \infty$.

We have that $|f(X_{U_{i+1}}) - f(X_{U_i})| \geq \varepsilon$ a.s. on $(U_{i+1} < \infty)$ by the strong Markov property and Proposition 6.9. It follows that $f(X_t)$ must be right continuous.

Fix t_0 and let $Z_t = X_{t_0-t} - X_0$. Z_t is a continuous Gaussian process with independent increments, and $\operatorname{Var}(Z_t - Z_s) = \operatorname{Var}(X_{t_0-t} - X_{t_0-s}) = (t - s)I$. Hence Z_t is also a Brownian motion. By the above, $f(Z_t)$ is right continuous, which implies $f(X_t)$ is left continuous. $\qquad\square$

(6.11) Corollary. *If f and g are excessive, then $f \wedge g$ is excessive.*

Proof. $P_t(f \wedge g) \leq P_t f \leq f$ and $P_t(f \wedge g) \leq P_t g = g$. Hence $P_t(f \wedge g) \leq f \wedge g$ for all t. $(f \wedge g)(X_t)$ is continuous, so $P_t(f \wedge g)(X_t) = \mathbb{E}^x(f \wedge g)(X_t) \to \mathbb{E}^x(f \wedge g)(X_0)$ as $t \to 0$ by dominated convergence. This is $(f \wedge g)(x)$. $\qquad\square$

Riesz decomposition

In Proposition 3.4 we observed that potentials $U\mu$ were excessive. We now show the converse, the Riesz decomposition theorem.

First, we have a definition.

(6.12) Definition. *If $f \geq 0$ is superharmonic on a domain D, a harmonic minorant is a harmonic function h such that $0 \leq h \leq f$ on D. The greatest harmonic minorant is a harmonic function h that is bigger than or equal to all other harmonic minorants of f but is itself a harmonic minorant.*

We need a technical compactness lemma.

(6.13) Lemma. *Suppose f is excessive and the greatest harmonic minorant of f is 0, C is compact, and $\varepsilon > 0$. Then there exists a compact set A such that if ν is a measure with $U\nu \leq f$, then*

$$\int_{A^c} U 1_C(x)\nu(dx) \leq \varepsilon.$$

Here A does not depend on ν.

Proof. Let $h_n(x) = \mathbb{E}^x f(X(\tau_{B(0,n)}))$. Then $f(x) \geq \mathbb{E}^x f(X(\tau_{B(0,n)})) = h_n(x)$ by Proposition 6.5. If $n \geq m$, by the strong Markov property,

$$h_n(x) = \mathbb{E}^x f(X(\tau_{B(0,n)})) = \mathbb{E}^x \mathbb{E}^{X(\tau_{B(0,m)})} f(X(\tau_{B(0,n)}))$$
$$= \mathbb{E}^x h(X_{\tau_{B(0,m)}}) \leq \mathbb{E}^x f(X_{\tau_{B(0,m)}}) = h_m(x).$$

Therefore $h_n(x)$ is decreasing in n. Let h be the limit. Since

$$h_n(x) = |B(x,r)|^{-1} \int_{B(x,r)} h_n(y)\, dy,$$

the same will be true with h_n replaced by h by letting $n \to \infty$ and using dominated convergence. Hence h is harmonic on $B(0,n)$ for all n, hence on $\mathbb{R}^d$. Since the greatest harmonic minorant of f is 0, h is 0, or $\mathbb{E}^x f(X(\tau_{B(0,n)})) \to 0$ as $n \to \infty$.

Take n_0 big so that $C \subseteq B(0, n_0/2)$. By Harnack's inequality, there exists c_1 such that $v(x) \leq c_1 v(y)$ for all v harmonic on $B(0, n_0)$, $x, y \in B(0, n_0/2)$. Take $x_0 \in C$ and take $n \geq n_0$ large enough so that

$$\mathbb{E}^{x_0} f(X(\tau_{B(0,n)})) \leq \varepsilon/(c_1|C|).$$

Then if $x \in C$, $\mathbb{E}^x f(X(\tau_{B(0,n)})) \leq \varepsilon/|C|$. Let $A = \overline{B(0,n)}$.

Suppose now that ν is a measure with $U\nu \leq f$. If $\widehat{\nu} = \nu|_{A^c}$ and $x \in B(0,n) \cap C$, then

$$U\widehat{\nu}(x) = \mathbb{E}^x U\widehat{\nu}(X(\tau_{B(0,n)}))$$
$$\leq \mathbb{E}^x U\nu(X(\tau_{B(0,n)})) \leq \mathbb{E}^x f(X(\tau_{B(0,n)})) \leq \varepsilon/|C|.$$

Hence

$$\int_{A^c} U1_C(x)\nu(dx) = \int\int 1_{A^c}(x)u(x,y)1_C(y)dy\, \nu(dx)$$
$$= \int_C U\widehat{\nu}(y)dy \leq (\varepsilon/|C|)|C| = \varepsilon.$$

$\square$

With this lemma in hand, we can now prove the following.

(6.14) Theorem. (Riesz decomposition) *If f is excessive and $\inf_{\mathbb{R}^d} f = 0$, then there exists μ such that $f = U\mu$.*

Proof. Let h be the greatest harmonic minorant of f. Since $h \geq 0$ and h is harmonic, it is constant. Since $\inf_{\mathbb{R}^d} f = 0$, h is 0.

Let $g_n = n(f - P_{1/n}f)$. Recall that $Ug_n \uparrow f$. Let $\mu_n(dy) = g_n(y)dy$. If B is any ball, for every x, $u(x, \cdot)$ is bounded below on B, so

$$\mu_n(B) \leq c \int_B u(x,y)\mu_n(dy) \leq cU\mu_n(x) \leq cf(x) < \infty.$$

Thus for each ball B, $\mu_n(B)$ is uniformly bounded in n. By a diagonalization procedure and Theorem I.7.4, there exists a subsequence n_j and a measure μ such that for each N, μ_{n_j} restricted to $B(0,N)$ converges weakly to μ restricted to $B(0,N)$. We claim $f = U\mu$.

To show $f \geq U\mu$, for each N let ψ_N be a function that is continuous with $1_{B(0,N)} \leq \psi_N \leq 1_{B(0,N+1)}$. Let $\varepsilon > 0$. If N is sufficiently large, $\int [u(x,y) \wedge N]\psi_N(y)\mu(dy) \geq U\mu(x) - \varepsilon$ by monotone convergence. Then

$$U\mu_{n_j}(x) \geq \int [u(x,y) \wedge N]\psi_N(y)\mu_{n_j}(dy)$$

$$\to \int [u(x,y) \wedge N]\psi_N(y)\mu(dy) \geq U\mu(x) - \varepsilon.$$

It follows that $U\mu_{n_j}(x) = Ug_{n_j}(x) \uparrow f(x)$, or $f(x) \geq U\mu(x)$ since ε is arbitrary. (The case when $U\mu(x)$ is infinite is similar.)

To show $f \leq U\mu$, let C be compact, $\varepsilon > 0$, and A as in Lemma 6.13. Let φ be continuous with compact support, equal to 1 on A, and bounded between 0 and 1. Then

$$(6.3) \qquad \int_C U\mu_{n_j}(x)dx = \int U1_C(x)\mu_{n_j}(dx)$$

$$\leq \int \varphi(x)U1_C(x)\mu_{n_j}(dx) + \varepsilon$$

by Lemma 6.13. Since $U\mu_{n_j} = Ug_{n_j} \uparrow f$, the left-hand side converges to $\int_C f(x)dx$ by monotone convergence. Since $U1_C$ is continuous (Exercise 18) and μ_{n_j} converges weakly to μ, the right-hand side of (6.3) converges to

$$\int \varphi(x)U1_C(x)\mu(dx) + \varepsilon \leq \int U1_C(x)\mu(dx) + \varepsilon = \int_C U\mu(x)dx + \varepsilon.$$

Since ε is arbitrary, $\int_C f(x)dx \leq \int_C U\mu(x)dx$.

Since $f \geq U\mu$ and C is any compact set, we must have $f \leq U\mu$, a.e. Hence $P_t f \leq P_t U\mu$ for all x. Letting $t \to 0$, we see that $f \leq U\mu$ for all x, since both are excessive. This completes the proof. □

Greatest harmonic minorants are only interesting in domains that have nonconstant harmonic functions. A function $f \geq 0$ is excessive on a domain $D \subseteq \mathbb{R}^d$ if $\mathbb{E}^x f(\widehat{X}_t) \leq f(x)$ for all t and x and $\mathbb{E}^x f(\widehat{X}_t) \to f(x)$ as $t \downarrow 0$ for all x, where $\widehat{X}_t$ is Brownian motion killed on exiting D.

(6.15) Theorem. *Suppose f is excessive on D. Let h be the greatest harmonic minorant of f on D. Then there exists μ such that $f = G_D\mu + h$.*

The proof is essentially the same as that of Theorem 6.14, and it will be left for the reader to do. The only significant change is to observe that $f - h$ is excessive on D and $\inf_D(f - h) = 0$; apply the preceding proof to $f - h$.

Balayage

We finish this section with some results about balayage ("sweeping out" in English) and réduite ("reduced function"). Suppose μ is a measure supported on B^c. The balayage problem is to find a measure ν supported on ∂B such that $U\mu = U\nu$ on B.

The electrostatic interpretation of balayage is the following. Suppose B is a conductor that is grounded. Suppose μ is some positive charge distribution in B^c. Since B is grounded, electrons will flow out of the ground onto the surface of B and form some charge distribution ν. Since B is grounded, the charge distribution inside B must be 0, or $U(\mu - \nu) = 0$ in B, or $U\mu = U\nu$ in B.

(6.16) Proposition. *Let μ be a finite measure supported on B^c with finite potential. Then there exists one and only one measure ν concentrated on B^r such that $U\mu = U\nu$ on B^r.*

Proof. Let $\nu(dy) = \int \mathbb{P}^z(X_{T_B} \in dy)\mu(dz)$. By Corollary 5.6, ν is concentrated on B^r. If x is regular for B, then

$$
U\nu(x) = \int \int u(x, y)\mathbb{P}^z(X_{T_B} \in dy)\mu(dz)
$$

$$
= \int \mathbb{E}^z u(x, X_{T_B})\mu(dz) = \int E^x u(z, X_{T_B})\mu(dz)
$$

by (5.2). This is $\mathbb{E}^x U\mu(X_{T_B})$, and since $x \in B^r$, this is just $U\mu(x)$.

As for uniqueness, suppose ν_1 and ν_2 are two such measures. If $y \in B^r$, $u(x, y) = \mathbb{E}^y u(x, X_{T_B})$. Since ν_1 is supported on B^r, we have

$$
U\nu_1(x) = \int \mathbb{E}^y u(x, X_{T_B})\nu_1(dy) = \int E^x u(y, X_{T_B})\nu_1(dy)
$$

$$
= \mathbb{E}^x U\nu_1(X_{T_B}) = \mathbb{E}^x U\mu(X_{T_B}).
$$

A similar equation holds for $U\nu_2(x)$, hence for all x, $U\nu_1(x) = U\nu_2(x)$, or by Proposition 3.5, $\nu_1 = \nu_2$. $\square$

(6.17) Definition. *Let f be excessive.*

$$
R_B(f) = \inf\{v : v \text{ excessive}, v \geq f \text{ on } B\}
$$

is the réduite of f with respect to B.

Let

$$(6.4) \qquad P_B f(x) = \mathbb{E}^x f(X_{T_B}).$$

(6.18) Proposition. *Suppose f is excessive. Then $P_B f$ is the smallest excessive function that dominates f on B^r.*

Proof. First we show $P_B f$ is excessive. Let $g_n = n(f - P_{1/n}f)$.

$$\begin{aligned}
P_B U g_n(x) &= \mathbb{E}^x U g_n(X_{T_B}) = \mathbb{E}^x \int_{T_B}^\infty g_n(X_s)ds \\
&= \mathbb{E}^x \int_0^\infty g_n(X_{s+T_B})ds = \mathbb{E}^x \int_0^\infty \mathbb{E}^{X_s} g_n(X_{T_B})ds \\
&= \mathbb{E}^x \int_0^\infty P_B g_n(X_s)ds = U(P_B g_n)(x).
\end{aligned}$$

Since potentials of functions are excessive, then the last expression is excessive, hence $P_B(U g_n)$ is excessive. Now let $n \to \infty$ and use monotone convergence. The increasing limit of excessive functions is easily checked to be excessive, hence $P_B f$ is excessive.

Clearly $P_B f = f$ on B^r. If v is excessive and $v \geq f$ on B^r, then

$$P_B f(x) = \mathbb{E}^x f(X_{T_B}) \leq \mathbb{E}^x v(X_{T_B}) \leq v(x).$$

$\square$

$P_B f$ is not the réduite of f since the réduite need not be excessive. It is supermeanvalued: $P_t(R_B f) \leq R_B f$. If we let $\overline{R}_B f$ be the excessive regularization of $R_B f$, namely, $\overline{R}_B f = \lim_{t \downarrow 0} P_t(R_B f)$, then $\overline{R}_B f$ is excessive and $\overline{R}_B f = P_B f$. Exercise 33 asks the reader to supply the proof of this.

The following representation of harmonic functions is often useful. In particular, we will use it in the next section on Martin boundary.

(6.19) Proposition. *Suppose h is positive and harmonic in D, B is open, and $\overline{B} \subseteq D$. Then there exists μ supported on ∂B such that $h = G_D \mu$ in B.*

Proof. Let $\widehat{X}_t$ be X_t killed on exiting D. Let $f(x) = \mathbb{E}^x h(\widehat{X}_{T_B})$. As in the proof of Proposition 6.18, f is excessive on D. $f = 0$ on the points of ∂D that are regular for D^c, so the greatest harmonic minorant of f in D is 0. By Theorem 6.14, $f = G_D \mu$ for some μ supported in D. Since $f = h$ on B, f is harmonic in B, and by Exercise 34, μ is supported in B^c. Similarly f is harmonic in the interior of B^c, hence μ is concentrated on the closure of B. Therefore μ is concentrated on ∂B. (Alternately, let μ' be the balayage of μ onto B. Then μ is supported on ∂B and for $x \in B$, $h = G_D \mu = G_D \mu'$.) $\square$

It would be interesting to understand the correspondence between h and μ better. Is there a more direct interpretation of μ in terms of h?

7 Martin boundary

Construction

We saw in Propositions 1.21 and 1.22 that every positive harmonic function on $D = B(0,1)$ could be written

$$h(x) = \int_{\partial D} P_B(x,y)\mu(dy)$$

for some positive measure μ on ∂D. We would like a similar formula for more general domains. In general the boundary one needs cannot be the Euclidean boundary. For example, if we take the set in the plane $D = B(0,1) - \{(x,0) : x \geq 0\}$, each point on the positive x-axis corresponds to two harmonic functions. We will not prove this here, but the intuitive notion is that Brownian motion could hit the positive x-axis from above or below. (To prove that there are two harmonic functions corresponding to each x on the positive axis, use a conformal mapping of the slit disk onto the unit ball; see Chap. V.) On the other hand, if $C_n = \{1/n\} \times [0, 1 - e^{-1/n}]$ and $D = (0,1)^2 - \cup_{n=1}^{\infty} C_n$, then the set of boundary points $\{0\} \times [0,1]$ corresponds to a single harmonic function. Again conformal mapping could be used to prove this; the intuitive notion is that if we start Brownian motion at $(0,0)$ and condition it to stay inside D, it must move up the y-axis infinitely fast and thus cannot distinguish between the points of the y-axis.

The first idea one might try is to let D_n be an increasing sequence of smooth domains with $\overline{D}_n \subseteq D_{n+1}$, $\overline{D}_n \subseteq D$, $D_n \uparrow D$. If h is harmonic in D, we know by Proposition 6.19 that $h(x) = \int g_D(x,y)\mu_n(dy)$ in D_n for a measure μ_n supported on ∂D_n. The hope is that μ_n does not become infinite as $n \to \infty$. Note that as $y \to \partial D$, $g_D(x,y) \to 0$. Since $h(x) \neq 0$, we must have $\mu_n(\mathbb{R}^d) \to \infty$, and our hope is not realized.

Martin's idea was the following. Let x_0 be fixed in D.

$$h(x) = \int g_D(x,y)\mu_n(dy) = \int \frac{g_D(x,y)}{g_D(x_0,y)} g_D(x_0,y)\mu_n(dy).$$

If we let $\nu_n(dy) = g_D(x_0,y)\mu_n(dy)$, then $\nu_n(\mathbb{R}^d) = \int g_D(x_0,y)\mu_n(dy) = h(x_0) < \infty$. Now we can hope that a subsequence of the ν_n converges weakly, say to ν. If the ratios $g_D(x,y)/g_D(x_0,y)$ were continuous for $y \in \overline{D} - \{x_0, x\}$, we would obtain in the limit $h(x) = \int g_D(x,y)/g_D(x_0,y)\,\nu(dy)$.

However, the ratios need not be continuous up to the boundary, and this is the first obstacle to overcome.

Let

$$(7.1) \qquad\qquad M(x,y) = \frac{g_D(x,y)}{g_D(x_0,y)}.$$

$M(x,y)$ is continuous for $y \in D - \{x\}$, but we need it to be continuous on a closed set containing D. To achieve this we compactify D appropriately.

Let us assume D is connected. Otherwise we deal with each component separately. A function $f : D \to [-\infty, \infty]$ is continuous in the extended value sense if it is continuous with respect to the topology on $[-\infty, \infty]$ generated by the open sets in $\mathbb{R}$ together with the sets of the form $[-\infty, -a)$ and $(a, \infty]$.

(7.1) Theorem. (Constantinescu-Cornea) *Let E be locally compact but not compact and $\mathcal{G}$ a family of functions continuous in the extended value sense. Then there exists a compact set E^*, unique up to homeomorphisms, such that E is a dense subset of E^*, each $f \in \mathcal{G}$ has a continuous extension to E^*, and these extended functions separate points of $E^* - E$. If E is Hausdorff, then so is E^*.*

If $E = (0, 1]$ and $\mathcal{G} = \{\sin(1/x)\}$, E^* is definitely not $[0, 1]$. Rather this compactification is similar to the Stone-Čech compactification.

Proof. Enlarge $\mathcal{G}$ by including all continuous functions on E that have compact support contained in E. Let R_f denote the range of f (R_f could include the points $-\infty$ and ∞). Let $A = \prod_{f \in \mathcal{G}} \overline{R}_f$. A is compact since it is the product of compact sets. Let I be the mapping of E into A such that the f-coordinate of $I(x)$ is $f(x)$, and let E^* be the closure of $I(E)$ in A. Note each $f \in \mathcal{G}$ is equal to $\pi_f \circ I$, where π_f is the projection onto the f-coordinate. Hence $\pi_f \circ I$ is the extension of $f \in \mathcal{G}$ to E^*. Since projections are continuous, the extension is continuous. $I(E)$ is clearly dense in E^*. If x and y are two distinct points in $E^* - I(E)$, there must be $f \in \mathcal{G}$ such that $\pi_f(x) \neq \pi_f(y)$. Then π_f is the extension of a function in $\mathcal{G}$ that separates x and y. If $x, y \in E^*$, $x \neq y$, and E is Hausdorff, there exists π_f separating x and y (here we are not assuming $x, y \notin I(E)$, which is why we need E to be Hausdorff). So if $a = (\pi_f(x) + \pi_f(y))/2$, then $(\pi_f > a)$ and $(\pi_f < a)$ are two open sets that separate x and y, hence E^* is Hausdorff. If we now identify E and $I(E)$, we have the existence of the desired compactification E^*.

As for the uniqueness (which, by the way, we do not need in what follows), that will be left for the reader to do as Exercise 41. $\square$

Now let $\mathcal{G} = \{M(x, \cdot) : x \in D\}$. Each $M(x, \cdot)$ is continuous in the extended sense on D. Embed D into D^* using Theorem 7.1. Let D_n be subdomains of D increasing up to D with $\overline{D}_n \subseteq D_{n+1}$ for all n.

(7.2) Theorem. *Every positive harmonic function on D can be represented as*

$$(7.2) \qquad h(x) = \int M(x,y)\nu(dy)$$

for some measure ν concentrated on $D^* - D$.

$D^* - D$ is called the Martin boundary, and we will denote it by $\partial_M D$. Since D^* is metrizable (see Exercise 37 for an example of a metric on D^*), the results of Sect. I.7 on weak convergence are applicable.

Proof. Fix x_0. If $x \in D_n$,

$$h(x) = \int g_D(x,y)\mu_n(dy) = \int M(x,y)\nu_n(dy),$$

where $\nu_n(dy) = g_D(x_0,y)\mu_n(dy)$ and μ_n is a measure supported on ∂D_n, by Proposition 6.19.

$$\nu_n(\mathbb{R}^d) = \int g_D(x_0,y)\mu_n(dy) = h(x_0) < \infty.$$

The ν_n are measures on D^* with total mass uniformly bounded by $h(x_0)$. D^* is compact, hence by Theorem I.7.4 (cf. proof of Proposition 1.22) there exists a subsequence that converges, say to ν. The measure ν is supported on D^* and clearly $\nu(D_n) = 0$ for each n. Hence ν is supported on $\partial_M D$. Since $M(x,y)$ is continuous on D^*, for n large so that $x \in D_n$,

$$h(x) = \int M(x,y)\nu_n(dy) \to \int M(x,y)\nu(dy).$$

$\square$

Conversely, we have the following proposition.

(7.3) Proposition. *If ν is a finite measure on $\partial_M D$ and we set $h(x) = \int M(x,y)\nu(dy)$, then h is harmonic in D.*

Proof. If $y \in D$, $M(x,y) = g_D(x,y)/g_D(x_0,y)$ is clearly a harmonic function of x in $D - \{y\}$. Now suppose $y \in \partial_M D$. Take $y_n \in D$ converging to y with respect to the topology of D^*. Fix $x \in D$ and let $r < \mathrm{dist}\,(x,\partial D)/4$. $M(x_0,y_n) = 1$ for all n, and by Harnack's inequality (Theorem 1.19), $M(\cdot,y_n)$ is bounded on $B(x,2r)$ by a constant independent of n, provided only that n is large enough so that $y_n \notin B(x,3r)$. Since $M(\cdot,y_n)$ is harmonic on $B(x,2r)$,

$$(7.3) \qquad M(x,y_n) = \frac{1}{|B(x,r)|} \int_{B(x,r)} M(z,y_n)\,dz.$$

Let $n \to \infty$. By the definition of D^*, $M(z, y_n) \to M(z, y)$. By the preceding paragraph, the convergence takes place boundedly over $z \in B(x, 2r)$. So using dominated convergence, we have (7.3) with y_n replaced with y. Hence $M(x, y)$ is a harmonic function of x.

If ν is a finite measure on $\partial_M D$ and h is defined as above, we have h satisfying (1.1). This follows by Fubini's theorem from (7.3) with y_n replaced by y. h is finite for all x by Harnack's inequality and the fact that $h(x_0) = \int M(x_0, y)\, \nu(dy) = \nu(\partial_M D) < \infty$. $\qquad\square$

Since $M(x_0, y) = 1$ for $y \in D^*$, Harnack's inequality tells us that $M(x, y) > 0$ for all $x \in D - \{y\}$.

Minimal Martin boundary

Although we have constructed $\partial_M D$, this actually is not the right boundary. The difficulty is that the representation (7.2) is not unique. A given h might be representable by more than one measure ν. To avoid this, we construct the minimal Martin boundary.

(7.4) Definition. *A positive harmonic function is minimal harmonic if v is harmonic and $0 \le v \le u$ implies that $v = cu$ for some constant c.*

(7.5) Proposition. *If u is minimal harmonic, then $u = cM(x, y)$ for some $y \in \partial_M D$.*

Proof. $u(x) = \int M(x, y)\mu(dy)$ for some measure μ. Let y_0 be a point of density for μ, that is, a point y_0 such that $\mu(V) \neq 0$ for every open neighborhood V of y_0. Since $\int M(x, y)\mu|_V(dy) \le u$, there exists λ such that $\int_V M(x, y)\mu(dy) = \lambda u(x)$ for all x.

From the definition, $M(x_0, y) = 1$, hence $\lambda u(x_0) = \int_V \mu(dy)$. Therefore

$$
\int_V \frac{M(x, y)}{M(x, y_0)}\mu(dy) = \frac{1}{M(x, y_0)} \int_V M(x, y)\mu(dy)
$$
$$
= \frac{\lambda u(x)}{M(x, y_0)} = \frac{\int_V \mu(dy)}{u(x_0)} \frac{u(x)}{M(x, y_0)}.
$$

Divide both sides by $\mu(V)$ and let $V \downarrow \{y_0\}$. Since $M(x, y)/M(x, y_0) \to 1$ as $y \to y_0$, on the left we obtain 1. Hence $u(x) = u(x_0)M(x, y_0)$. $\qquad\square$

The set of positive harmonic functions form a cone $\mathcal{C}$. ($\mathcal{C}$ is a cone if $h_1, h_2 \in \mathcal{C}$ implies $h_1 + h_2 \in \mathcal{C}$ and $ah_1 \in \mathcal{C}$ if $a > 0$.) The set $\mathcal{K}$ of positive harmonic functions with $h(x_0) = 1$ is the intersection of $\mathcal{C}$ with the hyperplane $\{h : Lh = 1\}$, where L is the linear functional mapping h to $h(x_0)$. Clearly $\mathcal{K}$ is convex.

(7.6) Proposition. *The extreme points of $\mathcal{K}$ are the set of $M(x, y)$ such that $M(\cdot, y)$ is minimal harmonic.*

Proof. Let us first show that if u is an extreme point of $\mathcal{K}$, then u is minimal harmonic. This follows because if $v \le u$, then $u - v$ is positive harmonic and

$$u(x) = t\frac{v(x)}{v(x_0)} + (1-t)\frac{(u-v)(x)}{(u-v)(x_0)},$$

where $t = v(x_0)$. Since u is an extreme point, we must have $v(x)/v(x_0) = u(x)$, or $v(x) = tu(x)$.

On the other hand, if u is minimal harmonic and $u = tv + (1-t)w$ for some $t \in (0, 1)$, then $tv \le u$, hence $tv = cu$. Since $v(x_0) = u(x_0) = 1$, $t = c$ and $u = v$. So $w = u$ and u is extremal. $\square$

We define the minimal Martin boundary, denoted $\partial_m D$, to be the set of $y \in \partial_M D$ such that $M(x, y)$ is minimal harmonic.

A lattice is a partially ordered set (ordered by $\le$) such that any two elements have a least upper bound and a greatest lower bound. A vector lattice is one where $x \le y$ implies $x + z \le y + z$ if $z \in \mathcal{C}$ and $ax \le ay$ if $a \ge 0$.

(7.7) Proposition. *$\mathcal{K}$ is compact and metrizable. $\mathcal{C}$ is a vector lattice.*

Proof. Let D_n be a sequence of domains increasing up to D, with $\overline{D}_n \subseteq D_{n+1}$. Let $\rho(h, g) = \sum_n 2^{-n} \sup_{D_n} (|h(x) - g(x)| \wedge 1)$. This is easily seen to be a metric on $\mathcal{K}$. Since $h(x_0) = 1$ if $h \in \mathcal{K}$, then for each n, h is bounded on D_{n+1} by Harnack's inequality and by Corollary 1.4, equicontinuous on D_n. Given a sequence $h_m \in \mathcal{K}$, there exists by diagonalization a subsequence that converges uniformly on each $\overline{D}_n$. The limit will be harmonic on each D_n (cf. the proof of Lemma 6.13), hence on D, will equal 1 at x_0, and is therefore a point of $\mathcal{K}$. Thus every sequence in $\mathcal{K}$ has a convergent subsequence. Since $\mathcal{K}$ is metrizable, this shows that $\mathcal{K}$ is compact.

If g and h are harmonic functions in $\mathcal{C}$, then Harnack tells us that $g + h$ is bounded on each D_n. Let $u_m(x) = \mathbb{E}^x (g \vee h)(X(\tau_{D_m}))$ for $x \in D_m$. As above, this is equicontinuous on D_n if $m > n+1$, hence there is a harmonic subsequential limit, say u. We claim u is the least upper bound for g and h. It is clear that u is bigger than both g and h on each D_m, hence on D. Suppose v is also an upper bound and $v(x) < u(x)$ for some x. Then for m sufficiently large, $v(x) < u_m(x)$. Since v is greater than or equal to both g and h on ∂D_m, then $v(x) \ge \mathbb{E}^x (g \vee h)(X(\tau_{D_m})) = u_m(x)$, a contradiction. The existence of a greatest lower bound is similar. The vector part of the definition is trivial. $\square$

(7.8) Definition. *Let us say that $h \in \mathcal{K}$ is the barycenter (or center of mass) of a probability measure μ on $\mathcal{K}$, written $h = b(\mu)$, if for all linear functionals L on $\mathcal{K}$,*

$$L(h) = \int_K L(k)\,\mu(dk).$$

(7.9) Proposition. *Every probability μ on $\mathcal{K}$ has one and only one barycenter. There exist probabilities μ_n on $\mathcal{K}$, converging weakly to μ, such that each has finite support and barycenter $b(\mu)$.*

Proof. Since $\mathcal{K}$ is compact, for each n we can cover $\mathcal{K}$ by finitely many balls B_i of radius less than $1/n$. Let $A_1 = B_1$ and for $i > 1$, let $A_i = B_i - \cup_{j<i} A_j$. Discard any A_i that is empty. Pick $k_i \in A_i$ and let $\nu_n = \sum_i \mu(A_i)\delta_{k_i}$. Each ν_n has finite support and it is easy to see that ν_n converges weakly to μ. Let $h_n = \sum_i \mu(A_i)k_i$. Then h_n is the barycenter of ν_n, and hence

$$(7.4) \qquad\qquad L(h_n) = \int L(k)\nu_n(dk)$$

for all linear functionals L. Replacing ν_n by a subsequence if necessary, we may assume h_n converges, say to h. Since L is continuous, letting $n \to \infty$ in (7.4) shows that h is the barycenter of μ.

For each n let $\mu_n = \sum_i \mu(A_i)\delta_{b_i}$, where b_i is the barycenter of $\mu(A_i)^{-1}\mu|_{A_i}$. Then

$$L(b(\mu_n)) = \int L(k)\mu_n(dk) = \sum_i \mu(A_i) \int L(k)\frac{\mu|_{A_i}(dk)}{\mu(A_i)}$$

$$= \int L(k)\mu(dk) = L(b(\mu)),$$

or the μ_n have the same barycenter as μ.

As for the uniqueness, suppose h_1, h_2 are distinct points of $\mathcal{K}$. Since $h_1(x_0) = h_2(x_0) = 1$, they must be linearly independent. Define a linear functional L on the subspace spanned by h_1, h_2 by using linearity and setting $L(h_1) = 1$ and $L(h_2) = 2$. By the Hahn-Banach theorem, we can extend L to all of $\mathcal{C} - \mathcal{C}$. Then h_1 and h_2 cannot both be barycenters of μ, since otherwise we would have $L(h_1) = \int L(k)\mu(dk) = L(h_2)$, a contradiction. $\qquad\square$

We can now apply Choquet's theorem. (The uniqueness part is often called the Choquet-Meyer theorem.)

(7.10) Theorem. (Choquet)
(a) *If $\mathcal{K}$ is convex, compact, and metrizable, and $x \in \mathcal{K}$, there exists a probability measure μ supported on the extreme points of $\mathcal{K}$ such that $x = b(\mu)$.*
(b) *If, in addition, $\mathcal{K}$ is the intersection of a cone $\mathcal{C}$ with a hyperplane and $\mathcal{C}$ is a vector lattice, then μ is unique.*

We will give a complete proof of Theorem 7.10 after the end of this subsection.

(7.11) Theorem. *There is a one-to-one correspondence between positive harmonic functions h with $h(x_0) = 1$ and measures μ supported on the minimal Martin boundary given by*

$$h(x) = \int_{\partial_m D} M(x, y)\mu(dy).$$

Proof. Since Proposition 7.7 shows that $\mathcal{K}$ and $\mathcal{C}$ satisfy the hypotheses of Theorem 7.10, each positive harmonic function h with $h(x_0) = 1$ is the barycenter of one and only one measure μ supported on the extreme points of $\mathcal{K}$. By Proposition 7.6, we saw that the extreme points are the minimal harmonic functions. For each x the functional L_x defined by $L_x(h) = h(x)$ is a linear functional, hence $h(x) = L_x(h) = \int L_x(M(\cdot, y))\mu(dy) = \int M(x, y)\mu(dy)$. □

The Martin boundary is, as we shall see in Chap. III, well suited to absorbing Brownian motion and the Dirichlet problem. For reflecting Brownian motion and the Neumann problem, the Kuramochi boundary is more appropriate.

Let D be a bounded domain, K a compact subset of D with nonempty interior, and $g^N(x, y)$ the Green function for reflecting Brownian motion that has normal reflection on ∂D and is killed at time T_K (it takes some work to construct this process and to prove the existence of g^N; see Fukushima [1]). The compactification of D so that each $g^N(x, \cdot)$ has a continuous extension is called the Kuramochi compactification. As with the Martin boundary, there is a minimal Kuramochi boundary and an integral representation of positive harmonic functions. See Bass and Hsu [1] or Ohtsuka [1] for details.

Existence of extremal measures

In this subsection we prove the existence part of Theorem 7.10. Let $\mathcal{S}$ be the set of continuous concave functions on $\mathcal{K}$. $\mathcal{S}$ is closed under the operation $\wedge$ (the operation of taking the greatest lower bound) and is closed under addition. By Exercise 39, $\mathcal{S} - \mathcal{S}$ is closed under $\wedge$ and $\vee$. $\mathcal{S} - \mathcal{S}$ contains constants, and since it contains linear functions, it separates points. By the Stone-Weierstrass theorem, $\mathcal{S} - \mathcal{S}$ is dense in $C(\mathcal{K})$.

Let us write $\lambda \prec \mu$ if $\int f d\lambda \geq \int f d\mu$ for all $f \in \mathcal{S}$, where μ and λ are probability measures on $\mathcal{K}$. The idea of the existence of an extremal measure is the following. If x is not extremal, it can be written as $\sum_i a_i x_i$. Any x_i that is not an extreme point has a similar representation. The measure $\sum_i a_i \delta_{x_i}$ is "closer" to the boundary than δ_x and we will see this means

$\sum_i a_i \delta_{x_i} \succ \delta_x$. The desired representation will come if we find the measure that is maximal with respect to $\prec$.

(7.12) Proposition. *There exists μ such that $\delta_x \prec \mu$ and $\mu \succ \lambda$ whenever $\lambda \succ \delta_x$.*

Proof. We use Zorn's lemma. Suppose I is a totally ordered set and μ_i is a probability measure on $\mathcal{K}$ for each $i \in I$ with $\mu_i \prec \mu_j$ if $i < j$. If $f \in \mathcal{S}$, then $\int f d\mu_i$ decreases as i increases. Let $L(f)$ denote the limit. In this context, this means that given $\varepsilon > 0$, there exists $i_0 \in I$ such that $|L(f) - \int f d\mu_i| < \varepsilon$ if $i > i_0$. Because $\int f d\mu_i$ decreases in i, it is easy to see that $L(f) = \inf_{i \in I} \int f d\mu_i$. Define L on $\mathcal{S} - \mathcal{S}$ by $L(f) = \lim_i \int f d\mu_i$. Because all the μ_i have total mass 1, L is a bounded linear operator, and we can then extend L to $C(\mathcal{K})$, the continuous functions on $\mathcal{K}$. By the Riesz representation theorem, there exists a measure μ such that $L(f) = \int f d\mu$ for all $f \in C(\mathcal{K})$. If $f \in \mathcal{S} - \mathcal{S}$ is nonnegative, $L(f) = \lim_i \int f d\mu_i \geq 0$, or μ is a positive measure. Since $L(1) = 1$, μ is a probability. Because $\int f d\mu = L(f) = \inf_i \int f d\mu_i$ if $f \in \mathcal{S}$, $\mu \succ \mu_i$ for all $i \in I$. Thus μ is an upper bound for the μ_i. By Zorn's lemma, then, $\{\lambda : \lambda \succ \delta_x\}$ has a maximal element. $\square$

We must now show that μ has barycenter x and that μ is supported on the extreme points of $\mathcal{K}$.

(7.13) Proposition. *If $\mu \succ \delta_x$, then $b(\mu) = x$.*

Proof. Suppose L is any linear functional on $\mathcal{K}$. Then $L \in \mathcal{S}$ and $L \in -\mathcal{S}$. Thus $\int L d\mu \leq \int L d\delta_x$ and $\int (-L) d\mu \leq \int (-L) d\delta_x$, or $\int L d\mu = \int L d\delta_x$ for all L linear. That implies $b(\mu) = x$. $\square$

For the verification that μ is supported on the extreme points, we need a version of the Hahn-Banach theorem.

(7.14) Definition. *P is a sublinear functional on E if $x, y \in E$ and $\lambda \geq 0$ implies $P(\lambda x) = \lambda P(x)$ and $P(x + y) \leq P(x) + P(y)$.*

(7.15) Theorem. *If P is sublinear on E and L is a linear functional on a subspace F of E such that $L(x) \leq P(x)$ for all $x \in F$, then L can be extended to a linear functional on E such that $L(x) \leq P(x)$ for all $x \in E$.*

Proof. Suppose L has been extended to a subspace H of E. If H is not E itself, there exists $a \in E - H$ and we show how to extend L to H', the subspace spanned by H and $\{a\}$. For $x, y \in H$,

$$L(x) + L(y) = L(x+y) \leq P(x+y) = P((x-a)+(y+a)) \leq P(x-a) + P(y+a).$$

So for all $x, y \in H$,

$$L(x) - P(x - a) \leq P(y + a) - L(y).$$

Choose λ so that

$$\sup_{x \in H} [L(x) - P(x - a)] \le \lambda \le \inf_{y \in H} [P(y + a) - L(y)].$$

Define $L(x + ra) = L(x) + r\lambda$ if $x \in H$. If we set $y = x/r$,

$$L(y + a) = L(y) + \lambda \le P(y + a),$$

so $L(x + ra) = rL(y + a) \le rP(y + a) = P(x + ra)$ if $r \ge 0$. Similarly, $L(y - a) = L(y) - \lambda \le P(y - a)$, so $L(x - ra) = rL(y - a) \le rP(y - a) = P(x - ra)$ if $r \ge 0$. We have thus extended L from H to H'.

We now use Zorn's lemma. Suppose I is a totally ordered subset and let $\{(H_i, L_i) : i \in I\}$ be a collection of pairs such that each H_i is a subspace containing F and each L_i is an extension of L to H_i and such that $H_i \subseteq H_j$ and L_j is an extension of L_i if $i \le j$. If we set $H_I = \cup_i H_i$ and define L_I by $L_I(x) = L_i(x)$ if $x \in H_i$, then (H_I, L_I) is an upper bound for our collection. By Zorn's lemma, there is a maximal subspace H on which an extension of L can be defined. By the above, H must be all of E or else we could find H' larger than H on which L can be defined. $\qquad \square$

If $f \in C(\mathcal{K})$, the continuous functions on $\mathcal{K}$, let

(7.5) $$\widetilde{f} = \inf\{g \in \mathcal{S}, g \ge f\}.$$

$\widetilde{f}$ is the least concave majorant of f. Note $f \le \widetilde{f}$ and that $\widetilde{f}$ is bounded, since the constant function $g \equiv \sup_{\mathcal{K}} f$ is continuous and dominates f.

The existence part of Theorem 7.10 is completed by the following.

(7.16) Proposition. *If μ is maximal with respect to $\succ$, then μ is supported on the extreme points of $\mathcal{K}$.*

Proof. Suppose μ is maximal. We first show

(7.6) $$\int f d\mu = \int \widetilde{f} d\mu$$

for all $f \in C(\mathcal{K})$. Let $E = C(\mathcal{K})$ and define

(7.7) $$P(f) = \int \widetilde{f} d\mu.$$

Since $\widetilde{f + g} \le \widetilde{f} + \widetilde{g}$, P is clearly sublinear. Suppose $\int f d\mu \ne \int \widetilde{f} d\mu$ for some $f \in C(\mathcal{K})$. Since $\mathcal{S} - \mathcal{S}$ is dense in $C(\mathcal{K})$, there exists $f \in \mathcal{S}$ such that $\int f d\mu \ne \int \widetilde{f} d\mu$. Let $F = \{cf : c \in \mathbb{R}\}$ and let $L(f) = P(f) = \int \widetilde{f} d\mu$.

$0 \le P(f) + P(-f)$ by sublinearity, so $L(-f) = -P(f) \le P(-f)$, or L is dominated by P on F. Use Theorem 7.15 to extend L to E. Since L is a linear functional on $C(\mathcal{K})$, there exists a measure ν such that $Lg = \int g d\nu$ for all $g \in C(\mathcal{K})$ by the Riesz representation theorem. We claim ν is a probability. $L(1) \leq P(1) = 1$, and $-L(1) = L(-1) \leq P(-1) = -1$, or $L(1) = 1$. If $g \geq 0$, $-L(g) = L(-g) \leq P(-g) = \int(-g)d\mu \leq 0$, or $L(g) \geq 0$. Thus ν is a positive measure with total mass 1, which proves the claim. If $h \in \mathcal{S}$, then $\tilde{h} = h$ and $\int h\, d\nu = L(h) \leq P(h) = \int \tilde{h}\, d\mu = \int h\, d\mu$, or $\nu \succ \mu$. Moreover, $\int f\, d\nu = L(f) = P(f) = \int \tilde{f}\, d\mu > \int f\, d\mu$, or $\nu \neq \mu$. This contradicts the maximality of μ. Therefore $\int f\, d\mu = \int \tilde{f}\, d\mu$ for all $f \in C(\mathcal{K})$.

Since (7.6) holds and $f \leq \tilde{f}$, μ must be concentrated on $B_f = \{x \in \mathcal{K} : f(x) = \tilde{f}(x)\}$ for all $f \in -\mathcal{S}$. Since $\mathcal{K}$ is metrizable, $C(\mathcal{K})$ has a countable dense subset. Since $\mathcal{S}-\mathcal{S}$ is dense in $C(\mathcal{K})$, we can find a countable sequence $f_n \in -\mathcal{S}$ that separate points. We may normalize so that $\|f_n\|_\infty \leq 1$. Let $f = \sum_n (f_n)^2/2^n$. Then f is convex, continuous, and also strictly convex. Since $\tilde{f}$ is concave, $\tilde{f}(x) > f(x)$ for all x in $\mathcal{K}$ that are not extremal. So B_f is contained in the set of extreme points of $\mathcal{K}$, which proves the proposition.

$\square$

Uniqueness of extremal measures

For the uniqueness part of Theorem 7.10, we assume $\mathcal{K}$ is the intersection of a cone $\mathcal{C}$ with a hyperplane. There is no loss of generality in making that assumption, for we can let $\mathcal{C} = \{(t, x) : t \in (0, \infty), x \in t\mathcal{K}\}$, and we can define L_0 on $\mathcal{C}$ by $L_0(t, x) = t$. Then $\mathcal{K}$ is the intersection of $\mathcal{C}$ with the hyperplane $\{y : L_0(y) = 1\}$. For $x, y \in \mathcal{C}$, we say $x \leq y$ if $y - x \in \mathcal{C}$. For the uniqueness part we need a decomposition theorem for vector lattices.

(7.17) Theorem. (Decomposition lemma) *Suppose $x + y = \sum_{i=1}^{n} z_i$, where $x, y, z_1, \ldots, z_n \in \mathcal{C}$. Then there exist $x_1, \ldots, x_n$, $y_1, \ldots, y_n \in \mathcal{C}$ with $z_i = x_i + y_i$ and $x = \sum_{i=1}^{n} x_i$, $y = \sum_{i=1}^{n} y_i$.*

Proof. Let us do the case $n = 2$ first. Let $x_1 = x \wedge z_1$. Since $x_1 \leq x$, then setting $x_2 = x - x_1$, we see that $x_2 \in \mathcal{C}$. Similarly, setting $y_1 = z_1 - x_1$, we have $y_1 \in \mathcal{C}$.

Suppose we show

(7.8) $$y \geq y_1.$$

Then $y - y_1 \in \mathcal{C}$, and if we let $y_2 = y - y_1$, we then have

$$x = x_1 + x_2, \qquad z_1 = x_1 + y_1, \qquad y = y_1 + y_2$$

and

$$x_2 + y_2 = (x - x_1) + (y - y_1) = (x + y) - (x_1 + y_1) = (x + y) - z_1 = z_2.$$

So we need to show (7.8).

Take $a = x$, $b = z_1$ and $c = y$ in Exercise 39. Then

$$(x \wedge z_1) + y = (x + y) \wedge (y + z_1) = z \wedge (y + z_1).$$

Since $z_1 \leq z$ and $z_1 \leq y + z_1$, then $z_1 \leq z \wedge (y + z_1) = (x \wedge z_1) + y$, or $y \geq z_1 - (x \wedge z_1) = z_1 - x_1 = y_1$, which is (7.8).

The general case follows by induction. If $x + y = z_1 + \cdots + z_{n+1}$, let $Z_1 = z_1 + \cdots + z_n$, $Z_2 = z_{n+1}$. Hence there exists $X_1, X_2, Y_1,$ and Y_2 such that $Z_1 = X_1 + Y_1$, $Z_2 = X_2 + Y_2$, $x = X_1 + X_2$, and $y = Y_1 + Y_2$. Let $x_{n+1} = X_2$, $y_{n+1} = Y_2$ and use the induction hypothesis on $X_1 + Y_1 = z_1 + \cdots + z_n$ to deduce the result for $n + 1$. □

We need two lemmas concerning $P(f)$. Recall the definition of $P(f)$ and $\tilde{f}$ given in (7.5) and (7.7).

(7.18) Lemma. $P(f) = \inf_{g \in \mathcal{S}, g \geq f} \int g d\mu$.

Proof. If $g \geq f$, then $P(f) = \int \tilde{f} d\mu \leq \int g d\mu$, so

$$P(f) \leq \inf_{g \in \mathcal{S}, g \geq f} \int g d\mu.$$

To prove the other direction, we need to show

$$(7.9) \qquad \inf \int g d\mu \leq \int (\inf g) d\mu$$

where the infimum is over $g \in \mathcal{S}, g \geq f$.

Let $\lambda > \int \tilde{f} d\mu$. $\tilde{f}$ is upper semicontinuous. Let φ be continuous, greater than or equal to $\tilde{f}$, with $\int \varphi d\mu \leq \lambda$ (see Exercise 40). Let $\varepsilon > 0$. For each $x \in \mathcal{K}$, there exists $g_x \in \mathcal{S}$ such that $g_x \geq f$ and $g_x(x) < \varphi(x) + \varepsilon$. Since φ is continuous and each g_x is continuous, there will be a neighborhood V_x of x such that $g_x(y) \leq \varepsilon + \varphi(y)$ if $y \in V_x$. Since $\mathcal{K}$ is compact, it will be covered by finitely many of the V_x, say, $V_{x_1}, \ldots, V_{x_n}$. Since the minimum of a finite collection of continuous concave functions is continuous and concave, $g' = g_{x_1} \wedge \ldots \wedge g_{x_n}$ is in $\mathcal{S}$ and dominates f. So $\inf \int g d\mu \leq \int g' d\mu \leq \int \varphi d\mu + \varepsilon \leq \lambda + \varepsilon$. Since ε is arbitrary, we have (7.9). □

(7.19) Lemma. *If $f \in C(\mathcal{K})$, $P(f) = \sup_{\nu \succ \mu} \int f d\nu$.*

Proof. Suppose $\nu \succ \mu$. If $g \in \mathcal{S}$ with $g \geq f$, then $\int f d\nu \leq \int g d\nu \leq \int g d\mu$. Taking the infimum over such gs,

$$\int f d\nu \leq P(f)$$

by Lemma 7.18. So $P(f)$ is larger than the supremum.

For the other direction, define a linear functional L on $\{cf\}$ by $L(cf) = P(cf)$. Extend L to a linear functional on $C(\mathcal{K})$ dominated by P by Theorem 7.15. By the Riesz representation theorem, there exists a measure ν such that $L(g) = \int g d\nu$. As in the proof of Proposition 7.16, ν is a probability. If $g \in \mathcal{S}$, then $g = \tilde{g}$, and $P(g) = \int \tilde{g} d\mu = \int g d\mu$. However, $\int g d\nu = L(g) \leq P(g)$, or $\mu \prec \nu$. Also $\int f d\nu = L(f) = P(f)$. Hence $P(f) \leq \sup_{\nu \succ \mu} \int f d\nu$. $\qquad\square$

In particular, if we take $\mu = \delta_x$, we have

$$(7.10) \qquad\qquad \tilde{f}(x) = \sup_{\nu \succ \delta_x} \int f d\nu.$$

Next we examine $\tilde{f}$ further.

(7.20) Proposition. *If $f \in C(\mathcal{K})$, then $\tilde{f}(x) = \sup\{\int f d\nu : b(\nu) = x\}$.*

Proof. Suppose $b(\nu) = x$. By Proposition 7.9, we can approximate ν by measures ν_n with finite support and barycenter x. If $g \in \mathcal{S}$ and $g \geq f$, then since g is concave,

$$g(x) \geq \int g d\nu_n \to \int g d\nu \geq \int f d\nu.$$

Taking the infimum over such gs shows that $\tilde{f}(x)$ is larger than the supremum.

Suppose $\nu \succ \delta_x$. Since any linear functional L is in $\mathcal{S}$ and $-\mathcal{S}$, then as in Proposition 7.13, $\int L d\nu = \int L d\delta_x = L(x)$, or $b(\nu) = x$. By (7.10), $\tilde{f}(x) = \sup\{\int f d\nu : \nu \succ \delta_x\} \leq \sup\{\int f d\nu : b(\nu) = x\}$. $\qquad\square$

(7.21) Proposition. *If μ is concentrated on the extreme points of $\mathcal{K}$, then μ is maximal with respect to $\prec$. Moreover, $\int f d\mu = \int \tilde{f} d\mu$ if $f \in -\mathcal{S}$.*

Proof. Suppose x is an extreme point of $\mathcal{K}$. If $\lambda \succ \delta_x$, as in the last paragraph of the proof of Proposition 7.20, $b(\lambda) = x$. If $\lambda \neq \delta_x$, there exists $y \neq x$ and a closed convex neighborhood V of y not containing x with $\lambda(V) > 0$. Let $\nu = (\lambda(V))^{-1} \lambda|_V$, $\nu' = (\lambda(\mathcal{K} - V))^{-1} \lambda|_{\mathcal{K} - V}$. Let $t = \lambda(V)$, $z = b(\nu)$, $z' = b(\nu')$. Since $x \notin V$, $z \neq x$, but $x = tz + (1-t)z'$, a contradiction to x being extreme. Therefore $\lambda = \delta_x$.

By (7.10), $\tilde{f}(x) = f(x)$. So if μ is concentrated on the extreme points of $\mathcal{K}$, then μ is concentrated on $\{x : f(x) = \tilde{f}(x)\}$ for all f continuous, and in particular,

$$(7.11) \qquad\qquad \int f d\mu = \int \tilde{f} d\mu$$

if $f \in -\mathcal{S}$.

Suppose $\mu \prec \nu$. Then $\int f d\nu \geq \int f d\mu$ for all $f \in -\mathcal{S}$. By Proposition 7.19, $\int f d\nu \leq P(f) = \int \tilde{f} d\mu = \int f d\mu$. Therefore $\int f d\nu = \int f d\mu$ for all

$f \in -\mathcal{S}$, hence for all $f \in \mathcal{S} - \mathcal{S}$, hence for all $f \in C(\mathcal{K})$, and hence $\nu = \mu$. Therefore μ is maximal. $\qquad\square$

(7.22) Lemma. *Suppose g is lower semicontinuous and convex. If $b(\mu) = x$, $g(x) \geq \int g d\mu$.*

Proof. Take μ_n with finite support, with barycenter x, and converging weakly to μ. Let $\varepsilon > 0$ and, using Exercise 40, pick h continuous such that $h \leq g$ and $\int h d\mu \geq \int g d\mu - \varepsilon$. Then since g is convex,

$$g(x) \geq \int g d\mu_n \geq \int h d\mu_n \to \int h d\mu \geq \int g d\mu - \varepsilon.$$

Since ε is arbitrary, this completes the proof. $\qquad\square$

(7.23) Lemma. *Suppose h is upper semicontinuous, bounded, and convex. If $b(\mu) = x$, then $\int h d\mu \leq h(x)$.*

Proof. Let $\varepsilon > 0$. h is bounded, say by B. Pick $t \in (h(x), h(x) + \varepsilon)$. Let G be the convex hull of $(\{B\} \times \mathcal{K}) \cup ((t, B) \times \{x\}))$ in $\mathbb{R} \times \mathcal{K}$. Let $g(y) = \inf\{s \in \mathbb{R} : (s, x) \in G\}$. Then g is convex, lower semicontinuous, $g \geq h$, and $g(x) \leq h(x) + \varepsilon$. Hence

$$h(x) + \varepsilon \geq g(x) \geq \int g d\mu \geq \int h d\mu$$

by Lemma 7.22. $\qquad\square$

We finally can finish the proof of Theorem 7.10. If f is any function on $\mathcal{K}$ and $\mathcal{K}$ is the intersection of $\mathcal{C}$ with $\{y : L_0(y) = 1\}$ for a linear functional L_0, extend f to $\mathcal{C}$ by setting $f(y) = L_0(y) f(y/L_0(y))$. Note that

$$(7.12) \qquad\qquad f(ry) = r f(y), \qquad r > 0, y \in \mathcal{C}.$$

Proof of Theorem 7.10, uniqueness. Our first step is to show that if $f \in -\mathcal{S}$, then $\tilde{f}$ is convex as well as concave. By Proposition 7.20, $\tilde{f}(x) = \sup\{\int f d\nu : b(\nu) = x\}$. By Proposition 7.9, we can approximate any ν such that $b(\nu) = x$ by probabilities with finite support, so

$$(7.13) \qquad \tilde{f}(x) = \sup\left\{\int f d\nu : b(\nu) = x, \nu \text{ has finite support}\right\}.$$

If $f \in -\mathcal{S}$, $\tilde{f}$ is concave. Clearly $\tilde{f}$ is upper semicontinuous. We want to show that $\tilde{f}$ is convex, or using Exercise 38, that

$$(7.14) \qquad\qquad \tilde{f}(x + y) \leq \tilde{f}(x) + \tilde{f}(y), \qquad x, y \in \mathcal{C}.$$

By (7.13),

$$\tilde{f}(x + y) = \sup\left\{\sum_i t_i f(w_i) : b\left(\sum_i t_i \delta_{w_i}\right) = x + y\right\}.$$

Setting $z_i = t_i w_i$,

$$(7.15) \qquad \widetilde{f}(x+y) = \sup\left\{ \sum_i f(z_i) : \sum_i z_i = x+y \right\}.$$

By Theorem 7.17, $z_i = x_i + y_i$ with $x = \sum_i x_i$, $y = \sum_i y_i$. Since f is convex, then

$$(7.16) \qquad \sum_i f(z_i) \leq \sum_i [f(x_i) + f(y_i)].$$

Let $\nu_x = (L_0(x))^{-1} \sum_i L_0(x_i) \delta_{x_i/L_0(x_i)}$. Then $b(\nu_x) = \sum_i x_i/L_0(x) = x/L_0(x)$ and

$$\sum_i f(x_i) = L_0(x) \int f d\nu_x \leq L_0(x) \widetilde{f}(x/L_0(x)) = \widetilde{f}(x)$$

by (7.12) and (7.13). Similarly $\sum_i f(y_i) \leq \widetilde{f}(y)$, This, (7.15), and (7.16) yield (7.14).

We now complete the proof. We continue to suppose that $f \in \mathcal{S}$, and hence by what we have just shown, $\widetilde{f}$ is convex. If $b(\mu) = x$, $g \in \mathcal{S}$, and $g \geq f$, then by Proposition 7.22, $g(x) \geq \int g d\mu \geq \int \widetilde{f} d\mu$. Taking the infimum over such g that dominate f, $\widetilde{f}(x) \geq \int \widetilde{f} d\mu$. By Proposition 7.23, $\int \widetilde{f} d\mu \leq \widetilde{f}(x)$. We therefore have $\int \widetilde{f} d\mu = \widetilde{f}(x)$.

Now suppose μ_1, μ_2 are two measures concentrated on the extreme points of $\mathcal{K}$ with barycenter x. By Proposition 7.21 they are both maximal, and $\int f d\mu_i = \int \widetilde{f} d\mu_i$, $i = 1, 2$, if $f \in -\mathcal{S}$. Hence $\int f d\mu_1 = \widetilde{f}(x) = \int f d\mu_2$ for each $f \in -\mathcal{S}$, hence for all $f \in \mathcal{S} - \mathcal{S}$, and hence for all $f \in C(\mathcal{K})$. Therefore $\mu_1 = \mu_2$, $\qquad \square$

The assumption that $\mathcal{C}$ is a lattice is quite stringent. For example, if $\mathcal{K}$ is the unit square in $\mathbb{R}^2$ and x is the center, the representation of x as a linear combination of the extreme points is clearly not unique. If $\mathcal{K}$ is a triangle, it is. One way to see what is going on is to look at what it means for $x, y \in \mathcal{C}$ to have a least upper bound.

$v \geq x$ means $v - x \in \mathcal{C}$, or $v \in x + \mathcal{C}$. Similarly, if $v \geq y$, then $v \in y + \mathcal{C}$. So any upper bound of x and y must be in $(x + \mathcal{C}) \cap (y + \mathcal{C})$. Conversely, if $v \in (x + \mathcal{C}) \cap (y + \mathcal{C})$, then $v - x \in \mathcal{C}$ or $v \geq x$, and similarly $v \geq y$. If v is the least upper bound of x and y, and if w is any other upper bound for $\{x, y\}$, then $v \leq w$, or $w - v \in \mathcal{C}$, or $w \in v + \mathcal{C}$. If $w \in v + \mathcal{C}$, then $w \geq v \geq x$ and $w \geq v \geq y$. So what is required is that the set of upper bounds, namely, $(x + \mathcal{C}) \cap (y + \mathcal{C})$, must equal $v + \mathcal{C}$ for some v.

If $\mathcal{K}$ is a triangle, then the intersection of two translates of $\mathcal{C}$ is again a translate of $\mathcal{C}$ and a least upper bound exists. However, if $\mathcal{K}$ is a rectangle, the intersection of two translates of $\mathcal{C}$ is not necessarily a translate of $\mathcal{C}$. This is illustrated by Fig. 7.1. On the left, the two larger triangles represent the intersection of $x+\mathcal{C}$ and $y+\mathcal{C}$ with a fixed hyperplane, and their intersection (the shaded area) is the intersection of $v+\mathcal{C}$ with this hyperplane. On the right, we see that the intersection of $x+\mathcal{C}$ and $y+\mathcal{C}$ need not be of the form $v+\mathcal{C}$ when $\mathcal{K}$ is a square.

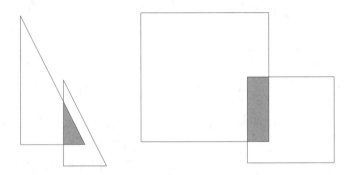

FIGURE 7.1. The intersection of $x+\mathcal{C}$ and $y+\mathcal{C}$ with a fixed hyperplane

8 Exercises and further results

Exercise 1. Show that a point is regular for the complement of a domain D in the probabilistic sense if and only if it is regular for D^c in the analytic sense.

Exercise 2. Give an example of a bounded domain D and f continuous on ∂D such that the Dirichlet problem has no solution.

Exercise 3. Show that if f is continuous, it is lower semicontinuous. Show that if f_n are lower semicontinuous and $f_n \uparrow f$, then f is lower semicontinuous. Show that if f is lower semicontinuous, then $\liminf_{x\to y} f(x) \geq f(y)$.

Exercise 4. Prove Liouville's theorem: if h is harmonic on $\mathbb{R}^d$, $d \geq 1$, and h is bounded below, then h is constant.

Exercise 5. Suppose D is a bounded domain (not necessarily regular for the Dirichlet problem). If f is continuous on ∂D, let

$$H(f) = \sup\{w : w \text{ is superharmonic on } D,$$
$$\limsup_{x\to y} w(x) \leq f(y) \text{ for all } y \in \partial D\}.$$

The Dirichlet problem for D with boundary values f is resolutive if $H(f) = -H(-f)$. Show that if the Dirichlet problem for D with boundary values f is resolutive, then $H(f)(x) = \mathbb{E}^x f(X_{\tau_D})$.

Exercise 6. Suppose $d = 2$, D is a domain, and $y \in \partial D$. Show that if there exists a line segment contained in D^c with one endpoint at y, then y is regular for D^c.

Exercise 7. The generalized Laplacian $\widetilde{\Delta}$ is defined by

$$\widetilde{\Delta} f(x) = \lim_{r \to 0} r^{-2} \int_{\partial B(0,r)} [f(x+y) - f(x)] \, \sigma_r(dy),$$

provided the limit exists, where σ_r is normalized surface measure on $\partial B(0,r)$. Show that if $f \in C^2$, then $\widetilde{\Delta} f(x) = \Delta f(x)$.

Exercise 8. Show that if h is a Lipschitz function, then the partial derivatives of h exist for a.e. point x.

Hint: Hold $x^1, \ldots, x^{i-1}, x^i, \ldots, x^d$ fixed and use the one dimensional result.

Exercise 9. If μ is a finite measure on the boundary of a domain D, show there exist functions f_n that are continuous on ∂D such that $f_n(y) \, dy$ converges weakly to μ.

Hint: Cf. construction in Proposition 7.9.

Exercise 10. If $(\mathbb{P}^x, X_t)$ is d-dimensional Brownian motion, A is a Borel set, and $t > 0$, show there exists $B \in \mathcal{F}^{00}_{t+}$ (defined in Sect. I.3) such that the symmetric difference of $(T_A \leq t)$ and B is $\mathbb{P}^x$-null for every x.

Hint: Verify that it suffices to show that $(D_A \circ \theta_s \leq t)$ differs from a $\mathcal{F}^{00}_{t+}$ set by a set that is $\mathbb{P}^x$-null for all x (cf. the proof of Theorem 2.8). Define

$$\mathbb{P}^{\mu}(\cdot) = \int \mathbb{P}^y(\cdot)(2\pi)^{-d/2} e^{-|y|^2/2} dy.$$

Show that a consequence of Theorem 2.8 is that there exist B_1 and B_2 in $\mathcal{F}^{00}_{t+}$ such that $B_1 \circ \theta_s \subseteq (D_A \circ \theta_s \leq t) \subseteq B_2 \circ \theta_s$ and $\mathbb{P}^{\mu}(B_2 - B_1) = 0$. Then

$$\mathbb{P}^x(B_2 \circ \theta_s - B_1 \circ \theta_s) = \mathbb{E}^x \mathbb{P}^{X_s}(B_2 - B_1)$$
$$= \int \mathbb{P}^y(B_2 - B_1)(2\pi s)^{-d/2} e^{-|y-x|^2/2s} dy.$$

Now use the fact that $\int \mathbb{P}^y(\cdot)(2\pi s)^{-d/2} e^{-|y-x|^2/2s} dy$ is absolutely continuous with respect to $\mathbb{P}^{\mu}$.

Exercise 11. Show that if B is a Borel set and $(\mathbb{P}^x, X_t)$ is d-dimensional Brownian motion, then $\mathbb{P}^x(T_B = t) = 0$ for each x and each $t > 0$.

Exercise 12. Show that if B is a Borel set, then B^r is also a Borel set.

Exercise 13. Give examples of ν_1 and ν_2 such that (a) ν_1 has density with respect to Lebesgue measure, but $U\nu_1(0) = \infty$, and (b) $U\nu_2$ is bounded but not continuous.

Exercise 14. Let X_t be one-dimensional Brownian motion killed on exiting $D = (0, 1)$. Let

$$g(x, y) = 2x(1 - y) \wedge 2y(1 - x).$$

Show that if f is bounded and measurable on D, then

$$\int_D f(y)g(x, y)\,dy = \mathbb{E}^x \int_0^{\tau_D} f(X_s)ds, \qquad x \in D.$$

Exercise 15. Suppose D is a domain and fix $x \in D$. Suppose $g(y)$ is 0 if $y \in (B^c)^r$ and $g(y)$ is equal to $u(x, y)$ minus a positive harmonic function. Show g must be equal to the Green function for D with pole at x.

Exercise 16. Show that if $A \subseteq B$, then $g_A(x, y) \leq g_B(x, y)$ for all x and y.

Exercise 17. Suppose $d \geq 2$ and $(\mathbb{P}^x, X_t)$ is d-dimensional Brownian motion. Show there exists a constant c such that if $x \in B(0, r)$, then $\mathbb{E}^x \tau_{B(0,r)} \leq cr^2$. Show there exists c such that if $x \in B(0, r/2)$, then $\mathbb{E}^x \tau_{B(0,r)} \geq cr^2$.

Exercise 18. Show that if f is bounded with compact support, then Uf is continuous.

Exercise 19. If C is Newtonian capacity, show that $C(B) = 0$ implies $|B| = 0$. An event that occurs everywhere except for a set of capacity 0 is said to occur quasi–everywhere, written "q.e."

Exercise 20. A set A is thin at x if $x \notin A^r$. A set A is thin if it is thin at every point of A. A set A is semipolar if it is the countable union of thin sets. Show semipolar sets are polar.

Exercise 21. Prove Proposition 5.11.

Exercise 22. Show a set B is polar if and only if there exists a measure μ such that $U\mu$ is infinite on B but not identically infinite.

Exercise 23. Show that if B is a bounded Borel set, ν is concentrated on B, and $U\nu \leq 1$ on B^r, then $C(B) \geq \nu(B)$.

Hint: As in the proof of Proposition 5.10, $U\nu \leq U\mu_B$ on $\mathbb{R}^d$, where μ_B is the equilibrium measure on B. As in the proof of Proposition 5.8,

$$\nu(\mathbb{R}^d) = \int U\gamma_r \, d\nu = \int U\nu \, d\gamma_r \leq \int U\mu_B \, d\gamma_r = C(B).$$

Exercise 24. Show that if ν gives positive mass to a polar set B, then $U\nu(x)$ must be infinite for at least one x.

Exercise 25. Prove the domination principle: if μ is a measure concentrated on B^r and f is nonnegative, superharmonic, and $f \geq U\mu$ q.e. on B, then $f \geq U\mu$ for all $x \in \mathbb{R}^d$. (q.e. means except for a set of capacity 0; see Exercise 19.)

Exercise 26. If f is nonnegative, $e^{-\alpha t} P_t f \leq f$ for all t, and $e^{-\alpha t} P_t f \to f$ pointwise as t tends to 0, f is said to be α-excessive. Show α-excessive functions are β-excessive if $\alpha < \beta$. Show that if f is α-excessive, there exist $g_m \geq 0$ such that $U^\alpha g_m \uparrow f$. Show $f(x) = \mathbb{E}^x e^{-\alpha T_B}$ is α-excessive for every Borel set B.

Exercise 27. The fine topology is the coarsest topology with respect to which all the α–excessive functions are continuous. Show G is open in the fine topology if and only if for all $x \in G$, x is not regular for G^c. Give an example of a finely open set in $\mathbb{R}^d$, $d \geq 3$, that is not open.

Exercise 28. Suppose $f \geq 0$ is superharmonic on a domain D. Show that the greatest harmonic minorant of f exists.

Exercise 29. Suppose $U\mu$ is finite for all x. Show there exists a continuous additive functional A_t such that $\mathbb{E}^x A_\infty = U\mu(x)$ for all x.

Hint: $U\mu(X_t)$ is a continuous supermartingale. Use the Doob-Meyer decomposition. (Some care is needed because here we have a family of probabilities $\{\mathbb{P}^x\}$, not just one probability.)

Exercise 30. Suppose A_t is a continuous additive functional with $f(x) = \mathbb{E}^x A_\infty$ finite for all x. Show f is excessive.

Exercise 31. Suppose A_t and B_t are two additive functionals with $\mathbb{E}^x A_\infty = \mathbb{E}^x B_\infty < \infty$ for all x. Show $A_t \equiv B_t$ for all t.

Hint: Use the uniqueness part of the Doob–Meyer decomposition.

Exercise 32. Suppose A_t and B_t are two continuous additive functionals. Suppose $\int_0^\infty h(X_s) dA_s = 0$ whenever $h \geq 0$ and $\int_0^\infty h(X_s) dB_s = 0$. Show there exists $f \geq 0$ such that $B_t = \int_0^t f(X_s) dA_s$ for all t.

Hint: Suppose first that $A_t \leq B_t$ and B_t is strictly increasing. Let

$$Z_s = \liminf_{r \to 0} \frac{B_{s+r} - B_s}{A_{s+r} - A_s},$$

and let $f(x) = \mathbb{E}^x Z_0$. In the general case, consider the densities of A_t and B_t with respect to $C_t = A_t + B_t + t$.

Exercise 33. In the notation of (6.4), show $P_B f = \overline{R}_B f$.

Exercise 34. Show that if $U\mu$ is harmonic on an open set B, then μ is supported on B^c.

Exercise 35. Let A and B be disjoint compact subsets of $\mathbb{R}^d$, $d \geq 3$. Solve the condenser problem: find a signed measure ν concentrated on $A \cup B$ such that $U\nu \geq 0$ on $\mathbb{R}^d$, $U\nu = 1$, on B^r and $U\nu = 0$ on A^r.

Hint: Imitate Proposition 5.8 to find ν such that

$$U\nu(x) = \mathbb{P}^x(T_B < T_A, T_{A \cup B} < \infty).$$

Exercise 36. Show that the definition of the Martin boundary is independent of the choice of the approximating sequence D_n. Show that the Martin boundaries that arise from two different choices of x_0 are homeomorphic.

Exercise 37. Find a metric for the Martin compactification D^*.

Hint: Let x_n be a countable dense subset of D and let

$$\rho(y, y') = \sum_n 2^{-n}[(M(x_n, y) - M(x_n, y')) \wedge 1].$$

Exercise 38. Suppose C is a cone and $f(ry) = rf(y)$ for all $r > 0$, $y \in C$. Show f is convex if and only if $f(x + y) \le f(x) + f(y)$ for all $x, y \in C$.

Exercise 39. Let C be a vector lattice. Show $(a \wedge b) + c = (a + c) \wedge (b + c)$. Show that if S is a subset of C that is closed under addition and under the operation $\wedge$, then $S - S$ is closed under the operations $\wedge$ and $\vee$. ($S - S = \{a - b : a, b \in S\}$.)

Hint: For the second assertion, show $((a - b) \wedge (c - d)) + (b + d) = (a + d) \wedge (c + b)$.

Exercise 40. Suppose f is lower semicontinuous and bounded on a metric space (S, ρ). Show there exist continuous functions f_n such that $f_n \uparrow f$.

Hint: Let $f_n(x) = \inf\{f(z) + n\rho(x, z) : z \in S\}$.

Exercise 41. Prove the uniqueness assertion of the Constantinescu-Cornea theorem.

Exercise 42. Prove (5.12).

Exercise 43. Show that if H is the upper half-space in $\mathbb{R}^d$, $d \ge 1$, then the transition densities for Brownian motion killed on exiting H is

$$p_H(t, x, y) = p(t, x, y) - p(t, \tilde{x}, y), \qquad x, y \in H,$$

where $\tilde{x} = (x^1, \dots, x^{d-1}, -x^d)$.

Hint: By the independence of X_t^d and $(X_t^1, \dots, X_t^{d-1})$, it suffices to consider the case $d = 1$. If X_t is one-dimensional Brownian motion, by translation invariance and symmetry,

$$\mathbb{P}^x(X_t > y, \inf_{s \le t} X_s > 0) = \mathbb{P}^0(X_t < x - y, \sup_{s \le t} X_s < x)$$

$$= \mathbb{P}^0(X_t < x - y) - \mathbb{P}^0(X_t < x - y, \sup_{s \le t} X_s \ge x).$$

By Theorem I.3.8, this is $\mathbb{P}^0(X_t < x - y) - \mathbb{P}^0(X_t > 2x - (x - y)) = \mathbb{P}^x(X_t > y) - \mathbb{P}^{-x}(X_t > y)$. Now differentiate with respect to y.

Exercise 44. Suppose $d = 2$, X_t is two-dimensional Brownian motion, and H is the upper half-plane. Let

$$g(x,y) = \frac{1}{\pi} \log(1/|x-y|) - \frac{1}{\pi} \log(1/|\tilde{x}-y|), \qquad x,y \in H,$$

where $\tilde{x} = (x^1, -x^2)$ if $x = (x^1, x^2)$. Show that if f is bounded and measurable on H, then

$$\int_H f(y)g(x,y)\,dy = \mathbb{E}^x \int_0^{\tau_H} f(X_s)ds, \qquad x \in H.$$

This says that $g(x,y)$ can be considered the Green function for H.

Hint: Integrate the conclusion of Exercise 43 over t from 0 to ∞, using the idea of (3.23).

Exercise 45. Suppose $d = 2$ and X_t is two-dimensional Brownian motion. Let

$$g(x,y) = \begin{cases} -\frac{1}{\pi}\log|x-y| + \frac{1}{\pi}\log\left|\frac{x}{|x|} - |x|y\right|, & x \neq 0, \\ -\frac{1}{\pi}\log|y|, & x = 0, \end{cases} \qquad x,y \in B(0,1).$$

Show that if f is bounded and measurable on $B(0,1)$, then

$$\int_{B(0,1)} f(y)g(x,y)\,dy = \mathbb{E}^x \int_0^{\tau_{B(0,1)}} f(X_s)ds, \qquad x \in B(0,1).$$

This says that g can be considered the Green function for $B(0,1)$. If $g(x,y)$ is defined by the same formula for $x,y \in B(0,1)^c$ and f is bounded on $B(0,1)^c$, show that

$$\int_{B(0,1^c)} f(y)g(x,y)\,dy = \mathbb{E}^x \int_0^{\tau_{B(0,1)}} f(X_s)ds, \qquad x \in B(0,1)^c.$$

Hint: Use Exercise 44 and the Kelvin transform.

Exercise 46. Show that if D is a bounded set and P_t^D is defined by (4.13), then there exists c independent of t such that

$$\|P_t^D f\|_{L^2(D)} \leq c\|f\|_{L^2(D)}.$$

Notes

Much of Sect. 1, 3, 4, 5, and 6 was adapted from Port and Stone [1]. See that reference for further information and for some of the history of these results. Another good reference for this material is the encyclopedic work of Doob [2]. Helms [1] is a reference to the analytic approach to potential theory, as is Wermer [1]. The proof of the Choquet capacity theorem in Sect. 2 was taken from Dellacherie and Meyer [1], and the applications to hitting times from Blumenthal and Getoor [1]. Our approach to the Feynman-Kac formula is that of Williams [1], while our proof of the smoothness of potentials is from Gilbarg and Trudinger [1]. The proof of Theorem 3.16 is taken from Bass and Khoshnevisan [1]. Yosida [1] provided the basis for the proof of the Hilbert-Schmidt expansion theorem. Our construction of the Martin boundary follows the outline given in Brelot [1]. Our proof of the Choquet theorem (or Choquet-Meyer theorem) used Dellacherie and Meyer [3]; the older book by Meyer [1] was also very useful. The idea of Fig. 7.1 came from Phelps [1]. Many of the exercises in Sect. 8 originate from material in Port and Stone [1]. Exercises 27, 29, 30, 31, and 32 are based on results from Blumenthal and Getoor [1]; see Benveniste and Jacod [1] for a solution to Exercise 32.

III
LIPSCHITZ DOMAINS

1 The boundary Harnack principle

Formulation

Many theorems that can be proved for smooth domains become much harder to prove as one makes the domains less and less regular. For many results Lipschitz domains seem to be the borderline case between where the theorem is true but hard to prove and where the assertion no longer continues to hold. In this chapter we look at quite a number of results that have been proved for Lipschitz domains, the earliest in 1968, the latest in 1991.

A key tool in obtaining many of these results is the boundary Harnack principle. Suppose D is a domain and u and v are two positive harmonic functions on D that vanish on the same nice subset A of ∂D, the boundary of D. If points on the boundary of D are regular for the complement of D, then $u(x)$ and $v(x)$ will tend to 0 as x tends to a point $z \in A$. The boundary Harnack principle says, roughly, that u and v tend to 0 at exactly the same rate. Slightly more precisely, if we pick $x_0 \in D$ and normalize u and v so that $u(x_0) = v(x_0) = 1$, then $u(x)/v(x)$ stays bounded above and below away from 0 as $x \to z$.

We will give two proofs of the boundary Harnack principle for Lipschitz domains in this section.

First we need to specify exactly what we mean by a Lipschitz domain. A function $\Gamma : \mathbb{R}^{d-1} \to \mathbb{R}$ is Lipschitz if there exists a constant M_Γ such that

(1.1) $\qquad |\Gamma(z_1) - \Gamma(z_2)| \leq M_\Gamma |z_1 - z_2|, \qquad z_1, z_2 \in \mathbb{R}^{d-1}.$

The smallest such constant M_Γ is called the Lipschitz constant of Γ.

(1.1) Definition. *A bounded domain $D \subseteq \mathbb{R}^d$ is a Lipschitz domain if for every $x \in D$ there exist $r \in (0, \infty)$, a Lipschitz function Γ, and an orthonormal coordinate system CS, all possibly depending on x, such that*

(1.2) $\qquad D \cap B(x, r) = \{y = (y^1, \dots, y^d) \text{ in } CS :$
$$y^d > \Gamma(y^1, \dots, y^{d-1})\} \cap B(x, r).$$

All this says is that locally D looks like the region above the graph of a Lipschitz function Γ.

One bothersome feature of Definition 1.1 is that it does not preclude $D \cap B(x, r)$ consisting of more than one component or containing narrow channels (see Fig. 1.1). However, it is not hard to see that one can choose r (depending on x) to avoid these possibilities. In any case this is a very minor point since for most of the theorems in this chapter our first step is to reduce the problem to the situation where the domain is the region above a Lipschitz function.

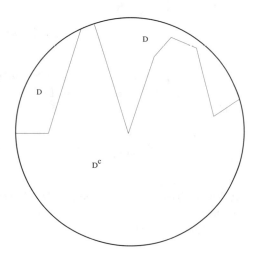

FIGURE 1.1. A Lipschitz domain such that $D \cap B(x, r)$ has more than one component.

Let us make the observation that if D is the region above the graph of a Lipschitz function Γ and $a > 0$, then $aD = \{ay : y \in D\}$ is the region above the function Γ_a, where $\Gamma_a(x) = a\Gamma(x/a)$. Note

$$|\Gamma_a(z_1) - \Gamma_a(z_2)| = a|\Gamma(z_1/a) - \Gamma(z_2/a)| \leq aM_\Gamma|z_1/a - z_2/a|,$$

and so Γ_a has the same Lipschitz constant as Γ. We will call this property "scaling."

We can now give a precise statement of the boundary Harnack principle for Lipschitz domains.

(1.2) Theorem. *Suppose D is a connected Lipschitz domain, $x_0 \in D$. Suppose V is open, K is compact, and $K \subseteq V$. Then there exists a constant c, depending only on K, V, and D, such that if u and v are two positive harmonic functions on D that both vanish continuously on $V \cap \partial D$, then*

$$(1.3) \qquad u(x)/v(x) \leq cu(x_0)/v(x_0), \qquad x \in K \cap D.$$

Saying that a harmonic function u vanishes continuously on $V \cap \partial D$ means that $u(x)$ tends to 0 as x tends to z uniformly over z in compact subsets of $V \cap \partial D$.

Reversing the roles of u and v, $v(y)/u(y) \leq c_1 v(x_0)/u(x_0)$ for $y \in K \cap D$. Combining, we obtain the following corollary of Theorem 1.2.

(1.3) Corollary. *Suppose K, V, and D are as in Theorem 1.2. There exists c such that if u, v are positive harmonic on D with u, v vanishing continuously on $V \cap \partial D$, then*

$$(1.4) \qquad u(x)/v(x) \leq cu(y)/v(y), \qquad x, y \in K \cap D.$$

The function u/v is "harmonic" with respect to the probability measures P_v^x (see (I.6.16) for the definition of $\mathbb{P}_v^x$). That is, if S is a stopping time less than τ_D, then since u is harmonic,

$$\mathbb{E}_v^x \left(\frac{u}{v} \right)(X_{t \wedge S}) = \mathbb{E}^x \left(\frac{u}{v} \right)(X_{t \wedge S}) \frac{v(X_{t \wedge S})}{v(x)} = \mathbb{E}^x u(X_{t \wedge S})/v(x) = u(x)/v(x).$$

One way to think of the boundary Harnack principle, then, is as the ordinary Harnack inequality for positive functions that are L_v-harmonic, i.e., $L_v u = 0$, where L_v is the operator corresponding to Brownian motion conditioned by the function v. This means that

$$(1.5) \qquad L_v u = \frac{1}{2} \Delta u + \frac{\nabla v}{v} \cdot \nabla u.$$

The box method

It will turn out that to prove the boundary Harnack principle, the key situation to consider is the case of a domain above the graph of a Lipschitz function. For now suppose Γ is a bounded Lipschitz function on $\mathbb{R}^{d-1}$ and

$$D = \{x : x^d > \Gamma(x^1, \ldots, x^{d-1})\}.$$

Define the vertical distance from x to ∂D by

(1.6) $$\delta(x) = x^d - \Gamma(x^1, \ldots, x^{d-1}).$$

Note that since this is a Lipschitz domain, there exists a constant c depending only on M_Γ such that

(1.7) $$c\delta(x) \leq \text{dist}(x, \partial D) \leq \delta(x).$$

Define the "box"

(1.8) $$Q(x, a, R) = \{y \in D : \delta(y) < a,$$
$$|(y^1, \ldots, y^{d-1}) - (x^1, \ldots, x^{d-1})| < R\}.$$

We will let U be the upper boundary of Q:

(1.9) $$U(x, a, R) = \{y \in \partial Q(x, a, R) : \delta(y) = a\},$$

and S the sides:

(1.10) $$S(x, a, R) = \{y \in \partial Q(x, a, R) :$$
$$|(y^1, \ldots, y^{d-1}) - (x^1, \ldots, x^{d-1})| = R\}.$$

For any two points x and y we say that y is "above" x or x is "below" y if $(y^1, \ldots, y^{d-1}) = (x^1, \ldots, x^{d-1})$ and $y^d \geq x^d$.

As we will see (Theorem 1.7), what we need to show is that the probability of exiting $Q(x, a, R)$ through the sides is less than a constant times the probability of exiting through the top, provided we do not start too near the sides. The point is that the process does *not* exit $Q(x, a, R)$ by creeping along the boundary.

Since D is the region above a Lipschitz function, clearly it satisfies the exterior cone condition, and hence every point of ∂D is regular for D^c (see Proposition II.1.13). We want to make a stronger uniform statement.

(1.4) Lemma. *There exists $\rho < 1$ such that if $z \in D$ and $a = \delta(z)$, then*

$$\frac{|\partial B(z, 2a) \cap D|}{|\partial B(z, 2a)|} \leq \rho.$$

Proof. By a scaling argument, we may assume $a = 1$. Without loss of generality, make a change of coordinates so that $z = (0, \ldots, 0, 1)$. By the definition of $\delta(z)$, the point $0 \in \partial D$. Since Γ is Lipschitz, there exists c_1 such that the set $A = \{x \in \partial B(z, 2) : |x - (0, \ldots, 0, -1)| \leq 1/c_1\} \subseteq D^c$. Note there exists c_2 such that $|A| \geq c_2|\partial B(z, 2)|$. The result now follows by taking $\rho = 1 - c_2$. $\square$

Since the distribution of $X_{\tau(B(z,2a))}$, started at z, is uniform on the boundary of $\partial B(z,2a)$, we conclude

$$(1.11) \qquad \mathbb{P}^z(X_{\tau(B(z,2a))} \in D) \le \rho.$$

We use this to find an upper bound on the probability of exiting $Q(x, a, R)$ through the sides.

(1.5) Lemma. *There exist c_1 and c_2 such that if $y \in Q(x, a, R/2)$, then*

$$\mathbb{P}^y(X_{\tau(Q(x,a,R))} \in S(x, a, R)) \le c_1 \exp(-c_2 R/a).$$

Proof. Suppose first that $R > 24a$. Let

$$V_1 = \inf\{t > 0 : |X_t - X_0| \ge 2a\}, \qquad V_{i+1} = V_i + V_1 \circ \theta_{V_i}, \qquad i = 1, 2, \dots .$$

The V_i are the successive times that the process X_t moves $2a$ (cf. Sect. I.3).

Let us write τ_Q for $\tau_{Q(x,a,R)}$. Note that $|X_{V_i} - X_0| \le 2ai$. Thus, if $X_{\tau_Q} \in S(x, a, R)$, we must have $\tau_Q > V_{[R/8a]}$, $\mathbb{P}^y$ a.s. if $y \in Q(x, a, R/2)$. By the strong Markov property,

$$\mathbb{P}^y(\tau_Q > V_{i+1}) \le \mathbb{P}^y(\tau_Q > V_i, \tau_Q \circ \theta_{V_i} > V_1 \circ \theta_{V_i})$$
$$= \mathbb{E}^y[\mathbb{P}^{X_{V_i}}(\tau_Q > V_1); \tau_Q > V_i].$$

If $i \le R/8a$ and $\tau_Q > V_i$, then $|X_{V_i} - X_0| \le R/4$ and $X_{V_i} \in Q(x, a, 3R/4)$, $\mathbb{P}^y$–a.s. Hence $\mathbb{P}^{X_{V_i}}(\tau_Q > V_1) \le \rho$ by Lemma 1.4. So

$$\mathbb{P}^y(\tau_Q > V_{i+1}) \le \rho \mathbb{P}^y(\tau_Q > V_i).$$

By induction,

$$\mathbb{P}^y(X_{\tau_Q} \in S(x, a, R)) \le \mathbb{P}^y(\tau_Q > V_{[R/8a]}) \le \rho^{[R/8a]}$$
$$= \exp([R/8a] \log \rho) \le c_1 \exp(-c_2 R/a).$$

(Since $\rho < 1$, $\log \rho < 0$.) This gives our result when $R > 24a$. We have the result for $R \le 24a$ by taking c_1 large enough. $\qquad\square$

Not surprisingly, we now want a lower bound on leaving $Q(x, a, R)$ through the top.

(1.6) Lemma. *There exist c and $\beta > 0$ such that if $a \le 1/2$, $y \in Q(x, a, a/2)$, and $b = \delta(y) \le a$, then*

$$\mathbb{P}^y(X_{\tau(Q(x,1,4))} \in U(x, 1, 4)) \ge cb^\beta.$$

Proof. Let $y_0 = y = (y^1, \dots, y^{d-1}, y^d)$. We will construct a sequence of points $y_i = (y^1, \dots, y^{d-1}, y_i^d)$ above y for $i = 1, \dots, n$ for some n by choosing y_i^d so that

$$|y_i - y_{i-1}| = r_i = \text{dist}\,(y_{i-1}, \partial D)/4$$

and $\delta(y_n) \geq 2$. Once we have this sequence, we let $B_i = B(y_i, 2r_i)$ and $B = \cup_{i=0}^n B_i$. By our construction $B \subseteq Q(x, 4, 4)$. We define

$$h(z) = \mathbb{P}^z(X_{\tau_B} \in \partial B - Q(x, 1, 4)).$$

So h is the harmonic function on B with boundary value 1 on $\partial B - Q(x, 1, 4)$ and 0 on the rest of ∂B.

By the support theorem (Theorem I.6.6), there exists c_1 such that

$$\mathbb{P}^0(\sup_{s \leq 2} |X_s - \psi(s)| < 1/8) \geq c_1,$$

where $\psi(s), 0 \leq s \leq 2$, is the curve that moves upward at constant unit speed: $\psi(s) = (0, \ldots, 0, s)$. If $A = \{x \in \partial B(0, 1) : x^d > 0, |(x^1, \ldots, x^{d-1}| < 1/8\}$ then $\mathbb{P}^0(X_{\tau(B(0,1))} \in A) \geq c_1$. By translation invariance and scaling,

$$\mathbb{P}^{y_n}(X_{\tau(B(y_n, r_n))} \in Q(x, 1, 4)) \geq c_1,$$

and therefore $h(y_n) \geq c_1$. By the usual Harnack inequality (Theorem II.1.19), there exists c_2 such that $h(y_i) \geq c_2 h(y_{i+1})$ for each i. By induction, $h(y_0) \geq c_1 c_2^n = c_1 \exp(n \log c_2)$. Since $c_2 < 1$, this is equal to $c_1 \exp(-c_3 n)$ for some $c_3 > 0$. If X_t exits B through $\partial B - Q(x, 1, 4)$, then X_t exits $Q(x, 1, 4)$ through $U(x, 1, 4)$. Hence $\mathbb{P}^y(X_{\tau(Q(x,1,4))} \in U(x, 1, 4)) \geq \mathbb{P}^y(X_{\tau_B} \in \partial B - Q(x, 1, 4))$, and we will have our result once we figure out what n must be.

Now $r_i \geq \text{dist}\,(y_{i-1}, \partial D)/4$ and by our construction, is increasing in i. So $r_1 \geq \delta(y_0)/4M_\Gamma \geq b/4M_\Gamma$. Let $m = [8M_\Gamma] + 1$. Then $\delta(y_m) \geq mr_1 \geq 2b$. So $r_m \geq 2b/4M_\Gamma$. Repeating, we see that $\delta(y_{2m}) \geq mr_m \geq 4b$, and by induction, $\delta(y_{km}) \geq 2^k b$. We choose n to be km where k is the first integer such that $2^k b > 2$. So $k = [\log(2/b)/\log 2] + 1 \leq -c_4 \log b$. Hence $n \leq -c_5 \log b$, or

$$c_1 \exp(-c_3 n) \geq c_1 \exp(c_3 c_5 \log b) = c_1 b^\beta$$

with $\beta = c_3 c_4$. $\qquad\square$

We finally can make precise the notion that the probability that the process leaves $Q(x, a, R)$ through the sides is not too high. We suppose $a \leq 1/2, R \geq 8a$. Define

$$(1.12) \qquad H_1 = \big(X_{\tau(Q(x,a,R))} \in S(x, a, R)\big),$$
$$H_2 = \big(X_{\tau(Q(x,a,R))} \in U(x, a, R)\big).$$

(1.7) Theorem. *There exists c such that if $y \in Q(x, a, R/2)$, then $\mathbb{P}^y(H_1) \leq c\mathbb{P}^y(H_2)$.*

Proof. By a scaling argument, it suffices to show our result for the case $a = 1/2$. Since increasing R increases $\mathbb{P}^y(H_2)$ but decreases $\mathbb{P}^y(H_1)$, it is enough to prove our result when $R = 4$. Let

$$(1.13) \qquad r_i = R\left(\frac{3}{4} - \frac{1}{40}\sum_{j=1}^{i}\frac{1}{j^2}\right).$$

Note $R/2 \le r_1 \le R$ for all i. (Recall that $\sum_{j=1}^{\infty} j^{-2} = \pi^2/6 \approx 1.5$; see any book on Fourier series, e.g., Folland [2], for a proof.) Let

$$J_i = Q(x, 2^{-i}, r_i) - Q(x, 2^{-i-1}, r_i)$$

and let

$$d_i = \sup_{y \in J_i} \mathbb{P}^y(H_1)/\mathbb{P}^y(H_2).$$

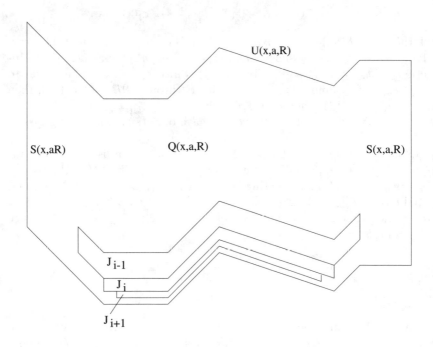

FIGURE 1.2. The sets J_i.

For $y \in Q(x, a, R/2) - \cup_{i=2}^{\infty} J_i$, we have by Lemma 1.6 that $\mathbb{P}^y(H_2) \ge c$. So

$$\mathbb{P}^y(H_1)/\mathbb{P}^y(H_2) \le 1/c < \infty$$

for y in this set, or $d_1 \le 1/c < \infty$.

Our aim is to bound d_i in terms of d_{i-1}. Suppose $y \in J_i$, write τ_i for $\tau_{Q(x,2^{-i},r_{i-1})}$, and write τ_Q for $\tau_{Q(x,a,R)}$. If X_t started at y exits $Q(x,a,R)$ through its sides $S(x,a,R)$, either it exits $Q(x,2^{-i},r_{i-1})$ through $S(x,2^{-i},r_{i-1})$ or it exits $Q(x,2^{-i},r_{i-1})$ through the top $U(x,2^{-i},r_{i-1})$ and then exits $Q(x,a,R)$ through $S(x,a,R)$. So we have

$$(1.14) \quad \mathbb{P}^y(H_1) \leq \mathbb{P}^y\big(X_{\tau_i} \in S(x,2^{-i},r_{i-1})\big)$$
$$+ \mathbb{P}^y\big(X_{\tau_i} \in U(x,2^{-i},r_{i-1}); X_{\tau_Q} \in S(x,a,R)\big).$$

By Lemma 1.5 and some algebra, the first term on the right-hand side of (1.14) is less than or equal to

$$(1.15) \qquad c_1 e^{-c_2(r_{i-1}-r_i)/2^{-i}} \leq c_1 e^{-c_3 2^i/i^2} \leq c(2^{-i})^\beta/i^2.$$

Here we used (1.13). Since $X_{\tau_Q} = X_{\tau_Q} \circ \theta_{\tau_i}$, the strong Markov property at time τ_i bounds the second term on the right in (1.14) by

$$(1.16) \qquad \mathbb{E}^y\big[\mathbb{P}^{X_{\tau_i}}(H_1); X_{\tau_i} \in U(x,2^{-i},r_{i-1})\big].$$

Since $X_{\tau_i} \in U(x,2^{-i},r_{i-1})$ implies $X_{\tau_i} \in J_{i-1}$, (1.16) in turn is bounded by

$$(1.17) \quad d_{i-1}\mathbb{E}^y\big[\mathbb{P}^{X_{\tau_i}}(H_2); X_{\tau_i} \in U(x,2^{-i},r_{i-1})\big]$$
$$\leq d_{i-1}\mathbb{P}^y\big(X_{\tau_i} \in U(x,2^{-i},r_{i-1}); X_{\tau_Q} \in U(x,a,R)\big)$$
$$\leq d_{i-1}\mathbb{P}^y\big(X_{\tau_Q} \in U(x,a,R)\big)$$
$$\leq d_{i-1}\mathbb{P}^y(H_2).$$

Combining (1.14), (1.15), and (1.17),

$$\mathbb{P}^y(H_1) \leq c2^{-i\beta}/i^2 + d_{i-1}\mathbb{P}^y(H_2).$$

By Lemma 1.6, for $y \in J_i$,

$$\mathbb{P}^y(H_2) \geq c(2^{-i})^\beta.$$

Hence

$$\mathbb{P}^y(H_1) \leq (c/i^2)\mathbb{P}^y(H_2) + d_{i-1}\mathbb{P}^y(H_2).$$

Taking the supremum over ys in J_i,

$$d_i \leq d_{i-1} + c/i^2.$$

Therefore

$$d_i \leq d_1 + c\sum_{j=1}^\infty \frac{1}{j^2}$$

or $\sup_i d_i < \infty$. This is exactly what we needed to prove. $\qquad\square$

The other key step is the following proposition, known as a Carleson estimate.

(1.8) Theorem. *Suppose u is positive and harmonic in D, the region above the graph of a Lipschitz function Γ. Let $x \in D$, suppose $u(x) = 1$, and suppose u vanishes continuously on $B(x, 4\delta(x)) \cap \partial D$. There exists c, not depending on x or u, such that $u(y) \leq c$ for all $y \in D \cap B(x, 2\delta(x))$.*

Proof. By a scaling argument, we may suppose $\delta(x) = 1$. By the usual Harnack inequality, $u \leq c$ on $U(x, 1, 4)$. If $y_1 \in U(x, 1, 4)$ and y lies directly below y_1, then by Exercise 4 we see that there exists β such that

$$(1.18) \qquad\qquad u(y) \leq c_1 \delta(y)^{-\beta}.$$

So if $u(y) \geq M$, then $\delta(y) \leq (c_1/M)^{1/\beta}$.

Take N large. Suppose there exists $x_1 \in B(x, 2) \cap D$ with $u(x_1) \geq N$. We will construct a sequence x_i tending to a point in $B(x, 4) \cap \partial D$ with $u(x_i) \to \infty$, which is a contradiction. Let $r_1 = \operatorname{dist}(x_1, \partial D)$. Using (1.11), there exists $\rho < 1$ not depending on x_1 such that

$$
\begin{aligned}
N \leq u(x_1) &= \mathbb{E}^{x_1} u(X_{\tau(B(x_1, 2r_1))}) \\
&= \mathbb{E}^{x_1}[u(X_{\tau(B(x_1, 2r_1))}); \tau_{B(x_1, 2r_1)} < \tau_D] \\
&\leq \rho \sup_{\partial B(x_1, 2r_1)} u.
\end{aligned}
$$

Therefore there must exist x_2 in $\partial B(x_1, 2r_1) \cap D$ with $u(x_2) \geq N/\rho$. Keep repeating the above argument, with x_1 and r_1 replaced by x_i and $r_i = \operatorname{dist}(x_i, D)$. As long as $B(x_i, 2r_i) \subseteq B(x, 3)$, there will exist $x_{i+1} \in \partial B(x_i, 2r_i) \cap D$ with $u(x_{i+1}) \geq u(x_i)/\rho$. This implies $u(x_i) \geq N/\rho^{i-1}$. In turn, by (1.7) and (1.18),

$$r_i \leq \delta(x_i) \leq c u(x_i)^{-1/\beta} \leq c \frac{(\rho^{i-1})^{1/\beta}}{N^{1/\beta}}.$$

The sum $\sum r_i$ converges, and if we take N large enough, the sum will be less than $1/4$. Since $x_{i+1} \in \partial B(x_i, 2r_i) \cap D$, $|x_{i+1} - x_i| = 2r_i$ and

$$|x_i - x| \leq |x_i - x_1| + |x_1 - x| \leq \sum_{j=1}^{i} 2r_j + 2 < 5/2,$$

hence $B(x_i, 2r_i) \subseteq B(x, 3)$. With this choice of N, we thus obtain a sequence $x_1, x_2, \ldots$ such that $\operatorname{dist}(x_i, \partial D) \to 0$, $u(x_i) \to \infty$, and $B(x_i, 2r_i) \subseteq B(x, 3)$. Since every point on the boundary of D is regular for D^c, this contradicts the fact that $u = 0$ on $B(x, 4) \cap \partial D$. Therefore we conclude that u is bounded on $B(x, 2) \cap D$ by N. $\qquad\square$

Finally we can put Theorems 1.7 and 1.8 together to prove Theorem 1.2.

Proof of Theorem 1.2. For the moment, let us continue to suppose that D is the region above the graph of a bounded Lipschitz function. Suppose

$\delta(x_0) \in [2,4]$. Write just Q for $Q(x_0, 1/2, 4)$ and similarly U, S. Then, by the usual Harnack inequality, $\sup_U u \le cu(x_0)$ and $\inf_U v \ge cv(x_0)$. If $x \in Q(x_0, 1/2, 2)$, we write

$$(1.19) \qquad u(x) = \mathbb{E}^x[u(X_{\tau_Q}); X_{\tau_Q} \in S] + \mathbb{E}^x[u(X_{\tau_Q}); X_{\tau_Q} \in U]$$
$$\le c(\sup_U u)\mathbb{P}^x(X_{\tau_Q} \in S) + cu(x_0)\mathbb{P}^x(X_{\tau_Q} \in U)$$
$$\le cu(x_0)\mathbb{P}^x(X_{\tau_Q} \in S) + cu(x_0)\mathbb{P}^x(X_{\tau_Q} \in U)$$

by Theorem 1.8. Using Theorem 1.7, this is less than or equal to

$$cu(x_0)\mathbb{P}^x(X_{\tau_Q} \in U) \le c\frac{u(x_0)}{v(x_0)}\mathbb{E}^x[v(X_{\tau_Q}); X_{\tau_Q} \in U]$$
$$\le c\frac{u(x_0)}{v(x_0)}v(x).$$

This gives our result in this special case. By scaling, we have the boundary Harnack principle for domains above the graph of a Lipschitz function, without assuming anything on $\delta(x_0)$.

Finally, cover K by a finite number of balls contained in V whose radii are small enough that the intersection of D with any one of these balls is the same as the intersection of the region above the graph of some Lipschitz function with the same ball (recall the definition of a Lipschitz domain). For each small ball B_i, take a point x_i. We have shown that $u(y)/v(y) \le cu(x_i)/v(x_i)$ for all $y \in B_i$. By the usual Harnack inequality, $u(x_i) \le cu(x_0)$ and $v(x_i) \ge cv(x_0)$, so $u(x_i)/v(x_i) \le c^2 u(x_0)/v(x_0)$. Putting this all together and remembering there are only finitely many B_i, we deduce $u(y)/v(y) \le cu(x_0)/v(x_0)$, which is what we wanted to prove.

□

The SDE method

We now proceed to give another proof of the boundary Harnack principle in Lipschitz domains, the "SDE" method. SDE is an abbreviation for "stochastic differential equations" (cf. Sect. I.6). Although this method gives the boundary Harnack principle for many of the same domains for which the box method works (see the extensions subsection), the box method seems to be the more versatile of the two. Nevertheless, the SDE method is more useful for proving Harnack inequalities for operators similar to ones like L_v in (1.5) but that are not given by an h-path transform. As part of our discussion of the SDE method, we will obtain some estimates on harmonic functions that are of considerable interest in themselves.

(1.9) Lemma. *Suppose D is the region above the graph of a Lipschitz function Γ and $x_1 \in D$. There exist c_1, c_2 and $\beta_1, \beta_2 > 0$ such that if u is positive and harmonic in D and $u(x_1) = 1$, then for x below x_1,*

(1.20) $$u(x) \leq c_1 \delta(x)^{-\beta_1}$$

and

(1.21) $$u(x) \geq c_2 \delta(x)^{\beta_2}.$$

In addition, if u vanishes continuously on $B(x_1, 4\delta(x_1)) \cap \partial D$, then there exists c_3 and $\beta_3 > 0$ such that

(1.22) $$u(x) \leq c_3 \delta(x)^{\beta_3}.$$

Proof. The first and second inequalities were essentially proved in Lemma 1.6; see Exercise 4. The third inequality we deduce from the Carleson estimate as follows. Without loss of generality, we may assume $\delta(x_1) = 1$ (scaling). By the usual Harnack inequality and Carleson's estimate, there exists K such that $u(y) \leq K$ for all $y \in Q(x_1, 1, 2)$. Let $M = 4M_\Gamma \vee 16$. Now suppose $z \in Q(x_1, M^{-1}, 3/2)$. By (1.11) there exist $\rho < 1$ so that

$$u(z) = \mathbb{E}^z u(X_{\tau(B(z,2M^{-1})) \wedge \tau_D})$$
$$\leq \left(\sup_{Q(x_1,1,2)} u \right) \mathbb{P}^z(X_{\tau_{B(z,2M^{-1})} \wedge \tau_D} \in D)$$
$$\leq \rho K.$$

Repeat: if $z \in Q(x_1, M^{-2}, 5/4)$,

$$u(z) = \mathbb{E}^z u(X_{\tau(B(z,2M^{-2})) \wedge \tau_D})$$
$$\leq \left(\sup_{Q(x_1,M^{-1},3/2)} u \right) \mathbb{P}^z(X_{\tau_{(B(z,M^{-2}))} \wedge \tau_D} \in D)$$
$$\leq \rho(\rho K).$$

Continuing, if $z \in Q(x_1, M^{-n}, 1 + (1/2)^n)$, then $u(z) \leq K\rho^n$. (1.22) follows from this with some algebra. $\qquad\square$

Let h be the positive harmonic function in $Q(x, a, R)$ that has boundary value 1 on $S(x, a, R)$ and $U(x, a, R)$ and boundary value 0 on $\partial D \cap \partial Q(x, a, R)$. The reason we look at this particular harmonic function is that

(1.23) $$\mathbb{P}_h^y(X_{\tau(Q(x,a,R))} \in S(x, a, R))$$
$$= \mathbb{P}^y(X_{\tau(Q(x,a,R))} \in S(x, a, R))/h(y)$$

since h is 1 on $S(x, a, R)$, with a similar equation for $U(x, a, R)$. So if we prove

(1.24) $$\mathbb{P}_h^y(X_{\tau(Q(x,a,R))} \in S(x, a, R)) \leq c\mathbb{P}_h^y(X_{\tau(Q(x,a,R))} \in U(x, a, R)),$$

we can then multiply by $h(y)$ to obtain the conclusion of Theorem 1.7. From then on the proof via the SDE method proceeds the same as the proof via the box method.

For this h we have the following.

(1.10) Lemma. *There exist c such that if $y \in Q(x, a/2, R/2)$, then*

$$\frac{|\nabla h(y)|}{h(y)} \geq \frac{c}{\delta(y)}.$$

Note that this inequality is sharp, since by Corollary II.1.4 and Harnack's inequality,

$$|\nabla h(y)| \leq \frac{c}{\text{dist}\,(y, \partial D)} \Big(\sup_{B(y, \text{dist}\,(y, \partial D)/2)} h \Big) \leq \frac{c}{\delta(y)} h(y).$$

Proof. Fix y. If y_1 is directly below y, then by Lemma 1.9 and scaling, there exists β such that

$$h(y_1) \leq c_1 h(y)[\delta(y_1)/\delta(y)]^\beta.$$

Choose y_1 below y so that

$$\delta(y_1) = (2c_1)^{-1/\beta} \delta(y).$$

Hence $h(y_1) \leq h(y)/2$. By the mean value theorem, there exists y_2 between y_1 and y such that

$$\frac{1}{2} h(y) \leq h(y) - h(y_1) = \frac{\partial h}{\partial x^d}(y_2)|\delta(y) - \delta(y_1)|,$$

or

$$\frac{\partial h}{\partial x^d}(y_2) \geq \frac{ch(y)}{\delta(y)}.$$

Now consider $\partial h/\partial x^d$. We want to show it is nonnegative. Since $h = 1$ on $S(x, a, R)$, then $\partial h/\partial x^d = 0$ on $S(x, a, R)$. Since $h = 0$ on ∂D and $h \geq 0$ in D, then $\partial h/\partial x^d \geq 0$ on $\partial D \cap \overline{Q(x, a, R)}$. Since $h = 1$ on $U(x, a, R)$ and $h \leq 1$ in $Q(x, a, R)$, then $\partial h/\partial x^d \geq 0$ on $U(x, a, R)$. Therefore the boundary values of $\partial h/\partial x^d$ are nonnegative, and hence $\partial h/\partial x^d$ is a nonnegative harmonic function in $Q(x, a, R)$.

Since $\text{dist}\,(y_2, \partial D) \geq c\delta(y_2) \geq c\delta(y_1) \geq c\delta(y)$ and $|y_2 - y_1| \leq c\delta(y)$, it follows by the Harnack inequality that

$$\frac{\partial h}{\partial x^d}(y) \geq c\frac{\partial h}{\partial x^d}(y_2) \geq c\frac{h(y)}{\delta(y)}.$$

Since $|\nabla h(y)| \geq (\partial h/\partial x^d)(y)$, our result follows. $\qquad \square$

We have seen in (I.6.19) that under P_h^x, X_t satisfies the stochastic differential equation

(1.25) $$dX_t = dW_t + (\nabla h/h)(X_t)dt,$$

where W_t is a d-dimensional Brownian motion. So an easy calculation using Itô's lemma tells us that if $\alpha \in (0,1)$ and $Y_t = h^\alpha(X_t)$, then

(1.26) $$dY_t = \alpha h^{\alpha-1}(X_t)\nabla h(X_t) \cdot dW_t$$
$$+ (\alpha(\alpha+1)/2)h^{\alpha-2}(X_t)|\nabla h(X_t)|^2\, dt, \qquad t < \tau_D,$$

using the fact that $\Delta h \equiv 0$. We let M_t denote the martingale part of Y_t, that is,

$$M_t = \int_0^{t\wedge\tau_D} \alpha h^{\alpha-1}(X_t)\nabla h(X_t) \cdot dW_t,$$

and we let B_t denote the bounded variation part of Y_t, that is,

$$B_t = \int_0^{t\wedge\tau_D} (\alpha(\alpha+1)/2)h^{\alpha-2}(X_t)|\nabla h(X_t)|^2\, dt.$$

Let

$$A_t = \int_0^t \frac{|\nabla h|}{h}(X_s)\, ds.$$

The idea of the SDE proof is as follows. In a short time t_0, a Brownian motion will not travel far. So in order for X_t to go a large distance in time t_0, A_{t_0} must be large. Either $h(X_t)$ also becomes large in that time, or else by comparing B_t to A_t, we see that B_{t_0} is large. If B_{t_0} is large, then provided we can control the martingale term in (1.26), $Y_t = h^\alpha(X_t)$ must become large. In either case, $h(X_t)$ is not too small, which says that X_t does not stay too close to the boundary.

Proof of Theorem 1.1, SDE method. Let X_t be the solution to (1.25). Let s be a number to be chosen in a moment and let

$$S = \inf\{t : |X_t - x_0| \geq 1/2\}, \qquad \sigma_s = \inf\{t : h(X_t) > s\},$$
$$\rho = \inf\{t : B_t \geq 1\}.$$

Let $b = \beta_2$ be the exponent from Lemma 1.9 and set $\alpha = 1/2b$. First we compare $\langle M\rangle_t$ and A_t to B_t. Note

(1.27) $$\frac{|\nabla h|}{h}(x) \geq \frac{c}{\delta(x)} \geq \frac{c}{h(x)^{1/b}}$$

from Lemma 1.10, and so

(1.28) $$\frac{|\nabla h|}{h}\frac{|\nabla h|}{h}h^\alpha \geq \frac{|\nabla h|}{h}\frac{1}{h^\alpha}.$$

Hence

(1.29) $$B_{t\wedge\sigma_s} \geq K s^{-\alpha} A_{t\wedge\sigma_s}$$

for some $K > 0$. Also,

$$(1.30) \qquad \langle M \rangle_{t \wedge \sigma_s} = \int_0^{t \wedge \sigma_s} \alpha^2 h^{2\alpha - 2}(X_t) |\nabla h(X_t)|^2 \, dt \leq c_2 s^\alpha B_{t \wedge \sigma_s}.$$

Next we partition our probability space into five sets $D_1, \ldots, D_5$. Let

$$D_1 = (\sigma_s \leq S), \qquad D_2 = (\sup_{u \leq t_0} |W_u| \geq 1/4),$$

where t_0 is also a number to be chosen in a moment. If ω is not in $D_1 \cup D_2$, then $\sigma_s > S$ and either $S \leq t_0 \wedge \rho$ or $S > t_0 \wedge \rho$. If $S > t_0 \wedge \rho$, either $t_0 < \rho$ or $t_0 \geq \rho$. Hence if we let

$$D_3 = (\sigma_s > S, S \leq t_0 \wedge \rho) \cap D_2^c, \qquad D_4 = (\sigma_s > S > t_0, \rho > t_0),$$

and

$$D_5 = (\sigma_s > S > \rho, t_0 \geq \rho),$$

then $\Omega = D_1 \cup D_2 \cup D_3 \cup D_4 \cup D_5$.

Now we show that the probabilities of D_2 and D_5 are small, while D_3 and D_4 must be empty. Choose t_0 small so that $2 \exp(-1/32t_0) \leq 1/4$. Hence $\mathbb{P}_h^x(D_2) \leq 1/4$. Choose $s < 4^{-1/\alpha}$ small so that $Ks^{-\alpha}/4 > 1$, $c_1^2 t_0 / s^{3\alpha} > 1$, and $2 \exp(-1/32c_2 s^\alpha) \leq 1/4$.

If $\omega \in D_3$, since $|X_S - x_0| \leq |W_S| + A_S$, then $A_S \geq 1/4$. Then since $S \leq \rho$, we have

$$1 \geq B_S \geq Ks^{-\alpha} A_S \geq Ks^{-\alpha}/4,$$

which is impossible. Therefore $D_3 = \emptyset$.

If $\omega \in D_4$ and $t < \sigma_s$, then by (1.27)

$$h^\alpha \left(\frac{|\nabla h|}{h} \right)^2 (X_t) \geq h^\alpha \left(\frac{c_1}{h^{1/b}} \right)^2 (X_t) = \frac{c_1^2}{h^{3\alpha}} (X_t).$$

Then we have

$$1 \geq B_{t_0} \geq \frac{c_1^2 t_0}{s^{3\alpha}},$$

which is impossible. So $D_4 = \emptyset$.

Since $Y_t = M_t + B_t$ and $\sigma_s > \rho$ on D_5, then $|Y_\rho| \leq s^\alpha$ and so $\sup_{r \leq \rho} |M_r| > 1/4$. However, $\langle M \rangle_\rho \leq c_2 s^\alpha B_\rho = c_2 s^\alpha$ by (1.30). Hence by Exercise I.8.13,

$$\mathbb{P}_h^x(D_5) \leq \mathbb{P}(M_\rho^* > 1/4, \langle M \rangle_\rho \leq c_2 s^\alpha) \leq 2 \exp \left(-\frac{(1/4)^2}{2c_2 s^\alpha} \right) \leq 1/4.$$

Finally, we bound the probability of D_1 from below and show that this leads to the boundary Harnack principle. We have

$$1 \leq \mathbb{P}_h^x(D_1) + \mathbb{P}_h^x(D_2) + \mathbb{P}_h^x(D_5) \leq \mathbb{P}_h^x(D_1) + 1/2,$$

or

$$\mathbb{P}_h^x(D_1) \geq 1/2.$$

On $\sigma_s \leq S$, $\delta(X_{\sigma_s}) \geq (s/c)^{1/\beta_3}$, where β_3 is given in (1.22). So there exists d such that

$$\mathbb{P}_h^x(X_{\tau(Q(x,d,1))} \in U(x,d,1/2)) \geq 1/2.$$

By the support theorem, if $z \in U(x,d,1/2)$, then $\mathbb{P}^z(X_{\tau(Q(x,a,1))} \in U(x,a,1)) \geq c$. Lemma 1.9 tells us that h is bounded below on $Q(x,d,1/2)$. Hence

$$\mathbb{P}_h^x(X_{\tau(Q(x,a,1))} \in U(x,a,1)) \geq c.$$

Thus by the strong Markov property, $\mathbb{P}_h^x(X_{\tau(Q(x,a,1))} \in U(x,a,1)) \geq c/2$.

With this estimate, we now proceed as in the first proof of the boundary Harnack principle.   □

Extensions

One can also consider domains that cannot be represented locally as the region above the graph of a function. For example, Bass and Burdzy [2] proved the boundary Harnack principle for what are called John domains. (Jerison and Kenig [3] had earlier considered a special case of John domains, the class of nontangentially accessible domains.)

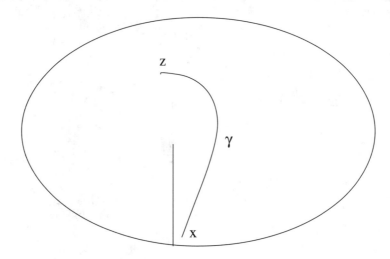

FIGURE 1.3. A John domain.

(1.11) Definition. *A domain D is a John domain if there exist $z \in D$ and $c > 0$ such that for every $x \in D$ there is a rectifiable curve γ connecting x and z in D satisfying*

$$(1.31) \qquad \text{dist}\,(y, \partial D) \geq c\ell(\gamma(x,y)) \qquad \text{for all } y \in \gamma,$$

where $\ell(\gamma(x,y))$ is the length of the piece of the curve γ between x and y.

A simple example of a John domain that is not a Lipschitz domain is simply the unit ball about 0 minus the line segment from 0 to $(0,\ldots,0,-1)$.

Returning to domains that are given locally as the region above the graph of a function, the boundary Harnack principle holds for Hölder domains of order α if $\alpha \in (0,1]$ (Bañuelos, Bass, and Burdzy [1]). A Hölder domain is one where the domain looks locally like the region above the graph of a Hölder function of order α; such a function is one where

$$(1.32) \qquad |\Gamma(z_1) - \Gamma(z_2)| \le M_\Gamma |z_1 - z_2|^\alpha.$$

In addition, in Hölder domains Brownian motion can be replaced by other diffusions, namely, ones corresponding to uniformly elliptic operators of either divergence type or nondivergence type (Bass and Burdzy [2, 5]) (in the latter case, some regularity of the boundary is necessary when $\alpha \le 1/2$). Even some degenerate elliptic operators are possible (Gao [1]).

There are, however, examples of domains that are given locally as the region above the graph of a continuous function where the boundary Harnack principle does not hold (Bass and Burdzy [2]). It would be interesting to characterize those domains for which it holds.

The Fatou theorem (see Sect. III.4) says that in a Lipschitz domain a positive harmonic function has nontangential limits at almost every boundary point. This is not necessarily true in arbitrary domains. Yet the martingale convergence theorem says that $\lim_{t \to \tau_D} u(X_t)$ exists a.s. if u is a positive harmonic function in D. This is, in some sense, a stochastic Fatou theorem valid in arbitrary domains. Might there be a stochastic version of the boundary Harnack principle that holds in every domain? A possible formulation would be: suppose V is open, K is compact contained in V, and u and v are two positive harmonic functions on D that vanish on the regular points of $V \cap \partial D$ and are bounded in a neighborhood of $V \cap \partial D$. (See Bass and Burdzy [2] for a discussion of why these extra assumptions are necessary.) Does $\lim_{t \to \tau_D}(u/v)(X_t)$ remain bounded on the set $(X(\tau_D) \in K)$? To state this result for arbitrary domains, one really has to replace ∂D by the Martin boundary of D. Doob [1] has shown that the limit exists P_v^x a.s., but here we need it to hold P_z^x a.s. for every $z \in K \cap \partial D$.

2 The Martin boundary for Lipschitz domains

Martin boundary

In this section we show that for a Lipschitz domain the Martin boundary is the same as the Euclidean boundary. This implies, for example, that Lipschitz domains have "Poisson kernels": there exists $M(x, z)$ such that every positive harmonic function h in D can be written $h(x) = \int_{\partial D} M(x, z)\mu(dz)$ for some μ supported on ∂D.

The main result can be stated as follows.

(2.1) Theorem. *Suppose D is a Lipschitz domain. There exists a one-to-one mapping of the Martin boundary onto the Euclidean boundary. Each harmonic function corresponding to a Martin boundary point z is minimal harmonic.*

The first step is to show that the Martin boundary can be regarded as a subset of the Euclidean boundary. Fix $x_0 \in D$ and recall from Sect. II.7 that $M(x, y) = g_D(x, y)/g_D(x_0, y)$, $x, y \in D$. The Martin boundary is the set $\partial_M D = D^* - D$, where D^* is the smallest compact set for which $M(x, y)$ is continuous in the extended sense in y. If we show $g_D(x, y)/g_D(x_0, y)$ is uniformly continuous in y near the boundary of D, this will imply $D^* \subseteq \overline{D}$, and so we can identify the Martin boundary with a subset of the Euclidean boundary.

The oscillation of a function f on a set A is defined by

$$(2.1) \qquad \underset{A}{\mathrm{Osc}}\, f = \sup_A f - \inf_A f.$$

(2.2) Proposition. *$M(x, y)$ is uniformly continuous in y for y in a neighborhood of ∂D.*

Proof. Fix x_0, $x \in D$. Pick $w \in \partial D$ and pick r small enough so that $D \cap B(w, r)$ is the intersection of $B(w, r)$ with the region above the graph of a Lipschitz function. r depends on D but can be chosen to be independent of w. Fix a coordinate system as in the definition of Lipschitz domain. Pick k_0 large enough so that $Q(w, 2^{-k_0}, 2^{-k_0}) \subseteq B(w, r)$ and $x, x_0 \notin Q(w, 2^{-k_0}, 2^{-k_0})$.

Write $Q_k = Q(w, 2^{-k}, 2^{-k})$ and $h(y) = g_D(x_0, y)$. We want to show the oscillation of $g_D(x, \cdot)/h(\cdot)$ on Q_{k+1} is controlled by the oscillation of $g_D(x, \cdot)/h(\cdot)$ on Q_k. Both h and $g_D(x, \cdot)$ are harmonic functions on $D - \{x, x_0\}$ that vanish on ∂D. Moreover $g_D(x, y)$ and $h(y)$ are nonzero and finite for any $y \in Q_{k_0}$ by Harnack's inequality. By the boundary Harnack principle in Q_k (see Exercise 11), $g_D(x, \cdot)/h$ is bounded above and below by positive constants.

Fix k. Let $u(y) = Ag_D(x, y) + Bh(y)$, where A and B are real numbers so that

$$\sup_{Q_k}(u/h) = 1, \qquad \inf_{Q_k}(u/h) = 0.$$

Clearly u is harmonic in Q_k, and

$$\operatorname*{Osc}_{Q_k}(u/h) = \sup_{Q_k}(u/h) - \inf_{Q_k}(u/h) = 1.$$

Pick $z_k \in U(w, 2^{-k}, 2^{-k})$. If $u(z_k)/h(z_k) \leq 1/2$, replace u by $h - u$. So we may suppose $u(z_k)/h(z_k) \geq 1/2$ and still have $\sup_{Q_k}(u/h) = 1$, $\inf_{Q_k}(u/h) = 0$, and u harmonic in Q_k.

By the boundary Harnack principle, if $y \in Q_{k+1}$,

$$(u/h)(y) \geq c_1(u/h)(z_k) \geq c_1/2.$$

So

$$\operatorname*{Osc}_{Q_{k+1}}(u/h) \leq 1 - c_1/2 = \rho.$$

Letting $\rho = 1 - c_1/2$ and undoing the algebra,

$$\operatorname*{Osc}_{Q_{k+1}}\big(g_D(x,y)/h(y)\big) \leq \rho \operatorname*{Osc}_{Q_k}\big(g_D(x,y)/h(y)\big),$$

and $\rho < 1$.

$g_D(x,y)/h(y)$ is bounded by c_2 on Q_{k_0} by the boundary Harnack principle. So $\operatorname{Osc}_{Q_{k+1}}(g_D(x,y)/h(y)) \leq c_2\rho^{k-k_0}$, or

$$\left| \frac{g_D(x,y)}{h(y)} - \frac{g_D(x,y')}{h(y')} \right| \leq c_3\rho^k$$

if $y, y' \in Q(w, 2^{-k}, 2^{-k})$. This and Exercise 12 imply the uniform Hölder continuity of $g_D(x,y)/h(y)$, which implies our result. $\qquad \square$

As a consequence, if $y \to z \in \partial D$, $g_D(x,y)/g_D(x_0,y) = M(x,y)$ converges. Call the limit $M(x,z)$. We next want to show that if $z_1 \neq z_2$, then $M(\cdot, z_1) \neq M(\cdot, z_2)$. This will imply that each Euclidean boundary point corresponds to a different positive harmonic function, hence the Martin boundary cannot be identified with a proper subset of the Euclidean boundary.

(2.3) Proposition. *If $M(\cdot, z_1) \equiv M(\cdot, z_2)$ for $z_1, z_2 \in \partial D$, then $z_1 = z_2$.*

Proof. First we want to show $M(\cdot, z)$ vanishes on $D - \{z_1\}$. Fix $y_0 \in D$, $y_0 \neq x_0$. Suppose $w \in \partial D$, $w \neq z_1$. Since $g_D(x, y_0) \to 0$ as $x \to w$, given ε there exists $\delta < |w - z_1|/4$ such that $g_D(x, y_0) \leq \varepsilon$ if $|x - w| \leq \delta$. Then by the boundary Harnack principle in $D - (B(x, \delta/2) \cup B(x_0, \delta/2))$ (see Exercise 9),

$$\frac{g_D(x,y)}{g_D(x_0,y)} \leq \frac{cg_D(x,y_0)}{g_D(x_0,y_0)} \leq \frac{c\varepsilon}{g_D(x_0,y_0)}$$

if $|x - w| \leq \delta$. Letting $y \to z_1$, we see that $M(x, z_1) \leq c\varepsilon/g_D(x_0, y_0)$. Since ε is arbitrary, $M(x, z_1) \to 0$ as $x \to w \neq z_1$. Moreover the convergence is uniform on compact subsets of $\partial D - \{z_1\}$.

The next step is to show that there is only one μ supported on ∂D such that $M(x, z_1) = \int M(x, w)\mu(dw)$, namely, $\mu = \delta_{z_1}$. If we show this, then the same argument for z_2 will tell us that if $M(x, z_1) \equiv M(x, z_2)$, then $\delta_{z_1} = \mu = \delta_{z_2}$, or $z_1 = z_2$.

So suppose $M(x, z_1) = \int_{\partial D} M(x, w)\mu(dw)$, $\mu \neq \delta_{z_1}$. Since $M(x_0, w) = 1$ for all w, $1 = M(x_0, z_1) = \mu(\partial D)$. If $\mu \neq \delta_{z_1}$, there exists ε such that $\mu_\varepsilon = \mu|_{B(z_1, \varepsilon)^c} \neq 0$. Then $\int M(x, w)\mu_\varepsilon(dw)$ is a harmonic function bounded above by $M(x, z_1)$.

Now the first paragraph of the proof showed that if $|w - z_1| \geq \varepsilon$, then $M(x, w) \to 0$ (uniformly) as $x \to w' \in \partial D \cap B(z_1, \varepsilon/2)$. So $\int M(x, w)\mu_\varepsilon(dw) \to 0$ as $x \to w' \in \partial D \cap B(z_1, \varepsilon/2)$. On the other hand, if $x \to w' \in \partial D \cap B(z_1, \varepsilon/2)^c$, then

$$\int M(x, w)\mu_\varepsilon(dw) \leq \int M(x, w)\mu(dw) = M(x, z_1) \to 0.$$

By the maximum principle, $\int M(x, w)\mu_\varepsilon(dw) = 0$. The boundary Harnack principle tells us that $g_D(x, y)/g_D(x_0, y)$ stays bounded below by a positive constant as $y \to \partial D$. Therefore $M(x, w)$ is positive for all w, hence $\mu_\varepsilon = 0$. Since ε is arbitrary, $\mu(\{z_1\}^c) = 0$. Because $\mu(\partial D) = 1$, we have $\mu = \delta_{z_1}$ as desired. □

The proof of Theorem 2.1 will be complete once we prove the following.

(2.4) Proposition. *For each $z \in \partial D$, $M(x, z)$ is minimal harmonic.*

Proof. Suppose $h(x) \leq M(x, z)$ in D. We can write $h(x) - \int M(x, w)\mu(dw)$ for some μ supported on ∂D, and by exactly the same argument as the one in the proof of Proposition 2.3, we deduce that $\mu(\{z\}^c) = 0$. Therefore $\mu = c\delta_z$ for some c. □

Harmonic measure

We want to show that the Martin kernel is the density of harmonic measure with respect to two different starting points.

(2.5) Definition. *The harmonic measure of a set $A \subseteq \partial D$ is defined by*

$$(2.2) \qquad \omega^x(A) = \omega(x, A) = \mathbb{P}^x(X_{\tau_D} \in A).$$

For each A, $\omega(x, A)$ is a harmonic function, and so by the usual Harnack inequality, $\omega(x, A) \leq c\omega(x_0, A)$ for some c independent of A. Hence $\omega(x, \cdot)$

is absolutely continuous with respect to $\omega(x_0, \cdot)$ with a Radon–Nikodym derivative that is bounded by c.

(2.6) Theorem.

$$\omega(x, A) = \int_A M(x, w)\, \omega(x_0, dw).$$

Proof. Let $R(x, y)$ be the Radon-Nikodym derivative of $\omega(x, dy)$ with respect to $\omega(x_0, dy)$. Fix $y \in \partial D$. For each k, let Q_{yk} denote the cube of the form

$$[j_1/2^k, (j_1 + 1)/2^k) \times \cdots \times [j_d/2^k, (j_d + 1)/2^k)$$

that contains y, where $j_1, \ldots, j_d$ are integers.

$$h_k(x) = \frac{\omega\big(x, Q_{yk} \cap \partial D\big)}{\omega\big(x_0, Q_{yk} \cap \partial D\big)}$$

(with the convention $0/0 = 0$) is a harmonic function on $D - Q_{yk}$. Since $h_k(x_0) = 1$, the Harnack inequality shows that for any k_0, if $k > k_0$, then h_k is bounded on any subdomain D' of $D - Q_{y,k_0}$, uniformly over $k > k_0$. By Corollary II.1.4, the h_k, $k > k_0$, are equicontinuous on subdomains D'' of D', so a subsequence converges to a harmonic function on D''. By a diagonalization procedure, we can find a subsequence of h_k that converges to a harmonic function, call it h, uniformly on compact subsets of D.

$h(x) = \int M(x, w)\mu(dw)$ for some μ supported on ∂D. Now we know each h_k is bounded on subdomains D' and clearly it vanishes continuously on $\partial D - Q_{y,k-1}$. So by the Carleson estimate (Theorem 1.8) and Lemma 1.9, we see that $h_k(x) \to 0$ as $x \to z \in \partial D - Q_{y,k-1}$, uniformly over such z. Hence $h(x) \to 0$ as $x \to z \in \partial D - \{y\}$. By the argument of Proposition 2.3, $\mu(\{y\}^c) = 0$, or $\mu = c\delta_y$. Since $h_k(x_0) = 1$, then $c = 1$. Therefore $h(x) = M(x, y)$.

Except for a set of ys in a $\omega(x_0, \cdot)$ null set,

$$(2.3) \qquad \lim_{k \to 0} \frac{\omega(x, Q_{yk} \cap \partial D)}{\omega(x_0, Q_{yk} \cap \partial D))} = R(x, y)$$

by Exercise 13. So $M(x, y) = R(x, y)$, a.e. Therefore $M(x, y)$ is also a Radon-Nikodym derivative of $\omega(x, \cdot)$ with respect to $\omega(x_0, \cdot)$, and the theorem follows. $\qquad \square$

h-path transforms

Let $(\mathbb{P}_z^x, X_t)$ denote the h-path transform of Brownian motion killed on exiting D by the harmonic function $M(\cdot, z)$, that is,

$$(2.4) \qquad \mathbb{P}_z^x(\cdot) = \mathbb{P}_{M(\cdot, z)}^x(\cdot).$$

The next proposition shows that $(\mathbb{P}_z^x, X_t)$ can be interpreted as Brownian motion conditioned to exit D at z. For the reader interested only in the case where D is a ball or half-space, replace $M(\cdot, z)$ everywhere by the Poisson kernel with pole at z.

(2.7) Proposition. *If $A \in \mathcal{F}_{\tau_D}$,*

$$\int_B \mathbb{P}_z^{x_0}(A)\mathbb{P}^{x_0}(X_{\tau_D} \in dz) = \mathbb{P}^{x_0}(X_{\tau_D} \in B; A).$$

Proof. Suppose D_n are subdomains increasing up to D (that is, $\overline{D}_n \subseteq D_{n+1}$, $\cup_n D_n = D$) and $A \in \mathcal{F}_{\tau_{D_n}}$. Then

$$\int_B \mathbb{P}_z^{x_0}(A)\mathbb{P}^{x_0}(X_{\tau_D} \in dz) = \int_B \frac{\mathbb{E}^{x_0}[M(X_{\tau_{D_n}}, z); A]}{M(x_0, z)}\mathbb{P}^{x_0}(X_{\tau_D} \in dz)$$

$$= \mathbb{E}^{x_0}\left[\int_B M(X_{\tau_{D_n}}, z)\mathbb{P}^{x_0}(X_{\tau_D} \in dz); A\right]$$

since $M(x_0, z) = 1$. By Theorem 2.6 and the strong Markov property, this is equal to

$$\mathbb{E}^{x_0}\left[\mathbb{P}^{X(\tau_{D_n})}(X_{\tau_D} \in B); A\right] = \mathbb{P}^{x_0}(X_{\tau_D} \in B; A).$$

This proves the lemma for $A \in \mathcal{F}_{\tau_{D_n}}$, and a linearity and limiting argument gives it for $A \in \mathcal{F}_{\tau_D}$. $\square$

We will need a zero-one law due to Brossard; see Durrett [1], pp. 102-104.

(2.8) Definition. *A is shift invariant if whenever $S < \tau_D$ is a stopping time, $1_A \circ \theta_S = 1_A$, $\mathbb{P}^x$-a.s. for every $x \in D$.*

Roughly, A is shift invariant if for each ε, A depends on X_t, $\tau_D - \varepsilon < t < \tau_D$. For example, $(\lim_{t \to \tau_D} u(X_t)$ exists$)$ and $(\lim_{t \to \tau_D} u(X_t) > B)$ are shift invariant, but $(\sup_{t < \tau_D} u(X_t) > B)$ is not.

(2.9) Theorem. *If A is shift invariant, then $x \to \mathbb{P}_z^x(A)$ is constant and is either identically 0 or identically 1.*

It is not true, in general, that $z \to \mathbb{P}_z^x(A)$ is constant. For example, the second example of a shift invariant set given above does not have this property.

Proof. Let $k(x) = \mathbb{P}_z^x(A)$, $S < \tau_D$. Then

$$k(x) = \mathbb{P}_z^x(A) = \mathbb{P}_z^x(A \circ \theta_S) = \mathbb{E}_z^x \mathbb{E}_z^{X_S}(A) = \mathbb{E}_z^x k(X_S).$$

So $k(x) = \mathbb{E}^x M(X_S, z)k(X_S)/M(x, z)$, or $M(\cdot, z)k(\cdot)$ is harmonic. It is clearly nonnegative and bounded by $M(\cdot, z)$ since $k \leq 1$. Since $M(\cdot, z)$ is a minimal harmonic function, $M(\cdot, z)k$ is a constant multiple of $M(\cdot, z)$, or k is identically constant.

If D_n are subdomains increasing to D and $B \in \mathcal{F}_{\tau_{D_n}}$,

$$
\begin{aligned}
\mathbb{P}_z^x(A \cap B) = \mathbb{P}_z^x(A \circ \theta_{\tau_{D_n}}; B) &= \mathbb{E}_z^x\big[\mathbb{P}_z^{X(\tau_{D_n})}(A); B\big] \\
&= \mathbb{E}_z^x[k(X_{\tau_{D_n}}); B] \\
&= k(x)\mathbb{P}_z^x(B) \\
&= \mathbb{P}_z^x(A)\mathbb{P}_z^x(B),
\end{aligned}
$$

since k is constant. So $\mathbb{P}_z^x(A \cap B) = \mathbb{P}_z^x(A)\mathbb{P}_z^x(B)$ for $B \in \mathcal{F}_{\tau_{D_n}}$. This is true for all n, so it is true for $B \in \mathcal{F}_{\tau_D}$. Now let $B = A$, or $\mathbb{P}_z^x(A) = [\mathbb{P}_z^x(A)]^2$, hence $\mathbb{P}_z^x(A)$ is 0 or 1. $\qquad\square$

The zero-one law is not surprising if it is viewed as follows. The law of $X_{\tau_D - t}$ under $\mathbb{P}_z^x$ can be shown to be the same as the law of Brownian motion started at z and conditioned to go to x before exiting D (see Williams [1], Chung and Walsh [1], and Exercise 14). Shift invariant events are in $\mathcal{F}_{0+}$ for the time-reversed process. The analog of Corollary I.3.6 holds for h-path transforms of Brownian motion (the same proof works), so events in $\mathcal{F}_{0+}$ have probability 0 or 1.

Extensions

Since the equivalence of the Martin boundary and the Euclidean boundary follows from the boundary Harnack principle, and since the boundary Harnack principle holds in Hölder domains, one might expect that the results of this section hold for Hölder domains. That is not the case. The problem is that the constant c_1 in Proposition 2.2 now depends on k. One can show that the Martin boundary equals the Euclidean boundary for domains a little less regular than Lipschitz domains, namely, C^γ domains, where $\gamma(x) \leq ax \log\log(1/x)/\log\log\log(1/x)$, provided a is sufficiently small (Bass and Burdzy [4]). A C^γ domain is one that looks locally like the region above the graph of a C^γ functions, and a C^γ function Γ is one where $|\Gamma(x) - \Gamma(y)| \leq \gamma(|x - y|)$.

A probabilist would see the above expression for γ and assume that the "right" theorem is one where $\gamma_c(x) = cx \log\log(1/x)$ for some c. That also is incorrect: for any c there exist domains above the graph of a C^{γ_c} function for which the Martin boundary is larger than the Euclidean boundary (Bass and Burdzy [4]).

Besides domains above the graphs of functions, there are other situations where the Martin boundary has been determined. See Ancona [2].

A Denjoy domain D is a subset of $\mathbb{R}^2$ such that D^c is a subset of the x-axis. Each point on ∂D is known to be either one or two Martin boundary points. A criterion is known for such domains together with results for similar sectorial domains (Cranston and Salisbury [1]). Given an arbitrary domain D and $z \in \partial D$, can one give a condition on whether z corresponds to more than one Martin boundary point? What about domains above the graph of a function? Presumably any such condition would be in terms of capacities or the solutions to certain simpler Dirichlet problems. Bishop [1] discusses some related problems.

3 The conditional lifetime problem

Formulation

The conditional lifetime problem is this: if one has a bounded domain D and h is positive and harmonic on D, is $\mathbb{E}^x_h \tau_D < \infty$? Here $\mathbb{E}^x_h$ is the expectation of Brownian motion h-path transformed by the harmonic function h and τ_D is the time of exiting D. We will show that the answer is yes if D is a bounded domain in $\mathbb{R}^2$ or if D is a Lipschitz domain in $\mathbb{R}^d$.

This problem looks like a purely probabilistic problem with little interest to analysts. Although we give an equivalent analytic formulation, strictly speaking this is true. However, the proof we give, with a few minor modifications, also shows that Lipschitz domains are intrinsically ultracontractive (which implies all sorts of interesting things about the fundamental solution to the heat equation on D with Dirichlet boundary conditions) and that the parabolic boundary Harnack principle holds (which is the boundary Harnack principle for the heat equation); see the Extensions subsection.

The condition $\mathbb{E}^x_h \tau_D < \infty$ can be rewritten as follows. Let D_n be subdomains increasing to D. This means that D_n are domains such that $\overline{D}_n \subseteq D_{n+1}$ and $\cup_n D_n = D$. If $\mathbb{E}^x_h \tau_D < \infty$,

$$\infty > c \geq \mathbb{E}^x_h \int_0^{\tau_{D_n}} 1_{D_n}(X_s)ds$$

$$= \mathbb{E}^x \left[h(X(\tau_{D_n})) \int_0^{\tau_{D_n}} 1_{D_n}(X_s)ds \right] / h(x)$$

$$= \mathbb{E}^x \left[\int_0^\infty 1_{(s<\tau_{D_n})} \mathbb{E}^x[h(X_{\tau_{D_n}})|\mathcal{F}_s] 1_{D_n}(X_s)\,ds \right] / h(x)$$

$$= \mathbb{E}^x \left[\int_0^{\tau_{D_n}} h(X_s) 1_{D_n}(X_s)ds \right] / h(x)$$

$$= \int_{D_n} g_{D_n}(x,y)h(y)dy/h(x).$$

To obtain the next to the last inequality, we used the fact that $h(X_{s \wedge \tau_D})$ is a martingale. Letting $n \to \infty$,

$$\int_{D_n} g_{D_n}(x, y) h(y) dy \uparrow \int_D g_D(x, y) h(y) dy,$$

and so the condition $\mathbb{E}_h^x \tau_D < \infty$ is equivalent to

$$\int_D g_D(x, y) h(y) dy / h(x) < \infty.$$

Before proceeding to the proof of the main results, let us prove a simple estimate on simple random walks, well known to probabilists (Feller [1]). Suppose $Y_i = +1$ with probability p, -1 with probability $1 - p$, the Y_i are independent, and $Z_n = Z_0 + \sum_{i=1}^n Y_i$. Let $\mathbb{P}^j$ denote the law of Z given that $Z_0 = j$.

(3.1) Proposition. *If $p > 1/2$, then $\mathbb{E}^j \sum_{n=0}^\infty 1_{\{k\}}(Z_n) \leq c < \infty$, c independent of j and k.*

Proof. By Exercise 26, $\mathbb{P}^1(Z_k$ ever hits $0) \leq \rho < 1$. So

$$\mathbb{P}^0(Z_k \text{ hits } 0 \text{ for some } k > 0) = \mathbb{P}^0(Z_1 = 1, Z_k \text{ hits } 0 \text{ for some } k > 0)$$
$$+ \mathbb{P}^0(Z_1 = -1)$$
$$\leq p\rho + (1 - p) = \rho'.$$

By the strong Markov property,

$$\mathbb{P}^0(Z_k \text{ hits } 0 \text{ at least twice}) \leq (\rho')^2,$$

and similarly for hitting 0 three times, etc. Therefore

$$\mathbb{E}^0 \sum_{k=0}^\infty 1_{\{0\}}(Z_k) \leq 1 + \mathbb{E}^0 \sum_{i=1}^\infty i \mathbb{P}^0(Z_k \text{ hits } 0 \text{ at least } i \text{ times })$$

$$\leq 1 + \sum_{i=1}^\infty i(\rho')^i < \infty.$$

By the strong Markov property,

$$\mathbb{E}^{j-k} \sum_{n=0}^\infty 1_{\{0\}}(Z_n) \leq \mathbb{E}^0 \sum_{n=0}^\infty 1_{\{0\}}(Z_n),$$

and our result follows by translation invariance. $\qquad \square$

Lipschitz domains

Let D be a Lipschitz domain. Fix $x_0 \in D$. Let h be a positive harmonic function on D. Since $\mathbb{P}^x_{ch} = \mathbb{P}^x_h$, we may normalize h so that $h(x_0) = 1$. For $n \in \mathbb{Z}$, let

$$(3.1) \quad L_n = \{x \in D : h(x) = 2^n\}, \quad R_n = \{x \in D : 2^{n-1} \leq h(x) \leq 2^{n+1}\}.$$

The main estimate we need is the following.

(3.2) Theorem. *There exist $c > 0$ and $\alpha > 1$ (depending on D but not n or x) such that*

$$\mathbb{E}^x \tau_{R_n} \leq c(1 + |n|)^{-\alpha}.$$

Proof. Since D is bounded, $D \subseteq B(x, M)$ for $M = 2 \operatorname{diam}(D)$, hence

$$(3.2) \quad \mathbb{E}^x \tau_{R_n} \leq \mathbb{E}^x \tau_{B(x,M)} \leq c(\operatorname{diam}(D))^2$$

(see Exercise II.8.17). So it suffices to obtain the estimate for $|n|$ large.

Suppose $|n|$ is large but n is negative. Let $S_n = \{x : h(x) \leq 2^{n+2}\}$. Clearly $R_n \subseteq S_n$, so $\tau_{R_n} \leq \tau_{S_n}$. Recall $h(x_0) = 1$. Now by Lemma 1.9, there exists β such that if $h(x) \leq 2^{n+2}$, then $\operatorname{dist}(x, \partial D) \leq c(2^{n+2})^\beta$. Let $r_n = c2^{(n+2)\beta}$. Hence $S_n \subseteq \{x : \operatorname{dist}(x, \partial D) \leq r_n\}$.

S_n is typically a thin strip and we argue that Brownian motion cannot go very long before exiting this strip. If $V_0 = 0$, $V_1 = \inf\{t : |X_t - X_0| \geq cr_n\}$ and $V_{i+1} = V_i + V_1 \circ \theta_{V_i}$, then

$$\sup_{z \in S_n} \mathbb{P}^z(V_1 \leq \tau_{S_n}) \leq \rho < 1,$$

just as in Lemma 1.5. Hence, as in Lemma 1.5, $\mathbb{P}^y(V_j < \tau_{S_n}) \leq \rho^j$. Now

$$(3.3) \quad \mathbb{E}^x \tau_{S_n} = \sum_{j=0}^\infty \mathbb{E}^x[\tau_{S_n}; V_{j+1} > \tau_{S_n} \geq V_j]$$

$$\leq \sum_{j=0}^\infty \mathbb{E}^x[V_{j+1}; V_{j+1} > \tau_{S_n} \geq V_j]$$

$$\leq \sum_{j=0}^\infty \sum_{i=0}^j \mathbb{E}^x[V_{i+1} - V_i; V_{j+1} > \tau_{S_n} \geq V_j]$$

$$= \sum_{i=0}^\infty \sum_{j=i}^\infty \mathbb{E}^x[V_{i+1} - V_i; V_{j+1} > \tau_{S_n} \geq V_j]$$

$$\leq \sum_{i=0}^\infty \mathbb{E}^x[V_1 \circ \theta_i; \tau_{S_n} \geq V_i] = \sum_{i=0}^\infty \mathbb{E}^x[\mathbb{E}^{X_{V_i}} V_1; \tau_{S_n} \geq V_i]$$

$$\leq c \sum_{i=0}^\infty cr_n^2 \rho^i \leq cr_n^2.$$

This gives our result for such n with a great deal to spare.

For n large and positive, the proof is similar after observing that by Lemma 1.9. there exists β such that $h(x) \geq 2^{n-2}$ implies that dist $(x, \partial D) \leq c2^{-(n-2)\beta}$. $\qquad\Box$

We can now prove the following.

(3.3) Theorem. *There exists c, depending only on D and not h or x, such that $\mathbb{E}_h^x \tau_D \leq c$.*

Proof. Let $S_0 = \inf\{t : X_t \in \cup_n L_n\}$, W_i the integer n such that $X_{S_i} \in L_n$, and $S_{i+1} = \inf\{t > S_i : X_t \in \cup_n L_n - L_{W_i}\}$. The S_is are successive hits to different L_ns and the W_is are which L_ns are hit.

Let $v_n = \sup_{y \in L_n} \mathbb{E}_h^y S_1$. Let us use Proposition 3.2 to obtain a bound on v_n. We have

$$v_n = \sup_{y \in L_n} \frac{\mathbb{E}^y[S_1 h(X_{S_1})]}{h(y)}.$$

If $y \in L_n$, then $h(y) = 2^n$ and $X_{S_1} \in L_{n+1}$ or L_{n-1}; hence $h(X_{S_1}) \leq 2^{n+1}$. Therefore

$$v_n \leq \sup_{y \in L_n} 2\mathbb{E}^y S_1 \leq \sup_{y \in R_n} 2\mathbb{E}^y \tau_{R_n} \leq c(1 + |n|)^{-\alpha}.$$

Next we bound $E_h^x \tau_D$ in terms of v_n and the number of visits to L_n. Since $X_t \to \partial D$, $\mathbb{P}_h^x$-a.s.,

$$\mathbb{E}_h^x \tau_D = \mathbb{E}_h^x \sum_{i=0}^{\infty} (S_{i+1} - S_i).$$

We then write

$$\mathbb{E}_h^x \tau_D = \mathbb{E}_h^x \sum_{i=0}^{\infty} \mathbb{E}_h^{X_{S_i}}(S_1) \leq \mathbb{E}_h^x \sum_{i=0}^{\infty} v_{W_i}$$

(3.4) $$= \mathbb{E}_h^x \sum_{i=0}^{\infty} \sum_{n=-\infty}^{\infty} v_n 1_{L_n}(X_{S_i}) = \mathbb{E}_h^x \sum_{n=-\infty}^{\infty} v_n N_n,$$

where $N_n = \sum_{i=0}^{\infty} 1_{L_n}(X_{S_i})$ is the number of times that $X_{S_i} \in L_n$.

Since h is harmonic, $h(X_t)$ is a martingale. Hence if $x \in L_n$, the $\mathbb{P}^x$ probability of hitting L_{n+1} before L_{n-1} is, by Corollary I.4.10,

$$(2^n - 2^{n-1})/(2^{n+1} - 2^{n-1}) = 1/3.$$

Therefore

$$\mathbb{P}_h^x(T_{L_{n+1}} < T_{L_{n-1}}) = \frac{\mathbb{E}^x\left[h(X(T_{L_{n+1}} \wedge T_{L_{n-1}}); T_{L_{n+1}} < T_{L_{n-1}}\right]}{h(x)}$$

$$= \frac{2^{n+1}\mathbb{P}^x(T_{L_{n+1}} < T_{L_{n-1}})}{2^n} = \frac{2}{3}.$$

Similarly $\mathbb{P}_h^x(T_{L_{n-1}} < T_{L_{n+1}}) = 1/3$ for $x \in L_n$. So, under $\mathbb{P}_h^x$, W_i is a simple random walk with $p = 2/3$, $1-p = 1/3$. By Proposition 3.1, $\mathbb{E}_h^x N_n \leq c < \infty$. Substituting in (3.4),

$$\mathbb{E}_h^x \tau_D \leq \sum_{n=-\infty}^{\infty} v_n \mathbb{E}_h^x N_n \leq c \sum_{n=-\infty}^{\infty} (1 + |n|)^{-\alpha} < \infty.$$

This completes the proof. $\qquad\qquad\square$

The two-dimensional case

For the $d = 2$ result, what is different is Proposition 3.2. As we shall see, once we have the following replacement, the Cranston-McConnell result follows quickly.

(3.4) Theorem. *Suppose $D \subseteq \mathbb{R}^2$ is bounded. Then there exists c depending on D but not n or x such that $\mathbb{E}^x \tau_{R_n} \leq c|R_n|$.*

Proof. Suppose $|R_n| = r^2$. Fix $x \in R_n$, and note $|B(x, 2r) - R_n| \geq r^2/2$. Let $B = B(x, 3r)$, $A = B(x, 2r) - R_n$. By the explicit formula for g_B (Exercise II.8.45), there exists c_1 such that $g_B(x, y) \geq c_1$ on $B(x, 2r)$. So

$$c_1 r^2/2 \leq c_1|A| \leq \int g_B(x, y) 1_A(y) dy = \mathbb{E}^x \int_0^{\tau_B} 1_A(X_s) ds$$

$$= \mathbb{E}^x \left[\mathbb{E}^{X_{T_A}} \int_0^{\tau_B} 1_A(X_s) ds ; T_A < \tau_B \right]$$

$$\leq \left(\sup_{y \in B} \mathbb{E}^y \tau_B \right) \mathbb{P}^x(T_A < \tau_B)$$

$$\leq c_2 r^2 \mathbb{P}^x(T_A < \tau_B).$$

Therefore $\mathbb{P}^x(T_A < \tau_B) \geq c_1/2c_2$. To leave B before leaving R_n, the process must leave B before hitting A, or $\mathbb{P}^x(\tau_{R_n} > \tau_B) \leq 1 - c_1/2c_2 = \rho$.

We now write:

$$\mathbb{E}^x \tau_{R_n} \leq \mathbb{E}^x \tau_B + \mathbb{E}^x[\tau_{R_n} - \tau_B ; \tau_{R_n} > \tau_B]$$

$$\leq cr^2 + \mathbb{E}^x \left[\mathbb{E}^{X_{\tau(B)}} \tau_{R_n} ; \tau_{R_n} > \tau_B \right]$$

$$\leq cr^2 + \left(\sup_y \mathbb{E}^y \tau_{R_n} \right) \mathbb{P}^x(\tau_{R_n} > \tau_B)$$

$$\leq cr^2 + M\rho,$$

where $M = \sup_y \mathbb{E}^y \tau(R_n)$. Taking a supremum over x, we see $M \leq cr^2 + M\rho$. Since D is bounded, $M < \infty$, and therefore

$$M \leq cr^2/(1 - \rho) \leq c|R_n|.$$

$\qquad\qquad\square$

(3.5) Theorem. *If $D \subseteq R^2$ is bounded, there exists c independent of x and h such that $\mathbb{E}_h^x \tau_D < c$.*

Proof. We follow the proof of Theorem 3.3 almost exactly, except that here $v_n \leq c|R_n|$. Since

$$\sum_{n=-\infty}^{\infty} v_n \leq c \sum_{n=-\infty}^{\infty} |R_n| \leq c|D| < \infty,$$

our result follows. □

Extensions

It is clear from the estimate $(2^{-|n|})^\beta \leq c(1 + |n|)^{-\alpha}$ that things can be improved greatly. In fact, for $d \geq 3$, the expected lifetime can be bounded by a constant depending only on D for domains that are given locally as the region above a continuous function, and even much worse (see Davis [4], Bañuelos [4, 5], and Bass and Burdzy [3]). It is not true, however, that the expected lifetime of conditional Brownian motion is finite for every bounded domain in $\mathbb{R}^d$, $d \geq 3$; see Cranston and McConnell [1]. In $d = 2$, domains with infinite area can still have finite expected lifetime; see, e.g., Xu [1], Bañuelos and Davis [2].

In addition to conditional Brownian motion, one can also work with conditioned diffusions, where the diffusions are associated to various sorts of elliptic operators (Bass and Burdzy [3], Bañuelos [4, 5], Gao [2]).

Since $\tau_D = \int_0^{\tau_d} 1(X_s)ds$ is an additive functional with bounded potentials, then by Theorem I.6.11, τ_D actually has moments of all orders and even exponential moments. So by Chebyshev's inequality, $\mathbb{P}_h^x(\tau_D > t) \leq c_1 \exp(-c_2 t)$. An explicit value for c_2 is known; see Bañuelos and Davis [1].

These exponential estimates are just one example of what one can say if the domain D is intrinsically ultracontractive. A domain D is intrinsically ultracontractive if there exists a positive eigenfunction φ of the Laplacian in D with Dirichlet boundary conditions (i.e., $\varphi = 0$ on the points of ∂D regular for D^c, $\varphi \geq 0$ on D, and $\Delta\varphi = \lambda\varphi$ in D for some λ) and constants $c_1(t)$ and $c_2(t)$ such that

$$c_1(t)\varphi(x)\varphi(y) \leq p_D(t, x, y) < c_2(t)\varphi(x)\varphi(y).$$

This concept, introduced by Davies and Simon [1], leads to very strong estimates of $p_D(t, x, y)$, the transition density of Brownian motion killed on exiting D (equivalently, the fundamental solution of the heat equation on D with Dirichlet boundary conditions).

It turns out that being intrinsically ultracontractive is equivalent to the parabolic boundary Harnack principle holding. The parabolic boundary Harnack principle holds if for all u, there exists c depending on D and u such that for all $s, t \geq u$ and for all $v, x, y, z \in D$,

$$\frac{p_D(t,x,y)}{p_D(t,x,z)} \geq c\frac{p_D(s,v,y)}{p_D(s,v,z)}.$$

This equivalence was shown in Davis [4]. The parabolic boundary Harnack principle was first shown for Lipschitz domains by Caffarelli, Fabes, Mortola, and Salsa [1]. See Bañuelos [4], Davis [4], and Bass and Burdzy [3] for results on other domains and other operators.

Conditional gauge

The finite expected lifetime result, $\mathbb{E}_h^x \tau_D < \infty$, and the related result, $\mathbb{E}_h^x(e^{a\tau_D}) < \infty$ for some a, can be generalized by asking: for what functions q is it true that

$$(3.5) \qquad \sup_{x,h} \mathbb{E}_h^x \int_0^{\tau_D} q(X_s)ds < \infty$$

and

$$(3.6) \qquad \sup_{x,h} \mathbb{E}_h^x \exp\left(\int_0^{\tau_D} q(X_s)ds\right) < \infty?$$

Here the supremum is over $x \in \overline{D}$ and h positive and harmonic on D. Quite good results are known in d dimensions, $d \geq 3$. We present, when D is a Lipschitz domain and $q \geq 0$, a necessary and sufficient condition for (3.5) to hold and a sufficient condition for (3.6) to hold.

Recall that we write $\mathbb{E}_z^x$ for $\mathbb{E}_{M(\cdot,z)}^x$. The conditional gauge is the expression $e_q(x,z) = \mathbb{E}_z^x \exp(\int_0^{\tau_D} q(X_s)ds)$, and the conditional gauge theorem (see Cranston, Fabes, and Zhao [1]) is the assertion that for D a Lipschitz domain and q in the Kato class, if $e_q(x,z)$ is finite for one pair $(x,z) \in \overline{D} \times \overline{D}$, $x \neq z$, then the conditional gauge is bounded above and below by positive constants on $\overline{D} \times \overline{D}$. q is said to be in the Kato class if

$$(3.7) \qquad \limsup_{\varepsilon \to 0} \sup_x \int_{D \cap B(x,\varepsilon)} \frac{|q(y)|}{|x-y|^{d-2}} dy = 0.$$

One reason for looking at the conditional gauge is that when it is finite, then many results that are known for the Laplacian Δ on D, e.g., the boundary Harnack principle, the equivalence of the Martin boundary with the Euclidean one, etc., are then also true for the Schrödinger operator $Lu(x) = \Delta u(x) + q(x)u(x)$.

Let us recall the estimates

$$(3.8) \qquad g_D(v,w) \leq u(v,w) \leq c|v-w|^{2-d},$$

and

$$(3.9) \qquad g_D(v,w) \geq g_{B(v_0,\text{dist }(v_0,\partial D))}(v,w) \geq c|v-w|^{2-d},$$

$v_0 \in D$, $v, w \in B(v_0, \text{dist}\,(v_0, \partial D)/2)$. In (3.8), u is the Newtonian potential density. In (3.9), the second inequality follows from the explicit formula for the Green function in the ball, while the first follows from the fact that $B(v_0, \text{dist}\,(v_0, \partial D)) \subseteq D$ and Exercise II.8.16.

The first step, both to resolve (3.5) and toward proving the conditional gauge theorem, is to prove what is known as the "three G" theorem.

(3.6) Theorem. *Let D be a bounded Lipschitz domain in $\mathbb{R}^d$, $d \geq 3$. There exists c such that if x, y, z are three distinct points in D, then*

$$(3.10) \qquad \frac{g_D(x, y) g_D(y, z)}{g_D(x, z)} \leq c[g_D(x, y) + g_D(y, z)].$$

Proof. Let $x, y, z \in D$ and let r_x, r_y, r_z denote dist $(x, \partial D)$, dist $(y, \partial D)$, and dist $(z, \partial D)$, respectively.

To prove (3.10), we consider three cases. Case 1 is when $z \in B(y, r_y/8)$. Suppose $x \in B(y, r_y/4)$. If y is closer to x than to z, then $|y - z| \geq |x - z|/2$, as a diagram shows. Then by (3.8) and (3.9) with $v_0 = y$,

$$(3.11) \qquad \frac{g_D(y, z)}{g_D(x, z)} \leq \frac{c|y - z|^{2-d}}{|x - z|^{2-d}} \leq c,$$

which gives (3.10). If y is closer to z than to x, the exact same argument shows that $g_D(x, y)/g_D(x, z) \leq c$.

Suppose next that we are still in Case 1, that is, $z \in B(y, r_y/8)$, but now suppose $x \notin B(y, r_y/4)$. $g_D(\cdot, y)$ and $g_D(\cdot, z)$ are both harmonic functions away from y and z. Let x_1 be a point on $\partial B(y, r_y/6)$. Then by the boundary Harnack principle in $D - B(y, r_y/6)$ (see Exercise 9),

$$(3.12) \qquad \frac{g_D(x, y)}{g_D(x, z)} \leq \frac{c g_D(x_1, y)}{g_D(x_1, z)}.$$

Just as in deriving (3.11), we have $g_D(x_1, y)/g_D(x_1, z) \leq c$. Combining with (3.12) gives (3.10) again.

The second case is when $x \in B(y, r_y/8)$. Reversing the roles of x and z, this follows from Case 1.

The third case is when neither x nor z is in $B(y, r_y/8)$. Suppose first that $z \in B(x, (r_x \wedge r_y)/8)$. By (3.9), $g_D(x, z) \geq |x - z|^{2-d} \geq c r_y^{2-d}$. Since $x \notin B(y, r_y/8)$, by (3.8) $g_D(x, y) \leq c|x - y|^{2-d} \leq c r_y^{2-d}$. Therefore $g_D(x, y)/g_D(x, z) \leq c$, which gives (3.10).

Finally, suppose we are still in Case 3, but $z \notin B(x, (r_x \wedge r_y)/8)$. Let z_1 be a point in $\partial B(y, r_y/16) - B(x, (r_x \wedge r_y)/8)$. By the boundary Harnack principle in $D - B(y, r_y/16) - B(x, (r_x \wedge r_y)/8)$ (Exercise 9) for the functions $g_D(\cdot, z)$ and $g_D(y, \cdot)$,

$$\frac{g_D(y, z)}{g_D(x, z)} \leq \frac{c g_D(y, z_1)}{g_D(x, z_1)},$$

and we need to show that $g_D(y, z_1)/g_D(x, z_1)$ can be bounded by a constant independent of x, y, and z. This is just Case 1 if we replace z there by z_1.

□

We now can prove the following.

(3.7) Theorem. *Suppose D is a Lipschitz domain in $\mathbb{R}^d$, $d \geq 3$, and $q \geq 0$.*
(a) A necessary and sufficient condition for (3.5) to hold is that

$$(3.13) \qquad \sup_x \mathbb{E}^x \int_0^{\tau_D} q(X_s)\,ds < \infty.$$

(b) There exists c such that if $\sup_x \mathbb{E}^x \int_0^{\tau_D} q(X_s)ds < c$, then (3.6) holds.

Proof. (a) First of all, if (3.5) holds, let $h \equiv 1$ on D. Then

$$\sup_{x,h} \mathbb{E}^x_h \int_0^{\tau_D} q(X_s)ds \geq \sup_x \mathbb{E}^x \left[h(X_{\tau_D}) \int_0^{\tau_D} q(X_s)ds\right]/h(x)$$

$$= \sup_x \mathbb{E}^x \int_0^{\tau_D} q(X_s)ds,$$

which shows that (3.13) is necessary for (3.5) to hold.

We will show that if (3.13) holds, then

$$(3.14) \qquad \sup_{x,z} \mathbb{E}^x_z \int_0^{\tau_D} q(X_s)ds < \infty.$$

Once we have (3.14), this will imply (3.5). To see this, suppose h is harmonic and D_n are subdomains increasing up to D. $h(x) = \int_{\partial D} M(x, z)\mu(dz)$ for some measure μ, and

$$\mathbb{E}^x \left[h(X_{\tau_{D_n}}) \int_0^{\tau_{D_n}} q(X_s)ds\right] = \int \mathbb{E}^x \left[M(X_{\tau_{D_n}}, z) \int_0^{\tau_{D_n}} q(X_s)ds\right]\mu(dz)$$

$$= \int M(x, z)\mathbb{E}^x_z \int_0^{\tau_{D_n}} q(X_s)ds\,\mu(dz)$$

$$\leq \left[\sup_{x,z} \mathbb{E}^x_z \int_0^{\tau_D} q(X_s)ds\right] \int M(x, z)\mu(dz).$$

Dividing both sides by $h(x)$,

$$\mathbb{E}^x_h \int_0^{\tau_{D_n}} q(X_s)ds \leq \sup_{x,z} \mathbb{E}^x_z \int_0^{\tau_D} q(X_s)ds.$$

Letting $n \to \infty$ and using monotone convergence shows (3.5) if the supremum is over $x \in D$. A continuity argument (see Exercise 16) shows (3.5) where the supremum is over $x \in \overline{D}$.

We now turn to (3.14). Pick $x_0 \in D$. Dividing the numerator and denominator of the left-hand side of (3.10) by $g_D(x_0, z)$ gives

$$(3.15) \qquad \frac{g_D(x,y)M(y,z)}{M(x,z)} \leq c_1[g_D(x,y) + g_D(z,y)], \qquad x, y, z \in D.$$

Taking a limit as z tends to the boundary, we have (3.15) for $x, y \in D$, $z \in \overline{D}$. The left-hand side of (3.15) is the Green function for the domain D with respect to $\mathbb{P}_z^x$, that is,

$$\mathbb{E}_z^x \int_0^{\tau_D} q(X_s)ds = \int \frac{q(y)g_D(x,y)M(y,z)}{M(x,z)}dy$$
$$\leq c_1\left[\int_D q(y)g_D(x,y)dy + \int_D q(y)g_D(z,y)dy\right]$$
$$= c_1\left[\mathbb{E}^x \int_0^{\tau_D} q(X_s)ds + \mathbb{E}^z \int_0^{\tau_D} q(X_s)ds\right],$$

using Theorem 3.6. Hence if (3.13) holds, we have (3.14).

(b) As in part (a), we can restrict ourselves to bounding the quantity $\sup_{x,z} \mathbb{E}_z^x \exp(\int_0^{\tau_D} q(X_s)ds)$, $x \in D$, $z \in \partial D$. Let $A_t = \int_0^{\tau_D \wedge t} q(X_s)ds$. By the strong Markov property for $(\mathbb{P}_z^x, X_t)$,

$$\mathbb{E}_z^x[A_\infty - A_S|\mathcal{F}_S] = \mathbb{E}_z^x[A_\infty \circ \theta_S|\mathcal{F}_S] = \mathbb{E}_z^{X_S}A_\infty \leq \sup_{w,z} \mathbb{E}_z^w A_\infty,$$

where S is a stopping time less than or equal to τ_D. By part (a), this is less than or equal to $2c_1 \sup_w \mathbb{E}^w \int_0^{\tau_D} q(X_s)ds < \infty$. By Theorem I.6.11, we get (b), provided c is small enough. $\qquad\square$

For some theorems on the conditional gauge in two dimensions, see Cranston [2], Zhao [1], McConnell [2], and Bass and Burdzy [6].

4 Fatou theorems

Preliminary estimates

The Fatou theorem for the ball says that a positive harmonic function u on $B(0, 1)$ has nontangential limits, a.e. What is a nontangential limit? If $\varepsilon \in (0, 1)$ and $z \in \partial B(0, 1)$, let $C(z)$ be the interior of the smallest convex set containing $\{z\}$ and $B(0, \varepsilon)$. $C(z)$ is called a Stolz domain and looks a bit like an ice cream cone with a scoop of ice cream in it. The assertion is that for almost every $z \in \partial B(0, 1)$, $\lim_{x \to z, x \in C(z)} u(x)$ exists.

This theorem has been extended to Lipschitz domains and we prove that nontangential limits exist for a.e. boundary point, where a.e. refers to harmonic measure instead of surface measure. By the results of Sect. 5, we see that this amounts to the same thing. In addition we obtain some results on the maximal function.

All the results are essentially local ones, so we will restrict ourselves to the case where D is the region above a bounded Lipschitz function. We will leave it to the reader to make the easy modifications for bounded Lipschitz domains. We will also assume for simplicity that $d \geq 3$.

Suppose D is the region above the graph of a bounded Lipschitz function. What is the substitute for a Stolz domain? Let $b < M_\Gamma$. For $z \in \partial D$, define

$$(4.1) \quad C(z) = C_b(z) = \Big\{ x \in D :$$
$$|(x^1, \ldots, x^{d-1}) - (z^1, \ldots, z^{d-1})| < b|x^d - z^d| \Big\}.$$

Fix $z \in \partial D$ and pick $x_0 \in D$. Fix $a < 1/2$, and if $y \in C(z)$, let

$$B_y = B(y, a \operatorname{dist}(y, \partial D)).$$

(4.1) Proposition. *There exist c_1 and c_2 such that if $y \in D$, $r = \operatorname{dist}(y, \partial D)$, and $|y - x_0| \geq 2r$, then*

$$(4.2) \qquad c_1 g_D(x_0, y) r^{d-2} \leq \mathbb{P}^{x_0}(T_{B_y} < \tau_D) \leq c_2 g_D(x_0, y) r^{d-2}.$$

Proof. The idea behind these inequalities is that if Brownian motion hits B_y, it will spend about r^2 time there. Let us make this rigorous. $g_D(x_0, \cdot)$ is harmonic on $D - \{x_0\}$ and $B(y, 2r) \subseteq D$. By Harnack's inequality,

$$(4.3) \qquad G_D 1_{B_y}(x_0) = \int g_D(x_0, w) 1_{B_y}(w) dw$$
$$\geq c g_D(x_0, y)|B_y| = c g_D(x_0, y) r^d$$

and

$$(4.4) \qquad G_D 1_{B_y}(x_0) \leq c g_D(x_0, y)|B_y| = c g_D(x_0, y) r^d.$$

On the other hand,

$$(4.5) \quad G_D 1_{B_y}(x_0) = \mathbb{E}^{x_0} \int_0^{\tau_D} 1_{B_y}(X_s) ds$$
$$= \mathbb{E}^{x_0} \Big[\mathbb{E}^{X(T_{B_y})} \int_0^{\tau_D} 1_{B_y}(X_s) ds; T_{B_y} < \tau_D \Big]$$
$$\leq \mathbb{P}^{x_0}(T_{B_y} < \tau_D) \sup_{w \in \overline{B_y}} \mathbb{E}^w \int_0^{\tau_D} 1_{B_y}(X_s) ds.$$

Now

$$\mathbb{E}^w \int_0^{\tau_D} 1_{B_y}(X_s) ds \leq \mathbb{E}^w \int_0^\infty 1_{B_y}(X_s) ds = \int_{B_y} u(w, v) dv = c r^2.$$

Substituting this in (4.5), $G_D 1_{B_y}(x_0) \leq c \mathbb{P}^{x_0}(T_{B_y} < \tau_D)r^2$, and combining with (4.3) gives the left-hand inequality in (4.2).

As in (4.5),

$$(4.6) \qquad G_D 1_{B_y}(x_0) = \mathbb{E}^{x_0} \int_0^{\tau_D} 1_{B_y}(X_s)ds$$

$$\geq \mathbb{P}^{x_0}(T_{B_y} < \tau_D) \inf_{w \in \overline{B_y}} \mathbb{E}^w \int_0^{\tau_D} 1_{B_y}(X_s)ds.$$

Also,

$$\mathbb{E}^w \int_0^{\tau_D} 1_{B_y}(X_s)ds \geq \mathbb{E}^w \int_0^{\tau(B(y,2r))} 1_{B_y}(X_s)ds$$

$$= \int_{B_y} g_{B(y,2r)}(w,v)dv \geq cr^2$$

by Proposition II.3.9. This, (4.6), and (4.4) give the right-hand inequality. $\qquad \square$

The key estimate we want in this section is the following. Let $z \in \partial D$ and let $h(x) = M(x,z)$, the harmonic function with pole at z, that is 0 on $\partial D - \{z\}$, and $h(x_0) = 1$ (see Sect. 2). Throughout this section we denote P_h^x by P_z^x.

(4.2) Proposition. *There exists c, depending on a but not y, such that if $y \in C(z)$ and $|y - x_0| \geq 2 \operatorname{dist}(y, \partial D)$, then*

$$\mathbb{P}_z^{x_0}(T_{B_y} < \tau_D) > c.$$

Proof. We need to bound from below the quantity

$$\mathbb{P}_z^{x_0}(T_{B_y} < \tau_D) = \frac{\mathbb{E}^{x_0}[h(X_{T_{B_y}}); T_{B_y} < \tau_D]}{h(x_0)}.$$

Let $r = \operatorname{dist}(y, \partial D)$. Since h is harmonic and $B(y, 2r) \subseteq D$, $h(w) \geq ch(y)$ for $w \in B_y$. Recall that $h(x_0) = 1$ and $h(y) = \lim_{z' \to z} g_D(y,z')/g_D(x,z')$. So it suffices to bound from below

$$(4.7) \qquad I = \frac{g_D(y,z')}{g_D(x_0,z')} \mathbb{P}^{x_0}(T_{B_y} < \tau_D)$$

for z' close to z, $z' \in D$. Suppose $z' \in C(z)$ so that $\delta(z') \leq \delta(y)/8$ and pick $z_0 \in \partial B(y, r/8)$.

The domain $D - B(y, r/16)$ contains z_0 and z' and by the boundary Harnack principle for this domain (see Exercise 9),

$$\frac{g_D(y,z')}{g_D(x_0,z')} \geq c \frac{g_D(y,z_0)}{g_D(x_0,z_0)}.$$

By Harnack's inequality, $g_D(x_0, z_0) \leq cg_D(x_0, y)$. On the other hand, using (3.9),

$$g_D(y, z_0) \geq g_{B(y,2r)}(y, z_0) \geq c|y - z_0|^{2-d} \geq cr^{2-d}.$$

Substituting all this in (4.7) and using Proposition 4.1,

$$I \geq \frac{cr^{2-d}}{g_D(x_0, y)} cg_D(x_0, y)r^{d-2} \geq c,$$

as required. □

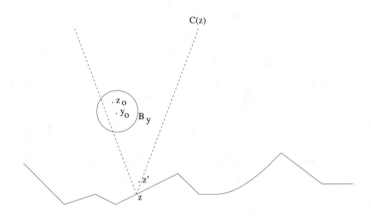

FIGURE 4.1. Diagram for the proof of Proposition 4.2.

Fatou theorems

Define $w^{x_0}(dz)$ by (2.2). We now can prove the following.

(4.3) Theorem. *Suppose u is a nonnegative harmonic function in D. Then $\lim_{y \to z, y \in C(z)} u(y)$ exists for almost every $z \in \partial D$ with respect to $w^{x_0}(dz)$.*

Proof. Let $U_t = u(X_{t \wedge \tau_D})$. For $t < \tau_D$, U_t is a martingale that is bounded below by 0, and by the martingale convergence theorem (Corollary I.4.14), $\lim_{t \to \tau_D} U_t$ exists, a.s.

$$1 = \mathbb{P}^{x_0}(\lim_{t \to \tau_D} U_t \text{ exists}) = \int \mathbb{P}_z^{x_0}(\lim_{t \to \tau_D} U_t \text{ exists}) w^{x_0}(dz)$$

by Proposition 2.7. So for almost every z,

$$\mathbb{P}_z^{x_0}(\lim_{t \to \tau_D} U_t \text{ exists}) = 1.$$

Let N be the null set of zs for which the $P_z^{x_0}$ probability is not 1 and fix a $z \notin N$. We will show $\lim_{y \to z, y \in C(z)} u(y)$ exists for this z.

By the zero-one law Theorem 2.9, there exists L such that $U_t \to L$ as $t \to \tau_D$, $\mathbb{P}_z^{x_0}$-a.s. (L depends on z.) If we have $\limsup_{y \to z, y \in C(z)} u(y) > L$, there exists ε and $y_1, y_2, \ldots \in C(z)$ such that $y_n \to z$ and $u(y_n) > L + \varepsilon$. By Harnack's inequality, there exists c_1 small such that if $B_n = B(y_n, c_1 \mathrm{dist}\,(y_n, \partial D))$, then $u \geq L + \varepsilon/2$ on B_n. By Proposition 4.2, there exists c_2 such that under $\mathbb{P}_z^{x_0}$, X_t hits each B_n with probability at least c_2. So if $A_n = (T_{B_n} < \tau_D)$, then

$$\mathbb{P}_z^{x_0}(A_n \text{ i.o.}) = \mathbb{P}_z^{x_0}(\cap_{j=1}^\infty \cup_{n=j}^\infty A_n) = \lim_{j \to \infty} \mathbb{P}_z^{x_0}(\cup_{n=j}^\infty A_n)$$

$$\geq \liminf_{j \to \infty} \mathbb{P}_z^{x_0}(A_j) \geq c_2.$$

Therefore $\mathbb{P}_z^{x_0}(T_{B_n} < \tau_D \text{ i.o.}) \geq c_2$, and hence $\mathbb{P}_z^{x_0}(U_t \geq L + \varepsilon/2 \text{ i.o.}) \geq c_2$, contradicting $U_t \to L$. Thus $\limsup_{y \to z, y \in C(z)} u(y) \leq L$. Similarly, $\liminf_{y \to z, y \in C(z)} u(y) \geq L$, which proves the theorem. $\qquad \square$

The following theorem is also known as a Fatou theorem.

(4.4) Theorem. *If $f \in L^p(\partial D, \omega^{x_0})$, $1 \leq p \leq \infty$, and $u(y) = \mathbb{E}^y f(X_{\tau_D})$, then*

$$\lim_{y \to z, y \in C(z)} u(y) = f(z)$$

for almost every $z \in \partial D$ with respect to ω^{x_0}.

Proof. Writing $f = f^+ - f^-$, where f^+, f^- are the positive and negative parts of f, respectively, it suffices to prove the theorem for $f \geq 0$. Now by the martingale convergence theorem, $U_t \to f(X_{\tau_D})$, a.s. as $t \to \tau_D$. So for almost every z, $U_t \to f(X_{\tau_D}) = f(z)$, $\mathbb{P}_z^{x_0}$ a.s. Now proceed as in the above proof with $L = f(z)$. $\qquad \square$

Maximal functions

Closely related to the Fatou theorems are maximal function inequalities. Let $U_t = u(X_{t \wedge \tau_D})$, $U^* = \sup_{s < \tau_D} |U_s|$, and $N(f)(z) = \sup_{y \in C(z)} |u(y)|$.

(4.5) Theorem. *Suppose $f \in L^p(\partial D, \omega^{x_0})$ and let $u(y) = \mathbb{E}^y f(X_{\tau_D})$. Then there exists c depending only on p such that*

(a)
$$\omega^{x_0}\left(\{z : N(f)(z) > \lambda\}\right) \leq c\mathbb{P}^{x_0}(U^* > \lambda/4);$$

(b)
$$\int |N(f)(z)|^p \omega^{x_0}(dz) \leq c \int |f(z)|^p \omega^{x_0}(dz) \qquad \text{if } 1 < p < \infty.$$

We will still restrict ourselves to the case when D is the region above the graph of a bounded Lipschitz function. Exercise 17 asks the reader to consider the case of bounded Lipschitz domains.

Proof. Since $f^+, f^- \leq |f|$ and $N(f)(z) \leq N(f^+)(z) + N(f^-)(z)$, we may again suppose $f \geq 0$. Let $A = \{z : N(f)(z) > \lambda\}$. If $z \in A$, there must exist $y \in C(z)$ with $u(y) > \lambda$. By Harnack's inequality, there exists c_1 independent of y such that $u > \lambda/2$ on $B = B(y, c_1 r)$, $r = \text{dist}(y, \partial D)$. Then $\mathbb{P}_z^{x_0}(T_B < \tau_D) \geq c_2$, or $\mathbb{P}_z^{x_0}(U^* > \lambda/2) \geq c_2$. So by Proposition 2.7

$$(4.8) \qquad c_2 \omega^{x_0}(A) \leq \int_A \mathbb{P}_z^{x_0}(U^* > \lambda/2)\, \omega^{x_0}(dz)$$

$$= \int_A \mathbb{P}_z^{x_0}(U^* > \lambda/2)\mathbb{P}^{x_0}(X_{\tau_D} \in dz)$$

$$= \mathbb{P}^{x_0}(U^* > \lambda/2, X_{\tau_D} \in A) \leq \mathbb{P}^{x_0}(U^* > \lambda/2),$$

which is (a).

For (b), multiply (a) by $p\lambda^{p-1}$ and integrate over λ from 0 to ∞. Using Proposition I.1.5 and Doob's inequality,

$$\int |N(f)(z)|^p \omega^{x_0}(dz) \leq c\mathbb{E}^x[(U^*)^p] \leq c\mathbb{E}^{x_0}[(f(X_{\tau_D}))^p]$$

$$= c\int f(z)^p \omega^{x_0}(dz).$$

This gives (b). $\qquad\qquad\qquad\qquad\qquad\qquad\qquad\qquad\qquad\qquad$ □

By taking D equal to the ball, we can recover the usual maximal inequality and Fatou theorems (see Sect. IV.1). For D equal to the half-space, recall

$$\omega^{x_0}(dz) = \frac{cx_0^d}{(|x_0^d|^2 + |x_0 - z|^2)^{d/2}}dz,$$

where x_0^d is the dth coordinate of x_0. Multiplying by $(x_0^d)^{d-1}$ and letting $x_0^d \to \infty$, we obtain the corresponding theorems for the half-space. See Chap. IV, however, where things are done more directly.

Extensions

Note that the only ingredients needed in all of the above were scale invariance, the boundary Harnack principle, and Harnack's inequality. Suppose D is a John domain with (1.31) holding, not only for $z \in D$, but for $z \in \partial D$. If $z \in \partial D$, let γ_z be the arc connecting x_0 and z. For suitably small b, define

$$C(z) = \cup_{y \in \gamma_z} B(y, br_y),$$

where $r_y = \ell(\gamma_z(y, z))$ and $\ell(\gamma_z(y, z))$ is the length of the piece of arc between y and z. Then slight modifications of the above proofs show that the Fatou theorems hold for John domains.

The usual analytic proofs of Fatou theorems proceed by first proving the maximal inequality Theorem 4.5. One key step is to prove the doubling property of harmonic measure: there exists c such that if $z \in \partial D$, then $\omega^{x_0}(B(z, 2r)) \leq c\omega^{x_0}(B(z, r))$ for all r (cf. Corollary 5.5). One feature of the boundary Harnack principle proof is that it works for domains that might not have the doubling property (for example, it is not hard to find a John domain that does not have the doubling property: take a domain that has a sharp enough cusp pointing into it).

Can Fatou theorems be proved for more general domains than Lipschitz ones? If D can be written as the union of countably many Lipschitz domains, the results can be derived for D from the results above, but we have something else in mind. If D is a simply connected domain in $\mathbb{R}^2$, then a conformal invariance argument shows that $\lim_{y \to z, y \in L_z} u(y)$ exists a.e. with respect to harmonic measure if u is positive and harmonic in D and L_z is the flow line through z determined by the vector field $\nabla g_D(x_0, \cdot)$. This means that if $y \in L_z$, then L_z is parallel to $\nabla g_D(x_0, y)$, or alternately, L_z is the image under φ of the radius connecting $\varphi^{-1}(z)$ to 0, where φ is the conformal map mapping the unit disk onto D and taking the origin to x_0. The question is, can one make some sort of similar assertion for $d \geq 3$?

5 Support of harmonic measure

Smooth domains

If one starts a Brownian motion at $x_0 \in D$, where D is a Lipschitz domain, what can one say about the distribution of X_{τ_D}? The sort of result one would like is that the $\mathbb{P}^{x_0}$ distribution of X_{τ_D} is mutually absolutely continuous with respect to surface measure on ∂D. Actually more can be said. In this section we establish the results obtained by Dahlberg [1] and reproved by different means in Jerison and Kenig [2]. Our method is a blend of the two. We also give a probabilistic demonstration of Gehring's inequality, which is needed to obtain the sharpest results.

Let σ denote surface measure on ∂D and ω^{x_0} harmonic measure starting from x_0. That is, $\omega^{x_0}(A) = \mathbb{P}^{x_0}(X_{\tau_D} \in A)$, $A \subseteq \partial D$. The main result is the following.

(5.1) Theorem. *Let D be either a bounded Lipschitz domain or else the region above the graph of a bounded Lipschitz function. Let $x_0 \in D$.*
(a) ω is absolutely continuous with respect to σ;
(b) if K is the Radon-Nikodym derivative of ω^{x_0} with respect to σ, there exists $\varepsilon > 0$ such that K is locally in $L^{2+\varepsilon}(d\sigma)$;
(c) σ is absolutely continuous with respect to ω^{x_0}.

The theorem says that harmonic measure and surface measure are mutually absolutely continuous, and the density is locally in $L^{2+\varepsilon}$ with respect to surface measure. Easy examples (see Dahlberg [1]) show that the $2+\varepsilon$ cannot be improved upon in general. K locally in $L^{2+\varepsilon}(d\sigma)$ means that for each $M > 0$, $K1_{B(0,M)} \in L^{2+\varepsilon}(d\sigma)$. For bounded domains, K locally in $L^{2+\varepsilon}(d\sigma)$ is precisely the same as $K \in L^{2+\varepsilon}(d\sigma)$; we phrase the theorem this way to cover the case of regions above the graph of bounded Lipschitz functions as well.

Observe that Theorem 5.1 is essentially a local result. If Theorem 5.1 holds for regions D above the graph of a Lipschitz function, we can obtain Theorem 5.1 easily for bounded domains. Let us therefore suppose we are in the situation that D is the region above the graph of a Lipschitz function.

A set of the form $A = B(w,r) \cap \partial D$ for some $w \in \partial D$ will be called a surface ball.

Let $\partial u/\partial n = \nabla u \cdot n$ denote the normal derivative, where $n(x)$ is the unit vector normal to the boundary at x pointing inwards. For Lipschitz domains, n is only defined for a.e. x. (See Exercise II.8.8.)

Set $v = g_D(x_0, \cdot)$. Since $v = 0$ on ∂D, it is clear by the Lipschitz nature of D that there exist two constants c_1, c_2 such that

$$(5.1) \qquad c_1 \frac{\partial v}{\partial(x^d)} \leq \frac{\partial v}{\partial n} \leq c_2 \frac{\partial v}{\partial(x^d)}.$$

Let us suppose at first that D is the region above the graph of a bounded C^∞ function Γ. We will derive estimates on harmonic measure that do not depend on the smoothness of Γ (beyond its Lipschitz constant), and we can then take a limit.

We begin by showing that for smooth domains, harmonic measure has a bounded density with respect to surface measure.

(5.2) Theorem. *Let Γ be a C^2 function such that Γ and its first and second derivatives are bounded. Suppose D is the region above the graph of Γ, $x_0 \in D$. Then $\omega^{x_0}(dy) = \mathbb{P}^{x_0}(X_{\tau_D} \in dy)$ has a density with respect to surface measure on ∂D that is bounded.*

This theorem could be obtained as an easy consequence of some results from partial differential equations. We give a proof that uses only the tools we have already developed.

A domain D satisfies a uniform exterior sphere condition if there exist $r > 0$ such that for all $z \in \partial D$, there exists a ball B_z of radius r contained in D^c with $z \in B_z \cap \partial D$.

Proof. For simplicity we will assume $d \geq 3$. We start with an estimate on $g_D(x_0, y)$ for y near the boundary of D. We show first that

(5.2) *There exists c and a such that $g_D(x_0, y) \leq c \operatorname{dist}(y, \partial D)$ if $\operatorname{dist}(y, \partial D)$ $\leq a$.*

It is easy to see that the region above Γ satisfies a uniform exterior sphere condition. Suppose $y \in D$. Let z be a point of ∂D that minimizes $|y - z|$ and let B_z be a ball of radius r contained in D^c touching ∂D at z. Then $g_D(x_0, y) \leq g_{B_z^c}(x_0, y)$ since $D \subseteq B_z^c$. (5.2) follows by the explicit formula for $g_{B_z^c}$ (Proposition II.3.9).

Next is an estimate on $\nabla g_D(x_0, y)$. Suppose $y \in D$ and $r = \text{dist}\,(y, \partial D) < a/2$. Since (5.2) implies that $g_D(x_0, \cdot)$ is bounded by $2cr$ on $B(y, r)$, then from Corollary II.1.4 we have

$$(5.3) \qquad |\nabla g_D(x_0, y)| \leq \frac{c}{r} \sup_{B(y,r)} |g_D(x_0, \cdot)| \leq c.$$

Define D_{+b} to be the region above $\Gamma(x^1, \ldots, x^{d-1}) + b$ and similarly D_{-b}. Let f be a smooth nonnegative function with compact support defined on ∂D. Extend f to all of $\mathbb{R}^d$ by defining $f(y) = f(w)$ if $w \in \partial D$ and y is above or below w. Fix b for the moment and let $v_b(x) = \mathbb{E}^x f(X_{\tau(D_{-b})})$.

Let $M > 1$ and let us apply Green's second identity to the functions v_b and g_D in $D_{+\delta} \cap B(0, M) - B(x_0, \varepsilon)$. Both v_b and g_D are harmonic and smooth in this region. We let $M \to \infty$ and argue that the terms involving $\partial B(0, M)$ go to 0. $g_D(x_0, y) \leq u(x_0, y)$, the Newtonian potential, which is bounded by a constant if y is within $(\delta \wedge \varepsilon)/2$ of the boundary of $D \cap B(0, M) - B(x_0, \varepsilon)$, and $g_D(x_0, y) \leq cM^{2-d}$ if M is large enough and y is within $(\delta \wedge \varepsilon)/2$ of $D \cap \partial B(0, M)$. By Corollary II.1.4, $|\nabla g_D(x_0, y)|$ is bounded by $c(\delta, \varepsilon)M^{2-d}$ for $y \in D_{+\delta} \cap \partial B(0, M)$. Since f is bounded, v_b is bounded, and again using Corollary II.1.4, $|\nabla v_b|$ is bounded on $D_{+\delta} \cap B(0, M)$. Since the surface area of $B(0, M)$ is less than cM^{d-1}, it follows that the terms involving the integral over $D_{+\delta} \cap B(0, M)$ tend to 0 as $M \to \infty$.

We are thus left with

$$\int_{\partial D_{+\delta}} v_b \frac{\partial g_D}{\partial n} + \int_{\partial B(x_0, \varepsilon)} v_b \frac{\partial g_D}{\partial n} = \int_{\partial D_{+\delta}} g_D \frac{\partial v_b}{\partial n} + \int_{\partial B(x_0, \varepsilon)} g_D \frac{\partial v_b}{\partial n}.$$

Precisely as in the proof of Proposition II.3.11, we can let $\varepsilon \to 0$ and see that

$$(5.4) \qquad cv_b(x_0) + \int_{\partial D_{+\delta}} g_D \frac{\partial v_b}{\partial n} = \int_{\partial D_{+\delta}} v_b \frac{\partial g_D}{\partial n}.$$

(Recall that here n denotes the inward normal while Green's identities are formulated in terms of the outward pointing normal.) Using (5.3), the right-hand side of (5.4) is bounded by $c \int_{\partial D_{+\delta}} v_b(y)\sigma(dy)$, which converges to $c \int_{\partial D} v_b(y)\sigma(dy)$ as $\delta \to 0$. By (5.2) the left-hand side of (5.4) converges to $cv_b(x_0)$.

Let e_d be the unit vector in the dth coordinate direction. By translation invariance, $v_b(y) = v_0(y + be_d)$. Every point of ∂D is regular for D^c; so if $y \in \partial D$, $v_b(y) = v_0(y + be_d) \to f(y)$ as $b \downarrow 0$. Also $v_b(x_0) = v_0(x_0 + be_d) \to v_0(x_0)$. Therefore we end up with

$$(5.5) \qquad \int_{\partial D} f(y)\, \omega^{x_0}(dy) = \mathbb{E}^{x_0} f(X_{\tau_D}) = v_0(x_0) = c \int_{\partial D} f(y)\sigma(dy).$$

That ω^{x_0} has a bounded Radon-Nikodym derivative with respect to σ follows immediately from (5.5). $\qquad\square$

L^2 estimates

We are now ready to show that $d\omega/d\sigma \in L^2(\sigma)$ for C^∞ domains with estimates only depending on M_Γ and not on any further smoothness. This will actually not take too long. The bulk of this section will be in deriving the $L^{2+\varepsilon}$ estimate.

(5.3) Proposition. *Suppose* $w \in \partial D$, $r > 0$, *and* $A = B(w, r) \cap \partial D$. *Let* y_0 *be above* w *with* $\mathrm{dist}\,(y_0, \partial D) = r$. *Then there exists* c *such that*

$$(5.6) \qquad \int_A \left(\frac{d\omega^{y_0}}{d\sigma}(w)\right)^2 \sigma(dw) \leq c/\sigma(A).$$

Proof. Pick z_0 above w with $\mathrm{dist}\,(z_0, \partial D) = r/2$. By Harnack's inequality in D, there exists c_1 such that if u is harmonic and positive in D, then $u(z_0)/u(y_0) \leq c_1$ and $u(y_0)/u(z_0) \leq c_1$. Take $\varepsilon < 1/8$ small enough so that if $x \in B(z_0, \varepsilon r)$, then $u(x)/u(z_0) \leq 2$. That this can be done follows by the expression for the constant in Harnack's inequality in Theorem II.1.19. Note if $x \in B(y_0, r/8)$, then $|x - z_0| \geq r/4$ and by Proposition I.5.8

$$\mathbb{P}^x(T_{B(z_0, \varepsilon r)} < \infty) \leq \frac{(\varepsilon r)^{d-2}}{(r/4)^{d-2}}.$$

So take ε even smaller if necessary so that

$$\sup_{x \in B(y_0, r/8)} \mathbb{P}^x(T_{B(z_0, \varepsilon r)} < \infty) \leq \frac{1}{4c_1}.$$

Write B for $B(z_0, \varepsilon r)$.

Let $h(x) = g_D(x, z_0)$, $k(x) = (\partial h/\partial(x^d))(x)$, and for $x \in \partial D$, $L(x) = (\partial h/\partial n)(x)$. We want to show

$$(5.7) \qquad \int_A L^2(w)\sigma(dw) \leq c_3/\sigma(A).$$

The key observation is that the left-hand side of (5.7) is equal to $\int_A L(w)\, \omega^{z_0}(dw) = \mathbb{E}^{z_0}[L(X_{\tau_D}); X_{\tau_D} \in A]$. We use our choices of ε, z_0, etc. to bound this in terms of $\mathbb{E}^{y_0} k(X_{\tau_D \wedge T_B})$.

Since $x \to \mathbb{E}^x(L1_A)(X_{\tau_D})$ is harmonic in D, then

$$(5.8) \qquad \mathbb{E}^{z_0}[L(X_{\tau_D}); X_{\tau_D} \in A] \leq c_1 \mathbb{E}^{y_0}[L(X_{\tau_D}); X_{\tau_D} \in A].$$

Now

$$\mathbb{E}^{y_0}[L(X_{\tau_D})] = \mathbb{E}^{y_0}[L(X_{\tau_D}); \tau_D < T_B] + \mathbb{E}^{y_0}[L(X_{\tau_D}); \tau_D > T_B]$$
$$= \mathbb{E}^{y_0}[L(X_{\tau_D}); \tau_D < T_B] + \mathbb{E}^{y_0}[\mathbb{E}^{X_{T_B}}L(X_{\tau_D}); \tau_D > T_B].$$

Again, $\mathbb{E}^z L(X_{\tau_D})$ is a positive harmonic function. For $v \in B$,

$$\mathbb{E}^v L(X_{\tau_D}) \le 2\mathbb{E}^{z_0} L(X_{\tau_D}) \le 2c_1 \mathbb{E}^{y_0} L(X_{\tau_D})$$

by our choice of ε. So

$$\mathbb{E}^{y_0} L(X_{\tau_D}) \le \mathbb{E}^{y_0}[L(X_{\tau_D}); \tau_D < T_B] + 2c_1 \mathbb{E}^{y_0} L(X_{\tau_D}) \mathbb{P}^{y_0}(\tau_D > T_B)$$
$$\le \mathbb{E}^{y_0}[L(X_{\tau_D}); \tau_D < T_B] + 2c_1 \mathbb{E}^{y_0} L(X_{\tau_D}) \mathbb{P}^{y_0}(T_B < \infty)$$
$$\le \mathbb{E}^{y_0}[L(X_{\tau_D}); \tau_D < T_B] + \mathbb{E}^{y_0} L(X_{\tau_D})/2.$$

Since our domain is smooth, L is bounded. Hence

(5.9) $\quad \mathbb{E}^{y_0}[L(X_{\tau_D}); X_{\tau_D} \in A] \le \mathbb{E}^{y_0} L(X_{\tau_D}) \le 2\mathbb{E}^{y_0}[L(X_{\tau_D}); \tau_D < T_B].$

By (5.1) this is less than or equal to

(5.10) $\qquad 2c\mathbb{E}^{y_0}[k(X_{\tau_D}); \tau_D < T_B] \le 2c\mathbb{E}^{y_0}[k(X_{\tau_D \wedge T_B})].$

Note k is harmonic on $D - B$. So $\mathbb{E}^{y_0}[k(X_{\tau_D \wedge T_B})] = k(y_0)$. Since

$$|\nabla h(y_0)| \le cr^{-1} \sup_{B(y_0, r/8)} h,$$

then

(5.11) $\qquad\qquad k(y_0) \le \frac{c}{r} r^{2-d} = cr^{1-d} \le c/\sigma(A).$

Combining (5.8), (5.9), (5.10), and (5.11), we obtain (5.7), or

(5.12) $\qquad\qquad \int_A \left(\frac{d\omega^{z_0}}{d\sigma}(w)\right)^2 \sigma(dw) \le c/\sigma(A).$

Finally, by Harnack's inequality (cf. the remarks following Definition 2.5), $\omega^{y_0}(B) \le c\omega^{z_0}(B)$ for any Borel set B, so

$$d\omega^{y_0}/d\sigma \le cd\omega^{z_0}/d\sigma.$$

$\qquad\qquad\qquad\qquad\qquad\qquad\qquad\qquad\qquad\qquad\qquad\qquad\qquad \square$

A suitable approximation scheme gives us Theorem 5.1(a). We approximate D by smooth domains D_n. One way to construct the D_ns is as follows. Let φ be a nonnegative C^∞ function with compact support that is radially symmetric and has integral 1. Let $\varphi_\varepsilon(x) = \varepsilon^{-d}\varphi(x/\varepsilon)$. Let $\Gamma_n(x) = \varphi_{1/n} * \Gamma(x) - \alpha_n$, where α_n tends to 0 slowly compared to $1/n$. Let D_n be the region above Γ_n. Since $\alpha_n \to 0$, $D_n \to D$, and moreover the

Lipschitz constants M_{Γ_n} are uniformly bounded. Since α_n is chosen to tend to 0 sufficiently slowly, $D \subseteq D_n$.

Proof of Theorem 5.1(a). Let the D_n be constructed as above. Fix y_0. Let A be a surface ball in ∂D and let $\widetilde{A}$ be the set of points in D that are above points of A. Let f be a nonnegative C^∞ function with support in A with $\int_{\partial D} f^2 \sigma_D(dz) \le 1$. Extend f to D by defining $f(y) = f(w)$ if y is above $w \in A$. Let σ_{D_n} be surface measure for the boundary of D_n and let L_n be the Radon-Nikodym derivative of $\mathbb{P}^{y_0}(X_{\tau(D_n)} \in \cdot)$ with respect to σ_{D_n}. Then

$$\mathbb{E}^{y_0} f(X(\tau_{D_n})) = \int_{\partial D_n} f(w)\mathbb{P}^{y_0}(X_{\tau_{D_n}} \in dw) = \int_{\partial D_n} f(w)L_n(w)\sigma_{D_n}(dw)$$

$$\le \left(\int_{\partial D_n} f^2(w)\sigma_{D_n}(dw) \right)^{1/2} \left(\int_{\partial D_n \cap \widetilde{A}} L_n^2(w)\sigma_{D_n}(dw) \right)^{1/2}$$

$$\le c\|f\|_{L^2(\sigma_{D_n})}/\sigma_{D_n}(\widetilde{A})^{1/2}.$$

We used the Cauchy-Schwarz inequality in the second line.

It should be clear that $\|f\|_{L^2(\sigma_{D_n})} \le c\|f\|_{L^2(\sigma)}$ and $\sigma_{D_n}(\widetilde{A}) \ge c\sigma(A)$. Since X_t is continuous, f is continuous, and $\tau_{D_n} \downarrow \tau_D$,

$$\int f(w)\omega^{y_0}(dw) = \mathbb{E}^{y_0} f(X_{\tau_D}) \le c\|f\|_{L^2(\sigma)}/\sigma_D(A)^{1/2}.$$

If $E \subseteq A$ is closed with $\sigma_D(E) = 0$, taking suitable f decreasing to 1_E shows $\omega^{y_0}(E) = 0$. So ω^{y_0} restricted to A is absolutely continuous with respect to σ restricted to A. This proves (a), but let us obtain a further estimate. Let L be the density of ω^{y_0} with respect to σ. Then

$$\int f(w)L(w)\sigma_D(dw) \le c\|f\|_{L^2(\sigma)}/\sigma_D(A)^{1/2}.$$

Taking the supremum over all such nonnegative C^∞ functions f with support in A and with $\|f\|_{L^2(d\sigma)} \le 1$ implies

(5.13)
$$\left(\int L^2(w)\sigma_D(dw) \right)^{1/2} \le c/\sigma_D(A)^{1/2}.$$

$\square$

Surface balls

To prove Theorem 5.1(b), the $L^{2+\varepsilon}$ estimate, we need to work harder. First we need some estimates on the harmonic measure of surface balls.

(5.4) Theorem. *Suppose $w \in \partial D$ and $A = B(w, r) \cap \partial D$. Let y be above w with $\mathrm{dist}\,(y, \partial D) = r$. Then there exist c_1 and c_2 such that*

$$c_1 g_D(x_0, y) r^{d-2} \leq \mathbb{P}^{x_0}(X_{\tau_D} \in A) \leq c_2 g_D(x_0, y) r^{d-2}.$$

Proof. Let $B = B(y, r/2)$. By the support theorem (Theorem I.6.6), if $z \in B$, $\mathbb{P}^z(X_{\tau_D} \in A) \geq c_3$. (We just draw a narrow tube centered about the line segment from z to w.) Then by Proposition 4.1,

$$\begin{aligned}
\mathbb{P}^{x_0}(X_{\tau_D} \in A) &\geq \mathbb{P}^{x_0}(X_{\tau_D} \in A; T_B < \tau_D) \\
&= \mathbb{E}^{x_0}\left[\mathbb{P}^{X_{T_B}}(X_{\tau_D} \in A); T_B < \tau_D\right] \\
&\geq c\mathbb{P}^{x_0}(T_B < \tau_D) \geq cg_D(x_0, y) r^{d-2}.
\end{aligned}$$

For the other direction, pick z above w with $\text{dist}\,(z, \partial D) = 7r/8$. If $x \in A$, then by the boundary Harnack principle in $D - B(z, r/16)$,

$$M(z, x) \geq \frac{cg_D(z, y)}{g_D(x_0, y)}.$$

Since $g_D(z, y) \geq g_{B(z, r/4)}(z, y) \geq cr^{2-d}$ by Proposition II.3.9, it follows that $M(z, x) \geq cr^{2-d}/g_D(x_0, y)$. Thus by Theorem 2.6,

$$1 \geq \mathbb{P}^z(X_{\tau_D} \in A) = \int_A M(z, x)\omega^{x_0}(dx) \geq c\frac{r^{2-d}}{g_D(x_0, y)}\omega^{x_0}(A),$$

or $\omega^{x_0}(A) \leq cg_D(x_0, y) r^{d-2}$ as required. $\qquad\square$

The following corollary, the doubling property of harmonic measure, follows easily from Theorem 5.4 and Harnack's inequality.

(5.5) Corollary. *There exists c such that for all $w \in \partial D$ and $r > 0$, $\omega^{x_0}(B(w, 2r) \cap \partial D) \leq c\omega^{x_0}(B(w, r) \cap \partial D)$.*

(5.6) Proposition. *Suppose D is the region above the graph of a C^∞ function. Let $w \in \partial D$, $r > 0$, $A = B(w, r) \cap \partial D$, and choose y above w with $\text{dist}\,(y, \partial D) = r$. There exists c such that if $E \subseteq A$, then*

$$\mathbb{P}^{x_0}(X_{\tau_D} \in E) \leq c\mathbb{P}^{x_0}(X_{\tau_D} \in A)\mathbb{P}^y(X_{\tau_D} \in E).$$

Proof. If we show $M(y, z)$ is bigger than $c/\mathbb{P}^{x_0}(X_{\tau_D} \in A)$ for $z \in A$, we will then have our result, since by Theorem 2.6,

$$\mathbb{P}^y(X_{\tau_D} \in E) = \int_E M(y, z)\omega^{x_0}(dz) \geq \frac{c}{\mathbb{P}^{x_0}(X_{\tau_D} \in A)}\omega^{x_0}(E).$$

Pick z_0 above w with $\text{dist}\,(z_0, \partial D) = \text{dist}\,(y, \partial D)/2$. By the boundary Harnack principle in $D - B(y, r/2)$, if $z \in A$, then

$$M(y, z) \geq cg_D(y, z_0)/g_D(x_0, z_0).$$

As we have seen several times before,

$$g_D(y, z_0) \geq g_{B(y,r/2)}(y, z_0) \geq cr^{2-d}.$$

By Harnack's inequality, $g_D(x_0, z_0) \leq cg_D(x_0, y)$. So

$$M(y, z) \geq cr^{2-d}/g_D(x_0, y).$$

By Theorem 5.4, this is larger than $c/\mathbb{P}^{x_0}(X_{\tau_D} \in A)$.   □

If $K_y(z)$ is the density of ω^y with respect to σ, we have from Proposition 5.6 that

$$(5.14) \qquad cK_y(z) \geq K_{x_0}(z)/\omega^{x_0}(A).$$

(5.7) Proposition. *Let A be a surface ball. Then*

$$\frac{1}{\sigma(A)} \int_A K_{x_0}^2(z)\sigma(dz) \leq c\left[\frac{1}{\sigma(A)} \int_A K_{x_0}(z)\sigma(dz)\right]^2.$$

Proof. Let y be above w as before. By Proposition 5.3,

$$(5.15) \qquad \int_A K_y^2(z)\sigma(dz) \leq c/\sigma(A).$$

As in the proof of Theorem 5.1(a), we take a limit to see that (5.15) is still valid in Lipschitz domains, not just C^∞ domains. By (5.14),

$$\frac{1}{[\omega^{x_0}(A)]^2} \int_A K_{x_0}^2(z)\sigma(dz) \leq c/\sigma(A),$$

or

$$\frac{1}{\sigma(A)} \int_A K_{x_0}^2(z)\sigma(dz) \leq c\left[\frac{\omega^{x_0}(A)}{\sigma(A)}\right]^2.$$

Since $\omega^{x_0}(A) = \int_A K_{x_0}(z)\sigma(dz)$, this gives what we wanted.   □

Reverse Hölder inequalities

The inequality in Proposition 5.7 is known as a reverse Hölder inequality because the converse inequality is an immediate consequence of Hölder's inequality. We now prove Gehring's inequality, which when combined with Proposition 5.7 will yield Theorem 5.1(b). Our proof is the standard proof suitably modified into martingale terms. (See also Dolèans-Dade and Meyer [1] for further results and references related to this topic.)

(5.8) Lemma. *Suppose X is a nonnegative integrable martingale, $Y = X^p$ for some $p > 1$, and Y is integrable. Let $X_n = \mathbb{E}[X|\mathcal{F}_n]$, $Y_n = E[Y|\mathcal{F}_n]$. Suppose*

(a) there exists c_1 such that $Y_n \le c_1 Y_{n-1}$, a.s., for all n;
(b) there exists c_2 such that $Y_n \le c_2 X_n^p$, a.s., for all n.
   Then there exists c_3 such that

$$(5.16) \qquad \mathbb{E}\,[X^p; X > \lambda] \le c_3 \lambda^{p-1} \mathbb{E}\,[X; X > \lambda].$$

Proof. Without loss of generality we may assume $c_2 \ge 1$. Let $\beta = 2c_2^{1/p}\lambda$. Let $N = \inf\{n : Y_n \ge \beta^p\}$. Then

$$(5.17) \qquad Y_N \le c_1 Y_{N-1} \le c_1 \beta^p.$$

Also, $\beta^p \le Y_N \le c_2 X_N^p$, or

$$(5.18) \qquad X_N \ge c_2^{-1/p}\beta = 2\lambda.$$

Let $c_4 = 2c_2^{1/p}$. If $X > c_4\lambda$, then $Y = X^p > c_4^p \lambda^p = \beta^p$. Since $Y_n \to Y$, a.s., then $Y_n > \beta^p$ for some n, or $N < \infty$ on the set where $X > c_4\lambda$.

We now bound $\mathbb{E}\,[X^p; X > c_4\lambda]$ in terms of $\mathbb{P}(N < \infty)$ and bound that in turn in terms of $\mathbb{E}\,[X; X > \lambda]$.

$$(5.19) \qquad \mathbb{E}\,[X^p; X > c_4\lambda] = \mathbb{E}\,[Y; X > c_4\lambda] \le \mathbb{E}\,[Y; N < \infty]$$
$$= \mathbb{E}\,[Y_N; N < \infty] \le c_1\beta^p \mathbb{P}(N < \infty)$$

by (5.17). By (5.18)

$$(5.20) \quad \mathbb{P}(N < \infty) \le \frac{\mathbb{E}\,[X_N; N < \infty]}{2\lambda} = \frac{\mathbb{E}\,[X; N < \infty]}{2\lambda}$$

$$\le \frac{E[X; N < \infty, X > \lambda]}{2\lambda} + \frac{\mathbb{E}\,[X; N < \infty, X \le \lambda]}{2\lambda}$$

$$\le \frac{\mathbb{E}\,[X; X > \lambda]}{2\lambda} + \frac{\lambda\mathbb{P}(N < \infty)}{2\lambda}.$$

So

$$\mathbb{P}(N < \infty) \le 2\mathbb{E}\,[X; X > \lambda]/2\lambda = \mathbb{E}\,[X; X > \lambda]/\lambda.$$

Combining (5.19) and (5.20),

$$\mathbb{E}\,[X^p; X > c_4\lambda] \le c_5 \lambda^{p-1} \mathbb{E}\,[X; X > \lambda].$$

Observing that

$$\mathbb{E}\,[X^p; X > \lambda] = \mathbb{E}\,[X^p; X > c_4\lambda] + \mathbb{E}\,[X^p; c_4\lambda \ge X > \lambda]$$
$$\le \mathbb{E}\,[X^p; X > c_4\lambda] + c_4^{p-1}\lambda^{p-1}\mathbb{E}\,[X; c_4\lambda \ge X > \lambda]$$
$$\le (c_5 + c_4^{p-1})\lambda^{p-1}\mathbb{E}\,[X; X > \lambda]$$

completes the proof. $\qquad\qquad\qquad\qquad\qquad\qquad\qquad\qquad\qquad\qquad\square$

(5.9) Theorem. (Gehring) *Let X and Y be as above. Then there exists $r > p$ and $c > 0$ such that $\mathbb{E}X^r \leq c\mathbb{E}X$.*

r and c depend on p, c_1, and c_2.

Proof. Suppose first that $\mathbb{E}X^r < \infty$. We have

$$\mathbb{E}\left[X^r; X > 1\right] = \mathbb{E}\left[X^p X^{r-p}; X > 1\right]$$

$$= \mathbb{E}\left[X^p\left(1 + (r-p)\int_1^X \lambda^{r-p-1}d\lambda\right); X > 1\right]$$

$$= \mathbb{E}\left[X^p; X > 1\right] + (r-p)\int_1^\infty \lambda^{r-p-1}\mathbb{E}\left[X^p; X > \lambda\right]d\lambda$$

$$\leq \mathbb{E}\left[X^p; X > 1\right] + c(r-p)\int_1^\infty \lambda^{r-p-1}\lambda^{p-1}\mathbb{E}\left[X; X > \lambda\right]d\lambda$$

by Lemma 5.8. This in turn is equal to

$$\mathbb{E}\left[X^p; X > 1\right] + c(r-p)\mathbb{E}\left[X\int_1^X \lambda^{r-2}d\lambda; X > 1\right]$$

$$\leq \mathbb{E}\left[X^p; X > 1\right] + \frac{c(r-p)}{r-1}\mathbb{E}\left[X^r; X > 1\right].$$

If $r > p$ but $r - p$ is small enough so that $c(r-p)/(r-1) < 1/2$, then

$$\mathbb{E}\left[X^r; X > 1\right] \leq 2\mathbb{E}\left[X^p; X > 1\right] \leq c\mathbb{E}\left[X; X > 1\right].$$

Finally
$$\mathbb{E}X^r = \mathbb{E}\left[X^r; X > 1\right] + \mathbb{E}\left[X^r; X \leq 1\right]$$
$$\leq c\mathbb{E}\left[X; X > 1\right] + \mathbb{E}\left[X; X \leq 1\right]$$
$$\leq (c+1)\mathbb{E}X.$$

To remove the restriction that $\mathbb{E}X^r < \infty$, let $X_N = X \wedge N$. By Exercise 19, we have inequality (5.16) with c_3 replaced by some c_3' (independent of N). So $\mathbb{E}X_N^r \leq c\mathbb{E}X_N \leq c\mathbb{E}X$. Now let $N \to \infty$ and use Fatou's lemma. $\square$

$L^{2+\varepsilon}$ estimates

We can now apply Gehring's inequality to prove part (b) of Theorem 5.1.

Proof of Theorem 5.1(b). Let $w \in \partial D$ and let Q be a cube of side length r centered at w. It suffices to show

$$(5.21) \qquad \int_{Q \cap \partial D} K^{2+\varepsilon}(z)\sigma(dz) < \infty,$$

where $K = K_{x_0} = d\omega^{x_0}/d\sigma$.

Define a probability measure $\mathbb{P}$ on $Q \cap \partial D$ by $\mathbb{P}(E) = \sigma(E \cap Q)/\sigma(Q \cap \partial D)$. We partition Q into 2nd equal subcubes, R_i, $i = 1, 2, \ldots, 2^{nd}$, and let $\mathcal{F}_n$ be the σ-field generated by the sets $\partial D \cap R_i$, $i = 1, 2, \ldots, 2^{nd}$. Then if X is a function on Q,

$$\mathbb{E}\left[X|\mathcal{F}_n\right](x) = \frac{\int_R X(y)\sigma(dy)}{\sigma(R)}$$

if $x \in R$ and R is one of the 2nd subcubes (cf. (I.4.2)).

If B is the surface ball with the same center as R and radius $\sqrt{d}R$, so that B circumscribes R, then by Proposition 5.7,

$$\frac{1}{\sigma(R)}\int_R K^2(y)\sigma(dy) \leq c \int_B K^2(y)\sigma(dy)/\sigma(R)$$
$$\leq c\frac{\sigma(B)}{\sigma(R)}\left(\frac{\omega^{x_0}(B)}{\sigma(B)}\right)^2.$$

Since $\sigma(B)/\sigma(R)$ is bounded by a constant and $\omega^{x_0}(B) \leq c\omega^{x_0}(R)$ by the doubling property of harmonic measure, we have

$$(5.22) \qquad \frac{1}{\sigma(R)}\int_R K^2(y)\sigma(dy) \leq c\left(\frac{\int_R K(y)\sigma(dy)}{\sigma(R)}\right)^2.$$

Now let $X(y) = K_{x_0}(y)$. We need to check the other hypotheses of Gehring's inequality. If $x \in R$, (5.22) tells us that

$$(5.23) \qquad \mathbb{E}\left[X^2|\mathcal{F}_n\right](x) = \frac{\int_R X^2(y)\sigma(dy)}{\sigma(R)}$$
$$\leq c\left(\frac{\int_R X(y)\sigma(dy)}{\sigma(R)}\right)^2 = c\left[\mathbb{E}\left[X|\mathcal{F}_n\right](x)\right]^2.$$

Also,

$$\mathbb{E}\left[X^2|\mathcal{F}_{n+1}\right](x) = \frac{\int_{R'} X^2(y)\sigma(dy)}{\sigma(R')} \leq \frac{\sigma(R)}{\sigma(R')}\frac{\int_R X^2(y)\sigma(dy)}{\sigma(R)}$$
$$\leq c\mathbb{E}\left[X^2|\mathcal{F}_n\right](x)$$

if $x \in R' \subseteq R$, $R' \in \mathcal{F}_{n+1}$, and $R \in \mathcal{F}_n$. Then by Gehring's inequality, $\mathbb{E}X^{2+\varepsilon} < \infty$ for some $\varepsilon > 0$. That is (5.21) and we are done. $\qquad \square$

Gehring's inequality is closely tied with the theory of A_p weights. Let ν be a measure on a cube Q_0 with the doubling property. A positive function w satisfies the A_p condition for some $p \in (1, \infty)$ if

$$\sup\left(\frac{1}{\nu(Q)}\int_Q w\,d\nu\right)\left(\frac{1}{\nu(Q)}\int_Q \left(\frac{1}{w}\right)^{1/(p-1)}d\nu\right)^{p-1} < \infty,$$

where the supremum is over all subcubes Q of Q_0.

The relevance to us is the following.

(5.10) Theorem. $K(z)$ *is an* A_2 *weight with respect to* $\omega^{x_0}(dz)$.

Proof. Let $\nu = \omega^{x_0}$. We have

$$\frac{1}{\sigma(R)} \int_R K^2(z)\sigma(dz) \leq c\Big(\frac{1}{\sigma(R)} \int_R K(z)\sigma(dz)\Big)^2.$$

Let $w(x) = K(x)$. Then we write

$$\sigma(R) \int_R K(z)\nu(dz) \leq c\Big(\int_R \nu(dz)\Big)^2,$$

or

$$\Big(\frac{1}{\nu(R)} \int_R \frac{1}{K(z)} \nu(dz)\Big)\Big(\frac{1}{\nu(R)} \int_R K(z)\nu(dz)\Big) \leq c.$$

Thus w is an A_2 weight with respect to ν. $\qquad\square$

See Exercise 20, Garnett [1], or Coifman and Fefferman [1] for more on A_p weights.

Mutual absolute continuity

We now prove Theorem 5.1(c). Define μ on ∂D by

(5.24) $\quad \mu(E) = \big|\{(x^1, \ldots, x^{d-1}) : (x^1, \ldots, x^{d-1}, \Gamma(x^1, \ldots, x^{d-1})) \in E\}\big|.$

Note there exist c_1 and c_2 such that $c_1 \leq d\mu/d\sigma \leq c_2$.

Proof of Theorem 5.1(c). Suppose $\sigma(E) > 0$ but $\omega^{y_0}(E) = 0$. Let Q_{zk} denote the cube of the form $[j_1/2^k, (j_1 + 1)/2^k) \times \cdots \times [j_d/2^k, (j_d + 1)/2^k)$ which contains z, where $j_1, \ldots, j_d$ are integers. By Exercise 13, there exists $z \in E$ such that

(5.25) $$\frac{\mu(Q_{zk} \cap E)}{\mu(Q_{zk})} \to 1.$$

(In fact, this is true for almost every point of E.)
   Let

$$I_k = Q_{zk} \cap \partial D \qquad \text{and} \qquad E_k = I_k \cap E.$$

Let z_k be a point above z with $\delta(z_k) = 2^{-k}$. [Recall the definition of $\delta(x)$ from (1.6).]

   By scaling and the support theorem (Theorem I.6.6), $\omega^{z_k}(I_k) = \mathbb{P}^{z_k}(X_{\tau_D} \in I_k) \geq c$ (cf. the proof of Theorem 5.4). Since $\omega^{y_0}(E) = 0$, then $\omega^{z_k}(E) = 0$ by Harnack's inequality. If w_k is the center of Q_{zk}, let $\widehat{w}_k$ be the point above w_k with $\delta(\widehat{w}_k) = 2^{-k}$. Then by Harnack's inequality again, there exists c such that $d\omega^{z_k}/d\omega^{\widehat{w}_k} \leq c$. Hence $d\omega^{z_k}/d\sigma$ has a density H_k in $L^2(\sigma)$ with $\int_{I_k} H_k^2(w)\sigma(dw) \leq c/\sigma(I_k)$.

So

$$\omega^{z_k}(I_k - E_k) = \int_{I_k - E_k} H_k(w)\sigma(dw)$$

$$\leq c\left(\sigma(I_k)^{-1}\right)^{1/2}\left(\sigma(I_k - E_k)\right)^{1/2}$$

by the Cauchy-Schwarz inequality. This is less than or equal to

$$c\left[\frac{\mu(I_k - E_k)}{\mu(I_k)}\right]^{1/2} = c\left[1 - \frac{\mu(I_k \cap E_k)}{\mu(I_k)}\right]^{1/2} \to 0.$$

This is a contradiction to $\omega^{z_k}(E_k) = 0$. □

We could actually have done this proof immediately following the proof of Theorem 5.1(a), except for the minor details of switching from balls to cubes. If we had looked ahead and used Theorem IV.1.2 and let Q_{zk} be the ball about z of radius 2^{-n}, we could have avoided using cubes altogether.

Extensions

A Lipschitz function is one where $\nabla\Gamma \in L^\infty$. We can define L_1^p domains to be ones that are given locally as the region above the graph of a function Γ with $\nabla\Gamma \in L^p$. Jerison and Kenig [1] have obtained the analog of Theorem 5.1 for L_1^p domains when $p > d - 1$.

Some of the results concerning the Dirichlet problem in Lipschitz domains can also be approached via layer potentials. See Verchota [1], Fabes, Jodeit, and Rivière [1], and Jerison and Kenig [2].

Suppose one considers regions above the graph of a function that is not necessarily Lipschitz. What can one say about the support of harmonic measure? It is not true in general that $\omega^{x_0} \ll \mu$, where μ is defined by (5.24). One can even find domains with finite surface measure for which this is not true. One might assert that harmonic measure is supported on those points where Γ is differentiable. This also is not precisely true because one can perturb the domain by sets of small capacity without affecting things too much. Can one give a more exact condition?

6 Exercises and further results

Exercise 1. Suppose Γ is a continuous (not necessarily Lipschitz) bounded function and D is the region above the graph of Γ. Suppose every point of ∂D is regular for D^c. Suppose $x_1 \in \partial D$ and $x_0 \in D$ with $\delta(x_0) = 1$. Suppose that the conclusion of Theorem 1.2 holds with $K = \overline{B(x_1, 2)}$ and $V = B(x_0, 4)$. Show there exists c such that if u is positive and harmonic in D, u vanishes continuously on $B(x_0, 4) \cap \partial D$, and $u(x_0) = 1$, then u

is bounded by c on $B(x_0, 2) \cap D$. In other words, the boundary Harnack principle implies Carleson's estimate.

Exercise 2. Let D be the region above the graph of a continuous (not necessarily Lipschitz) region and let $F_a = \{x \in D : \delta(x) < a\}$. Show there exists $\rho < 1$ such that $|\partial B(z, 2a) \cap F_a| \leq \rho |\partial B(z, 2a)|$ whenever $z \in F_a$.

Exercise 3. Suppose D is the region above the graph of a Hölder function [see (1.32)]. State and prove the appropriate analog to Lemma 1.6.

Exercise 4. Prove (1.20) and (1.21).

Exercise 5. Let D_1 and D_2 be two Lipschitz domains and let $r > 0$ so that $B(0, r) \cap D_1 = B(0, r) \cap D_2 \neq \emptyset$. Suppose g_{D_1} has a finite strictly positive normal derivative on $B(0, r/2) \cap \partial D$. Prove that g_{D_2} has a finite strictly positive normal derivative on $B(0, r/2) \cap \partial D$.

Exercise 6. Suppose K is a compact set contained in an open set V and there exists c such that $g_D(x, y)/g_D(x', y) \leq c$ whenever $x, x' \in K \cap D$, $y \in D \cap \partial V$. Show the boundary Harnack inequality holds for the triple (K, V, D).

Exercise 7. Let $\{G_i\}$ be a finite collection of bounded open sets covering a compact set $F \subseteq \mathbb{R}^d$. A C^∞ partition of unity subordinate to $\{G_i\}$ is a collection of C_K^∞ functions h_i such that the support of h_i is contained in G_i for each i and $\sum_i h_i = 1$ on F. Prove a partition of unity exists.

Exercise 8. Suppose D is a bounded Lipschitz domain, f is a function defined on ∂D, and for each $x \in \partial D$, there exists a C^∞ function $h_x : \mathbb{R}^d \to \mathbb{R}$ and $r_x > 0$ such that f agrees with h_x in $B(x, r_x)$. Show there exists a C^∞ function $h : \mathbb{R}^d \to \mathbb{R}$ such that f is the restriction of h to ∂D.

   Hint: Use a partition of unity.

Exercise 9. Let D be a Lipschitz domain. Show there exists a constant c depending only on D such that if $y \in D$, then we have the boundary Harnack principle in $D - B(y, r)$ (i.e., the conclusion of Theorem 1.2 holds) with constant c provided $r < \text{dist}(y, \partial D)$.

Exercise 10. Suppose D is a Lipschitz domain, u and v are positive and continuous, and u and v vanish continuously on $V \cap \partial D$ for some open set V. Show u/v is Hölder continuous on $K \cap D$ if K is a compact subset of V.

Exercise 11. Let Γ be a Lipschitz function and D the region above the graph of Γ. Show there exists a constant c such that for all r and x, we have the boundary Harnack principle in $Q(x, r, r)$ (i.e., the conclusion of Theorem 1.2 holds) with constant c.

Exercise 12. Suppose f is a bounded function and there exists $\rho < 1$ such that $\text{Osc}_{B(x,r)} f \leq \rho \, \text{Osc}_{B(x,2r)} f$ for all x and all $r < 1$. Show f is Hölder continuous.

Hint: The hypothesis implies that if $y \in B(x, 2^{-n})$, then $|f(x) - f(y)| \leq \rho^n \|f\|_\infty$.

Exercise 13. (a) Let $\mathcal{F}_n$ be an increasing collection of σ–fields on $\mathbb{R}^d$ such that each $\mathcal{F}_n$ is generated by finitely many disjoint sets $A_{n1}, \ldots, A_{nm_n}$. Show that if ν is a finite measure, $f \geq 0$ is integrable, and

$$g_n(x) = \sum_{i=1}^{m_n} \frac{\int_{A_{ni}} f \, d\nu}{\nu(A_{ni})} 1_{A_{ni}}(x),$$

(with the convention $0/0 = 0$), then $g_n(x) \to f(x)$ for almost every x.
(b) Use (a) to prove (2.3).
(c) Use (a) to prove (5.25).

Hint: Cf. Exercise I.8.16.

Exercise 14. Suppose D is a Lipschitz domain, $z \in \partial D$, and $(\mathbb{P}_z^x, X_t)$ is Brownian motion h-path transformed by the harmonic function $M(\cdot, z)$. If $x_0 \in \partial D$, show $\mathbb{P}_z^x$ has a weak limit as $x \to x_0$ with $x \in D$. We call the weak limit $\mathbb{P}_z^{x_0}$.

Hint: Use the boundary Harnack principle to show tightness. Use the arguments of Sect. 2 to show the limit exists.

Exercise 15. Show that q is in the Kato class if and only if

$$\lim_{K \to \infty} \sup_{x \in D} \int \frac{|q(y)| 1_{(|q(y)| \geq K)}}{|x - y|^{d-2}} dy = 0.$$

Exercise 16. If $q \geq 0$ and $\sup_{x \in D} \mathbb{E}_z^x \int_0^{\tau_D} q(X_s) ds < \infty$, show

$$\sup_{x \in \overline{D}} \mathbb{E}_z^x \int_0^{\tau_D} q(X_s) ds < \infty.$$

Exercise 17. State and prove the analog of Theorem 4.5 for bounded Lipschitz domains.

Hint: The definition of $C(z)$ needs to be modified.

Exercise 18. Suppose f is a nonnegative function, $p > 1$, $c > 1$, and

$$\frac{1}{|Q|} \int_Q f(x)^p dx \leq \left(\frac{c}{|Q|} \int_Q f(x) dx \right)^p$$

for all cubes Q. Show $f > 0$, a.e.

Exercise 19. Suppose $X \geq 0$, $p > 1$, and there exists c_1 such that for all λ,

$$\mathbb{E}[X^p; X > \lambda] \leq c_1 \lambda^{p-1} \mathbb{E}[X; X > \lambda].$$

Let $X_N = X \wedge N$. Show there exists c_2, independent of N and X, such that

$$\mathbb{E}\left[X_N^p; X_N > \lambda\right] \le c_2 \lambda^{p-1} \mathbb{E}\left[X_N; X_N > \lambda\right].$$

Hint: Choose c_3 large but fixed. For λ such that $N < \lambda$ or $\lambda \le N < c_3\lambda$, the result is easy. For $\lambda \le c_3^{-1}N$, use the inequality

$$\mathbb{E}\left[X^p - N^p; X > N\right] \ge pN^{p-1}\mathbb{E}\left[X - N; X > N\right]$$
$$\ge pc_3^{p-1}\lambda^{p-1}\mathbb{E}\left[X - N; X > N\right].$$

Exercise 20. Suppose w is an A_p weight. (a) Show that w is an A_r weight for all $r > p$. (b) Show there exists $\varepsilon > 0$ such that w is an $A_{p-\varepsilon}$ weight.

Hint: Use Gehring's inequality.

Exercise 21. Let D be a region as in Theorem 5.2. (a) Show that D satisfies a uniform interior sphere condition (i.e., the uniform exterior sphere condition for D^c). (b) Use (a) to prove there exists $c > 0$ such that $d\omega^{x_0}/d\sigma \ge c$.

Exercise 22. For each $\varepsilon > 0$ construct an example where $d\omega^{x_0}/d\sigma$ is not locally in $L^{2+\varepsilon}(d\sigma)$.

Hint: Look at $d = 2$ and look at a wedge with the appropriate aperture.

Exercise 23. Give an example of a John domain where harmonic measure does not have the doubling property.

Hint: Let $d = 2$ and let D be the unit ball with the interval $[0,1] \times \{0\}$ removed.

Exercise 24. Show that harmonic functions are real analytic. That is, if x is in a domain D and h is harmonic in D, then h can be expanded in a Taylor series about x with radius of convergence equal to dist $(x, \partial D)$.

Hint: Show it first for the Poisson kernel in a ball.

Exercise 25. Suppose h is a nonconstant harmonic function in a domain D. Show there cannot exist a subdomain on which h is constant.

Hint: Use Exercise 24.

Exercise 26. Let p, Y_i, and Z_n be as in Proposition 3.1. Show

$$\mathbb{P}^1(Z_k \text{ ever hits } 0) < 1.$$

Hint: Cf. Exercise I.8.18.

Notes

The boundary Harnack principle for Lipschitz domains was proved by Dahlberg [1], Ancona [1], and Wu [1]. Later, another proof was given by Jerison and Kenig [2] and still another, a probabilistic one, was given in Bass and Burdzy [1]. A proof can be given using heat kernel estimates

(Bañuelos [5]). The proof using the box method follows Bass and Burdzy [1]. The proof using the SDE method is new.

The results on the Martin boundary were first established by Hunt and Wheeden [2] in 1970, well before the boundary Harnack principle. Other proofs have been given by Jerison and Kenig [2] and Bass and Burdzy [1]. We follow the proof of Bass and Burdzy [1].

The conditional lifetime problem was posed by K.L. Chung around 1983 and solved in domains in $\mathbb{R}^2$ shortly thereafter by Cranston and Mc-Connell [1]. For Lipschitz domains in $\mathbb{R}^d$ the results were first obtained in Cranston [1]. The proof given here follows Bass and Burdzy [3]; see that paper for many other references to the literature. Our account of the conditional gauge material and the three G theorem follows Cranston, Fabes, and Zhao [1].

The original Fatou theorems are classical. See, for example, Doob [2]. Some of our approach comes from Durrett [1]. The Fatou theorem for Lipschitz domains was first proved by Hunt and Wheeden [1]. The idea of proving a Fatou theorem using the boundary Harnack principle has its roots in Doob [2].

Theorem 5.1 was first proved by Dahlberg [1], with a later proof by Jerison and Kenig [2].

IV
SINGULAR INTEGRALS

1 Maximal functions

Formulation

Let Z_t be a Brownian motion in $\mathbb{R}^2$, let u be a function that is harmonic in the upper half-plane, and let v be the conjugate harmonic function to u. The basic connection between Brownian motion and singular integrals comes about through the observation that $v(Z_t)$ is a martingale transform of $u(Z_t)$. In this chapter we will exploit this relationship and similar ones to obtain many results about the Hilbert transform, Riesz transform, Littlewood–Paley functionals, and other singular integrals.

This section contains some basic properties of the maximal function. Certain of these results are consequences of what was proved in Chap. III, but here the derivation is much more direct.

In this section we will be working in the half-space $D = \mathbb{R}^d \times [0, \infty)$. We will denote elements of D by $z = (x, y)$ with $x \in \mathbb{R}^d$, $y \in [0, \infty)$. Let X_t be a d-dimensional Brownian motion, Y_t a one-dimensional Brownian motion independent of X_t, and set $Z_t = (X_t, Y_t)$. Instead of τ_D we will just write τ. We will write $x = (x^1, \ldots, x^d)$, $\partial_j u$ for $\partial u / \partial x^j$, and $\partial_y u$ for $\partial u / \partial y$.

Given a reasonable function $f : \mathbb{R}^d \to \mathbb{R}$, (for example, $f \in L^p$ for some $p \in [1, \infty]$) we define the harmonic extension u of f by

$$u(z) = \mathbb{E}^z f(X_\tau).$$

Of course, u is harmonic. From Theorem II.1.16, we can also write

$$u(z) = u(x, y) = P_y f(x) = \int f(z) P_y(x - z) dz,$$

where

(1.1) $\qquad P_y(t) = \dfrac{c_d y}{(y^2 + |t|^2)^{(d+1)/2}}, \qquad c_d = \Gamma((d+1)/2)/\pi^{(d+1)/2}$

is the Poisson kernel for the half-space. $P_y(t)$ can also be expressed as $c_d y / |(y, t)|^{d+1}$. When h_w is the positive harmonic function defined by $h_w(t, y) = P_y(t - w)$, we write $\mathbb{P}_w^z$ instead of $\mathbb{P}_{h_w}^z$ for the h-path transform of Brownian motion by the harmonic function h_w.

We make the observation that if $|t| \leq y$, then $P_y(t) \geq c y^{-d}$, hence

(1.2) $$P_y(t) \geq \frac{c}{|B(0, y)|} 1_{B(0, y)}(t).$$

Given $f : \mathbb{R}^d \to \mathbb{R}$, let

(1.3) $$A_r f(x) = \frac{1}{|B(0, y)|} \int_{B(0, r)} f(x + h) \, dh,$$

the average of f over the ball of radius r centered at x. The function

(1.4) $$M f(x) = \sup_r \frac{1}{|B(0, r)|} \int_{B(0, r)} |f(x + h)| \, dh$$

is called the Hardy–Littlewood maximal function, and one of the main results of this section is the following.

(1.1) Theorem.
(a) $|\{x : M f(x) > \lambda\}| \leq c \|f\|_1 / \lambda$.
(b) If $1 < p \leq \infty$, then $\|M f\|_p \leq c_p \|f\|_p$.

Part (a) is known as a weak (1-1) inequality. If $f = 1_{B(0,1)}$, then $M f(x) \sim c / |x|^d$ for $|x|$ large, and we see (b) cannot hold for $p = 1$.

As a corollary to Theorem 1.1, we will obtain the Lebesgue density theorem.

(1.2) Theorem.
(a) If $1 < p < \infty$ and $f \in L^p$, then $\|A_r f - f\|_p \to 0$ as $r \to 0$.
(b) If $1 \leq p \leq \infty$ and $f \in L^p$, then $\lim_{r \to 0} A_r f = f$, a.e.

The standard proofs of Theorems 1.1 and 1.2 use a covering lemma such as the one of Vitali. We will prove Theorem 1.1 probabilistically (cf. Sect. III.4).

In the literature there are a number of ways of overcoming the fact that probabilities are finite measures while Lebesgue measure is not. One is to define $\mathbb{P}^\mu(A) = \int \mathbb{P}^z(A)\mu(dz)$, set $\mu = \nu \times \delta_a$, where ν is Lebesgue measure on $\mathbb{R}^d$ and δ_a is point mass at a, and then let $a \to \infty$. Another is to define the so-called background radiation, which is Brownian motion started at ∞ and run on the time scale $(-\infty, 0]$. What we will do is look at $c_d^{-1} s^d \mathbb{P}^{(0,s)}(X_\tau \in dy) = s^{d+1}/(s^2 + y^2)^{(d+1)/2} dy$, and then let $s \to \infty$. Since $\mathbb{P}^{(0,s)}(X_\tau \in dy)$ is a genuine probability measure, we can still use the material of Chap. I and II without checking to see whether they remain valid for measures with infinite mass.

Proof of main results

The key to Theorem 1.1 is the following. Let

$$(1.5) \qquad C_b(x) = \{(w, y) \in D : |w - x| < by\}.$$

When $b = 1$ we will write just $C(x)$. Let

$$(1.6) \qquad N_b(f)(x) = \sup_{z \in C_b(x)} |u(z)|,$$

$$N_b^A(f)(x) = \sup\{|u(w, y)| : (w, y) \in C_b(x), y < A\}.$$

Again we drop the b from the notation if $b = 1$. Let

$$(1.7) \qquad U_t = u(Z_{t \wedge \tau}), \qquad U^* = \sup_{t < \tau} |U_t|.$$

(1.3) Proposition. *Fix A and R and suppose $f \geq 0$. There exists c, not depending on A and R, and s_0, depending on A and R, such that if $\lambda > 0$ and $s > s_0$, then*

$$|\{x : N_b^A(f)(x) > \lambda\} \cap B(0, R)| \leq cs^d \mathbb{P}^{(0,s)}(U^* > \lambda/2).$$

Proof. The first step is to show that for s sufficiently large,

$$(1.8) \quad \text{If } N_b^A(f)(x) > \lambda \text{ and } |x| \leq R, \text{ then } \mathbb{P}_x^{(0,s)}(U^* > \lambda/2) \geq c.$$

Suppose $N_b^A(f)(x) > \lambda$ for some $|x| < R$. Then $u(w, y) > \lambda$ for some $(w, y) \in C_b(x)$ with $y \leq A$. By Harnack's inequality, there exists a ball B_y of radius cy about $z = (w, y)$ such that $u > \lambda/2$ on B_y. We want to show

$$(1.9) \qquad \mathbb{P}_x^{(0,s)}(T_{B_y} < \tau) \geq c$$

for s sufficiently large. Let $S_y = \inf\{t : Y_t \leq y\}$. $S_y < \tau$, so

$$\mathbb{P}_x^{(0,s)}(T_{B_y} < \tau) \geq \mathbb{P}_x^{(0,s)}(Z_{S_y} \in B_y).$$

By the definition of $\mathbb{P}_z^{(0,s)}$, the right-hand side is

$$\frac{\mathbb{E}^{(0,s)}[h_x(Z_{S_y}); Z_{S_y} \in B_y]}{h_x((0, s))}.$$

Since h_x is nonnegative and harmonic, by Harnack's inequality, $h_x(v) \geq ch_x(w, y)$ for $v \in B_y$. So it suffices to bound from below

$$(1.10) \qquad \frac{h_x((w, y))\mathbb{P}^{(0,s)}(Z_{S_y} \in B_y)}{h_x((0, s))} = \frac{ch_x((w, y))}{h_x((0, s))} \int_{B_y \cap \partial H_y} \frac{s - y}{|v - s|^{d+1}} dv.$$

Here $H_y = \mathbb{R}^d \times [y, \infty)$. We have used the fact that the distribution of Z_{S_y} is given by the Poisson kernel. As $s \to \infty$, $h_x((0, s)) \sim c/s^d$ and $(s - y)/|v - s|^{d+1} \sim c/s^d$. Since $(w, y) \in C_b(x)$, then $0 \leq |w - x| \leq by$, hence $h_x((w, y)) \geq cy/y^{d+1} = c/y^d$. The volume of $B_y \cap \partial H_y$ is cy^d. Hence the right-hand side of (1.10) is larger than

$$c\frac{y^{-d}}{s^{-d}}s^{-d}y^d = c$$

for s large.

Substituting in (1.10) then gives (1.9). Since $u > \lambda/2$ on B_y, then $(U^* > \lambda/2)$ on the set $(T_{B_y} < \tau)$ and we obtain (1.8).

For the second and final step of the proof, let $E = \{x : N_b^A(f)(x) > \lambda\}$. Then by Proposition III.2.7

$$s^d \mathbb{P}^{(0,s)}(U^* > \lambda/2) \geq \int_{B(0,R) \cap E} s^d \mathbb{P}_x^{(0,s)}(U^* > \lambda/2)\mathbb{P}^{(0,s)}(X_\tau \in dx)$$

$$\geq c \int_{B(0,R) \cap E} s^d \mathbb{P}^{(0,s)}(X_\tau \in dx) \geq c|B(0, R) \cap E|.$$

$\square$

The lemma is all we need to prove Theorem 1.1.

Proof of Theorem 1.1. By writing $f = f^+ - f^-$, it suffices to assume $f \geq 0$. To prove (a), we use Doob's inequality (Theorem I.4.6) to see

$$\mathbb{P}^{(0,s)}(U^* > \lambda/2) \leq 2\mathbb{E}^{(0,s)}U_\tau/\lambda.$$

Combining with Proposition 1.3,

$$|B(0, R) \cap \{x : N_b^A(f)(x) > \lambda\}| \leq cs^d \mathbb{E}^{(0,s)}U_\tau/\lambda,$$

if s is large enough. Let $s \to \infty$, and note that the right-hand side converges to $c\|f\|_1/\lambda$, since $U_\tau = f(X_\tau)$. Now let $R \to \infty$ and then $A \to \infty$. We thus have

$$|\{x : N_b(f)(x) > \lambda\}| \leq c\|f\|_1/\lambda.$$

Since by (1.2) $Mf(x) \leq cN_b(f)(x)$, (a) follows.

Part (b) is similar. Multiplying the statement of Lemma 1.3 by $p\lambda^{p-1}$ and integrating over λ from 0 to ∞, we have

$$\int_{B(0,R)} [N_b^A(f)(x)]^p dx \leq cs^d \mathbb{E}^{(0,s)}(U^*)^p.$$

By Doob's inequality (Theorem I.4.7), the right-hand side is bounded by $cs^d\mathbb{E}^{(0,s)}|U_\tau|^p$, which converges to $c\|f\|_p^p$ as $s \to \infty$. Now let $R \to \infty$, then $A \to \infty$. Thus $\|N_b(f)\|_p \le c\|f\|_p$. (b) follows. $\qquad\square$

(1.4) Corollary. *If $b > 0$, there exists c such that*
(a) $|\{x : N_b(f)(x) > \lambda\}| \le c\|f\|_1/\lambda.$
(b) *If $1 < p \le \infty$, then $\|N_b(f)\|_p \le c_p\|f\|_p$.*

We obtain Theorem 1.2 as a corollary to Theorem 1.1.

Proof of Theorem 1.2. Again writing $f = f^+ - f^-$, we may suppose $f \ge 0$. Let $\varepsilon > 0$ and write $f = g + h$, where g is continuous with compact support and $\|h\|_p < \varepsilon$. It is easy to see that $A_r g \to g$ uniformly. So

$$\limsup_{r \to 0} \|A_r f - f\|_p \le \limsup_{r \to 0} \|A_r f - A_r g\|_p + \|f - g\|_p$$
$$\le \|Mh\|_p + \varepsilon \le c\varepsilon.$$

Since ε is arbitrary, this gives (a).

For (b), note that the result is a local one, i.e., it suffices to show the result for almost every x in $B(0, R)$ for each R. So it is enough to look at $f 1_{B(0,2R)}$, and we therefore restrict attention to $f \in L^1$.

Pick g, h as above. Let $\delta > 0$. We have $\limsup_{r \to 0} |A_r f(x) - A_r g(x)| \le Mh(x)$. So

$$\{x : \limsup |A_r f(x) - f(x)| > \delta\}$$
$$\subseteq \{x : Mh(x) > \delta/2\} \cup \{x : |h(x)| > \delta/2\}.$$

By Theorem 1.1 and Chebyshev's inequality,

$$|\{x : \limsup |A_r f(x) - f(x)| > \delta\}| \le c\|h\|_1/\delta + 2\|h\|_1/\delta \le c\varepsilon/\delta.$$

Since ε is arbitrary, $|\{x : \limsup |A_r f(x) - f(x)| > \delta\}| = 0$. Since δ is arbitrary, we have $\limsup |A_r f(x) - f(x)| = 0$, a.e. $\qquad\square$

Extensions

There is a local version of the nontangential limit theorems that applies when u is not necessarily positive. Suppose u is harmonic in D. u is nontangentially bounded at $x \in \partial D$ if $\sup_{z \in C_b(x)} |u(z)| < \infty$. u has a nontangential limit at $x \in \partial D$ if $\lim_{z \to x, z \in C_a(x)} u(z)$ exists.

(1.5) Theorem. *Suppose $a < b$ and suppose u is nontangentially bounded on $E \subseteq \partial D$. Then u has nontangential limits at almost every point of E.*

Proof. By looking at $E \cap B(0, M)$ and letting $M \to \infty$, we may suppose E is bounded. By looking at $E_N = \{x \in E : \sup_{z \in C(x)} |u(z)| \leq N\}$ and letting $N \to \infty$, we may suppose u is nontangentially bounded by N on E.

Let $G = \cup_{x \in E} C(x)$. G is sometimes called a sawtooth domain. (cf. Fig. 6.1) It is clear that G is the region above the graph of a Lipschitz function with Lipschitz constant $1/b$. u is bounded in absolute value by N on G, so $u + N$ is nonnegative on G. By Theorem III.4.3, the nontangential limit of $u + N$, hence of u, exists at almost every point of ∂G. However, $E \subseteq \partial G$, hence the nontangential limit of u exists at almost every point of E. □

In the above theorem we used cones $C_b(x)$ for the definition of nontangentially bounded and cones $C_a(x)$ for the definition of nontangential limits with $a < b$. The apertures are actually unimportant and the theorem is still true if $a \geq b$. See Durrett [1].

In Theorem 1.1, can Lebesgue measure be replaced by something else? If we define

$$M_\mu f(x) = \sup_r \frac{1}{\mu(B(x,r))} \int_{B(x,r)} |f(w)| \mu(dw),$$

then the analog of Theorem 1.1 holds provided that μ satisfies a doubling condition: there exists c such that for all x and r we have $\mu(B(x,2r)) \leq c\mu(B(x,r))$. See Garnett [1]. More interesting is the result that the inequality

$$\|Mf\|_{L^p(\mu)} \leq c\|f\|_{L^p(\mu)}, \qquad 1 < p < \infty$$

[here M is the usual maximal function as defined by (1.4)] holds if and only if μ has a density with respect to Lebesgue measure and the density is an A_p weight (cf. Theorem III.5.10). See Garnett [1] for a proof of this.

Approximations to the identity

The last thing we want to do in this section is give a converse to (1.2). Let φ be nonnegative and integrable on $\mathbb{R}^d$ and let $\varphi_r(x) = r^{-d}\varphi(x/r)$. Suppose φ is radially symmetric, decreasing as a function of $|x|$, and $c_1 = \int |\varphi(x)| dx < \infty$. A typical φ would be $1_{B(0,1)}(x)$. Another would be $P_1(x)$.

(1.6) Theorem. (Approximation to the identity)
(a) If $f \in L^p$, $1 \leq p \leq \infty$, then

$$\sup_{r>0} |f * \varphi_r(x)| \leq c_1 Mf(x).$$

(b) If $f \in L^p$, $1 \leq p < \infty$, and $\int \varphi(x)dx = 1$, then $\|f * \varphi_r - f\|_p \to 0$ as $r \to 0$.
(c) If $f \in L^p$, $1 \leq p \leq \infty$ and $\int \varphi(x)dx = 1$, then $\lim_{r \to 0}(f * \varphi_r)(x) = f(x)$, a.e.

Proof. In proving (a), by translation invariance and scaling, we need only show $f * \varphi(0) \leq c_1 M f(0)$. First suppose φ is piecewise constant on annuli: there exist $a_1 \leq a_2 \leq \cdots \leq a_n$ and $A_1 \geq A_2 \geq \cdots \geq A_n$ such that $\varphi(x) = A_1$ on $|x| \leq a_1$, $\varphi(x) = A_i$ if $a_{i-1} < |x| \leq a_i$, and $\varphi(x) = 0$ if $|x| > a_n$. Then

$$
\begin{aligned}
f * \varphi(0) &= \int f(x)\varphi(x)\,dx \\
&= A_1 \int_{B(0,a_1)} f + A_2 \int_{B(0,a_2)-B(0,a_1)} f + \cdots + A_n \int_{B(0,a_n)-B(0,a_{n-1})} f \\
&= (A_1 - A_2) \int_{B(0,a_1)} f + (A_2 - A_3) \int_{B(0,a_2)} f + \cdots + A_n \int_{B(0,a_n)} f \\
&\leq \left[(A_1 - A_2)|B(0,a_1)| + \cdots + A_n|B(0,a_n)|\right] M f(0) \\
&= \left[A_1|B(0,a_1)| + A_2|B(0,a_2) - B(0,a_1)| + \cdots \right. \\
&\qquad \left. + A_n|B(0,a_n) - B(0,a_{n-1})|\right] M f(0).
\end{aligned}
$$

Note the coefficient of $Mf(0)$ in the last expression is just $\int \varphi(x)dx$. To handle the general case, just approximate φ by φ_n of the above form and take a limit.

To prove (b),

$$
f * \varphi_r(x) - f(x) = \int [f(x-y) - f(x)]\varphi_r(y)dy,
$$

so by Jensen's inequality

$$
\|f * \varphi_r - f\|_p^p \leq \int \|f(\cdot - y) - f(\cdot)\|_p^p \varphi_r(y)dy = \int \|f(\cdot - ry) - f(\cdot)\|_p^p \varphi(y)dy.
$$

Let $\varepsilon > 0$ and write $f = g + h$ where g is continuous with compact support and $\|h\|_p < \varepsilon$. Then $\|g(\cdot - ry) - g(\cdot)\|_p \to 0$ and $\|h(\cdot - ry) - h(\cdot)\|_p \leq 2\varepsilon$. Hence $\limsup \|f * \varphi_r - f\|_p \leq 2\varepsilon$. Since ε is arbitrary, we have (b).

Finally there is (c). If $p < \infty$, we proceed exactly as in the proof of Theorem 1.2(b), using part (a). So there remains the case $p = \infty$. Let R be arbitrary, and we need to show that $f * \varphi_r(x) \to f(x)$ a.e. for $x \in B(0, R)$. Write $f = f1_{B(0,2R)} + f1_{B(0,2R)^c}$. Since f is bounded, $f1_{B(0,2R)}$ is in L^1 and we obtain our result for this function by the $p = 1$ result. Set $h = f1_{B(0,2R)^c}$. If $x \in B(0, R)$, then $h(x) = 0$, and

$$
|h * \varphi_r(x)| = \left| \int h(x-y)\varphi_r(y)dy \right| \leq \int_{|y| \geq R} \varphi_r(y)dy \, \|h\|_\infty,
$$

since $h(x-y) = 0$ if $x \in B(0, R)$ and $|y| < R$. Note now that $\|h\|_\infty \leq \|f\|_\infty$ and

$$\int_{|y|\geq R} \varphi_r(y)dy = \int_{|y|\geq R/r} \varphi(y)dy \to 0$$

as $r \to 0$. □

The assumptions of nonnegativity and of radial symmetry may be dispensed with as long as $\int \psi(x)dx < \infty$, where $\psi(x) = \sup_{|y|\geq|x|} |\varphi(y)|$. This follows since

$$|f * \varphi_r(x)| \leq |f| * \psi_r(x),$$

where $\psi_r(x) = r^{-d}\psi(x/r)$.

2 Hilbert transforms

Basic properties

The Hilbert transform is an operator on functions defined by

$$(2.1) \qquad Hf(x) = \lim_{\varepsilon \to 0, N \to \infty} \frac{1}{\pi} \int_{N>|y|>\varepsilon} \frac{f(x-y)}{y} dy.$$

Of course, $1/y$ is not absolutely integrable, so even for $f \equiv 1$, $\int_{N_1>y>\varepsilon_1} dy/y$ will not have a limit as $\varepsilon_1 \to 0$ or $N_1 \to \infty$. If we take integrals over symmetric intervals instead, however, cancelation takes place. To see how this works, let us show the limit exists for each x if $f \in C_K^1$, where C_K^1 denotes the C^1 functions with compact support.

(2.1) Proposition. *If $f \in C_K^1$,*

$$\lim_{\varepsilon \to 0, N \to \infty} \int_{N>|y|>\varepsilon} \frac{f(x-y)}{y} dy$$

exists for every x.

Proof. Fix x. Since f has compact support, $f(x-y)$ will be 0 for $|y|$ large. There is thus no problem with the limit as $N \to \infty$, and we may concentrate on the limit as $\varepsilon \to 0$.

$$\int_{\varepsilon_2 \geq |y| > \varepsilon_1} \frac{f(x-y)}{y} dy = \int_{\varepsilon_2 \geq |y| > \varepsilon_1} \frac{f(x-y) - f(x)}{y} dy$$

since $1/y$ is odd. Now $|f(x-y) - f(x)| \leq \|f'\|_\infty |y|$, and so

$$\left| \int_{|y|>\varepsilon_1} \frac{f(x-y)}{y} \, dy - \int_{|y|>\varepsilon_2} \frac{f(x-y)}{y} \, dy \right|$$

$$\leq \int_{\varepsilon_2 \geq |y| > \varepsilon_1} \frac{|f(x-y) - f(x)|}{|y|} \, dy$$

$$\leq \|f'\|_\infty \int_{\varepsilon_2 \geq |y| > \varepsilon_1} dy$$

$$\leq 2|\varepsilon_2 - \varepsilon_1| \, \|f'\|_\infty.$$

Hence $\int_{|y|>\varepsilon} f(x-y)/y \, dy$ is a Cauchy sequence in ε, and therefore the limit exists. $\qquad\square$

What is more interesting is that Hf exists for almost every x if $f \in L^p$, $1 \leq p < \infty$. We have the following.

(2.2) Theorem. (M. Riesz)
(a) If $1 < p < \infty$ and $f \in C_K^1 \cap L^p$, then $\|Hf\|_p \leq c_p \|f\|_p$;
(b) $|\{x : |Hf(x)| > \lambda\}| \leq c \|f\|_1 / \lambda$ if $f \in C_K^1$.

Note the M. Riesz inequalities are only for f in C_K^1. Once we have the inequalities for those fs, (a) and (b) allow us to define Hf for all $f \in L^p$, since C_K^1 is a dense subset of L^p. Later we shall see that variants of the Riesz inequalities allow us to show that the limit in (2.1) exists if $f \in L^p$ and agrees with the extension given by Theorem 2.2.

Hf turns out to be related to conjugate harmonic functions. To see the connection, let us calculate the Fourier transform of Hf. The Fourier transform of a function will be denoted $\widehat{f}$. Let

(2.2) $$H_{\varepsilon N}(x) = \frac{1}{\pi x} 1_{(N>|x|>\varepsilon)}.$$

(2.3) Proposition. If $f \in C_K^1$, then $\widehat{Hf}(\xi) = i \operatorname{sgn}(\xi) \widehat{f}(\xi)$, a.e.

Proof. Let us estimate $\widehat{H}_{\varepsilon N}$. Since $1/x$ is odd,

$$\int_{N>|x|>\varepsilon} \frac{e^{i\xi x}}{x} \, dx = 2i \int_{N>x>\varepsilon} \frac{\sin(\xi x)}{x} \, dx.$$

This is 0 if ξ is 0, and is equal to $-2i \int_{N>x>\varepsilon} \sin(|\xi|x)/x \, dx$ if $\xi < 0$. Also

$$\int_{N>x>\varepsilon} \frac{\sin(|\xi|x)}{x} \, dx = \int_{|\xi|N>x>|\xi|\varepsilon} \frac{\sin x}{x} \, dx,$$

which is well known to converge boundedly to the value $\pi/2$ as $N \to \infty$ and $\varepsilon \to 0$. Therefore $\widehat{H}_{\varepsilon N}(\xi) \to i \operatorname{sgn}(\xi)$ pointwise and boundedly.

By Plancherel's identity,

$$\|H_{\varepsilon_1 N_1} f - H_{\varepsilon_2 N_2} f\|_2^2 = c \int |\widehat{H}_{\varepsilon_1 N_1}(\xi) - \widehat{H}_{\varepsilon_2 N_2}(\xi)|^2 |\widehat{f}(\xi)|^2 d\xi.$$

This tends to 0 as $\varepsilon_1, \varepsilon_2 \to 0$ and $N_1, N_2 \to \infty$ by dominated convergence and the fact that $\|\widehat{f}\|_2 = c\|f\|_2 < \infty$. Therefore $H_{\varepsilon N} f$ converges in L^2 as $\varepsilon \to 0$ and $N \to \infty$. Since $H_{\varepsilon N} f$ converges pointwise to Hf by Proposition 2.1, it converges to Hf in L^2. By Plancherel's identity again, the Fourier transform of $H_{\varepsilon N} f$ converges in L^2 to the Fourier transform of Hf. The Fourier transform of $H_{\varepsilon N} f$ is $\widehat{H}_{\varepsilon N}(\xi)\widehat{f}(\xi)$, which converges pointwise to $i \operatorname{sgn}(\xi)\widehat{f}(\xi)$. $\square$

(2.4) Proposition. *Suppose $f \in C_K^1$. Let u be the harmonic extension of f and let v be the harmonic extension of Hf. Then u and v are conjugate harmonic functions.*

Proof. We will show that u and v satisfy the Cauchy-Riemann conditions by looking at their Fourier transforms. Recall from Theorem II.1.16 that $\widehat{P}_y(\xi)$, the Fourier transform of the Poisson kernel in x with y held fixed, is $e^{-y|\xi|}$.

$u(x, y) = [P_y(\cdot) * f](x)$. Then the Fourier transform of u (in each of the formulas below the Fourier transform is in the x variable only, with y considered to be fixed) is

$$(2.3) \qquad \widehat{u}(\xi, y) = \widehat{P}_y(\xi)\widehat{f}(\xi) = e^{-y|\xi|}\widehat{f}(\xi).$$

Also,

$$(2.4) \qquad \widehat{\partial_x u}(\xi, y) = i\xi \widehat{u}(\xi, y) = i\xi e^{-y|\xi|}\widehat{f}(\xi)$$

and

$$(2.5) \qquad \widehat{\partial_y u}(\xi, y) = -|\xi| e^{-y|\xi|}\widehat{f}(\xi).$$

We obtain (2.5) simply by differentiating (2.3). Similarly,

$$\widehat{v}(\xi, y) = e^{-y|\xi|}\widehat{Hf}(\xi) = i \operatorname{sgn}(\xi)e^{-y|\xi|}\widehat{f}(\xi),$$

hence

$$(2.6) \qquad \widehat{\partial_x v}(\xi, y) = i\xi e^{-y|\xi|} i \operatorname{sgn}(\xi)\widehat{f}(\xi)$$

and

$$(2.7) \qquad \widehat{\partial_y v}(\xi, y) = -|\xi| e^{-y|\xi|} i \operatorname{sgn}(\xi)\widehat{f}(\xi).$$

Comparing (2.4) and (2.7) and comparing (2.5) and (2.6), we see that the Cauchy-Riemann equations hold for almost all pairs (x, y). Since u and v are both harmonic in D, they are both continuous, and hence the Cauchy-Riemann equations hold everywhere. $\square$

Riesz' theorem

Recall that if u and v are conjugate harmonic functions, then the Cauchy-Riemann equations imply that $|\nabla u| = |\nabla v|$. We now have the following.

Proof of Theorem 2.2(a). Let u and v be the harmonic extensions of f and Hf, respectively. Let $U_t = u(Z_{t \wedge \tau})$ and $V_t = v(Z_{t \wedge \tau})$. Note

$$\langle U \rangle_t = \int_0^{t \wedge \tau} |\nabla u|^2(Z_r)\,dr = \int_0^{t \wedge \tau} |\nabla v|^2(Z_r)\,dr = \langle V \rangle_t.$$

By Theorem I.6.8,

$$
\begin{aligned}
(2.8) \quad s\mathbb{E}^{(0,s)}|V_\tau|^p &\le cs\mathbb{E}^{(0,s)}|V_0|^p + cs\mathbb{E}^{(0,s)}\langle V \rangle_\tau^{p/2} \\
&= cs\mathbb{E}^{(0,s)}|V_0|^p + cs\mathbb{E}^{(0,s)}\langle U \rangle_\tau^{p/2} \\
&\le cs\mathbb{E}^{(0,s)}|V_0|^p + cs\mathbb{E}^{(0,s)}|U_\tau|^p + cs\mathbb{E}^{(0,s)}|U_0|^p.
\end{aligned}
$$

Let $s \to \infty$. In a moment we will show

$$(2.9) \qquad\qquad s\mathbb{E}^{(0,s)}|U_0|^p \to 0$$

and similarly with U_0 replaced by V_0. Then (2.8) yields

$$\int |Hf(x)|^p dx \le c \int |f(x)|^p dx,$$

as required. (Recall $\lim s\mathbb{E}^{(0,s)} h(X_\tau) = c \int h(x)\,dx$.)

Let us show (2.9). Under $\mathbb{P}^{(0,s)}$,

$$U_0 \equiv u(0,s) = c \int \frac{s}{z^2 + s^2} f(z)\,dz \le \frac{c}{s}\|f\|_1.$$

So if $p > 1$, $s\mathbb{E}^{(0,s)}|U_0|^p \le cs(s^{-p}\|f\|_1^p) \to 0$ as $s \to \infty$.

To take care of V_0, note by Exercise 7 that there exists c (depending on f) such that $|Hf(x)| \le c(1 + |x|)^{-1}$, since $f \in C_K^1$. So $Hf \in L^q$ for all $q > 1$. For any $r > 1$,

$$\int P_s(z)^r dz = c \int \left(\frac{s}{s^2 + y^2}\right)^r dy = cs^{1-r} \int \left(\frac{1}{1 + y^2}\right)^r dy \le cs^{1-r}.$$

Take $q < p$ and define r by $r^{-1} + q^{-1} = 1$. Under $\mathbb{P}^{(0,s)}$, $V_0 \equiv v(0,s)$. Then $|v(0,s)| = |P_s Hf(0)| \le \|P_s\|_r \|Hf\|_q$. Therefore

$$s\mathbb{E}^{(0,s)}|V_0|^p = s|v(0,s)|^p \le cs(s^{1/r-1})^p.$$

Since $q < p$, then $1 - 1/r > 1/p$, and the right-hand side tends to 0 as $s \to \infty$. $\qquad\square$

Weak (1–1) inequality

Now let us prove part (b) of Theorem 2.2.

Proof of Theorem 2.2(b). By Itô's formula,

$$U_t = u(Z_{\tau \wedge t}) = u(Z_0) + \int_0^{\tau \wedge t} \nabla u(Z_s) \cdot dZ_s.$$

A similar equation holds for v. By the Cauchy-Riemann equations, $\partial_x v = -\partial_y u$ and $\partial_y v = \partial_x u$. Therefore, if B is the matrix

$$(2.10) \qquad\qquad B = \begin{pmatrix} 0 & -1 \\ 1 & 0 \end{pmatrix},$$

then

$$V_t = v(Z_{\tau \wedge t}) = v(Z_0) + \int_0^{\tau \wedge t} (B\nabla u)(Z_r) \cdot dZ_r.$$

Let $\lambda > 0$ and $S = \inf\{t : |U_t| > \lambda\}$. Then

$$s\mathbb{P}^{(0,s)}(|V_\tau| > \lambda) \leq s\mathbb{P}^{(0,s)}(U_\tau^* \geq \lambda) + s\mathbb{P}^{(0,s)}(|V_\tau| > \lambda, U_\tau^* < \lambda).$$

On $(U_\tau^* < \lambda)$, $S \geq \tau$. If $U_t^\lambda = U_{t \wedge S}$ and $V_t^\lambda = V_{t \wedge S}$, then

$$s\mathbb{P}^{(0,s)}(|V_\tau| > \lambda, U_\tau^* < \lambda) = s\mathbb{P}^{(0,s)}(|V_\tau^\lambda| > \lambda) \leq \frac{s\mathbb{E}^{(0,s)}(V_\tau^\lambda)^2}{\lambda^2}$$

by Chebyshev's inequality. As in the proof of part (a),

$$s\mathbb{E}^{(0,s)}(V_\tau^\lambda)^2 \leq cs\mathbb{E}^{(0,s)}(V_0^\lambda)^2 + cs\mathbb{E}^{(0,s)}\langle V^\lambda \rangle_\tau,$$

while

$$\langle V^\lambda \rangle_\tau = \int_0^{\tau \wedge S} |B\nabla u(Z_r)|^2 dr = \int_0^{\tau \wedge S} |\nabla u(Z_r)|^2 dr = \langle U^\lambda \rangle_\tau$$

and

$$s\mathbb{E}^{(0,s)}\langle U^\lambda \rangle_\tau \leq cs\mathbb{E}^{(0,s)}(U_\tau^\lambda - U_0^\lambda)^2.$$

Hence

$$(2.11) \quad s\mathbb{P}^{(0,s)}(|V_\tau| > \lambda, U_\tau^* < \lambda) \leq \frac{c\mathbb{E}^{(0,s)}|U_\tau^\lambda - U_0^\lambda|^2}{\lambda^2} + cs\mathbb{E}^{(0,s)}V_0^2/\lambda^2$$

$$\leq \frac{c\lambda}{\lambda^2} s\mathbb{E}^{(0,s)}|U_\tau^\lambda| + cs\mathbb{E}^{(0,s)}V_0^2/\lambda^2$$

$$+ cs\mathbb{E}^{(0,s)}|U_0|/\lambda.$$

Since $|U_t|$ is a submartingale, the first term on the right in (2.11) can be bounded by $cs\mathbb{E}^{(0,s)}|U_\tau|/\lambda$. By Doob's inequality, $s\mathbb{P}^{(0,s)}(U_\tau^* \geq \lambda) \leq s\mathbb{E}^{(0,s)}|U_\tau|/\lambda$.

Letting $s \to \infty$, we have our result, except for the terms involving U_0, V_0. These are a little more delicate than in the proof of Theorem 2.2(a); the details are Exercise 8. $\qquad\square$

Extensions

As we discussed following the statement of Theorem 2.2, Theorem 2.2 allows us to define Hf for all $f \in L^p$. Let

$$(2.12) \qquad H^* f(x) = \sup_{N,\varepsilon} |H_{\varepsilon N} f(x)|.$$

We have the following.

(2.5) Proposition. $H^* f(x) \leq cMf(x) + M(Hf)(x).$

Proof. This is Exercise 32. □

As an immediate corollary of this proposition, we have that if $f \in L^p$,

$$(2.13) \qquad \|H^* f\|_p \leq c\|f\|_p, \qquad 1 < p < \infty.$$

Then, just as in the proof of Theorem 1.2, we see that

$$(2.14) \qquad \lim_{\varepsilon \to 0, N \to \infty} H_{\varepsilon N} f(x) = Hf(x), \qquad \text{a.e.,}$$

if $f \in L^p$, $1 < p < \infty$. The almost everywhere existence of the limit also holds if $f \in L^1$; see Stein [1].

Just as for the maximal function, it turns out that the inequality

$$\|Hf\|_{L^p(\mu)} \leq c\|f\|_{L^p(\mu)}$$

holds for all $f \in L^p(\mu)$ if and only if μ has a density that is an A_p weight. See Garnett [1] for a proof.

There is a notion of Hilbert transforms of Banach-space-valued functions. Using probabilistic techniques, Burkholder [2] gave a condition on the Banach space that is sufficient for this Hilbert transform to be bounded in L^p, while Bourgain [1] showed that this condition is also necessary.

Related to the Hilbert transform is the Cauchy transform, or Hilbert transform on Lipschitz curves. See Stein [3] and Coifman, Jones, and Semmes [1]. It would be nice to have a probabilistic proof of the boundedness of the Cauchy transform.

3 Riesz transforms

Fourier transforms

The higher-dimensional analogs of the Hilbert transform are the Riesz transforms. We define the jth Riesz transform by

$$(3.1) \qquad R_j f(x) = \lim_{\varepsilon \to 0, N \to \infty} \int_{N > |y| > \varepsilon} K_j(y) f(x - y)\, dy, \qquad j = 1, \ldots, d,$$

where

$$(3.2) \qquad K_j(y) = \frac{cy^j}{|y|^{d+1}}, \qquad c = \Gamma((d+1)/2)/\pi^{(d+1)/2}$$

(y^j is the jth coordinate of y). The value of c is chosen to make $\widehat{K}_j$ turn out nicely. This is again a singular integral, and the existence of the limit relies on cancelation. Just as in Proposition 2.1, since $\int_{N>|y|>\varepsilon} K_j(y)\,dy = 0$, $R_j f(x)$ exists and is well-defined if $f \in C_K^1$.

We will need the Fourier transform of R_j.

(3.1) Proposition. *If $f \in C_K^\infty$,*

$$\widehat{R_j f}(\xi) = \mathrm{i}\frac{\xi^j}{|\xi|}\widehat{f}(\xi).$$

Proof. As in Proposition 2.3, to calculate $\widehat{R_j f}$, we need to calculate

$$\lim_{\varepsilon \to 0, N \to \infty} \widehat{K_j 1}_{(\varepsilon < |\cdot| < N)}(\xi) = c \lim \int_{\varepsilon < |y| < N} e^{\mathrm{i}\xi \cdot y} K_j(y)\,dy.$$

Changing to polar coordinates $r = |y|, \theta = y/|y|, u = \xi/|\xi|, R = |\xi|$, this is

$$\lim_{\varepsilon \to 0, N \to \infty} c \int_{\partial B(0,1)} \int_\varepsilon^N e^{\mathrm{i}Rru\cdot\theta} \frac{\Omega_j(\theta)}{r}\,dr\,\sigma_1(d\theta),$$

where $\Omega_j(\theta) = \theta^j$ and σ_1 is normalized surface measure on $\partial B(0,1)$. Since Ω_j is odd, as in the proof of Proposition 2.3, this is

$$\lim c\mathrm{i} \int_{\partial B(0,1)} \int_\varepsilon^N \frac{\sin(Rru\cdot\theta)}{r}\Omega_j(\theta)\,dr\,\sigma_1(d\theta)$$

$$= c \int_{\partial B(0,1)} \mathrm{i}\,\mathrm{sgn}\,(Rru\cdot\theta)\Omega_j(\theta)\,\sigma_1(d\theta)$$

$$= c\mathrm{i} \int_{\partial B(0,1)} \mathrm{sgn}\,\left(\frac{\xi}{|\xi|}\cdot\theta\right)\theta^j\,\sigma_1(d\theta).$$

The interchange of the limit and the integral is justified by dominated convergence.

If $\xi = e_k$ is the unit vector in the kth coordinate direction, then

$$\int_{\partial B(0,1)} \mathrm{sgn}\,\left(\frac{\xi}{|\xi|}\cdot\theta\right)\theta^j\sigma_1(d\theta) = \int_{\partial B(0,1)} \mathrm{sgn}\,(\theta^k)\theta^j\sigma_1(d\theta).$$

By symmetry this is 0 if $k \neq j$ and a constant independent of k and j if $k = j$, and so $\widehat{K}_j(e_k) = c\mathrm{i}\delta_{kj}$.

From this we get a formula for $\widehat{K}_j(\xi)$ for any ξ with $|\xi| = 1$. The d–tuple $K(y) = (K_1(y), \ldots, K_d(y)) = cy/|y|^{d+1}$ preserves rotations: if A is an orthogonal matrix, then $K(Ay) = AK(y)$. Therefore the same is true of

$$\widehat{K}(\xi) = (\widehat{K}_1(\xi), \ldots, \widehat{K}_d(\xi)) = ci \int_{\partial B(0,1)} \operatorname{sgn}\left(\frac{\xi}{|\xi|} \cdot \theta\right) \theta \, \sigma_1(d\theta).$$

Given ξ, let A be an orthogonal matrix such that $\xi = Ae_k$. Then

$$\widehat{K}_j(\xi) = \widehat{K}(\xi) \cdot e_j = \widehat{K}(Ae_k) \cdot e_j = A\widehat{K}(e_k) \cdot e_j = ciA_{jk},$$

since $\widehat{K}(e_k) = cie_k$. However, $A_{jk} = \xi^j$. For general ξ, scaling gives us $\widehat{K}_j(\xi) = ci\xi^j/|\xi|$.

For the appropriate choice of the constant in the definition of K_j in (3.2), we have $\widehat{K}_j(\xi) = i\xi^j/|\xi|$. One can check that the c given in (3.2) is the correct choice, but the exact value is of no importance to us. □

L^p boundedness

We will prove the boundedness of the Riesz transforms by two related methods, the first due to Bennett [1]. Let $f : \mathbb{R}^d \to \mathbb{R}$ with $f \in L^p$, let u be the harmonic extension of f, and let

$$(3.3) \quad S(f) = \left(\int_0^\tau |\nabla u(Z_r)|^2 dr\right)^{1/2}, \qquad S_1(f) = \left(\int_0^\tau |\partial_y u(Z_r)|^2 dr\right)^{1/2}.$$

(3.2) Lemma. *Suppose $F \geq 0$, $\int \int_0^\infty yF(x,y) \, dy \, dx < \infty$, and there exists c_1 and $\beta > 0$ such that*

$$\sup_{(x,y)\in B((0,s),s/2)} F(x,y) \leq c_1 s^{-d-2-\beta}, \qquad s \geq 1.$$

Then there exists c_2 not depending on F such that

$$(3.4) \quad \lim_{s\to\infty} s^d \mathbb{E}^{(0,s)} \int_0^\tau F(X_r, Y_r) dr = c_2 \int \int_0^\infty yF(x,y) \, dy \, dx.$$

Proof. For simplicity we will assume $d \geq 3$; a very similar proof works for $d = 2$. Recall that the Green function for $D = \mathbb{R}^d \times (0, \infty) \subseteq \mathbb{R}^{d+1}$ is given by

$$g_D((x_1, y_1), (x_2, y_2))$$
$$= c\Big\{ \big[(x_1 - x_2)^2 + (y_1 - y_2)^2\big]^{-(d-1)/2}$$
$$- \big[(x_1 - x_2)^2 + (y_1 + y_2)^2\big]^{-(d-1)/2}\Big\}.$$

We will show in a little bit that there exists c_3 such that for each $z = (x, y)$,

(3.5) $$s^d g_D((0, s), z) \to c_3 y \qquad \text{as } s \to \infty,$$

and there exists c_4 such that if $|z - (0, s)| \geq s/2$, then

(3.6) $$s^d g_D((0, s), z) \leq c_4 y.$$

Assuming (3.5) and (3.6) for the moment, let us complete the proof. Let $\varepsilon > 0$ and let M be large so that $\int \int_0^\infty y F(x, y) 1_{B(0,M)^c}(x, y) dy\, dx < \varepsilon$. Then

$$s^d \mathbb{E}^{(0,s)} \int_0^\tau F(X_r, Y_r) dr \geq s^d \mathbb{E}^{(0,s)} \int_0^\tau F(X_r, Y_r) 1_{B(0,M)}(X_r, Y_r)\, dr$$

$$= s^d \int \int_0^\infty F(x, y) 1_{B(0,M)}(x, y) g_D((0, s), (x, y)) dy\, dx.$$

By (3.5), (3.6), and dominated convergence, for $s \geq 2M$, the right-hand side converges to

$$c_3 \int \int_0^\infty y F(x, y) 1_{B(0,M)}(x, y)\, dy\, dx.$$

Therefore

(3.7) $$\liminf_s s^d \mathbb{E}^{(0,s)} \int_0^\tau F(X_r, Y_r)\, dr \geq c_3 \int \int_0^\infty y F(x, y)\, dy\, dx - c_3 \varepsilon.$$

To prove the other direction of the inequality, consider s large enough so that $|F(x, y)| \leq c\varepsilon s^{-(2+d)}$ if $(x, y) \in B((0, s), s/2)$. Then

$$s^d \mathbb{E}^{(0,s)} \int_0^\tau F(X_r, Y_r) 1_{B((0,s),s/2)}(X_r, Y_r) dr$$

$$\leq c\varepsilon s^d s^{-(2+d)} \mathbb{E}^{(0,s)} \int_0^\tau 1_{B((0,s),s/2)}(X_r, Y_r) dr$$

$$\leq c\varepsilon s^{-2} \mathbb{E}^{(0,s)} \int_0^\infty 1_{B((0,s),s/2)}(X_r, Y_r) dr \leq c\varepsilon.$$

For such s,

$$s^d \mathbb{E}^{(0,s)} \int_0^\tau F(X_r, Y_r) dr$$

$$\leq c\varepsilon + s^d \mathbb{E}^{(0,s)} \int_0^\tau F(X_r, Y_r) 1_{B((0,s),s/2)^c}(X_r, Y_r) dr$$

$$= c\varepsilon + s^d \int \int_0^\infty F(x, y) 1_{B((0,s),s/2)^c}(x, y) g_D((0, s), (x, y)) dy\, dx.$$

By (3.5), (3.6), and dominated convergence,

$$(3.8) \quad \limsup_{s} s^d \mathbb{E}^{(0,s)} \int_0^\tau F(X_r, Y_r) dr \le c\varepsilon + c_3 \int \int_0^\infty yF(x,y) dy \, dx.$$

Since ε is arbitrary, (3.7) and (3.8) together give our result.

To prove (3.5), note that

$$(3.9) \; s^d g_D((0,s),(x,y))$$

$$= cs^d \frac{\left[x^2 + (y+s)^2\right]^{(d-1)/2} - \left[x^2 + (y-s)^2\right]^{(d-1)/2}}{\left[x^2 + (y+s)^2\right]^{(d-1)/2} \left[x^2 + (y-s)^2\right]^{(d-1)/2}}$$

$$= cs \frac{\left[1 + 2y/s + (x^2+y^2)/s^2\right]^{(d-1)/2} - \left[1 - 2y/s + (x^2+y^2)/s^2\right]^{(d-1)/2}}{\left[(x/s)^2 + (1 - y/s)^2\right]^{(d-1)/2} \left[(x/s)^2 + (1 + y/s)^2\right]^{(d-1)/2}}.$$

Using a Taylor expansion for $(1+x)^{(d-1)/2}$ near 0, we see that the numerator of the last line of (3.9) is asymptotically $2(d-1)y/s$ as $s \to \infty$, from which (3.5) follows easily.

To prove (3.6), we start with the inequality (Exercise 10)

$$(3.10) \qquad\qquad 1 - (1-z)^{(d-1)/2} \le (d-1)z/2, \qquad z \le 1.$$

If $|(x,y) - (0,s)| \ge s/2$, we write

$$s^d g_D((0,s),(x,y))$$

$$= cs^d \left\{ 1 - \left[\frac{x^2 + (y-s)^2}{x^2 + (y+s)^2}\right]^{(d-1)/2} \right\} \Big/ [x^2 + (y-s)^2]^{(d-1)/2}$$

$$\le cs \left\{ 1 - \left[\frac{x^2 + (y-s)^2}{x^2 + (y+s)^2}\right]^{(d-1)/2} \right\}$$

$$\le cs\left(4ys/[x^2 + (y+s)^2]\right) \le cy.$$

This proves (3.6). □

This result is not surprising. Starting with a uniform distribution at ∞, Z_t will be uniform at each level. Also, the Green function for one-dimensional Brownian motion killed at 0 is $g(y, y_0) = y \wedge y_0 \to y$ as $y_0 \to \infty$.

Usually the functions to which we will be applying Lemma 3.2 are of the form $|\nabla u(x,y)|^2$ or $\nabla u(x,y) \cdot \nabla v(x,y)$, where u and v are the harmonic extensions of L^1 functions f and g, respectively.

(3.3) Lemma. *Suppose either that $f \in L^1$ and u is its harmonic extension or else that u is the Poisson kernel for some point $w \in \mathbb{R}^d$. Then*

$$\sup_{z \in B((0,s),s/2)} |\nabla u(z)|^2 \le cs^{-2d-2}.$$

Proof. In the first case,

$$|u(x,y)| = |P_y f(x)| = \left| \int P_y(x-t)f(t)\,dt \right|$$
$$\leq \sup_t P_y(t)\|f\|_1 \leq cy^{-d}\|f\|_1.$$

So in either case $|u(x,y)| \leq cy^{-d}$. By Corollary II.1.4 with $r = s/2$, $|\nabla u(x,y)| \leq s^{-d-1}$ if $(x,y) \in B((0,s), s/2)$. □

The situation $F = \nabla u \cdot \nabla v$ is similar.

(3.4) Lemma. *If* $f \in C_K^1$,

$$\lim_{s\to\infty} s^d \mathbb{E}^{(0,s)}[S_1(f)^2] = c \lim_{s\to\infty} s^d \mathbb{E}^{(0,s)}[S(f)^2] = c \int f(x)^2 dx.$$

Proof. By Lemma 3.2, the left-hand side is $\int_0^\infty y \int |\partial_y u(x,y)|^2 \, dx \, dy$. By Plancherel's identity this is [cf. (2.5)]

$$c \int_0^\infty y \int |\widehat{\partial_y u}(\xi,y)|^2 \, d\xi \, dy = c \int \int_0^\infty y|\xi|^2|\widehat{f}(\xi)|^2 e^{-2y|\xi|} \, dy \, d\xi$$
$$= c \int |\widehat{f}(\xi)|^2 d\xi = c \int |f(x)|^2 dx.$$

That $\lim s^d \mathbb{E}^{(0,s)}[S(f)^2] = c \int |f(x)|^2 dx$ is similar. □

(3.5) Theorem. *If* $f \in C_K^1$ *and* $1 < p < \infty$, *then there exists* c *such that*

$$\|R_j f\|_p \leq c\|f\|_p.$$

Proof. Let u_j be the harmonic extension of $R_j f$. Then

$$\widehat{u}_j(\xi,y) = e^{-|\xi|y} \widehat{R_j f}(\xi) = \frac{i\xi_j}{|\xi|} e^{-y|\xi|} \widehat{f}(\xi).$$

So

(3.11)
$$\widehat{\partial_y u_j}(\xi,y) = -i\xi_j e^{-y|\xi|} \widehat{f}(\xi) = -\widehat{\partial_j u}(\xi,y).$$

Hence

$$S_1(R_j f) = \left(\int_0^\tau |\partial_j u(Z_r)|^2 dr \right)^{1/2} \leq S(f).$$

Now suppose $g \in C_K^1$ and let v be its harmonic extension. We use polarization on Lemma 3.4. That means we apply Lemma 3.4 to $f + g$, f, and g, and we obtain [cf. (I.4.14)],

$$\lim_{s\to\infty} s^d \mathbb{E}^{(0,s)} \int_0^\tau \partial_y u(Z_r)\partial_y v(Z_r)dr = c \int fg.$$

Suppose $\|g\|_q \leq 1$. Since $|\partial_y u_j| = |\partial_j u| \leq |\nabla u|$, we can use Lemmas 3.2 and 3.3 to write

$$(3.12) \quad \left| \int (R_j f) g \right|$$

$$= c \lim s^d \left| \mathbb{E}^{(0,s)} \int_0^\tau \partial_y u_j(Z_r) \partial_y v(Z_r) dr \right|$$

$$= c \lim s^d \left| \mathbb{E}^{(0,s)} \int_0^\tau \partial_j u(Z_r) \partial_y v(Z_r) dr \right|$$

$$\leq c \lim \sup s^d \mathbb{E}^{(0,s)} \left[\left(\int_0^\tau |\partial_j u(Z_r)|^2 dr \right)^{1/2} \left(\int_0^\tau |\partial_y v(Z_r)|^2 dr \right)^{1/2} \right]$$

$$\leq c \left(\lim \sup s^d \mathbb{E}^{(0,s)}[S(f)]^p \right)^{1/p} \left(\lim \sup s^d \mathbb{E}^{(0,s)}[S(g)]^q \right)^{1/q}.$$

Using Theorems I.6.8 and I.4.7,

$$s^d \mathbb{E}^{(0,s)} S(f)^p \leq c s^d \mathbb{E}^{(0,s)} |f(X_\tau)|^p + c s^d \mathbb{E}^{(0,s)} |u(Z_0)|^p$$

$$\leq c \|f\|_p^p + c s^d |u(0,s)|^p,$$

and similarly with g. As in the proof of Theorem 2.2(a),

$$\lim \sup_{s \to \infty} s^d \mathbb{E}^{(0,s)}[S(f)^p] \leq c \|f\|_p^p,$$

similarly with g, and

$$\left| \int (R_j f) g \right| \leq c \|f\|_p \|g\|_q.$$

Now take the supremum over gs in C_K^1 with $\|g\|_q \leq 1$, and we conclude $\|R_j f\|_p \leq c \|f\|_p$. $\qquad \square$

Note c does not depend on the dimension.

The other method of proving the L^p boundedness of the Riesz transforms is the culmination of work by Gundy and Silverstein [1], Gundy and Varopoulos [1], Meyer [2, 3], and Bañuelos [1, 3]. Let u be the harmonic extension of f. Before proving the L^p boundedness we give (Corollary 3.7) a probabilistic interpretation of the Riesz transforms.

(3.6) Proposition. *If $\hat{f} \in L^1$ and there exists c such that $|\partial_y u(w,t)| \leq ct^{-1-d}$, then*

$$\lim_{s \to \infty} \mathbb{E}_x^{(0,s)} \left[\int_0^\tau \partial_y u(Z_r) dY_r \right] = cf(x).$$

Proof. Let $h(w,y) = P_y(w - x)$, the positive harmonic function with pole at x that is 0 on $\partial D - \{0\}$. Set $S_y = \inf\{t : Y_t \leq y\}$. Then

(3.13) $\quad \mathbb{E}_x^{(0,s)} \int_0^{S_y} \partial_y u(Z_r) dY_r$

$$= \mathbb{E}^{(0,s)} \Big[h(Z_{S_y}) \int_0^{S_y} \partial_y u(Z_r) dY_r \Big] / h((0,s)).$$

By Itô's formula,

(3.14) $$h(Z_{S_y}) = h(Z_0) + \int_0^{S_y} \nabla h(Z_r) \cdot dZ_r.$$

Since the stochastic integral is a mean zero martingale,

(3.15) $$\mathbb{E}^{(0,s)} \Big[h(Z_0) \int_0^{S_y} \partial_y u(Z_r) dY_r \Big] = 0.$$

Since X_t is independent of Y_t, then $\langle X, Y \rangle_t \equiv 0$, and

(3.16) $\mathbb{E}^{(0,s)} \Big[\int_0^{S_y} \nabla h(Z_r) \cdot dZ_r \int_0^{S_y} \partial_y u(Z_r) dY_r \Big]$

$$= \mathbb{E}^{(0,s)} \Big[\int_0^{S_y} \partial_y h(Z_r) \partial_y u(Z_r) dr \Big].$$

Using (3.13), (3.14), (3.15), and (3.16) we conclude

(3.17) $\quad \mathbb{E}_x^{(0,s)} \Big[\int_0^{S_y} \partial_y u(Z_r) dY_r \Big] = c s^d \mathbb{E}^{(0,s)} \Big[\int_0^{S_y} \partial_y h(Z_r) \partial_y u(Z_r) dr \Big].$

Next we let $y \to 0$. After an argument to justify the passage to the limit (Exercise 11) we end up with

(3.18) $\quad \mathbb{E}_x^{(0,s)} \Big[\int_0^\tau \partial_y u(Z_r) dY_r \Big] = c s^d \mathbb{E}^{(0,s)} \Big[\int_0^\tau \partial_y h(Z_r) \partial_y u(Z_r) dr \Big].$

By Lemmas 3.2 and 3.3, the right-hand side of (3.18) converges to

$$c \int \int_0^\infty y \partial_y h(w,y) \partial_y u(w,y) dy\, dw$$

as $s \to \infty$. By the Plancherel identity and polarization, this is equal to

$$c \int_0^\infty \int y \widehat{\partial_y u}(\xi, y) \overline{\widehat{\partial_y h}(\xi, y)} d\xi\, dy$$

$$= c \int \int_0^\infty y e^{-i\xi \cdot x} |\xi| e^{-y|\xi|} |\xi| e^{-y|\xi|} \widehat{f}(\xi) dy\, d\xi$$

$$= c \int e^{-i\xi \cdot x} \widehat{f}(\xi) d\xi = c f(x).$$

$\square$

Let A be the $(d+1) \times (d+1)$ matrix with A_{ik} zero unless $i = d+1$, $k = j$, in which case it is one. We have the following.

(3.7) Corollary. *Suppose $f \in C_K^\infty$. Then there exists c independent of f such that*

$$R_j f(x) = c \lim_{s \to \infty} \mathbb{E}_x^{(0,s)} \int_0^\tau A\nabla u(Z_r) \cdot dZ_r.$$

Proof.

$$A\nabla u(Z_r) \cdot dZ_r = \partial_j u(Z_r) dY_r = -\partial_y u_j(Z_r) dY_r,$$

where u_j is the harmonic extension of $R_j f$. Since $\partial_y u_j = -\partial_j u$ and $f \in C_K^\infty$, $|\widehat{R_j f}(\xi)| \leq |\hat{f}(\xi)| \in L^1$. So the hypotheses of Proposition 3.6 are satisfied, and the result follows from that proposition. □

An expression like $\int_0^t A\nabla u(Z_r) \cdot dZ_r$ is called the martingale transform of the martingale $\int_0^t \nabla u(Z_r) \cdot dZ_r$ by the matrix A.

The idea behind the second proof of the L^p boundedness of the Riesz transforms is this: if $\mathbb{E}^*$ represents "Brownian motion started with the distribution of Lebesgue measure at ∞," then

$$\int |R_j f(x)|^p dx = \mathbb{E}^* |R_j f(X_\tau)|^p = \mathbb{E}^* \left| \mathbb{E}^* \left[\int_0^\tau A\nabla u(Z_r) \cdot dZ_r | X_\tau \right] \right|^p$$

$$\leq \mathbb{E}^* \mathbb{E}^* \left[\left| \int_0^\tau A\nabla u(Z_r) \cdot dZ_r \right|^p | X_\tau \right]$$

$$= \mathbb{E}^* \left[\left| \int_0^\tau A\nabla u(Z_r) \cdot dZ_r \right|^p \right]$$

by Jensen's inequality. The Burkholder-Davis-Gundy inequalities then finish the job. That is what motivates the following, although it is hidden due to the fact that we will use the duality of L^p and L^q rather than Jensen's inequality.

Proof of Theorem 3.5, second method. Suppose $f, g \in C_K^1$. Then as in (3.12),

$$(3.19) \qquad \int (R_j f) g = c \lim s^d \mathbb{E}^{(0,s)} \int_0^\tau \partial_y u_j(Z_r) \partial_y v(Z_r) dr,$$

where v and u_j are the harmonic extensions of g and $R_j f$, respectively. By Itô's formula,

$$(3.20) \qquad g(X_\tau) = v(Z_0) + \int_0^\tau \nabla v(Z_r) \cdot dZ_r.$$

We have

$$(3.21) \qquad \mathbb{E}^{(0,s)} \left[v(Z_0) \int_0^\tau \partial_y u_j(Z_r) dY_r \right] = 0$$

since the stochastic integral has mean zero expectation. As in the proof of Proposition 3.6,

$$(3.22) \quad \mathbb{E}^{(0,s)}\left[\int_0^\tau \nabla v(Z_r) \cdot dZ_r \int_0^\tau \partial_y u_j(Z_r) dY_r\right]$$

$$= \mathbb{E}^{(0,s)} \int_0^\tau \partial_y v(Z_r) \partial_y u_j(Z_r) dr.$$

Combining (3.19), (3.20), (3.21), and (3.22),

$$(3.23) \quad \int (R_j f) g$$

$$= c \lim s^d \mathbb{E}^{(0,s)}\left[g(X_\tau) \int_0^\tau \partial_y u_j(Z_r) dY_r\right]$$

$$\le \left(\limsup s^d \mathbb{E}^{(0,s)} |g(X_\tau)|^q\right)^{1/q} \left(\limsup s^d \mathbb{E}^{(0,s)} \left|\int_0^\tau A\nabla u(Z_r) \cdot dZ_r\right|^p\right)^{1/p}.$$

By Theorem I.6.8,

$$\limsup s^d \mathbb{E}^{(0,s)} \left|\int_0^\tau A\nabla u(Z_r) \cdot dZ_r\right|^p$$

$$\le c \limsup s^d \mathbb{E}^{(0,s)}\left(\int_0^\tau |A\nabla u(Z_r)|^2 dr\right)^{p/2}$$

$$\le c \limsup s^d \mathbb{E}^{(0,s)}\left(\int_0^\tau |\nabla u(Z_r)|^2 dr\right)^{p/2}$$

$$\le c \limsup s^d \mathbb{E}^{(0,s)} \left|\int_0^\tau \nabla u(Z_r) \cdot dZ_r\right|^p$$

$$= c \lim s^d \mathbb{E}^{(0,s)} |f(X_\tau)|^p.$$

By this and (3.23),

$$\int (R_j f) g \le c\|g\|_q \|f\|_p.$$

Taking the supremum over gs with $\|g\|_q \le 1$ completes the proof. $\square$

Applications

As our first application, let us derive an inequality that is important in the theory of partial differential equations.

(3.8) Theorem. *Suppose $f \in C_K^\infty$. Then for $1 < p < \infty$, there exists c such that*

$$(3.24) \quad \|\partial_i \partial_j f\|_p \le c\|\Delta f\|_p.$$

Proof. We will do the case $d \geq 3$ for simplicity, although the other cases are not much harder. Suppose we show

$$(3.25) \qquad \|\partial_i \partial_j Ug\|_p \leq c\|g\|_p, \qquad g \in C_K^\infty.$$

If $f \in C_K^\infty$ and $g = -(1/2)\Delta f$, then $g \in C_K^\infty$ and $Ug = f$. Hence (3.24) will follow once we prove (3.25).

If P_t is defined by (I.3.2), then $\widehat{P_t g}(\xi) = \hat{P}_t(\xi)\hat{g}(\xi)$ and the Fourier transform of $(2\pi t)^{-d/2} e^{-|x|^2/2t}$ is $e^{-|\xi|^2 t/2}$ by (I.1.6). Integrating over t from 0 to ∞, $\widehat{Ug}(\xi) = 2|\xi|^{-2}\hat{g}(\xi)$. Then

$$\widehat{\partial_i \partial_j Ug}(\xi) = i\xi_i i\xi_j \frac{2}{|\xi|^2}\hat{g}(\xi) = -2\frac{\xi_i}{|\xi|}\frac{\xi_j}{|\xi|}\hat{g}(\xi).$$

By Proposition 3.1 this is cR_iR_jg. By Theorem 3.5,

$$\|R_iR_jg\|_p \leq c\|R_jg\|_p \leq c\|g\|_p.$$

$\square$

(3.9) Corollary. *Suppose $g \in L^p$. Then*

$$\|\partial_i \partial_j U^\lambda g\|_p \leq c\|g\|_p.$$

Proof. By the above theorem, $\|\partial_i \partial_j U^\lambda g\|_p \leq c\|\Delta(U^\lambda g)\|_p$. For nice g, $U^\lambda g = U(g - \lambda U^\lambda g)$ by (I.3.5). Then $(1/2)\Delta U^\lambda g = \lambda U^\lambda g - g$ by Proposition II.3.3. So provided $\|\lambda U^\lambda g\|_p \leq c\|g\|_p$, we are done. Recall that

$$\int \lambda u^\lambda(x,y)dx = \int_0^\infty \lambda e^{-\lambda t} \int p(t,x,y)dx\,dt = 1.$$

Using Jensen's inequality,

$$(3.26) \quad \int |\lambda U^\lambda g(x)|^p dx = \int \left|\int \lambda u^\lambda(x,y)g(y)dy\right|^p dx$$

$$\leq \int \int \lambda u^\lambda(x,y)|g(y)|^p dy\,dx = \int |g(y)|^p dy.$$

$\square$

Sobolev inequalities

Our second application of Riesz transforms is to give a proof of one of the Sobolev inequalities. Define the Riesz potentials by

$$I_\alpha(x) = \frac{c}{|x|^{d-\alpha}}, \qquad \alpha \in (0,d), \ c = \pi^{d/2}2^\alpha \Gamma(\alpha/2)/\Gamma((n-\alpha)/2).$$

We will show shortly (Theorem 3.11) that if $f \in L^p$, then $I_1 f \in L^q$, where $q^{-1} = p^{-1} - d^{-1}$ and (Proposition 3.12) that the Fourier transform of $I_1 f$ is essentially given by $(c/|\xi|)\widehat{f}(\xi)$. We then have the following.

(3.10) Theorem. (Sobolev) *If $f \in C_K^1$,*

$$\|f\|_{2d/(d-2)} \leq c\|\nabla f\|_2.$$

Proof. By a limiting procedure it suffices to prove these inequalities for $f \in C_K^\infty$. We claim

$$(3.27) \qquad f = c \sum_{j=1}^d R_j I_1 \partial_j f, \qquad \text{a.e.}$$

Once we have (3.27), if $\nabla f \in L^2$, then by Theorem 3.11 it follows that $I_1 \partial_j f \in L^{2d/(d-2)}$. Since $2d/(d-2) > 1$ and the Riesz transforms are bounded on L^p for $p > 1$, then $f \in L^{2d/(d-2)}$.

Since R_j and I_1 are both operators given by convolutions, they commute, and so it suffices to prove

$$(3.28) \qquad f = cI_1 \Big(\sum_{j=1}^d R_j \partial_j f \Big).$$

We will show (3.28) by using Fourier transforms. Let $H = \sum_{j=1}^d R_j \partial_j f$. Then

$$\widehat{H}(\xi) = \sum_{j=1}^d \frac{i\xi^j}{|\xi|} i\xi^j \widehat{f}(\xi) = -|\xi|\widehat{f}(\xi).$$

Suppose g is such that $\widehat{g} \in C_K^\infty$. Then by Proposition 3.12,

$$\int I_1 H(x)\overline{g(x)}dx = c \int |\xi|^{-1}\widehat{H}(\xi)\overline{\widehat{g}(\xi)}d\xi$$

$$= c \int \widehat{f}(\xi)\overline{\widehat{g}(\xi)}d\xi$$

$$= c \int f(x)\overline{g(x)}dx.$$

The first and third inequalities come from the polarization of Plancherel's identity. Since the class of such gs is dense in L^2, we have (3.28). □

(3.11) Theorem. (Hardy-Littlewood-Sobolev) *Suppose $p > 1$, $f \in L^p$, and $\alpha \in (0, d)$.*
(a) $I_\alpha f$ converges absolutely for almost every x;
(b) If $p < d/\alpha$ and $q^{-1} = p^{-1} - \alpha/d$, then

$$\|I_\alpha f\|_q \leq c\|f\|_p.$$

When $p > d/\alpha$, I_α maps L^p into the Hölder space C^β, where $\beta = \alpha - d/p$. (These spaces are defined in Definition 3.13.) See Exercise 13.

Proof. Let $a > 0$ and let

$$K_1(x) = \begin{cases} I_\alpha(x) & \text{if } |x| \le a, \\ 0 & \text{if } |x| > a. \end{cases}$$

Let $K_2(x) = I_\alpha(x) - K_1(x)$. $K_1 * f$ exists a.e. since it is the convolution of the L^1 function K_1 with the L^p function f. If p' is the conjugate exponent of p, i.e., $p^{-1} + p'^{-1} = 1$, then $K_2 * f$ exists everywhere since it is the convolution of the $L^{p'}$ function K_2 with the L^p function f. So (a) is proved.

For (b), let us normalize so that $\|f\|_p = 1$, and we must show there exists c such that $\|I_\alpha f\|_q \le c$.

$$\|K_2 * f\|_\infty \le \|K_2\|_{p'} \|f\|_p = \|K_2\|_{p'}$$

where

$$\|K_2\|_{p'} = c\left(\int_{|x|>a} (|x|^{\alpha-d})^{p'} dx \right)^{1/p'} = ca^{(d+p'(\alpha-d))/p'} = ca^{-d/q}.$$

By Theorem 1.6, note $K_1 * f(x) \le a^\alpha M f(x)$. Adding,

$$|I_\alpha(x)| \le ca^{-d/q} + ca^\alpha M f(x).$$

Optimizing, we choose $a = M f(x)^{-1/(\alpha+d/q)}$, and we conclude

$$|I_\alpha(x)| \le cM f(x)^{d/(\alpha q + d)}.$$

Since $dq/(\alpha q + d) = p$,

$$\|I_\alpha(x)f\|_q \le c\left(\int M f(x)^p \right)^{1/q} \le c\|f\|_p^{p/q} \le c$$

by Theorem 1.1. $\qquad\qquad\qquad\qquad\qquad\qquad\qquad\qquad\qquad\qquad \square$

The Schwartz class is the set of functions that are in C^∞ and all of whose derivatives remain bounded when multiplied by polynomials. It is not hard to see that if f is in the Schwartz class, so is $\hat{f}$ (Exercise 12). I_α is not in L^1, but its Fourier transform is in some sense given by $c|\xi|^{-\alpha}$. The precise statement is the following.

(3.12) Proposition. *Suppose f and g are in the Schwartz class. There exists c not depending on f or g such that*

$$\int I_\alpha f(x)\overline{g(x)}dx = c \int \hat{f}(\xi)|\xi|^{-\alpha}\overline{\hat{g}(\xi)}d\xi.$$

Proof. By polarization of Plancherel's identity,

(3.29)
$$\int h(x)\overline{k(x)}dx = c \int \widehat{h}(\xi)\overline{\widehat{k}(\xi)}d\xi.$$

Suppose k is in the Schwartz class. Let $h(x) = (2\pi t)^{-d/2}e^{-|x|^2/2t}$. Then by (I.1.6), $\widehat{h}(\xi) = e^{-t|\xi|^2/2}$. Let us multiply both sides of (3.29) by $t^{\alpha/2-1}$ and integrate over t from 0 to ∞. Since

$$\int_0^\infty t^{\alpha/2-1}e^{-t|\xi|^2/2}dt = c|\xi|^{-\alpha}\int_0^\infty u^{\alpha/2-1}e^{-u}du = c|\xi|^{-\alpha}$$

and

$$\int_0^\infty t^{-d/2}t^{\alpha/2-1}e^{-|x|^2/2t}dt = c|x|^{-d+\alpha}\int_0^\infty u^{(d-\alpha)/2-1}e^{-u}du$$
$$= c|x|^{-d+\alpha},$$

we see that

$$\int |x|^{-d+\alpha}\overline{k(x)}dx = c\int |\xi|^{-\alpha}\overline{\widehat{k}(\xi)}d\xi.$$

We use the fact that k is in the Schwartz class to justify the interchange in the order of integration.

Now let $k(y) = f(x-y)$, so

$$\int |y|^{-d+\alpha}f(x-y)dy = c\int \widehat{f}(-\xi)|\xi|^{-\alpha}e^{ix\cdot\xi}d\xi.$$

Multiplying by $\overline{g}(x)$ and integrating over x proves the proposition. $\quad\square$

Obtaining the I_α from Gaussian kernels as we did in the above proof has a probabilistic interpretation and goes under the name of subordination; see Feller [2].

The Hölder spaces

Let us define the space of Hölder continuous functions C^α.

(3.13) Definition. *Let $\alpha \in (0,1)$. f is in C^α if $f \in L^\infty$ and*

$$\|f\|_{C^\alpha} = \|f\|_\infty + \sup_x \sup_{|t|>0} \frac{|f(x+t) - f(x)|}{|t|^\alpha}$$

is finite.

Let us note the equality

(3.30)
$$\int \partial_y P_y(t)dt = \partial_y \int P_y(t)dt = \partial_y 1 = 0.$$

If u is the harmonic extension of f, we have the following.

(3.14) Proposition. *Suppose $0 < \alpha < 1$ and $f \in L^\infty$. Then $f \in C^\alpha$ if and only if*

$$\|\partial_y u(x, y)\|_\infty \leq cy^{-1+\alpha}$$

for all x and y.

Proof. Suppose $f \in C^\alpha$. Then

$$|\partial_y u(x, y)| = \int \partial_y P_y(t) f(x - t) dt = \int \partial_y P_y(t)[f(x - t) - f(x)] dt,$$

using (3.30). This is bounded by

$$\|f\|_{C^\alpha} \int |\partial_y P_y(t)| \, |t^\alpha| dt \leq cy^{-1+\alpha}\|f\|_{C^\alpha},$$

using scaling for the last inequality.

To go the other direction, suppose $\|\partial_y u(x, y)\|_\infty \leq cy^{-1+\alpha}$. We first want to show

(3.31) $$\|\partial_j u(x, y)\|_\infty \leq cy^{-1+\alpha},$$

for each j. To see this, note that $\|\partial_j P_{y/2}\|_1 \leq c/y$ by scaling. So

$$\partial_j \partial_y u(x, y) = \partial_j \partial_y P_{y/2} * P_{y/2} f(x)$$
$$= (\partial_j P_{y/2}) * (\partial_y P_{y/2} f)(x) \leq cy^{-2+\alpha}.$$

Since

$$\|\partial_j u(x, y)\|_\infty = \|\partial_j P_y * f\|_\infty \leq \|\partial_j P_y\|_1 \|f\|_\infty \leq cy^{-1}\|f\|_\infty,$$

then $\partial_j u(x, y) \to 0$ as $y \to \infty$. Therefore

$$\partial_j u(x, y) = -\int_y^\infty \partial_y \partial_j u(x, z) dz,$$

and so

$$|\partial_j u(x, y)| \leq c \int_y^\infty z^{-2+\alpha} dz = cy^{-1+\alpha}.$$

This proves (3.31).

We write

$$\left| f(x + t) - f(x) \right| \leq \left| f(x + t) - u(x + t, |t|) \right|$$
$$+ \left| u(x + t, |t|) - u(x, |t|) \right| + \left| f(x) - u(x, |t|) \right|.$$

Now

$$\left| f(x + t) - u(x + t, |t|) \right| = \left| \int_0^{|t|} \partial_y u(x + t, y) dy \right|$$
$$\leq c \int_0^{|t|} y^{-1+\alpha} dy = c|t|^\alpha.$$

The same bound holds for $|f(x) - u(x, |t|)|$, and

$$
\begin{aligned}
|u(x+t, |t|) - u(x, |t|)| &= \left| \int_0^{|t|} \partial_s u(x + st/|t|, |t|) ds \right| \\
&\leq \int_0^{|t|} |\nabla u(x + st/|t|, |t|)| ds \\
&\leq c \int_0^{|t|} |t|^{-1+\alpha} ds = c|t|^\alpha
\end{aligned}
$$

by (3.31). Summing gives our result. □

As a corollary, the Riesz transforms are almost bounded operators on C^α.

(3.15) Corollary. *Suppose $\alpha \in (0,1)$ and $f \in C^\alpha$. Let $M > 0$. There exists c_1 (independent of M) and c_2 and c_3 depending on M such that*
(a) $|R_j f(x+t) - R_j f(x)| \leq c_1 |t|^\alpha \|f\|_{C^\alpha}$.
(b) If f has support in $B(0, M)$, then $\|R_j f\|_\infty \leq c_1 \|f\|_{C^\alpha}$ and $|R_j f(x)| \leq c_2/|x|^d$ if $|x| \geq 2M$.
(c) If $|f(x)| \leq c_2/|x|^d$ for $|x| \geq 2M$, then $\|R_j f\|_\infty \leq c_3 \|f\|_{C^\alpha}$.

Proof. Let u_j be the harmonic extension of $R_j f$. Then

$$
|\partial_y u_j(x, y)| = |\partial_j u(x, y)| \leq c y^{-1+\alpha}
$$

by (3.31). By Proposition 3.14, this gives (a).
For (b),

$$
\begin{aligned}
R_j f(x) &= \int R_j(y) f(x - y) dy \\
&= \int R_j(y)[f(x - y) - f(x) 1_{(|y| \leq 3M)}] dy.
\end{aligned}
$$

If $|x| \leq 2M$, this is bounded by

$$
\int_{|y| \leq 3M} \frac{c}{|y|^d} \|f\|_{C^\alpha} |y|^\alpha dy \leq c \|f\|_{C^\alpha},
$$

since $f(x - y) = 0$ if $|y| \geq 3M$. If $|x| \geq 2M$, then $f(x - y)$ is nonzero only if $|y| \geq |x|/2$ and $|x - y| \leq M$, so

$$
|R_j f(x)| \leq c \int_{B(x, M)} |y|^{-d} \|f\|_\infty dy \leq c|x|^{-d}.
$$

The proof of part (c) is Exercise 14. □

As a consequence, if $f \in C^\alpha$ with support in $B(0, M)$ for some M, then $R_j R_k f$ is in C^α, and hence if $d \geq 3$ (cf. Theorem 3.8), $\partial_j \partial_k f$ is in C^α.

For $\alpha \geq 1$, let k be the smallest integer strictly greater than α and define

(3.32) $$C^\alpha = \{f \in L^\infty : \left\|\frac{\partial^k u(x,y)}{\partial y^k}\right\|_\infty \leq cy^{-k+\alpha}\}.$$

This turns out to be equivalent to saying $f \in C^\alpha$ if and only if $f \in L^\infty$ and $\partial_j f \in C^{\alpha-1}$, $j = 1, \cdots, d$, provided $\alpha \neq 1$. When $\alpha = 1$, the right definition of C^α is that $f \in L^\infty$ and $|f(x+t) + f(x-t) - 2f(x)| \leq c|t|^\alpha$ for all x and t. We will not need these facts; see Stein [1] for proofs.

Extensions

Another way to obtain the boundedness of the Riesz transforms and many other singular integral operators is the method of rotations. If v is a unit vector, it follows easily from the boundedness of the Hilbert transform that

$$H_v f(x) = \lim_{\varepsilon \to 0, N \to \infty} \frac{1}{\pi} \int_{\varepsilon < |t| < N} \frac{f(x - tv)}{t} dt$$

is a bounded operator on $L^p(\mathbb{R}^d)$. For suitable functions Ω_j on the unit sphere S^{d-1}, one can write

$$R_j f(x) = \int_{S^{d-1}} H_v f(x) \Omega_j(v) dv,$$

and conclude the L^p boundedness of the Riesz transforms; see Stein and Weiss [1].

The weak (1-1) inequality also holds for the Riesz transforms (Stein [1]), but no probabilistic proof is known. One reason for searching for a probabilistic proof is that it is likely that if one exists, it would show that the constants could be chosen independent of the dimension d, something that is currently not known.

There is a notion of Riesz transforms on infinite-dimensional spaces. Instead of Lebesgue measure, the appropriate substitute is a Gaussian measure, namely, Wiener measure (i.e., the law of a Brownian motion). These Riesz transforms are bounded in L^p (see Meyer [4, 5], Pisier [1], and Gundy [1]), and one approach is similar to the one given above, with the modification that the Brownian motion X_t is replaced by an Ornstein-Uhlenbeck process. Whether the weak (1-1) inequality holds for these Riesz transforms is still open.

4 Littlewood-Paley functions

Formulation

One way to approach other singular integrals is through the introduction of certain nonlinear functionals of f, the Littlewood-Paley functions. This is analogous to the use of $\langle M \rangle_t$ to study a martingale M_t. In this section we obtain some inequalities for the Littlewood-Paley functionals and in the next we give some applications.

Let $f \in L^p$, let u be its harmonic extension, and let us define the following nonlinear functionals:

$$(4.1) \quad g(f)(x) = \left(\int_0^\infty y |\nabla u(x,y)|^2 dy \right)^{1/2},$$

$$(4.2) \quad g_\lambda^*(f)(x) = \left(\int_0^\infty \int_{\mathbb{R}^d} \left(\frac{y}{|t|+y} \right)^{\lambda d} |\nabla u(x-t,y)|^2 y^{1-d} dt \, dy \right)^{1/2},$$

$$(4.3) \quad A(f)(x) = \left(\int_{C(x)} y^{1-d} |\nabla u(t,y)|^2 dy \, dt \right)^{1/2},$$

where $C(x)$ is defined by (1.5). We will also use in our proofs the functional

$$(4.4) \quad L^*(f)(x) = \left(\int_0^\infty \int \frac{y}{(|t|^2+y^2)^{(d+1)/2}} |\nabla u(x-t,y)|^2 dt \, y \, dy \right)^{1/2}.$$

$g(f)$ is known as the Littlewood-Paley g functional. $A(f)$ is called the area functional; the reason for the name is that when $d = 2$, $A(f)(x)$ represents the area of $u(C(x))$ (counting multiplicity).

Note $L^*(f)(x) = c \int_0^\infty y P_y * |\nabla u(\cdot, y)|^2 (x) dy$, where P_y is the Poisson kernel. Note also that the ratio of $g_{(d+1)/d}^*(f)(x)$ to $L^*(f)(x)$ is bounded above and below by positive constants not depending on f or x.

Our main result on these functions is the following.

(4.1) Theorem. *(a) If $\lambda > 1$, then $g(f)(x) \leq cA(f)(x) \leq cg_\lambda^*(f)(x)$ for each x;*
(b) $\|g(f)\|_p \leq c\|f\|_p$ if $1 < p < 2$;
(c) $\|g_{(d+1)/d}^*(f)\|_p \leq c\|f\|_p$ if $2 \leq p < \infty$;
(d) $\|f\|_p \leq c\|g(f)\|_p$ if $1 < p < \infty$.

Proofs of the inequalities

The proof of Theorem 4.1 will take most of the section, but we can give the proof of part (a) now.

Proof of Theorem 4.1(a). It suffices to assume $x = 0$. Write B_y for $B((0,y), cy)$, where c is chosen small enough so that $B_y \subseteq C(0)$. c can be chosen to be independent of y. Since $\partial_y u$ is harmonic,

$$\partial_y u(0,y) = \frac{1}{|B_y|} \int_{B_y} \partial_y u(x,s)\, dx\, ds.$$

By the Cauchy-Schwarz inequality,

(4.5) $$|\partial_y u(0,y)|^2 \le \frac{1}{|B_y|} \int_{B_y} |\partial_y u(x,s)|^2 dx\, ds.$$

Using (4.5),

$$\int_0^\infty y |\partial_y u(0,y)|^2 dy \le \int_0^\infty c y^{-d} \Big(\int_{B_y} |\partial_y u(x,s)|^2 dx\, ds \Big) dy$$

$$\le c \int_{C(0)} \Big(\int_{s/2}^{2s} y^{-d} dy \Big) |\partial_y u(x,s)|^2 dx\, ds$$

$$\le c \int_{C(0)} s^{1-d} |\partial_y u(x,s)|^2 dx\, ds.$$

Here we used the fact that if $(x,s) \in B_y$, then $s/2 \le y \le 2s$. We obtain a similar estimate for $\partial_j u$ for each j and sum:

(4.6) $$\int_0^\infty y |\nabla u(0,y)|^2 dy \le c \int_{C(0)} y^{1-d} |\nabla u(x,s)|^2 dx\, ds = c(A(f)(0))^2.$$

Finally, since $y/(|t|+y) \ge c$ in $C(0)$, the right-hand side of (4.6) is bounded by $c(g_\lambda^*(f)(0))^2$. $\square$

Before proving part (b) we need the following.

(4.2) Lemma. *Suppose $f \ge 0$ is C_K^∞ and $1 < p < 2$. Then*

$$\int f(x)^p dx = \frac{1}{2} \int \int_0^\infty y \Delta(u^p)(x,y)\, dy\, dx.$$

Proof. Let $U_t = u(Z_{t \wedge \tau})$. By Itô's formula,

$$(U_\tau + \varepsilon)^p = (U_0 + \varepsilon)^p + p \int_0^\tau (U_r + \varepsilon)^{p-1} dU_r$$

$$+ \frac{p(p-1)}{2} \int_0^\tau (U_r + \varepsilon)^{p-2} d\langle U \rangle_r.$$

Take the $\mathbb{E}^{(0,s)}$ expectation, multiply by s^d, and let $s \to \infty$ to see that

$$\lim s^d \mathbb{E}^{(0,s)} (U_\tau + \varepsilon)^p$$

$$= \varepsilon^p + \frac{p(p-1)}{2} \lim s^d \mathbb{E}^{(0,s)} \int_0^\tau (U_r + \varepsilon)^{p-2} |\nabla u(Z_r)|^2 dr,$$

or by Lemma 3.2,

$$\int (f(x) + \varepsilon)^p dx$$

$$= c\varepsilon^p + \frac{p(p-1)}{2} \int \int_0^\infty y(u(x,y) + \varepsilon)^{p-2} |\nabla u(x,y)|^2 dy\, dx.$$

By a direct calculation,

(4.7) $$\Delta(u^p) = p(p-1)u^{p-2}|\nabla u|^2.$$

Our result follows by letting $\varepsilon \to 0$. $\square$

Proof of Theorem 4.1(b). We may assume that $f \geq 0$ is smooth with compact support (Exercise 15). By (4.7) and Theorem 1.6

$$g(f)(x)^2 = \int_0^\infty y|\nabla u(x,y)|^2 dy = c \int_0^\infty yu^{2-p}\Delta(u^p)\, dy$$

$$\leq c(Mf(x))^{2-p} \int_0^\infty y\Delta(u^p)(x,y)\, dy.$$

If we set

$$I(x) = \int_0^\infty y\Delta(u^p)(x,y)dy,$$

then $\int I(x)dx = c\|f\|_p^p$ by Lemma 4.2. Thus

$$\int |g(f)(x)|^p dx \leq c \int Mf(x)^{p(2-p)/2} I(x)^{p/2} dx$$

$$\leq c\left(\int (Mf)^p\right)^{(2-p)/2} \left(\int I(x)\right)^{p/2}$$

$$\leq c(\|f\|_p^p)^{(2-p)/2}(\|f\|_p^p)^{p/2} = c\|f\|_p^p,$$

by Lemma 4.2 and Theorem 1.1. We used Hölder's inequality with exponents $2/p$ and $2/(2-p)$. $\square$

Part (c) is the interesting part from the probabilistic point of view.

Proof of Theorem 4.1(c). By Exercise 15 it suffices to assume $f \in C_K^1$. The case $p = 2$ is Exercise 16, so we will suppose $p > 2$. Since $g_{(d+1)/d}^*(f)(x)$ is comparable to $L^*(f)(x)$, it suffices to show

$$\|L^*(f)\|_p \leq c\|f\|_p.$$

We first show that if $h \in C_K^1$, then

(4.8) $$\lim s^d \mathbb{E}^{(0,s)}\left[h(X_\tau) \int_0^\tau |\nabla u(Z_r)|^2 dr\right] = \int (L^*(f)(x))^2 h(x)dx.$$

Let $v(z) = \mathbb{E}^z h(X_\tau)$ be the harmonic extension of h, so that

(4.9) $$v(z) = \int h(w)P_w(z)dw.$$

By the Markov property at time r,

$$s^d \mathbb{E}^{(0,s)} \left[\int_0^\tau |\nabla u(Z_r)|^2 dr \, h(X_\tau) \right]$$

$$= s^d \mathbb{E}^{(0,s)} \left[\int_0^\infty 1_{(r<\tau)} |\nabla u(Z_r)|^2 dr \, h(X_\tau) \right]$$

$$= s^d \mathbb{E}^{(0,s)} \left[\int_0^\infty 1_{(r<\tau)} |\nabla u(Z_r)|^2 \mathbb{E}\left[h(X_\tau) | \mathcal{F}_r \right] \right] dr$$

$$= s^d \mathbb{E}^{(0,s)} \int_0^\tau |\nabla u(Z_r)|^2 v(Z_r) \, dr$$

$$\to c \int \int_0^\infty y |\nabla u(x,y)|^2 v(x,y) \, dy \, dx.$$

Applying (4.9) proves (4.8).

Let $r = p/2$ and let q be the conjugate exponent to r. By Hölder's inequality and the Theorem I.6.8 (and the argument of the second proof of Theorem 3.5),

$$\limsup s^d \mathbb{E}^{(0,s)} \left[\int_0^\tau |\nabla u(Z_r)|^2 dr \, h(X_\tau) \right]$$

$$\leq c \left(\limsup s^d \mathbb{E}^{(0,s)} \left(\int_0^\tau |\nabla u(Z_r)|^2 dr \right)^r \right)^{1/r}$$

$$\times \left(\limsup s^d \mathbb{E}^{(0,s)} |h(X_\tau)|^q \right)^{1/q}$$

$$\leq c \|h\|_q \left(\limsup s^d \mathbb{E}^{(0,s)} \left| \int_0^\tau \nabla u(Z_r) \cdot dZ_r \right|^{2r} \right.$$

$$\left. + \limsup s^d u((0,s))^{2r} \right)^{1/r}$$

$$= c \|h\|_q \left(\limsup s^d \mathbb{E}^{(0,s)} |f(X_\tau)|^{2r} \right)^{1/r}$$

$$= c \|h\|_q \|f\|_{2r}^2.$$

Taking the supremum over $h \in C_K^1$ with $\|h\|_q \leq 1$ shows

$$\left(\int |L^*(f)(x)|^{2r} \right)^{1/r} \leq c \left(\int |f(x)|^{2r} \right)^{2/2r}.$$

Since $2r = p$, this completes the proof. $\qquad \square$

Finally we prove part (d).

Proof of Theorem 4.1(d). By Theorem 4.1(b), $\|g(f)\|_p \leq c\|f\|_p$, $1 < p < 2$. For $2 \leq p < \infty$, combining parts (a) and (c) with $\lambda = (d+1)/d$ shows $\|g(f)\|_p \leq c\|f\|_p$.

By Plancherel's identity,

$$\int \big(g(f)(x)\big)^2 dx = \int \int_0^\infty y|\nabla u(x,y)|^2 dy\, dx$$
$$= c\int_0^\infty \int y|\widehat{\nabla u}(\xi,y)|^2 d\xi\, dy$$
$$= c\int_0^\infty \int y|\xi|^2|\hat{f}(\xi)|^2 e^{-2|\xi|y} d\xi\, dy$$
$$= c\int |\hat{f}(\xi)|^2 d\xi = c\int |f(x)|^2 dx.$$

By polarization, if u_i is the harmonic extension of f_i for $i = 1, 2$,

$$c\int f_1 f_2 dx = \int \int_0^\infty y\nabla u_1(x,y)\cdot\nabla u_2(x,y)\, dy\, dx.$$

By the Cauchy-Schwarz inequality, the right-hand side is bounded by $\int g(f_1)(x)g(f_2)(x)dx$. If q is the conjugate exponent to p and $\|f_2\|_q \le 1$,

$$\int f_1 f_2 \le c\int g(f_1)g(f_2)$$
$$\le c\|g(f_1)\|_p\|g(f_2)\|_q$$
$$\le c\|g(f_1)\|_p\|f_2\|_q$$
$$\le c\|g(f_1)\|_p.$$

Taking the supremum over such f_2 gives the result. $\qquad\square$

We want to improve Theorem 4.1(c).

(4.3) Proposition. *Suppose $\lambda > 1$ and $p \ge 2$. There exists c depending only on p such that*

$$\|g_\lambda^*(f)\|_p \le c\|f\|_p.$$

Proof. Let

$$Q_y(t) = \Big(\frac{y}{|t| + y}\Big)^{\lambda d}.$$

Since

$$g_\lambda^*(f)(x)^2 = \int_0^\infty Q_y * |\nabla u(x,y)|^2 y^{1-d} dy,$$

then we have the inequality

$$\int h(x)\Big(g_\lambda^*(f)(x)\Big)^2 dx = \int_0^\infty Q_y h(x)|\nabla u(x,y)|^2 dx\, y^{1-d}\, dy$$
$$\le \int \int_0^\infty y|\nabla u(x,y)|^2 dy\, \Big(\sup_y \big(Q_y h(x)y^{-d}\big)\Big) dx$$
$$\le c\int \big(g(f)(x)\big)^2 Mh(x)dx$$

by Theorem 1.6. The case $p = 2$ follows by letting $h = 1$. For $p > 2$, let $r = p/2$ and let q be the conjugate exponent to r. Then

$$\int h(x)\big(g_\lambda^*(f)(x)\big)^2 dx \leq c\|g(f)^2\|_r \|Mh\|_q$$

$$= c\|g(f)\|_{2r}^2 \|Mh\|_q$$

$$\leq c\|f\|_{2r}^2 \|h\|_q.$$

Now taking a supremum over h with $\|h\|_q \leq 1$ shows $\|g_\lambda^*(f)^2\|_r \leq c\|f\|_{2r}^2$.

$\square$

In the next section we will also need the following. Let

(4.10) $$g_y(f)(x) = \Big(\int_0^\infty y|\partial_y u(x,y)|^2 dy \Big)^{1/2}.$$

Clearly $g_y(f)(x) \leq g(f)(x)$ for each x.

(4.4) Proposition. *Let $1 < p < \infty$. There exists c such that*

$$\|f\|_p \leq c\|g_y(f)\|_p.$$

Proof. This follows by a duality argument. By Plancherel's identity (cf. Exercise 16),

$$\int |f(x)|^2 dx = c \int |g_y(f)(x)|^2 dx.$$

If u_i is the harmonic extension of f_i for $i = 1, 2$, then by polarization and the Cauchy-Schwarz inequality,

$$\int f_1 f_2 = c \int \int_0^\infty y(\partial_y u_1(x,y))(\partial_y u_2(x,y))\, dy\, dx \leq c \int g_y(f_1) g_y(f_2).$$

Hölder's inequality shows that

(4.11) $$\Big| \int f_1 f_2 \Big| \leq c\|g_y(f_1)\|_p \|g_y(f_2)\|_q.$$

If $\|f_2\|_q \leq 1$, then $\|g_y(f_2)\|_q \leq \|g(f_2)\|_q \leq c\|f_2\|_q \leq c$ by Theorem 4.1. Taking the supremum over such f_2 proves the proposition. $\square$

If R_j is the jth Riesz transform and u_j is its harmonic extension, then $\partial_y u_j = -\partial_j u$, so $|g_y(R_j f)(x)| \leq |g(f)(x)|$ [cf. (3.11)]. Then Proposition 4.4 together with Theorem 4.1 provides yet another proof of the L^p boundedness of the Riesz transforms.

Area functionals

By Theorem 4.1(a) and (d) we have $\|f\|_p \leq c\|A(f)\|_p$ if $1 < p < \infty$. By Theorem 4.1(a) and (c) we have $\|A(f)\|_p \leq c\|f\|_p$ if $2 \leq p < \infty$. The loose end left over is to show the following.

(4.5) Proposition.

$$\|A(f)\|_p \leq c\|f\|_p, \qquad 1 < p < 2.$$

Proof. We will show this when f is in C_K^1, and then obtain the general case by appealing to Exercise 15. Recall the definitions of $C(x)$ and $N_b(f)(x)$. Take $b > 1$. We will show that

$$(4.12) \qquad \|A(f)\|_p \leq c\|N_b(f)\|_p.$$

By Corollary 1.4, this will prove our result.

Let $E = \{x : N_b(f)(x) \leq \lambda\}$ and let $G = \cup_{x \in E} C(x)$. The first step is to get a bound on the integral of $A(f)^2$ over E. We have

$$\int_E A(f)(x)^2 dx = \int_E \int_{C(x)} y^{1-d} |\nabla u(w,y)|^2 dw \, dy \, dx.$$

Since $(w,y) \in C(x)$ and $x \in E$ implies $(w,y) \in G$, using Fubini's theorem and integrating over dx first shows

$$\int_E A(f)(x)^2 dx = \int_G |\nabla u(w,y)|^2 y^{1-d} |\{x \in E : (w,y) \in C(x)\}| dw \, dy.$$

Since $(w,y) \in C(x)$ if and only if $|x - w| < y$, the above is bounded above by

$$c \int_G |\nabla u(w,y)|^2 y \, dw \, dy.$$

Let us apply Green's second identity (Theorem II.3.10) in $B(0, M) \cap G$ with one function being $u^2/2$ and the other being y, and then letting $M \to \infty$. Note $\Delta(u^2) = 2|\nabla u|^2$. Since u is the Poisson integral of a C_K^1 function, the $\int_{\partial B(0,M)}$ term tends to 0 as $M \to \infty$ and we are left with

$$(4.13) \qquad \int_E A(f)(x)^2 dx \leq c \left\{ \int_{\partial G} y \frac{\partial u^2}{\partial n} d\sigma - \int_{\partial G} u^2 \frac{\partial y}{\partial n} d\sigma \right\}.$$

Let us look at the first integral on the right. If $z = (w,y) \in \partial G \cap E$, this means $y = 0$. If $z = (w,y) \in \partial G \cap D$, then $|u(z)| \leq \lambda$ and $(w,y) \in C(v)$ for some $v \in E$. So there exists c independent of w and y such that $B(z, cy) \subseteq C_b(v)$. By Corollary II.1.4, $y|\nabla u(w,y)| \leq c\lambda$. Since $y|\partial u^2/\partial n| \leq 2y|u(z)||\nabla u(z)|$ and $d\sigma$ is comparable to dx,

$$\int_{\partial G \cap D} y \frac{\partial u^2}{\partial n} \leq c\lambda^2 |E^c|.$$

As for the second integral in (4.13), by our construction $|\partial y/\partial n|$ is bounded. $|u|^2 \leq \lambda^2$ on ∂G. So the second integral is bounded by

$$c\int_E N_b(f)^2 + c\lambda^2 |E^c|.$$

Some integrations will finish the proof. We have

$$|\{A(f) > \lambda\}| = |\{A(f) > \lambda\} \cap E| + |\{A(f) > \lambda\} \cap E^c|$$

$$\leq \lambda^{-2} \int_E A(f)(x)^2 dx + |E^c|$$

$$\leq c|E^c| + c\lambda^{-2} \int_E N_b(f)^2$$

$$= c|\{N_b(f) > \lambda\}| + c\lambda^{-2} \int_E N_b(f)^2.$$

We multiply by $p\lambda^{p-1}$ and integrate over λ from 0 to ∞. Since

$$\int_0^\infty p\lambda^{p-1}\lambda^{-2} \int_E N_b(f)^2 d\lambda$$

$$\leq c\int_0^\infty \lambda^{p-3} \Big[\int_0^\lambda \beta |\{N_b(f) > \beta\}| \, d\beta\Big] d\lambda$$

$$= c\int_0^\infty \beta |\{N_b(f) > \beta\}| \int_\beta^\infty \lambda^{p-3} d\lambda \, d\beta$$

$$= c\int_0^\infty \beta^{p-1} |\{N_b(f) > \beta\}| \, d\beta = c\|N_b(f)\|_p^p,$$

(4.12) follows.   □

Extensions

Sharper estimates relating the nontangential maximal function $N_b(f)$ and the area functional are possible and in more general domains D. See Bañuelos and Brossard [1], Bañuelos and Moore [3, 4], Burkholder and Gundy [1], and Davis [1].

Gundy and Silverstein [2] introduced the density of the area integral:

$$D_f^r(z) = \int_{C(z)} y^{1-d} |\Delta(|u - r|)(x, y)| \, dx \, dy.$$

The reason for the name is that

$$(4.14) \qquad \int_{-\infty}^{\infty} D_f^r(z)dr = A(f)(z)$$

(Exercise 17). The density of the area integral is the analog of local times of martingales. Motivated by the Barlow and Yor [1] results, Gundy and Silverstein [2] showed that the L^p norm of D_f^r is comparable to the L^p norm of f for all $p \in (1, \infty)$. See Bañuelos and Brossard [1] and Bañuelos and Moore [4] for further results. See also Brossard and Chevalier [1] for applications to H^1.

It is possible to define Littlewood-Paley functionals where both X_t and Y_t are replaced by other processes, and in many cases the analog of Theorem 4.1 holds. See Stein [2], Meyer [2, 3], and Marias [1].

The functional $g_\lambda^*(f)$ has a nice probabilistic interpretation when $\lambda = (d+1)/d$ [Theorem 4.1(c)]. Is there also a probabilistic interpretation for other λ?

Suppose in the definition of the Littlewood-Paley functionals we replace $u(x,t)$ by $v(x,t) = \int p(t,x,w)f(w)\,dw$. That is, we replace the Poisson integral by the heat semigroup. Several of the inequalities of Theorem 4.1 are still true (see Stein [2]). Can these be proven by techniques similar to ones used in this section?

5 Singular integral operators and Fourier multipliers

Nonsingular kernels

In the last section we obtained some results about the Littlewood–Paley g functional and some other quadratic functionals of f, but we did not say what they were used for. That is the purpose of this section. The problem we consider is: for what kernels K does the operator $Tf = K * f$ map L^p to L^p? The first proposition below gives a satisfactory and useful answer if K is not singular, but most of the Ks we will be interested in are similar to the Hilbert transform and the Riesz transforms in that cancelations are crucial to the existence.

First we handle the nonsingular case.

(5.1) Theorem. *Suppose μ is a measure, $K(x,y) \geq 0$, and there exists c_1 such that*

$$(5.1) \qquad \int K(x,y)\mu(dy) \leq c_1 \qquad \text{for all } x$$

and

$$(5.2) \qquad \int K(x,y)\mu(dx) \leq c_1 \qquad \text{for all } y.$$

Let $Tf(x) = \int K(x,y)f(y)\mu(dy)$. Then $\|Tf\|_p \leq c_1\|f\|_p$, $p \in [1,\infty]$.

Proof. If $p = \infty$, the result is immediate by (5.1). If $p = 1$, the result follows by an application of Fubini's theorem. So let us suppose $1 < p < \infty$ and let q be the conjugate exponent of p. Write $K = K^{1/p}K^{1/q}$ and use Hölder's inequality:

$$
\begin{aligned}
|Tf(x)|^p &= \left| \int K(x,y)f(y)\mu(dy) \right|^p \\
&\leq \left[\left(\int K(x,y)|f(y)|^p\mu(dy) \right)^{1/p} \left(\int K(x,y)\mu(dy) \right)^{1/q} \right]^p \\
&\leq c_1^{p/q} \int K(x,y)|f(y)|^p\mu(dy).
\end{aligned}
$$

So then

$$
\begin{aligned}
\int |Tf(x)|^p\mu(dx) &\leq c_1^{p/q} \int\int K(x,y)|f(y)|^p\mu(dy)\mu(dx) \\
&\leq c_1 c_1^{p/q} \int |f(y)|^p\mu(dy)
\end{aligned}
$$

by (5.1). $\qquad\qquad\qquad\qquad\qquad\qquad\qquad\qquad\qquad\qquad\qquad\qquad\quad$ ☐

Another way to prove this is to observe that (5.1) implies that T is a bounded operator on L^∞ and (5.2) and Fubini's theorem imply that T is a bounded operator on L^1. So by the Marcinkiewicz interpolation theorem (Theorem 7.21) T is a bounded operator on L^p.

Singular integrals

Let us suppose K is a function so that if the operator T is defined by $Tf(x) = \int K(y)f(x-y)\,dy$, it can be well defined for a large class of fs. For example, if K is an odd function and $|K(x)| \leq c/|x|$ for some c, then the proof of Proposition 2.11 shows that $\lim_{\varepsilon\to 0, N\to\infty} \int_{\varepsilon<|y|<N} K(y)f(x-y)dy$ exists if $f \in C_K^1$. By Exercise 18, TP_y also exists.

We will first give a criterion for Tf to be bounded on L^p and then later relate this to conditions on K.

(5.2) Theorem. *Suppose there exists $\lambda > 1$ such that*

$$
(5.3) \qquad\qquad |\partial_y P_y K(x)| \leq cy^{-d-1}\left(\frac{y}{y+|x|} \right)^{\lambda d}.
$$

Then for each $p \in (1,\infty)$, there exists c depending on p such that if $f \in C_K^1$,

$$
\|Tf\|_p \leq c\|f\|_p.
$$

Proof. Suppose first that $p \geq 2$. Let

$$Q_z(x) = z^{-d}\left(\frac{z}{z+|x|}\right)^{\lambda d}.$$

$P_y = P_{y/2}P_{y/2}$, and K, ∂y, and P_y commute (look at their Fourier transforms). By virtue of (5.3), $\partial_y P_y K f(x) \to 0$ as $y \to \infty$, so

$$\left(g_y(Tf)(x)\right)^2 = \int_0^\infty y|\partial_y P_y K f(x)|^2 dy$$

$$= \int_0^\infty y|P_{y/2}K\partial_y P_{y/2}f(x)|^2 dy$$

$$= \int_0^\infty y\left[\int_y^\infty \frac{z}{z}\partial_y P_{z/2}K\partial_y P_{z/2}f(x)dz\right]^2 dy.$$

By the Cauchy-Schwarz inequality the square of the integral inside the brackets is bounded by

$$\left(\int_y^\infty \frac{dz}{z^2}\right)\left(\int_y^\infty z^2(\partial_y P_{z/2}K\partial_y P_{z/2}f(x))^2 dz\right).$$

Therefore,

$$(5.4) \qquad \left(g_y(Tf)(x)\right)^2 \le \int_0^\infty \int_y^\infty z^2\left(\partial_y P_{z/2}K\partial_y P_{z/2}f(x)\right)^2 dz\, dy$$

$$= \int_0^\infty \int_0^z dy\, z^2(\partial_y P_{z/2}K\partial_y P_{z/2}f(x))^2 dz$$

$$= \int_0^\infty z^3\left(\partial_y P_{z/2}K\partial_y P_{z/2}f(x)\right)^2 dz$$

$$\le c\int_0^\infty z^3\left(z^{-1}Q_z\partial_y P_{z/2}f(x)\right)^2 dz$$

$$= c\int_0^\infty z\left(Q_z\partial_y P_{z/2}f(x)\right)^2 dz.$$

By Hölder's inequality, for any function k

$$(Q_z k(x))^2 = \left(\int k(x-y)Q_z(y)dy\right)^2$$

$$\le c\int k(x-y)^2 Q_z(y)dy = cQ_z(k^2)(x)$$

since $\int Q_z(t)dt = c$. So the right-hand side of (5.4) is bounded by

$$c\int_0^\infty zQ_z(\partial_y P_{z/2}f)^2(x)dz \le c(g_\lambda^*(f)(x))^2$$

or

$$|g_y(Tf)(x)| \le c|g_\lambda^*(f)(x)|.$$

This then tells us that

$$\|Tf\|_p \le c\|g_y(Tf)\|_p \le c\|g_\lambda^*(f)\|_p \le c\|f\|_p,$$

using Theorem 4.1 and the fact that $p \ge 2$.

Suppose next that $p < 2$. Let $K^*(x) = K(-x)$ and define T^* in terms of K^*. By the symmetry of the Poisson kernel, (5.3) holds when K is replaced by K^*. If $g \in C_K^1$ and q is the conjugate exponent to p,

$$\int g(Tf) = \int g(x) \int K(x-y)f(y)dy\,dx$$

$$= \int f(y) \int K^*(y-x)g(x)dx\,dy$$

$$= \int f(T^*g).$$

If $p < 2$, then $q > 2$, and so by the above

$$\left| \int g(Tf) \right| \le \|f\|_p \|T^*g\|_q \le c\|f\|_p\|g\|_q.$$

Now take the supremum over such g with $\|g\|_q \le 1$. $\square$

For this theorem actually to be useful, we need to be able to express (5.3) directly in terms of K. We give a theorem for $d = 1$, although the proof generalizes to the case $d > 1$ (Exercise 19).

(5.3) Theorem. *Suppose $d = 1$, K is odd, and there exists c_1 such that $|K(y)| \le c_1/|y|$, and $|K'| \le c_1/|y|^2$, $y \ne 0$. Then there exists c_2 depending only on p such that if $f \in C_K^1$, then $\|Tf\|_p \le c_2\|f\|_p$, $1 < p < \infty$.*

Proof. By our conditions on K, Tf is well defined for $f \in C_K^1$ and so is TP_y. Choose $\lambda \in (1,2)$. We will have our result from Theorem 5.2 if we show (5.3) holds. By scaling, it is enough to show (5.3) for $z = 1$, or that

$$|\partial_y P_1 K(x)| \le Q_1(x) = \frac{c}{(1 + |x|)^\lambda}.$$

Define

$$R(z) = \partial_y P_y(z)|_{y=1}.$$

Then

$$R(z) = c\frac{z^2 - 1}{(z^2 + 1)^2} \sim cz^{-2},$$

for z large and it is easy to check that $R'(z) \sim cz^{-3}$ if z is large. Also $\int R(z)dz = 0$ by (3.30).

Let $\alpha = \lambda - 1$. Since $|K'(y)| \le c/y^2$, it follows that

(5.5) $$|K(x-y) - K(x)| \le c\frac{|y|^\alpha}{|x|^{1+\alpha}}$$

if $|y| \leq |x|/2$.

Suppose that $|x| \leq 10$, and without loss of generality, that $x > 0$. Then using the fact that K is odd and the Cauchy-Schwarz inequality,

$$\partial_y P_1 K(x) = \int R(x - y)K(y)dy$$

$$= \int_{|y| \leq 10} [R(x - y) - R(x)]K(y)dy + \int_{|y| \geq 10} R(x - y)K(y)dy$$

$$\leq \int_{|y| < 10} \frac{|y|}{|y|} \sup_{|z| \leq 20} |R'(z)|dy$$

$$+ \left(\int_{|y| \geq 10} R(x - y)^2 dy \right)^{1/2} \left(\int_{|y| \geq 10} K(y)^2 dy \right)^{1/2}$$

$$\leq c \leq Q_1(x).$$

Now look at $|x| \geq 10$ and without loss of generality assume $x > 0$. We can write

$$(5.6) \quad \int R(x - y)K(y)dy = \int_{|y| \leq x/2} + \int_{y \geq 3x/2} + \int_{y < -x/2} + \int_{x/2 < y < 3x/2}.$$

For the first integral on the right-hand side of (5.6), using the fact that K is odd,

$$(5.7) \quad \left| \int_{|y| \leq x/2} R(x - y)K(y)dy \right| = \left| \int_{|y| \leq x/2} [R(x - y) - R(x)]K(y)dy \right|$$

$$\leq \int_{|y| \leq x/2} |y| \left[\sup_{|z-x| < x/2} |R'(z)| \right] \frac{dy}{|y|}$$

$$\leq \frac{c}{x^3} \int_{|y| \leq x/2} dy \leq cx^{-2}.$$

Let us now look at the second and third integrals. When $y \geq 3x/2$ or $y < -x/2$, $|R(x - y)| \leq c|R(y)|$, and

$$(5.8) \quad \int_{|y| > x/2} |R(y)| \, |K(y)|dy \leq \int_{|y| > x/2} \frac{c}{y^2} \frac{c}{|y|} dy \leq cx^{-2}.$$

For the fourth integral, we have

$$(5.9) \quad \left| \int_{x/2 < y < 3x/2} R(x - y)K(y)dy \right| = \left| \int_{|y| < x/2} R(y)K(x - y)dy \right|$$

$$\leq \left| \int_{|y| < x/2} R(y)[K(x - y) - K(x)]dy \right|$$

$$+ |K(x)| \left| \int_{|y| < x/2} R(y)dy \right|.$$

Using the hypotheses on K, the first integral on the last line of (5.9) is bounded by

$$\int_{|y|<x/2} |R(y)| \frac{|y|^\alpha}{|x|^{1+\alpha}} dy \le c/|x|^{1+\alpha}.$$

Using the fact that $\int R(z)\, dz = 0$,

$$\left| \int_{|y|<x/2} R(y) dy \right| = \left| \int_{|y|>x/2} R(y) dy \right| \le \int_{|y|>x/2} \frac{c}{y^2} = c/x.$$

So the last integral on the last line of (5.9) is bounded by c/x^2. Substituting these bounds in (5.9), we see that each of the integrals on the right-hand side of (5.6) is bounded by $c/|x|^{1+\alpha}$, as required. $\qquad\square$

Since the only place we used the condition $|K'(y)| \le c/y^2$ was to prove (5.5), we have the following.

(5.4) Corollary. *Theorem 5.3 remains true if the condition $|K'(y)| \le c/y^2$ is replaced by*
(5.10) *There exists $\alpha \in (0,1)$ such that*

$$|K(x-y) - K(x)| \le c \frac{|y|^\alpha}{|x|^{1+\alpha}}, \qquad |y| \le |x|/2.$$

The condition (5.10) is a very common hypothesis in the literature and is only very slightly weaker than the best condition known, Hörmander's condition. Hörmander's condition is that there exists c such that for all y

$$(5.11) \qquad \int_{|y| \le |x|/2} |K(x-y) - K(x)| dx \le c,$$

see also Theorem 7.22.

Another probabilistic approach to the proof of Theorem 5.3 can be found in Bass [1].

Fourier multipliers

Another important use of the Littlewood-Paley functionals is to prove the multiplier theorems of Mihlin and Marcinkiewicz. If T is a convolution operator and $f \in C_K^\infty$, then $\widehat{Tf} = \widehat{K * f} = \widehat{K}\widehat{f}$. So characterizing which operators T map L^p to L^p can be done either through conditions on K or by conditions on $\widehat{K}$. A Fourier multiplier is a bounded function m, and the corresponding convolution operator is T_m defined on all L^2 functions by $\widehat{T_m f}(\xi) = m(\xi)\widehat{f}(\xi)$. (In particular, $T_m f$ for $f \in C_K^1$ and $T_m P_y$ are well defined.) We will now work on giving sufficient conditions on a Fourier multiplier m so that T_m is a bounded operator on L^p.

(5.5) Theorem. (Mihlin) *Suppose $d = 1$, $c > 0$, and m is a bounded function that is C^1 on $\mathbb{R} - \{0\}$ with $|m'(x)| \leq c/|x|$, $x \neq 0$. Then T_m can be extended from a bounded operator on $L^p \cap L^2$ to a bounded operator on L^p, $1 < p < \infty$.*

Proof. From the proof of Theorem 5.2 we have

$$(g_y(Tf)(x))^2 \leq \int_0^\infty z^3 |\partial_y P_z K \partial_y P_z f(x)|^2 dz.$$

By the Cauchy-Schwarz inequality,

$$|\partial_y P_z K \partial_y P_z f(x)|^2 = \left| \int \partial_y P_z K(t) \partial_y P_z f(x - t) dt \right|^2$$

$$\leq \int (z + |t|)^2 |\partial_y P_z K(t)|^2 dt \int \frac{|\partial_y P_z f(x - t)|^2}{(z + |t|)^2} dt.$$

If we show

(5.12)
$$\int (z + |t|)^2 |\partial_y P_z K(t)|^2 dt \leq c/z$$

then

$$(g_y(Tf)(x))^2 \leq \int_0^\infty \frac{z^2}{(z + |t|)^2} |\partial_y P_z f(x - t)|^2 dt = c(g_\lambda^*(f)(x))^2$$

with $\lambda = 2$. Then just as in Theorem 5.2, the proof will be complete when $p \geq 2$. The case $p < 2$ will follow by a duality argument (Exercise 23).

So we must prove (5.12). When $|t| \leq z$, we have $(z + |t|)^2 \leq 4z^2$. Now the Fourier transform of $\partial_y P_z K$ is

$$-|\xi| e^{-|\xi| z} m(\xi).$$

Since m is bounded, the Fourier transform of $\partial_y P_z K$ is in L^1 with L^1 norm bounded by c/z^2. Therefore $\partial_y P_z K$ is bounded by c/z^2. Hence

(5.13)
$$\int_{|t| \leq z} (z + |t|)^2 |\partial_y P_z K(t)|^2 dt \leq c \int_{|t| \leq z} z^2 (c/z^2)^2 dt = c/z.$$

When $|t| > z$, we have $(z + |t|)^2 \leq 4|t|^2$. Let $H(t) = t \partial_y P_z K(t)$. By Plancherel's identity,

$$\int |t|^2 |\partial_z P_z K(t)|^2 dt = c \int |\widehat{H}(\xi)|^2 d\xi$$

$$= c \int \left[\left(|\xi| e^{-|\xi| z} m(\xi) \right)' \right]^2 d\xi.$$

Note

$$(|\xi| e^{-|\xi| z} m(\xi))' = (-z|\xi| e^{-|\xi| z} + e^{-|\xi| z}) \operatorname{sgn}(\xi) m(\xi) + |\xi| e^{-|\xi| z} m'(\xi).$$

By our assumptions on m, this is in L^2 with L^2 norm bounded by c/z. So

$$\int |t|^2 |\partial_z P_z K(t)|^2 dt \le c/z.$$

Combining with (5.13) proves (5.12). □

For the higher dimensional version, the conclusion holds if all the partial derivatives of order j are bounded by $c|x|^{-j}$ whenever $j \le k = [n/2]+1$. The proof is very similar, although it is necessary to look at $\partial_y^k P_z K f$; see Stein [1] for details.

Partial sum operators

From the Mihlin multiplier theorem one can obtain results about partial sum operators and the Marcinkiewicz multiplier theorem.

If I is an interval, define S_I by $\widehat{S_I f}(\xi) = 1_I(\xi)\widehat{f}(\xi)$. Note $S_{(-\infty,0)}f = (f + iHf)/2$ and $Hf = -2i(S_{(-\infty,0)}f - f/2)$. Also, since the Fourier transform of $f(x)e^{ixa}$ is $\widehat{f}(\xi + a)$, then the Fourier transform of $H(e^{ixa}f)$ is $i\operatorname{sgn}(\xi)\widehat{f}(\xi + a)$, and so

$$(5.14) \qquad S_{(-\infty,a)}f = \frac{f + ie^{-ixa}H(e^{ixa}f)}{2}.$$

Therefore each S_I is bounded on L^p. We can actually say more.

(5.6) Proposition. *Suppose* $1 < p < \infty$. *Let* f_i *be a sequence of functions,* I_i *a sequence of intervals. Then*

$$\left\| \left(\sum_i |S_{I_i}(f_i)|^2 \right)^{1/2} \right\|_p \le c \left\| \left(\sum_i |f_i|^2 \right)^{1/2} \right\|_p,$$

where c *depends only on* p.

Proof. If $I_i = (a_i, b_i)$, then $S_{I_i} = S_{(-\infty,b_i)} - S_{(-\infty,a_i]}$. If $I_i = (a_i, \infty)$, then $S_{I_i} = I - S_{(-\infty,a_i]}$. So without loss of generality we may assume each I_i is of the form $(-\infty, b_i)$. Let X_i be a sequence of independent random variables with $\mathbb{P}(X_i = 1) = \mathbb{P}(X_i = -1) = 1/2$. Note

$$\left(\sum |S_{I_i}(f_i)|^2 \right)^{1/2} \le c \left(\sum |f_i|^2 \right)^{1/2} + c \left(\sum |H(e^{ixb_i}f_i)|^2 \right)^{1/2}.$$

By Theorem I.6.13,

$$\left(\sum |H(e^{ixb_i}f_i)(x)|^2 \right)^{1/2} \le c \left(\mathbb{E} \left| \sum X_i H(e^{ixb_i}f_i)(x) \right|^p \right)^{1/p},$$

and so

$$\int \left(\sum |H(e^{ixb_i} f_i)(x)|^2 \right)^{p/2} dx \leq c\mathbb{E} \int \left| \sum X_i H(e^{ixb_i} f_i)(x) \right|^p dx.$$

Thus the pth power of the L^p norm of $(\sum |H(e^{ixb_i} f_i)|^2)^{1/2}$ is bounded by a constant times the expected value of the pth power of the L^p norm of $\sum X_i H(e^{ixb_i} f_i)$. However, $\sum X_i H(e^{ixb_i} f_i) = H(\sum X_i e^{ixb_i} f_i)$, which has L^p norm bounded by a constant times the L^p norm of $\sum X_i e^{ixb_i} f_i$. Hence

$$\mathbb{E} \int \left| \sum X_i H(e^{ixb_i} f_i)(x) \right|^p dx \leq c\mathbb{E} \int \left| \sum X_i e^{ixb_i} f_i(x) \right|^p dx.$$

By using Theorem I.6.13 as above, this in turn is bounded by

$$\int \left(\sum |e^{ixb_i} f_i(x)|^2 \right)^{p/2} dx = \int \left(\sum |f_i(x)|^2 \right)^{p/2}.$$

$\square$

Let I_i now be the interval $[2^i, 2^{i+1}]$, $i = \ldots, -2, -1, 0, 1, 2, \ldots$, and let $I_i^- = [-2^{i+1}, -2^i]$. These are called the dyadic intervals.

(5.7) Theorem. If $1 < p < \infty$, there exists c_1 and c_2 such that

$$(5.15) \quad \left\| \left(\sum_i [|S_{I_i} f|^2 + |S_{I_i^-} f|^2] \right)^{1/2} \right\|_p$$

$$\leq c_1 \|f\|_p \leq c_2 \left\| \left(\sum_i [|S_{I_i} f|^2 + |S_{I_i^-} f|^2] \right)^{1/2} \right\|_p.$$

Proof. The case $p = 2$ is Plancherel's identity.

$$(5.16) \quad \int \sum_i |S_{I_i} f|^2 + |S_{I_i^-} f|^2 = c \sum_i \int (1_{I_i}(\xi) + 1_{I_i^-}(\xi)) |\widehat{f}(\xi)|^2$$

$$= c \int |\widehat{f}(\xi)|^2 = c \int |f(x)|^2 dx.$$

We will show the first inequality in (5.15). Once we have that the second inequality follows by duality (Exercise 20). To show the first inequality, it suffices to show $\|(\sum |S_{I_i} f|^2)^{1/2}\|_p \leq c\|f\|_p$. By symmetry the same is true when I_i is replaced by I_i^-, and then the inequality in (5.15) follows by Minkowski's inequality.

Let φ be in C^∞ so that $\varphi = 1$ on $[1, 2]$ and has support in $[1/2, 4]$. Define V_i by $\widehat{V_i f}(\xi) = \varphi(2^{-i}\xi) \widehat{f}(\xi)$. Note $(\varphi(2^{-i}\xi))' = 2^{-i}\varphi'(2^{-i}\xi)$. Let X_i be a sequence of independent random variables with $\mathbb{P}(X_i = 1) = \mathbb{P}(X_i = -1) = 1/2$. Note $S_{I_i} V_i f = S_{I_i} f$. Set $f_i = V_i f$.

Let m_X be the multiplier $\sum X_i V_i$. Then $m_X(\xi) = \sum X_i \varphi(2^{-i}\xi)$. For any ξ, $\varphi(2^{-i}\xi)$ is nonzero for at most three is. So m_X is bounded since φ

is. Also $m'_X(\xi)$ is bounded by $c/|\xi|$. Therefore m_X satisfies the hypotheses of the Mihlin multiplier theorem, and we have

$$\left\|\sum_i X_i V_i f\right\|_p \le c\|f\|_p.$$

By Khintchine's inequality (Theorem I.6.13) and the argument of Proposition 5.6,

$$\left\|\left(\sum |f_i|^2\right)^{1/2}\right\|_p \le c\|f\|_p.$$

Combining with Proposition 5.6, $\|(\sum |S_{I_i} f_i|^2)^{1/2}\|_p \le c\|f\|_p$. The fact that $S_{I_i} f_i = S_i f$ finishes the proof. $\square$

We will use Theorem 5.7 to prove the Marcinkiewicz multiplier theorem.

(5.8) Theorem. *Suppose there exists c_1 such that m is bounded by c_1 and for each i, m is of bounded variation on I_i and on I_i^- with variation bounded by c_1. Then if $1 < p < \infty$, there exists c_2 such that $\|T_m f\|_p \le c_2\|f\|_p$, $f \in L^p \cap L^2$.*

Before proceeding with the proof, we first need a lemma.

(5.9) Lemma. *Suppose there exists c_1 such that for each i, γ_i is a positive measure on I_i with total mass bounded by c_1. Let $f_i = S_{I_i} f$. Then there exists c_2 such that*

$$\left\|\left(\sum_i \int_{2^i}^{2^{i+1}} |S_{[u,2^{i+1}]} f_i|^2 \gamma_i(du)\right)^{1/2}\right\|_p \le c_2 \left\|\left(\sum_i |f_i|^2\right)^{1/2}\right\|_p.$$

Proof. It suffices to prove the inequality for $f \in C_K^\infty$. It also suffices to prove the result for each γ_i having finite support; the general case follows from this by taking measures with finite support converging weakly to the γ_i. Since $S_{[u,2^{i+1}]}$ is continuous in u for $u \in (2^i, 2^{i+1})$ (Exercise 21), we can pass to the limit.

Suppose $\gamma_i = \sum_j a_{ij} \delta_{u_{ij}}$ with $u_{ij} \in [2^i, 2^{i+1}]$ and $\sum_j a_{ij} \le c_1$. Then

$$(5.17) \qquad \left(\sum_i \int_{2^i}^{2^{i+1}} |S_{[u,2^{i+1}]} f_i|^2 \gamma_{ij}(du)\right)^{1/2}$$

$$= \left(\sum_i \sum_j a_{ij} |S_{[u_{ij},2^{i+1}]} f_i|^2\right)^{1/2}$$

$$= \left(\sum_{i,j} |S_{[u_{ij},2^{i+1}]}(a_{ij}^{1/2} f_i)|^2\right)^{1/2}.$$

By Proposition 5.6, the right-hand side of (5.17) is bounded in L^p norm by

$$c\left\|\left(\sum_{i,j}|a_{ij}^{1/2}f_i|^2\right)^{1/2}\right\|_p \le cc_1\left\|\left(\sum_i|f_i|^2\right)^{1/2}\right\|_p.$$

$\square$

Proof of Theorem 5.8. Since $m = m1_{(-\infty,0]} + m1_{(0,\infty)}$, let us suppose that m is nonzero only for $\xi > 0$. If we apply the same argument for ms that are nonzero only for $\xi \le 0$ and add our estimates, we will obtain the general result.

We first claim

$$S_{I_i}(T_m f) = m(2^i)S_{I_i}f + \int_{2^i}^{2^{i+1}} m'(u)S_{[u,2^{i+1}]}S_{I_i}f\,du.$$

To prove this we look at the Fourier transform of both sides. If $\xi \notin [2^i, 2^{i+1}]$, both sides are zero. If $\xi \in [2^i, 2^{i+1}]$, the Fourier transform of the left-hand side is $m(\xi)\widehat{f}(\xi)$. The Fourier transform of the right-hand side is

$$m(2^i)\widehat{f}(\xi) + \int_{2^i}^{2^{i+1}} m'(u)1_{[u,2^{i+1}]}(\xi)\widehat{f}(\xi)du$$

$$= m(2^i)\widehat{f}(\xi) + \widehat{f}(\xi)\int_{2^i}^{\xi} m'(u)du = m(\xi)\widehat{f}(\xi).$$

By the Cauchy-Schwarz inequality with the functions $|m'|^{1/2}$ and $|m'|^{1/2}|S_{[u,2^{i+1}]}S_{I_i}f|$,

(5.18) $\qquad |S_{I_i}(T_m f)|^2 \le c|S_{I_i}f|^2 + c\int_{2^i}^{2^{i+1}} |m'(u)|\,|S_{[u,2^{i+1}]}S_{I_i}f|^2du.$

We do this for each interval I_i.

Let $f_i = S_{I_i}f$, $\gamma_i(du) = 1_{I_i}(u)|m'(u)|du$, sum (5.18) over i and apply Lemma 5.9. This tells us that

$$\left\|\left(\sum|S_{I_i}(T_m f)|^2\right)^{1/2}\right\|_p \le c\left\|\left(\sum|S_{I_i}f|\right)^{1/2}\right\|_p.$$

An application of Theorem 5.7 completes the proof. $\qquad\qquad\square$

Extensions

There are higher dimensional analogs of Theorem 5.8; see Stein [1] for details. Thus, for example, if m is the indicator of a polyhedron in $\mathbb{R}^d$, then T_m is a bounded operator on L^p for all p. Surprisingly, if m is the indicator of the unit ball in $\mathbb{R}^d$, $d \ge 2$, then T_m is a bounded operator on L^2 but is not a bounded operator on L^p for any other p; see Fefferman [1]. This has connections with the L^p convergence of multiple Fourier series.

There is an analog of the Mihlin multiplier theorem for operators that arise when the Brownian motions X_t and Y_t are replaced by other processes; see Stein [2], Meyer [2, 3], and Varopoulos [1]. For a Banach space version of Mihlin's theorem, see McConnell [1].

Inequalities of the form (5.15) when the I_i are not the dyadic intervals have been studied by Rubio de Francia [1].

There are a number of analogies between the partial sum operators and the partial sums of independent random variables. The question of whether $s_n = S_{(-n,n)}f$ converges a.e for $f \in L^p$ is essentially the situation that the Carleson-Hunt theorem (Carleson [2] and Hunt [1]) deals with. One could also ask about central limit theorems and laws of the iterated logarithms for the sequence s_n under various hypotheses on f.

6 The space H^1

Definitions

We have seen that the Hilbert transform is not bounded on L^∞ and by duality it is not bounded on L^1 either (see Exercise 9). However it is bounded on the Hardy space H^1, which is a substitute for L^1 in this context. (The substitute for L^∞ is the space of bounded mean oscillation, which is discussed in Sect. 7.)

There are a number of alternative definitions of the space H^1. Let us fix on one, and then show that all the others are equivalent.

Let $f : \mathbb{R}^d \to \mathbb{R}$ be a L^1 function and let u be its harmonic extension. Recall the definitions of $N(f)$ and $C(x)$ from (1.5) and (1.6).

(6.1) Definition. $H^1 = \{f \in L^1 : N(f) \text{ is in } L^1\}$ and we set $\|f\|_{H^1} = \|N(f)\|_1$.

Let $R_j f$ denote the jth Riesz transform of f, u_j is the harmonic extension of $R_j f$,

$$(6.1) \qquad F(z) = \left(u(z)^2 + \sum_{j=1}^{d} (u_j(z))^2 \right)^{1/2},$$

$U_t = u(Z_{t \wedge \tau})$, and $U^* = \sup_{s \le \tau} |U_t|$. Our main purpose in this section is to prove the following theorem.

(6.2) Theorem. *The following are equivalent.*
(a) $\|f\|_{H^1} < \infty$;

(b) $\limsup_{s \to \infty} s^d \mathbb{E}^{(0,s)} U^* < \infty$;
(c) $\|f\|_1 + \sum_{j=1}^d \|R_j f\|_1 < \infty$;
(d) $\sup_{y>0} \int |F(x,y)| dx < \infty$.
Moreover there exist constants c_1, c_2, c_3, and c_4 independent of f such that

$$\|f\|_{H^1} \leq c_1 \limsup_{s \to \infty} s^d \mathbb{E}^{(0,s)} U^* \leq c_2 \|f\|_1 + \sum_{j=1}^d \|R_j f\|_1$$

$$\leq c_3 \sup_{y>0} \int |F(x,y)| dx \leq c_4 \|f\|_{H^1}.$$

Equivalence to the probabilistic definition

The equivalence of (a) and (b) are due to Burkholder, Gundy, and Silverstein [1] in the $d = 2$ case and Burkholder and Gundy [2] in the $d \geq 3$ case. That will be our first goal. Then we will prove a subharmonicity lemma that will allow us to prove the remaining equivalences.

Let $E = \{x : N(f)(x) > \lambda\}$.

(6.3) Proposition. *There exists c such that*

$$\mathbb{P}^{(0,s)}(U^* > \lambda) \leq c\mathbb{P}^{(0,s)}(X_\tau \in E).$$

Proof. If $(x,y) \in D$, let $S(x,y) = \{w \in \mathbb{R}^d : |w - x| < y\}$. By scaling and the support theorem (Theorem I.6.6), $\mathbb{P}^{(x,y)}(X_\tau \in S(x,y)) \geq c$, with c independent of the point (x,y).

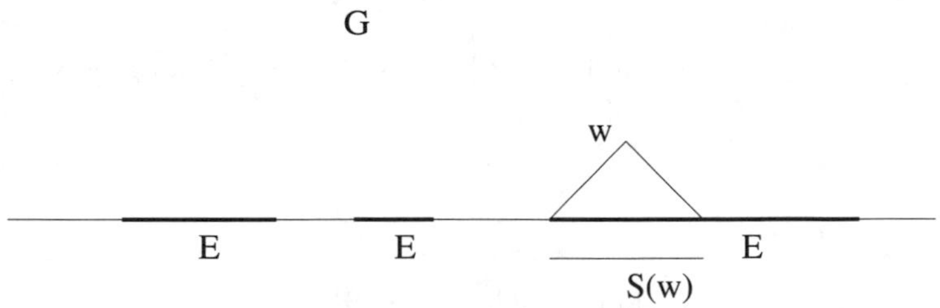

G

W

E   E   S(w)   E

FIGURE 6.1. Diagram for Proposition 6.3.

Let $G = \cup_{z \in E^c} C(z)$. If $w \in G^c$, then $S(w) \subseteq E$, so $\mathbb{P}^w(X_\tau \in E) > c$. Hence

$$\mathbb{P}^{(0,s)}(X_\tau \in E) \geq \mathbb{P}^{(0,s)}(X_\tau \in E, \tau_G < \tau)$$

$$= \mathbb{E}^{(0,s)}[\mathbb{P}^{X_{\tau_G}}(X_\tau \in E); \tau_G < \tau]$$

$$\geq c\mathbb{P}^{(0,s)}(\tau_G < \tau).$$

If $U^* > \lambda$, then the path of Z_t must enter G^c before τ. So

$$\mathbb{P}^{(0,s)}(\tau_G < \tau) \geq \mathbb{P}^{(0,s)}(U^* > \lambda).$$

□

Proof of Theorem 6.2, (a) implies (b). From Proposition 6.3,

$$s^d \mathbb{P}^{(0,s)}(U^* > \lambda) \leq cs^d \mathbb{P}^{(0,s)}(X_\tau \in E) \leq |E| = |\{N(f) > \lambda\}|.$$

Integrating over λ from 0 to ∞, $s^d \mathbb{E}^{(0,s)} U^* \leq \int N(f)(x)dx$. Now let $s \to \infty$.

□

We need the following lemmas. Suppose v is a nonnegative function on D and set $N_a(v)(x) = \sup_{z \in C_a(x)} v(z)$.

(6.4) Lemma. *If $b > a$,*

$$\left|\{x : N_b(v)(x) > \lambda\}\right| \leq (1 + (b/a)^d)\left|\{x : N_a(v)(x) > \lambda\}\right|.$$

Proof. Let $f(t) = 1_{(N_a(v) > \lambda)}(t)$. If $N_b(v)(w) > \lambda$, there exists (x, y) such that $|w - x| < by$ and $v(x, y) > \lambda$. So for any $t \in \mathbb{R}^d$ with $|t - x| < ay$, $N_a(v)(t) > \lambda$, or $f(t) = 1$. Hence $f = 1$ on $B(x, ay)$.

$B(x, ay) \subseteq B(w, ay + by)$, hence

$$Mf(x) \geq \frac{1}{|B(x, ay + by)|} \int_{B(w, ay+by)} f(t)\, dt$$

$$\geq \frac{|B(x, ay)|}{|B(w, ay + by)|} = \frac{a^d}{(a + b)^d}.$$

So if $N_b(v)(w) > \lambda$, $Mf(w) \geq \alpha = a^d/(a + b)^d$. Then

$$\left|\{N_b(v) > \lambda\}\right| \leq \left|\{Mf \geq \alpha\}\right| \leq \frac{c}{\alpha}\|f\|_1 = \frac{c}{\alpha}\left|\{N_a(v) > \lambda\}\right|.$$

□

(6.5) Corollary. *If $b > a$,*

$$\|N_b^h(f)\|_1 \leq (1 + (b/a)^d)\|N_a^h(f)\|_1.$$

$[N_b^h(f)$ is defined in (1.6).]

Proof. Applying Lemma 6.4 to $v(x, y) = u(x, y)1_{(y<h)}$, we see that

$$(6.2) \qquad |\{N_b^h(f) > \lambda\}| \leq (1 + (b/a)^d)|\{N_a^h(f) > \lambda\}|.$$

Integrating (6.2) over λ from 0 to ∞ proves the corollary.

□

(6.6) Lemma. *There exist a_0 and c depending only on the dimension such that if $a \le a_0$, then*

$$\|N_a^{2h}(f)\|_1 \le \|N_a^h(f)\|_1 + c \limsup s^d \mathbb{E}^{(0,s)} U^*.$$

Proof. We have $s^d \mathbb{E}^{(0,s)} |f(X_\tau)| \le s^d \mathbb{E}^{(0,s)} U^*$, and letting $s \to \infty$,

(6.3) $$\int |f(x)| dx \le \limsup s^d \mathbb{E}^{(0,s)} U^*.$$

It follows that

(6.4) $$\|P_y f\|_1 = \|P_y * f\|_1 \le \|P_y\|_1 \|f\|_1 \le \|f\|_1$$

or

$$\int |u(x,y)| dx \le \|f\|_1$$

for all y.

Let $u_h^*(w) = \sup\{|u(x,y)| : (x,y) \in C_a(w), h \le y < 2h\}$. Fix w. Provided a_0 is chosen small enough, if $(x,y) \in C_a(w)$ with $h \le y < 2h$, then $B((x,y), h/2) \subseteq B(w, 3h/2) \times [h/4, 11h/4]$. Since u is harmonic,

$$|u(x,y)| = ch^{-(d+1)} \left| \int_{B((x,y),h/2)} u(z) dz \right|$$

$$\le ch^{-d-1} \int_{B(w,3h/2)} \int_{h/4}^{11h/4} |u(t,r)| dr\, dt.$$

Integrating over $w \in \mathbb{R}^d$ and using Fubini's theorem,

(6.5) $$\int u_h^*(w) dw$$

$$\le ch^{-d-1} \int \int_{B(w,3h/2)} \int_{h/4}^{11h/4} |u(t,r)| dr\, dt\, dw$$

$$\le ch^{-1} \int_{h/4}^{11h/4} \int |u(t,r)| dt\, dr$$

$$\le c \limsup s^d \mathbb{E}^{(0,s)} U^*.$$

To finish,

$$|\{N_a^{2h}(f) > \lambda\}| \le |\{N_a^{2h}(f) > \lambda, N_a^h(f) \le \lambda\}| + |\{N_a^h(f) > \lambda\}|$$

$$\le |\{u_h^* > \lambda\}| + |\{N_a^h(f) > \lambda\}|.$$

Integrating over λ from 0 to ∞ gives

$$\|N_a^{2h}(f)\|_1 \le \|u_h^*\|_1 + \|N_a^h(f)\|_1.$$

Now apply (6.5). $\qquad\qquad\qquad\qquad\qquad\qquad\qquad\qquad\qquad\qquad\quad\square$

Very similar is the following lemma, whose proof is Exercise 26.

(6.7) Lemma. *Let $\varepsilon > 0$. There exists a_0 and c depending only on the dimension d such that if $a \le a_0$,*

$$\|N_a^{\varepsilon/2}(P_\varepsilon f)\|_1 \le c \limsup s^d \mathbb{E}^{(0,s)} U^*.$$

Proof of Theorem 6.2, (b) implies (a). Note $C(w) + (0,\varepsilon) \subseteq C(w)$. Let $S_\varepsilon = \inf\{t : Y_t < \varepsilon\}$. By the translation invariance of the Poisson kernel and of Brownian motion, $\sup_{(x,y)\in C(w)+(0,\varepsilon)} |u(x,y)| \uparrow N(f)(w)$ as $\varepsilon \to 0$ and $\limsup s^d \mathbb{E}^{(0,s)} \sup_{r \le S_\varepsilon} |U_r| \le \limsup s^d \mathbb{E}^{(0,s)} U^*$ for $\varepsilon > 0$. Therefore it suffices to prove

$$\|N_a(f_\varepsilon)\|_1 \le c \limsup s^d \mathbb{E}^{(0,s)} U^*$$

with c independent of ε, where $f_\varepsilon = P_\varepsilon f$, and then to let $\varepsilon \to 0$.

Take $b_0 = 2a_0$ and let $K = 2(1 + 2^d)$. Let

$$E(\lambda, h) = \{x : N_{a_0}^h(f_\varepsilon)(x) > \lambda, N_{b_0}^{2h}(f_\varepsilon)(x) \le K\lambda\}.$$

We write

(6.6) $$|\{N_{a_0}^h(f_\varepsilon) > \lambda\}| \le |E(\lambda, h)| + |\{N_{b_0}^{2h}(f_\varepsilon)/K > \lambda\}|.$$

We will show there exists c independent of h and ε such that

(6.7) $$\int_0^\infty |E(\lambda, h)| \, d\lambda \le c \limsup s^d \mathbb{E}^{(0,s)} U^*.$$

Suppose we have (6.7). Integrating (6.6) over λ from 0 to ∞, we have

$$\|N_{a_0}^h(f_\varepsilon)\|_1 \le \int_0^\infty |E(\lambda, h)| d\lambda + \|N_{b_0}^{2h}(f_\varepsilon)/K\|_1$$
$$\le c \limsup s^d \mathbb{E}^{(0,s)} U^* + K^{-1}\|N_{b_0}^{2h}(f_\varepsilon)\|_1$$
$$\le c \limsup s^d \mathbb{E}^{(0,s)} U^* + K^{-1}\|N_{b_0}^h(f_\varepsilon)\|_1.$$

Combining Lemma 6.7 and repeated uses of Lemma 6.6, $\|N_{a_0}^h(f_\varepsilon)\|_1 < \infty$. Using Corollary 6.5 we see that

$$\|N_{a_0}^h(f_\varepsilon)\|_1 \le 2c \limsup s^d \mathbb{E}^{(0,s)} U^*.$$

Using Corollary 6.5 again, for any $a > 0$,

$$\|N_a^h(f_\varepsilon)\|_1 \le c \limsup s^d \mathbb{E}^{(0,s)} U^*.$$

Letting $h \to \infty$ and then $\varepsilon \to 0$ completes the proof.

It remains to prove (6.7). Let $R > 2h > 0$. Suppose $w \in E(\lambda, h)$. There exists $z = (x, y) \in C_{a_0}(w)$ with $y < h$ and $|u(x,y)| > \lambda$. Moreover there exists c_1 such that $|u| \le K\lambda$ in $B(z, c_1 y)$. By Corollary II.1.4, $|\nabla u| \le c_2 K\lambda/y$

in $B(z, c_1 y/2)$. Hence there exists c_3 such that $|u| \geq \lambda/2$ in $B(z, c_3 y)$. As in (1.9), since $|w| \leq 2R$, if s is sufficiently large

$$\mathbb{P}_w^{(0,s)}(Z_t \text{ hits } B(z, c_3 y) \text{ before } \tau) \geq c.$$

If Z_t hits $B(z, c_3 y)$ before τ, $U^* \geq \lambda/2$.
    Hence

$$s^d \mathbb{P}^{(0,s)}(U^* > \lambda/2)$$

$$\geq \int_{E(\lambda,h) \cap B(0,2R)} s^d \mathbb{P}_w^{(0,s)}(U^* > \lambda/2)\mathbb{P}^{(0,s)}(X_\tau \in dw)$$

$$\geq c \int_{E(\lambda,h) \cap B(0,2R)} s^d \mathbb{P}^{(0,s)}(X_\tau \in dw).$$

For s sufficiently large, this is greater than

$$c|E(\lambda, h) \cap B(0, 2R)|.$$

Integrate λ from 0 to ∞ to get

$$s^d \mathbb{E}^{(0,s)}U^* \geq c \int |E(\lambda, h) \cap B(0, 2R)| \, d\lambda.$$

Let $s \to \infty$ to show

$$\limsup s^d \mathbb{E}^{(0,s)}U^* \geq c \int |E(\lambda, h) \cap B(0, 2R)| \, d\lambda.$$

Letting $R \to \infty$, we then have (6.7). $\qquad\qquad\qquad\qquad\qquad\qquad$ $\square$

Recall that u_j is the harmonic extension of $R_j f$. Let $u_0 = u$, write ∂_0 for ∂_y, and let $m_{ij} = \partial_i u_j$, $i, j = 0, 1, \ldots, d$. Then

$$\widehat{m}_{ij}(\xi) = i\xi^i \frac{\xi^j}{|\xi|} e^{-y|\xi|} \widehat{f}(\xi) = \widehat{m}_{ji}(\xi)$$

if $i, j = 1, 2, \ldots, d$, and a similar argument in the case when i or j is 0 shows m_{ij} is symmetric. Also

$$\text{trace } \widehat{m}(\xi) = \widehat{m}_{00}(\xi) + \sum_{i=1}^{d} \widehat{m}_{ii}(\xi) = |\xi|e^{-y|\xi|}\widehat{f}(\xi) + \sum_{i=1}^{d} \frac{-(\xi^j)^2}{|\xi|} e^{-y|\xi|} \widehat{f}(\xi) = 0,$$

so the trace of m is 0.

(6.8) Proposition. *If $x \in \mathbb{R}^{d+1}$,*

$$(6.8) \qquad \sum_i \left(\sum_j m_{ij} x^j \right)^2 \leq \frac{d}{d+1} \left(\sum_{i,j} m_{ij}^2 \right) \sum_k (x^k)^2.$$

Proof. Without loss of generality assume $\|x\|_{\ell^2}^2 = \sum_k (x^k)^2 = 1$. First suppose the matrix m is diagonal with diagonal entries $\lambda_0, \ldots, \lambda_d$ and that λ_{j_0} is the largest in absolute value. Since the trace of m is 0,

$$\lambda_{j_0} = -\sum_{j \neq j_0} \lambda_j,$$

or by the Cauchy-Schwarz inequality,

$$\lambda_{j_0}^2 \leq d \sum_{j \neq j_0} \lambda_j^2.$$

Hence $(d+1)\lambda_{j_0}^2 \leq d \sum_j \lambda_j^2$.

Now the left-hand side of (6.8) is just

$$(6.9) \qquad \|mx\|_{\ell^2}^2 \leq \lambda_{j_0}^2 \leq \frac{d}{d+1} \sum_j \lambda_j^2 = \frac{d}{d+1} \sum_{i,j} m_{ij}^2.$$

If m is not diagonal, there exists a unitary matrix P such that $n = P^t m P$ is diagonal. By Exercise 27, $\sum_{i,j} m_{ij}^2 = \sum_{i,j} n_{ij}^2$. By the same exercise,

$$(6.10) \qquad \sup_{\|x\|_{\ell^2} \leq 1} \sum_i \left(\sum_j m_{ij} x^j \right)^2 = \sup_{\|x\|_{\ell^2} \leq 1} \sum_i \left(\sum_j n_{ij} x^j \right)^2.$$

This with (6.9) completes the proof. $\qquad\qquad\qquad\qquad\qquad\qquad$ $\square$

Let $F_\varepsilon = (\varepsilon^2 + \sum_{j=0}^d u_j^2)^{1/2}$, so that $F_\varepsilon > 0$.

(6.9) Proposition. (Subharmonicity lemma) *If $q > (d-1)/d$, then $\Delta(F_\varepsilon^q) \geq 0$.*

Proof. Note

$$\partial_i(F_\varepsilon^q) = \frac{q}{2} \left(\varepsilon^2 + \sum_j u_j^2 \right)^{q/2-1} \left(\sum_j 2u_j \partial_i u_j \right)$$

and

$$\partial_{ii}^2(F_\varepsilon^q) = \frac{q}{2}\left(\frac{q}{2}-1\right) F_\varepsilon^{q-4} \left(\sum_j 2u_j \partial_i u_j \right)^2$$
$$+ \frac{q}{2} F_\varepsilon^{q-2} \sum_j \left[2(\partial_i u_j)^2 + 2u_j \partial_{ii}^2 u_j \right].$$

Since $\Delta u_j = 0$, summing over i gives

$$(6.11) \qquad \Delta(F_\varepsilon^q) = q F_\varepsilon^{q-4} \left[(q-2) \sum_i \left(\sum_j u_j \partial_i u_j \right)^2 + F_\varepsilon^2 \sum_{i,j} (\partial_i u_j)^2 \right].$$

If $q \geq 2$, this is trivially ≥ 0. Suppose $(d-1)/d < q < 2$. Let $m_{ij} = \partial_i u_j$, $x^j = u_j$. Then

$$\sum_i \left(\sum_j u_j \partial_i u_j \right)^2 = \sum_i \left(\sum_j m_{ij} x^j \right)^2$$

$$\leq \frac{d}{d+1} \sum_{i,j} (\partial_i u_j)^2 \sum_j u_j^2$$

$$\leq \frac{d}{d+1} \sum_{i,j} (\partial_i u_j)^2 F_\varepsilon^2.$$

Hence the term inside the brackets in (6.11) is greater than

$$(q-2)\frac{d}{d+1} F_\varepsilon^2 \sum_{i,j} (\partial_i u_j)^2 + F_\varepsilon^2 \sum_{i,j} (\partial_i u_j)^2.$$

Since $q > (d-1)/d$, then $1 + (q-2)d/(d+1) \geq 0$ and the last expression is greater than or equal to 0. $\qquad\square$

We can now complete the proof of Theorem 6.2.

Proof of Theorem 6.2. The equivalence of (a) and (b) was proved above. We show (d) implies (b). Let $(d-1)/d < q < 1$. Let $G_\varepsilon = F_\varepsilon^q$. Since $\Delta G_\varepsilon \geq 0$, Itô's formula tells us that

$$G_\varepsilon(Z_{t\wedge\tau}) - G_\varepsilon(Z_0) = \text{ martingale } + \frac{1}{2} \int_0^{t\wedge\tau} \Delta G_\varepsilon(Z_s)ds,$$

or $G_\varepsilon(Z_{t\wedge\tau})$ is a submartingale. $1/q > 1$, so by Doob's inequality,

$$(6.12) \qquad \mathbb{E}^{(0,s)} \sup_{t<\tau} F_\varepsilon(Z_t) = \mathbb{E}^{(0,s)} \sup_{t<\tau} G_\varepsilon(Z_t)^{1/q}$$

$$\leq c\mathbb{E}^{(0,s)} G_\varepsilon(Z_\tau)^{1/q} = c\mathbb{E}^{(0,s)} F_\varepsilon(Z_\tau).$$

Let $\varepsilon \to 0$. Then

$$s^d \mathbb{E}^{(0,s)} U^* \leq s^d \mathbb{E}^{(0,s)} \sup_{t<\tau} F(Z_t) \leq cs^d \mathbb{E}^{(0,s)} F(Z_\tau)$$

$$\leq c\sup_{y>0} \int |F(x,y)|dx.$$

Next we show (d) implies (c). Just as above,

$$s^d \mathbb{E}^{(0,s)} \sup_{t<\tau} F(Z_t) \leq c\sup_{y>0} \int |F(x,y)|dx.$$

However, $U_\tau^j \leq F(Z_\tau)$, so

$$\int |R_j f(x)|dx = \limsup cs^d \mathbb{E}^{(0,s)} U_\tau^j \leq c\sup_{y>0} \int |F(x,y)|dx.$$

This is true for $j = 1, 2, \ldots, d$, and if we write R_0 for the identity, it is also true for $j = 0$.

To see that (b) implies (c) we use an argument very similar to the second proof of Theorem 3.5. Let $g \in C_K^1$ with $\|g\|_\infty \leq 1$. Since $f \in H^1 \subseteq L^1$, arguing as in the first proof of Theorem 3.5 we obtain

$$\int (R_j f) g = \lim s^d \mathbb{E}^{(0,s)} \left[g(X_\tau) \int_0^\tau \partial_y u_j(Z_r) dY_r \right]$$

$$\leq \|g\|_\infty \limsup s^d \mathbb{E}^{(0,s)} \left| \int_0^\tau A \nabla u(Z_r) \cdot dZ_r \right|.$$

By Theorem I.6.8, this is bounded by

$$c \limsup s^d \mathbb{E}^{(0,s)} \left[\left(\int_0^\tau |\nabla u(Z_r)|^2 dr \right)^{1/2} \right]$$

$$\leq c \limsup s^d \mathbb{E}^{(0,s)} U^* + c \limsup s^d |u((0,s))|$$

$$\leq c \limsup s^d \mathbb{E}^{(0,s)} U^* + c \|f\|_1.$$

Taking the supremum over such gs, we see

(6.13) $$\|R_j f\|_1 \leq c \limsup s^d \mathbb{E}^{(0,s)} U^* + c \|f\|_1.$$

Since $|f(x)| \leq N(f)(x)$, by the equivalence of (a) and (b)

$$\|f\|_1 \leq \|N(f)\|_1 = \|f\|_{H^1} \leq \limsup s^d \mathbb{E}^{(0,s)} U^*.$$

This and (6.13) imply (c).

Finally, we want to see that (c) implies (d). Let $F_y(x, z) = F(x, y+z)$. So $F_y = P_y F$. We have

$$\int |F(x,y)| dx = \int F_y(x,0) dx$$

$$= \lim s^d \mathbb{E}^{(0,s)} F_y(Z_\tau)$$

$$\leq c \sum_{j=0}^d \lim s^d \mathbb{E}^{(0,s)} |P_y R_j f(Z_\tau)|$$

$$\leq c \sum \|P_y R_j f\|_1 \leq c \sum \|R_j f\|_1.$$

$\square$

As an immediate corollary we have the following.

(6.10) Theorem. *If $f \in H^1$, then f has nontangential limits almost everywhere.*

Proof. $f \in H^1$ implies that the harmonic extension of f, namely u, is such that $N(f) \in L^1$. Hence u is nontangentially bounded a.e. The result follows by Theorem 1.5. $\square$

In the case $d = 1$ we prove the boundedness of the Hilbert transform on H^1.

(6.11) Proposition. *If $d = 1$, the Hilbert transform is bounded on H^1.*

Proof. If $f \in H^1$, then $\limsup s^d \mathbb{E}^{(0,s)} U^* \leq c\|f\|_{H^1}$ by Theorem 6.2. By Theorem I.6.8,

$$\limsup s^d \mathbb{E}^{(0,s)} \langle U \rangle_\tau^{1/2} \leq c\|f\|_{H^1} + \limsup s^d |u(0,s)|$$
$$\leq c\|f\|_{H^1} + c\|f\|_1 \leq c\|f\|_{H^1}.$$

If v is the harmonic extension of Hf and $V_t = v(Z_{t \wedge \tau})$, then $\langle V \rangle_\tau = \langle U \rangle_\tau$. Using the inequalities of Theorem I.6.8 again,

$$\limsup s^d \mathbb{E}^{(0,s)} V^* \leq c\|f\|_{H^1} + c\limsup s^d |v(0,s)|$$
$$\leq c\|f\|_{H^1} + \|Hf\|_1 \leq c\|f\|_{H^1}$$

by Theorem 6.2. Finally, again by Theorem 6.2, $Hf \in H^1$ and $\|Hf\|_{H^1} \leq c\|f\|_{H^1}$. $\qquad\square$

Extensions

Let us define H^p as follows.

(6.12) Definition. *If $p \in (0, \infty)$, then f is in H^p if $N(f) \in L^p$ and we define $\|f\|_{H^p} = \|N(f)\|_p$.*

If $p > 1$, by Corollary 1.4 the H^p norm of f and the L^p norm of f are comparable. Note that for $p < 1$, H^p is no longer a Banach space because the triangle inequality fails to hold.

Much of the H^1 theory goes through for H^p, $p < 1$. For example, the equivalence of (a) and (b) in Theorem 6.2 still holds if we replace H^1 by H^p and $\lim s^d \mathbb{E}^{(0,s)} U^*$ by $\lim s^d \mathbb{E}^{(0,s)} (U^*)^p$. The equivalence of the analogs of (a) and (b) with (c) and (d) of Theorem 6.2 also continues to hold provided $p > (d-1)/d$. A suitable variation holds when $p \leq (d-1)/d$.

In Theorem 4.1 and Proposition 4.5 we saw that the area function was equivalent in L^p norm to f when $p > 1$. An examination of the proof of Proposition 4.5 shows that we actually proved that the L^p norm of the area function of f was bounded by a constant times the H^p norm of f for all p. It turns out that the inequality goes the other way as well. In fact, for all p, the H^p norm of f is equivalent to the L^p norm of $A(f)$, the L^p norm of $g(f)$, and the L^p norm of u^+, where $u^+(x) = \sup_{y>0} |u(x,y)|$; see Fefferman and Stein [1] for the proofs.

Results for H^p spaces can also be obtained through the use of atoms. For H^1 an atom is a function $a(x)$ having support in a cube Q that is bounded by $|Q|^{-1}$ and $\int a(x) = 0$. It turns out that if f is in H^1, then there exist constants b_i and atoms a_i such that $f(x) = \sum_i b_i a_i(x)$ and the H^1 norm of f is comparable to $\sum |b_i|$ (see Coifman [1] and Latter [1]).

7 Bounded mean oscillation

Introduction

The space BMO or space of functions of bounded mean oscillation is the appropriate substitute for L^∞ in many results concerning singular integrals. In this section we will consider the space $\mathcal{BMO}$ of martingales, the relationship between BMO and $\mathcal{BMO}$, prove that the dual of H^1 is BMO, and use BMO to prove some strong results on the L^p boundedness of singular integrals.

Let us start with the space $\mathcal{BMO}$ of martingales.

(7.1) Definition. *A continuous martingale M is in $\mathcal{BMO}$ if there exists c such that for all stopping times T*

$$\mathbb{E}\left[(M_\infty - M_T)^2 | \mathcal{F}_T\right] \leq c^2, \qquad \text{a.s.}$$

The smallest such c is the $\mathcal{BMO}$ norm of M.

Note that adding a $\mathcal{F}_0$ measurable random variable to M does not affect the $\mathcal{BMO}$ norm of M.

If M is a bounded martingale, then $M \in \mathcal{BMO}$. For a slightly less trivial example, let $M_t = X_{t \wedge 1}$, where X_t is Brownian motion. Then

$$
\begin{aligned}
\mathbb{E}\left[(M_\infty - M_T)^2 | \mathcal{F}_T\right] &= \mathbb{E}\left[(X_1 - X_{1 \wedge T})^2 | \mathcal{F}_T\right] \\
&\leq \mathbb{E}\left[\sup_{s \leq 1}(X_{T+s} - X_T)^2 | \mathcal{F}_T\right] \\
&= \mathbb{E}^{X_T}[\sup_{s \leq 1}(X_s - X_0)^2] = \mathbb{E}^0[\sup_{s \leq 1} X_s^2] \\
&\leq c\mathbb{E}^0 X_1^2 = c < \infty.
\end{aligned}
$$

If in the definition of $\mathcal{BMO}$ we integrate over $A = (T < \infty)$, we see that

(7.1) $$\mathbb{E}\left[(M_\infty - M_T)^2; T < \infty\right] \leq \|M\|_{\mathcal{BMO}}^2 \mathbb{P}(T < \infty).$$

In particular, if $M_0 \equiv 0$ and we take $T \equiv 0$, then $\mathbb{E}\, M_\infty^2 \leq \|M\|_{\mathcal{BMO}}^2$, or the $\mathcal{BMO}$ martingales can be considered as a subset of the square integrable ones.

Since $M_t^2 - \langle M \rangle_t$ is a martingale, then $M \in \mathcal{BMO}$ if and only if

(7.2) $$\mathbb{E}\left[\langle M \rangle_\infty - \langle M \rangle_T | \mathcal{F}_T\right] \leq c^2.$$

(7.2) Proposition. (John-Nirenberg) *There exist $\alpha > 0$ and $c > 0$ such that if $M \in \mathcal{BMO}$, then*

$$\mathbb{E}\, e^{\alpha M^*/\|M\|_{\mathcal{BMO}}} < c.$$

Proof. By looking at $M/\|M\|_{\mathcal{BMO}}$, we may assume that the $\mathcal{BMO}$ norm of M is 1. By Jensen's inequality,

$$\mathbb{E}\left[|M_\infty - M_T| \,|\mathcal{F}_T\right] \le \left(\mathbb{E}\left[(M_\infty - M_T)^2 |\mathcal{F}_T\right]\right)^{1/2} \le 1.$$

The result now follows by Theorem I.6.11. □

Let

(7.3) $\qquad \mathcal{M}^p = \big\{M : M$ is a continuous martingale,

$$M_0 = 0, \mathbb{E}\,(M_\infty^*)^p < \infty\big\}.$$

For a norm on $\mathcal{M}^p$ we use

$$\|M\|_{\mathcal{M}^p} = \mathbb{E}\left[(M_\infty^*)^p\right]^{1/p}.$$

By Theorem I.6.8, $\mathcal{M}^p$ also equals $\{M : M_0 \equiv 0, \mathbb{E}\,\langle M\rangle^{p/2} < \infty\}$. When $p > 1$, Doob's inequality implies that $\mathcal{M}^p = L^p(P)$, and in fact $\mathcal{M}^p$ is isomorphic to $L^p(\mathbb{P})$.

Duality for martingales

We will obtain the fact that BMO is the dual of H^1 as a consequence of the corresponding duality of $\mathcal{M}^1$ and $\mathcal{BMO}$.

(7.3) Proposition. *If φ is a continuous linear functional on $\mathcal{M}^1$, then there exists $Y \in \mathcal{BMO}$ such that for all $X \in \mathcal{M}^2$,*

$$\varphi(X) = \mathbb{E}\,X_\infty Y_\infty.$$

Proof. Let us assume $\|\varphi\| = 1$. Since $\mathcal{M}^2 \subseteq \mathcal{M}^1$, then φ is a continuous linear functional on $\mathcal{M}^2$. Since $\mathcal{M}^2$ is isomorphic to $L^2(\mathbb{P})$, the dual of $\mathcal{M}^2$ is $\mathcal{M}^2$ itself. Therefore there exists $Y \in \mathcal{M}^2$ such that $\varphi(X) = \mathbb{E}\,X_\infty Y_\infty$ for $X \in \mathcal{M}^2$. We need to show that $Y \in \mathcal{BMO}$.

If T is a stopping time, let $X_t = Y_t - Y_{t \wedge T}$, and let $Z = \langle X\rangle_\infty$. Note $X_t = \int_0^t 1_{[T,\infty)}(s)dY_s$, so

$$\langle X, Y\rangle_\infty = \int_0^\infty 1_{[T,\infty)}(s)d\langle Y\rangle_s = \langle X\rangle_\infty.$$

Hence

$$\mathbb{E}\,Z = \mathbb{E}\,\langle X, Y\rangle_\infty = \mathbb{E}\left[\langle Y\rangle_\infty - \langle Y\rangle_T\right].$$

On the other hand,

$$\mathbb{E}\, Z = \mathbb{E}\, \langle X, Y \rangle_\infty = \mathbb{E}\, X_\infty Y_\infty = \varphi(X) \le \|X\|_{\mathcal{M}^1} \le c\mathbb{E}\,[Z^{1/2}].$$

Since $Z = 0$ on $(T = \infty)$, by the Cauchy-Schwarz inequality,

$$\mathbb{E}\,[Z^{1/2}] \le (\mathbb{E}\, Z)^{1/2}\mathbb{P}(T < \infty)^{1/2},$$

or $\mathbb{E}\, Z \le c\mathbb{P}(T < \infty)$. So

$$\mathbb{E}\,[\langle Y \rangle_\infty - \langle Y \rangle_T] = \mathbb{E}\, Z \le c\mathbb{P}(T < \infty).$$

This is equivalent to $\mathbb{E}\,[\langle Y \rangle_\infty - \langle Y \rangle_T | \mathcal{F}_T] \le c$, or $Y \in \mathcal{BMO}$. □

(7.4) Proposition. (Fefferman's inequality) *There exists c such that if $X \in \mathcal{M}^1$ and $Y \in \mathcal{BMO}$, then*

$$\mathbb{E}\,\langle X, Y \rangle_\infty \le c\|X\|_{\mathcal{M}^1}\|Y\|_{\mathcal{BMO}}.$$

Proof. By stopping at suitable stopping times, we may assume that X, Y, $\langle X \rangle$, and $\langle Y \rangle$ are all bounded.

If we write

$$1 = \langle X \rangle_t^{-1/4}\langle X \rangle_t^{1/4},$$

then by the inequalities of Kunita and Watanabe (Exercise I.8.24) and the Cauchy-Schwarz inequality,

$$(7.4) \quad (\mathbb{E}\,\langle X, Y \rangle_\infty)^2 \le \left(\mathbb{E}\,\left[\int_0^\infty |d\langle X, Y \rangle_s|\right]\right)^2$$

$$\le \left(\mathbb{E}\,\int_0^\infty \langle X \rangle_t^{-1/2}d\langle X \rangle_t\right)\left(\mathbb{E}\,\int_0^\infty \langle X \rangle_t^{1/2}d\langle Y \rangle_t\right).$$

The first factor on the right in (7.4) is equal to

$$\mathbb{E}\,[2\langle X \rangle_\infty^{1/2}] \le 2c\|X\|_{\mathcal{M}^1}.$$

For the second factor, we integrate by parts (cf. Corollary I.5.7):

$$\mathbb{E}\,\int_0^\infty \langle X \rangle_t^{1/2}d\langle Y \rangle_t = \mathbb{E}\,\langle X \rangle_\infty^{1/2}\langle Y \rangle_\infty - \mathbb{E}\,\int_0^\infty \langle Y \rangle_t d\langle X \rangle_t^{1/2}$$

$$= \mathbb{E}\,\int_0^\infty [\langle Y \rangle_\infty - \langle Y \rangle_t]d\langle X \rangle_t^{1/2}$$

$$= \mathbb{E}\,\int_0^\infty \mathbb{E}\,[\langle Y \rangle_\infty - \langle Y \rangle_t | \mathcal{F}_t]\,d\langle X \rangle_t^{1/2}$$

$$\le \|Y\|_{\mathcal{BMO}}^2 \mathbb{E}\,\langle X \rangle_t^{1/2} \le c\|Y\|_{\mathcal{BMO}}^2\|X\|_{\mathcal{M}^1}.$$

Substituting these bounds in (7.4) proves the proposition. □

If we define $\varphi(X) = \mathbb{E}\, X_\infty Y_\infty$ for $X \in \mathcal{M}^2$, this inequality says that we can extend φ to a continuous linear functional on $\mathcal{M}^1$.

Bounded mean oscillation

Now let us turn to the analytic counterparts.

(7.5) Definition. *We say f is in BMO if there exists c such that*

$$\sup_Q \frac{1}{|Q|} \int_Q |f - f_Q| \le c,$$

where the supremum is over cubes in $\mathbb{R}^d$ (intervals if $d = 1$) and f_Q denotes $|Q|^{-1} \int_Q f \, dx$. The smallest such c is the BMO norm of f.

The space BMO is only defined up to additive constants; that is, f and $f + c$ have the same BMO norm. $f - f_Q$ is the oscillation of f about its mean (over Q), and the mean oscillation, namely, $|Q|^{-1} \int_Q |f - f_Q|$, is supposed to be bounded, which gives rise to the name BMO.

We have the analytic version of the John-Nirenberg inequality.

(7.6) Proposition. *There exists α such that for each cube Q_0*

$$(7.5) \qquad \left| \left\{ x \in Q_0 : \frac{|f - f_{Q_0}|}{\|f\|_{BMO}} > \lambda \right\} \right| \le e^{-\alpha \lambda}.$$

Proof. By scaling we may assume Q_0 is the unit cube and that $\|f\|_{BMO} = 1$. We consider f as a dyadic martingale. Let $\mathcal{F}_n$ be the 2nd subcubes of Q_0 of side length 2^{-n} and set $f = X_\infty$. Then $f_Q = |Q|^{-1} \int_Q f$ is the value of X_n for ω in the interior of Q, if $Q \in \mathcal{F}_n$.

Our inequality will follow from Corollary I.6.12 once we verify the hypotheses of that proposition. First of all, let T be a stopping time. On the set $(T = n)$, if Q is a cube in $\mathcal{F}_n$ such that $\omega \in Q$, then

$$(X_\infty - X_T)(\omega) = f - f_Q.$$

If $A \in \mathcal{F}_\infty$, then $A \cap (T = n) \in \mathcal{F}_n$, and so

$$\mathbb{E}\left[|X_\infty - X_T|; A \cap (T = n)\right] = \sum_{Q \in \mathcal{F}_n, Q \subseteq A \cap (T = n)} \mathbb{E}\left[|X_\infty - X_T|; Q\right]$$

$$= \sum_{Q \in \mathcal{F}_n, Q \subseteq A \cap (T = n)} \int_Q |f - f_Q|$$

$$\le \|f\|_{BMO} \sum_{Q \in \mathcal{F}_n, Q \subseteq A \cap (T = n)} |Q|$$

$$= \|f\|_{BMO} \mathbb{P}\left(A \cap (T = n)\right).$$

It follows that $\mathbb{E}\left[|X_\infty - X_T| \, | \mathcal{F}_T\right] \le \|f\|_{BMO}$.

Secondly we must check that the jumps of X_n are bounded. If $\omega \in Q \subseteq Q'$ with $Q \in \mathcal{F}_{n+1}$, $Q' \in \mathcal{F}_n$, then

$$(X_{n+1} - X_n)(\omega) = |Q|^{-1} \int_Q f - |Q'|^{-1} \int_{Q'} f = |Q|^{-1} \int_Q f - f_{Q'}.$$

Therefore

$$|(X_{n+1} - X_n)(\omega)| \leq \frac{1}{|Q|} \int_Q |f - f_{Q'}|$$

$$\leq 2^d \frac{1}{|Q'|} \int_{Q'} |f - f_{Q'}| \leq 2^d \|f\|_{BMO}.$$

We now apply Corollary I.6.12. □

(7.7) Corollary.

$$\sup_Q \frac{1}{|Q|} \int |f - f_Q|^2 \leq c\|f\|_{BMO}^2.$$

Proof. By scaling we may take Q to be a unit cube. Multiply (7.5) by 2λ, integrate over λ from 0 to ∞ and use Proposition I.1.5. □

Let u be the harmonic extension of f and let $U_t = u(Z_{t \wedge \tau})$. One might guess that $f \in BMO$ if and only if $U_t \in \mathcal{BMO}$, and that turns out to be true.

(7.8) Proposition. *If $U_t \in \mathcal{BMO}$, then $f \in BMO$ and $\|f\|_{BMO} \leq c\|U\|_{\mathcal{BMO}}$.*

Proof. For any t, by the strong Markov property

$$(7.6) \qquad \|U\|_{\mathcal{BMO}}^2 \geq \mathbb{E}\left[(U_\infty - U_t)^2 | \mathcal{F}_t\right] = \mathbb{E}^{Z_{t \wedge \tau}}\left[(U_\infty - U_0)^2\right].$$

Let us write $P_H f(z) = P_H f(x,y)$ for $P_y f(x)$ if $z = (x,y)$. Under $\mathbb{P}^z$ we have $U_0 = u(z)$ while $U_\infty = f(X_\tau)$ and $\mathbb{E}^z U_\infty = \mathbb{E}^z f(X_\tau) = u(z)$. Thus

$$\mathbb{E}^z\left[(U_\infty - U_0)^2\right] = \mathbb{E}^z U_\infty^2 - 2u(z)\mathbb{E}^z U_\infty + u(z)^2$$
$$= \mathbb{E}^z\left[(f(X_\tau))^2\right] - u(z)^2.$$

Therefore (7.6) says that

$$(7.7) \qquad P_H(f^2)(Z_{t \wedge \tau}) - (P_H f(Z_{t \wedge \tau}))^2 \leq \|U\|_{\mathcal{BMO}}^2, \qquad \text{a.s.}$$

for each t.

Since $P_H(f^2)$ and $P_H f$ are harmonic, they are continuous in D. We must have that

$$(7.8) \qquad P_H([f - P_H f(z)]^2)(z) = P_H(f^2)(z) - (P_H f(z))^2 \leq \|U\|_{\mathcal{BMO}}^2$$

for all $z \in D$. For if not, there is a neighborhood of z in which the left-hand side of (7.8) is strictly greater than the right-hand side. However, Z_t

will hit this neighborhood before leaving D with positive probability, which contradicts (7.7).

We want to obtain a bound on $|Q|^{-1}\int_Q |f - f_Q|^2$. By scaling and translation, we may assume Q is the unit cube in $\mathbb{R}^d$ centered at 0. Let $z_0 = (0,\dots,0,1)$. Note

(7.9)
$$|Q|^{-1}\int_Q (f - P_H f(z_0))^2$$

$$= |Q|^{-1}\int_Q f^2 - 2f_Q P_H f(z_0) + (P_H f(z_0))^2$$

$$= |Q|^{-1}\int_Q f^2 - f_Q^2 + (f_Q - P_H f(z_0))^2$$

$$\geq |Q|^{-1}\int_Q f^2 - f_Q^2$$

$$= |Q|^{-1}\int_Q (f - f_Q)^2.$$

Also, $|Q|^{-1}1_Q(y) \leq cP_H(z_0, y)$, so

$$|Q|^{-1}\int_Q (f - P_H f(z_0))^2 \leq cP_H\big([f - P_H f(z_0)]^2\big)(z_0).$$

Combining this, (7.8), and (7.9) shows

$$|Q|^{-1}\int_Q |f - f_Q|^2 \leq c\|U\|_{\mathcal{BMO}}^2.$$

An application of Jensen's inequality finishes the proof. $\qquad\square$

The other direction is a bit harder.

(7.9) Proposition. *If* $f \in BMO$, *then* $U_t \in \mathcal{BMO}$ *and* $\|U\|_{\mathcal{BMO}} \leq c\|f\|_{BMO}$.

Proof. By the proof of Proposition 7.8,

$$\mathbb{E}\left[|U_\infty - U_T|^2\big|\mathcal{F}_T\right] = P_H[(f - P_H f(Z_T))^2](Z_T).$$

So we want to show $P_H\big([f - P_H f(z)]^2\big)(z) \leq c\|f\|_{BMO}^2$ for each z. By scaling and translation, it is enough to show it for $z_0 = (0,\dots,0,1)$.

Let Q_0 be the unit cube in $\mathbb{R}^d$ centered at 0 and $Q_n = 2^n Q_0$. As in (7.9),

$$P_H\big([f - f_{Q_0}]^2\big)(z_0) = P_H\big([f - P_H f(z_0)]^2\big)(z_0) + \big(P_H f(z_0) - f_{Q_0}\big)^2$$
$$\geq P_H\big([f - P_H f(z_0)]^2\big)(z_0).$$

Now $P_H(z_0, x) \leq c\sum_{n=0}^{\infty} 2^{-n}|Q_n|^{-1}1_{Q_n}(x)$. So

$$(7.10) \qquad P_H\big([f - P_H f(z_0)]^2\big)(z_0) \le c \sum 2^{-n} \frac{1}{|Q_n|} \int_{Q_n} |f - f_{Q_0}|^2.$$

By the triangle inequality and the Cauchy-Schwarz inequality,

$$\frac{1}{|Q_n|} \int_{Q_n} |f - f_{Q_0}|^2 \le \frac{n}{|Q_n|} \int_{Q_n} |f - f_{Q_n}|^2 + n \sum_{i=0}^{n-1} |f_{Q_{i+1}} - f_{Q_i}|^2.$$

As in the proof of Proposition 7.6, $|f_{Q_{i+1}} - f_{Q_i}| \le c\|f\|_{BMO}$. So

$$\frac{1}{|Q_n|} \int_{Q_n} |f - f_{Q_0}|^2 \le cn\|f\|_{BMO}^2 + cn^2\|f\|_{BMO}^2.$$

Substituting in (7.10),

$$P_H\big([f - P_H f(z_0)]^2\big)(z_0) \le c \sum_{n=0}^{\infty} (1 + n^2) 2^{-n} \|f\|_{BMO}^2 \le c\|f\|_{BMO}^2.$$

$\square$

The dual of H^1

We now prove the $H^1 - BMO$ duality in the analytic case. The first fact we need is that R_j maps L^∞ to BMO. Later, we will want a similar result for more general kernels, so we take care of both at once.

(7.10) Proposition. *Suppose* $K : \mathbb{R}^d \to \mathbb{R}$ *satisfies (5.11) and there exists* c_1 *such that*

$$(7.11) \qquad |K(x)| \le c_1|x|^{-d}, \qquad x \ne 0$$

and

$$(7.12) \qquad \|Kf\|_2 \le c_1\|f\|_2.$$

Then there exists c_2 *such that if* $f \in L^\infty$ *with compact support, then* $Kf \in BMO$ *and* $\|Kf\|_{BMO} \le c_2\|f\|_\infty.$

By our results in Sect. 3, we see that the Riesz transforms R_j satisfy the hypotheses of this proposition.

Proof. Let Q be a cube and Q' the cube with the same center but twice the side length. Let $f = f_1 + f_2$ where $f_1 = f 1_{Q'}$. By scaling we may assume $|Q| = 1$ and that Q is centered at the origin.

$$\left(\frac{1}{|Q|} \int_Q |Kf_1(x)| dx\right)^2 \le \frac{1}{|Q|} \int_Q |Kf_1(x)|^2 dx \le \frac{1}{|Q|} \|Kf_1\|_2^2$$

$$\le \frac{1}{|Q|} c\|f_1\|_2^2 \le \frac{c\|f\|_\infty^2}{|Q|} |Q'|^2.$$

Let $a = \int K(-y)f_2(y)dy$. So

$$|Kf_2(x) - a| = \left| \int [K(x-y) - K(-y)]f_2(y)dy \right|$$

$$\leq \|f\|_\infty \int_{(Q')^c} |K(x-y) - K(-y)|dy \leq c\|f\|_\infty$$

by (5.11). Then

$$\frac{1}{|Q|} \int_Q |Kf_2(x) - a| \leq c\|f\|_\infty.$$

Combining,

$$\frac{1}{|Q|} \int_Q |Kf(x) - a| \leq c\|f\|_\infty.$$

Our result follows since

$$|Q|^{-1} \int_Q |Kf - (Kf)_Q| \leq |Q|^{-1} \int_Q |Kf - a| + |(Kf)_Q - a|$$

$$= 2|Q|^{-1} \int_Q |Kf - a| \leq c\|f\|_\infty.$$

$\square$

To verify (7.12) we have

(7.11) Proposition. *Suppose K satisfies (7.11), (5.11) and*

$$(7.13) \qquad \int_{R_1 < |x| < R_2} K(x) = 0 \qquad \text{for all } 0 < R_1 < R_2.$$

Then there exists c such that $|\widehat{Kf}(\xi)| \leq c|\hat{f}(\xi)|$ for all $f \in L^2$.

Proof. It suffices to show the inequality for $f \in C_K^\infty$. For such f it is not hard to see (cf. Proposition 2.3) that $\lim_{\varepsilon \to 0, N \to \infty} \widehat{K_{\varepsilon N}f}(\xi) \to \widehat{Kf}(\xi)$, where $K_{\varepsilon N}(x) = K(x)1_{\varepsilon < |x| < N}$, and so we need only look at $\widehat{K_{\varepsilon N}f}(\xi) = \widehat{K_{\varepsilon N}}(\xi)\hat{f}(\xi)$.

Fix ξ. If ε is small enough and N is large enough,

$$(7.14) \quad \widehat{K_{\varepsilon N}}(\xi) = \int_{\varepsilon < |x| < N} e^{i\xi \cdot x} K(x)dx = \int_{\varepsilon < |x| \leq 10/|\xi|} + \int_{10/|\xi| < |x| < N}.$$

Since K satisfies (7.11) and (7.13),

$$(7.15) \quad \int_{\varepsilon < |x| \leq 10/|\xi|} e^{i\xi \cdot x} K(x)dx = \int_{\varepsilon < |x| \leq 10/|\xi|} [e^{i\xi \cdot x} - 1]K(x)dx$$

$$\leq \int_{|x| \leq 10/|\xi|} \frac{c}{|x|^d} |\xi \cdot x|dx \leq c.$$

To handle the second integral in (7.14), let $z = \pi\xi/|\xi|^2$ so that $e^{iz\cdot\xi} = -1$ and $|z| = \pi/|\xi|$.

$$(7.16) \qquad \int K_{\varepsilon N}(x)e^{i\xi\cdot x}dx = -\int K_{\varepsilon N}(x-z)e^{i\xi\cdot x}dx$$

by a change of variables. So

$$\int K_{\varepsilon N}(x)e^{i\xi\cdot x}dx = \frac{1}{2}\int [K_{\varepsilon N}(x) - K_{\varepsilon N}(x-z)]e^{i\xi\cdot x}dx$$

and

$$(7.17) \quad \int_{10/|\xi|<|x|} K_{\varepsilon N}(x)e^{i\xi\cdot x}dx$$

$$= \frac{1}{2}\int_{10/|\xi|<|x|} [K_{\varepsilon N}(x) - K_{\varepsilon N}(x-z)]e^{i\xi\cdot x}dx$$

$$- \int_{|x|\le 10/|\xi|} K_{\varepsilon N}(x)e^{i\xi\cdot x}dx$$

$$+ \frac{1}{2}\int_{|x|\le 10/|\xi|} [K_{\varepsilon N}(x) - K_{\varepsilon N}(x-z)]e^{i\xi\cdot x}dx.$$

The second integral on the right in (7.17) we have already handled. Note $1/|\xi| = |z|/\pi$. If $|x| \ge 10|z|/\pi$, the only way for $|x|$ to be larger than N and $|x-z|$ less than N or vice versa is if $N/10 \le |x| \le 10N$. So the first integral on the right in (7.17) is bounded, using (5.11) and (7.11), by

$$\int_{|x|\ge 10|z|/\pi} |K_{\varepsilon N}(x) - K_{\varepsilon N}(x-z)|dx$$

$$\le \int_{|x|\ge 10|z|/\pi} |K(x) - K(x-z)|dx$$

$$+ 2\int_{N/10\le|x|\le 10N} |K(x)|dx \le c.$$

For the third integral on the right in (7.17), because of (7.15) we need to consider only

$$(7.18) \qquad \int_{|x|\le 10/|\xi|} K_{\varepsilon N}(x-z)e^{i\xi\cdot x}dx$$

$$= -\int_{|x+z|\le 10/|\xi|} K_{\varepsilon N}(x)e^{i\xi\cdot x}dx$$

$$= -\int_{\varepsilon<|x|<1/|\xi|} K(x)e^{i\xi\cdot x}dx$$

$$- \int_{1/|\xi|<|x|,|x+z|\le 10/|\xi|} K(x)e^{i\xi\cdot x}dx.$$

The first integral on the right in (7.18) is similar to what we did in (7.15), and

$$\left| \int_{1/|\xi|<|x|,|x+z|\leq 10/|\xi|} K(x)e^{i\xi \cdot x}dx \right| \leq c \int_{1/|\xi|\leq |x|\leq 15/|\xi|} |x|^{-d}dx \leq c.$$

$\square$

We now show that every BMO function gives rise to a linear functional on H^1.

(7.12) Proposition. *If $g \in BMO$ and $\varphi(f) = \int fg$ for $f \in C_K^\infty$, then φ has a bounded extension to H^1 with $\|\varphi\| \leq c\|g\|_{BMO}$.*

Proof. Let u and v be the harmonic extensions of f and g, respectively, and let $U_t = u(Z_{t \wedge \tau})$, $V_t = v(Z_{t \wedge \tau})$.

$$\varphi(f) = \lim s^d \mathbb{E}^{(0,s)} f(X_\tau)g(X_\tau) = \lim s^d \mathbb{E}^{(0,s)} U_\tau V_\tau$$
$$\leq (c \limsup s^d \mathbb{E}^{(0,s)} U_\tau^*) \|V\|_{BMO} \leq c\|f\|_{H^1}\|g\|_{BMO}.$$

This completes the proof. $\square$

As a consequence we have Fefferman's inequality for functions:

(7.19) $$\left| \int fg \right| \leq c\|f\|_{H^1}\|g\|_{BMO}.$$

We show that every linear functional on H^1 arises from a BMO function. This is the content of Proposition 7.15, but we need a few propositions first.

Let

$$H_0^1 = \{f \in H^1 : f \text{ is in the Schwartz class}\}.$$

(7.13) Proposition. H_0^1 *is dense in* H^1.

Proof. Let $A_1 = \{f \in H^1 : \widehat{f}$ has compact support not containing $0\}$ and $A_2 = \{f \in A_1 : \widehat{f} \in C^\infty\}$. We will show A_1 is dense in H^1 and A_2 is dense in A_1. That will prove the proposition, for if $f \in A_2$, then $\widehat{f}$ is in the Schwartz class, and hence so is f.

Let φ be a C^∞ radially symmetric function on $\mathbb{R}^d$ with compact support such that $\varphi(x) = 1$ if $|x| \leq 1$. Let Φ be the function such that $\widehat{\Phi} = \varphi$. Define T_δ by $\widehat{T_\delta f}(\xi) = \varphi(\xi/\delta)\widehat{f}(\xi)$. So

$$T_\delta f(x) = \delta^d \int f(x-y)\Phi(\delta y)dy.$$

φ is in the Schwartz class, hence so is Φ. This implies that Φ is integrable, and hence by scaling there exists c such that $\|T_\delta f\|_1 \leq c\|f\|_1$ for all δ. Note that $\int \Phi(x)dx = \varphi(0) = 1$.

Let $f \in H^1$ and let $f_k = T_k(I - T_{1/k})f$. It is not hard to see that if $g \in C_K^\infty$, then $(I - T_{1/k})g \to g$ uniformly and then $T_k(I - T_{1/k})g \to g$ in L^1. By the uniform boundedness of the T_δ and the fact that C_K^∞ is dense in L^1, we obtain that $f_k \to f$ in L^1. Since $f \in H^1$, then $R_j f \in L^1$. So $R_j f_k = T_k(I - T_{1/k})R_j f \to R_j f$ in L^1. Consequently $f_k \to f$ in H^1. Since $\widehat{f_k}(\xi) = \varphi(\xi/k)[\widehat{f}(\xi) - \varphi(k\xi)\widehat{f}(\xi)]$, $\widehat{f_k}$ has compact support disjoint from 0, or A_1 is dense in H^1.

Next suppose $f \in A_1$. Let ψ be in C_K^∞ with $\int \psi(\xi)d\xi = 1$. Let Ψ satisfy $\widehat{\Psi} = \psi$. Let $\psi_k(\xi) = k^d \psi(k\xi)$. Define f_k by $\widehat{f_k}(\xi) = (\widehat{f} * \psi_k)(\xi)$. Then $\widehat{f_k}$ is C^∞ with compact support, and there is a neighborhood V of the origin such that for k large enough, $\widehat{f_k}$ is 0 in V. $\widehat{f_k}$ is the Fourier transform of $f_k(x) = f(x)\Psi(x/k)$. It is not hard to see that $f_k \to f$ in L^1. Let M_j be a function such that $\widehat{M_j} = m_j = ix^j/|x|$ outside of V and $m_j \in C_K^\infty$. Then M_j is in the Schwartz class and hence integrable. Since $f, f_k \in A_1$,

$$R_j f_k = M_j * f_k \to M_j * f = R_j f$$

in L^1. Therefore $f_k \to f$ in H^1, and A_2 is dense in A_1. $\square$

(7.14) Proposition. *Suppose $g \in L^\infty$. Then there exists h with $\|h\|_{BMO} \leq c\|g\|_\infty$ such that*

$$\int (R_j f)g = \int fh$$

for all $f \in H^1$. c does not depend on g.

The idea is that one would like to write $\int(R_j f)g = -\int f(R_j g)$, and by Proposition 7.10, we should have $\| - R_j g\|_{BMO} \leq c\|g\|_\infty$. The difficulty is that $R_j g$ may not be defined when $g \in L^\infty$, and we have to go through some contortions to sidestep this.

Proof. Let

$$g_N(y) = g(y)1_{(|y| \leq N)},$$

$$h_N(x) = -\int g_N(y)\big[K_j(x-y) - 1_{(|y| \geq 1)}K_j(-y)\big]dy,$$

$$k_N(x) = -\int g_N(y)\big[K_j(x-y)1_{B(x,1)}(y)\big]dy,$$

and

$$\ell_N(x) = -\int g_N(y)\big[K_j(x-y)1_{B(x,1)^c}(y) - 1_{(|y| \geq 1)}K_j(-y)\big]dy,$$

where $K_j(z) = cz^j/|z|^{d+1}$ is the kernel for the jth Riesz transform.

Since $g_N \in L^p$ for all p, $R_j g_N(x)$ exists for almost every x. Since g_N has compact support, it follows that h_N, k_N, and ℓ_N exist for almost every x.

$h_N(x)$ differs from $-R_j g_N(x)$ by $\int_{|y| \geq 1} g_N(y) K_j(-y) dy$. This is a constant, and since BMO is defined only up to additive constants,

$$\|h_N\|_{BMO} = \|R_j g_N\|_{BMO} \leq c \|g_N\|_\infty$$

by Proposition 7.10.

Let us show $h_N(x)$ converges for almost every x as $N \to \infty$ and that the convergence is uniform on compacts. For $|x| \leq R$, $k_N(x) = -\int g(y) K_j(x - y) 1_{B(x,1)}(y) dy$ for all $N \geq R + 1$, so the k_Ns converge. Call the limit k. Provided $|x| \leq R$, $|y - x| \geq 1$, and $|y| \geq 1$,

$$(7.20) \qquad |K_j(x - y) - K_j(-y)| \leq c|x|/|y|^{d+1}$$

is absolutely integrable in y. So if $|x| \leq R$ and $N, M \geq R + 1$,

$$(7.21) \quad |\ell_N(x) - \ell_M(x)|$$

$$\leq 2\|g\|_\infty \int_{|y| \geq N \wedge M} |K_j(x - y) - K_j(-y)| dy$$

$$\leq \frac{c|x|\,\|g\|_\infty}{N \wedge M} \to 0$$

as $N, M \to \infty$. Let $\ell(x) = \lim_{N \to \infty} \ell_N(x)$. Since $h_N = k_N + \ell_N$, $h_N(x)$ converges, say to $h(x)$, uniformly on compacts.

If Q is a cube, then

$$|Q|^{-1} \int |h - h_Q| = \lim_N |Q|^{-1} \int_Q |h_N - (h_N)_Q| \leq c\|g\|_\infty,$$

by virtue of the uniform convergence. So $\|h\|_{BMO} \leq c\|g\|_\infty$.

Because $f \in H^1$, $\int f(x) dx = 0$ (Exercise 24).

$$\int (R_j f) g_N = - \int f(R_j g_N) = \int f(-R_j g_N - a_N)$$

for any constant a_N. Taking $a_N = \int_{|y| \geq 1} g_N(y) K_j(-y) dy$,

$$\int (R_j f) g_N = \int f h_N.$$

We must now let $N \to \infty$ and show we can pass to the limit. The left-hand side is easy. If $f \in H^1$, then $R_j f \in L^1$ and

$$\left| \int (R_j f) g_N - \int (R_j f) g \right| \leq \|g\|_\infty \int_{|y| \geq N} |R_j f(y)| dy \to 0$$

as $N \to \infty$.

It remains to show convergence of the right-hand side. Since H_0^1 is dense in H^1, by (7.19) it suffices to show

$$(7.22) \qquad \lim_{N \to \infty} \int f h_N = \int f h, \qquad f \in H_0^1.$$

Let $f \in H_0^1$, let $\varepsilon > 0$, and choose R such that $\int_{|x|>R} |f(x)|^2 |x|^{4d+4} dx < \varepsilon^2$. Since $h_N \to h$ uniformly on $B(0,R)$, we concentrate on $B(0,R)^c$ and show

$$(7.23) \qquad \sup_N \left| \int_{B(0,R)^c} f h_N \right| \le c\varepsilon, \qquad \left| \int_{B(0,R)^c} f h \right| < c\varepsilon.$$

$L(y) = K_j(y) 1_{(|y| \le 1)}$ is a kernel to which Proposition 7.11 applies. $k_N(x) = -L g_N(x)$, so

$$(7.24) \qquad \int |k_N|^2 \le c \int |g_N|^2 \le c \int_{B(0,N)} |g|^2 \le cN^d.$$

Hence if $N \le M + 1$,

$$(7.25) \qquad \int_{B(0,M)} |k_N|^2 \le cN^d \le cM^d.$$

If $|x| \le M$ and $N > M + 1$, then $k_N(x) = k_{M+1}(x)$, so

$$(7.26) \qquad \int_{B(0,M)} |k_N|^2 = \int_{B(0,M)} |k_{M+1}|^2 \le \int |k_{M+1}|^2 \le cM^d$$

by (7.24). Hence for any N

$$\int_{B(0,M)} |k_N|^2 \le cM^d.$$

We have

$$\int_{|x| \ge 1} |k_N(x)|^2 |x|^{-4-4d} \le \sum_{n=1}^{\infty} \int_{B(0,2^n)-B(0,2^{n-1})} |k_N(x)|^2 |x|^{-4-4d}$$

$$\le c \sum_{n=1}^{\infty} 2^{-(4+4d)n} \int_{B(0,2^n)} |k_N|^2$$

$$\le \sum_{n=1}^{\infty} 2^{-(4+4d)n} 2^{nd} \le c.$$

Then

$$(7.27) \qquad \left| \int_{|x| \ge R} f k_N \right|$$

$$\le \left(\int_{|x| \ge R} |f(x)|^2 |x|^{4+4d} \right)^{1/2} \left(\int_{|x| \ge 1} |k_N|^2 |x|^{-4-4d} \right)^{1/2}$$

$$\le c\varepsilon.$$

By Fatou's lemma, $\int_{|x|\geq 1}|k(x)|^2|x|^{-4-4d} \leq c$, and we conclude the same estimate for $|\int_{|x|\geq R} fk|$.

Note

$$\ell_N(x) = -R_j g_N(x) - k_N(x) - \int_{|y|\geq 1} g_N(y)K_j(-y)dy.$$

Since

$$\left|\int_{|y|\geq 1} g_N(y)|y|^{-d}dy\right| \leq c\log N,$$

then

$$\int_{B(0,M)} |\ell_N|^2 \leq c\|g_N\|_2^2 + c(\log N)^2 M^d \leq cN^d + c(\log N)^2 M^d.$$

If $N \leq M+1$, this shows

$$\int_{B(0,M)} |\ell_N|^2 \leq cM^{d+1}.$$

On the other hand, if $N > M+1$ and $|x| \leq M$, by (7.21)

$$|\ell_N(x) - \ell_M(x)| \leq c|x|/M,$$

so

$$\int_{B(0,M)} |\ell_N(x)|^2 \leq c\int_{B(0,M)} |\ell_{M+1}|^2 + c\int_{B(0,M)} |\ell_N - \ell_{M+1}|^2$$
$$\leq cM^{d+1} + cM^d \leq cM^{d+1}.$$

We now proceed to bound $\int_{|x|\geq R} f\ell_N$ and $\int_{|x|\geq R} f\ell$ by the same method we used with k_N and k. Combining this with (7.27) proves (7.23). $\qquad\square$

(7.15) Proposition. *If φ is a continuous linear functional on H^1, there exists $g \in BMO$ with $\|g\|_{BMO} \leq c\|\varphi\|$ such that $\varphi(f) = \int fg$ for all $f \in H^1$.*

Proof. By Theorem 6.2, H^1 can be identified with a subset of $L^1 \times \cdots \times L^1$ and $\|f\|_{H^1} \leq \|f\|_1 + \sum \|R_j f\|_1$. So φ can be extended to a linear functional on $L^1 \times \cdots \times L^1$ with the same norm. The dual of this latter space is $L^\infty \times \cdots \times L^\infty$, and there exist $g_0, g_1, \ldots, g_d \in L^\infty$ with $\sum_{j=0}^d \|g_j\|_\infty \leq c\|\varphi\|$ such that

$$\varphi(f) = \int fg_0 + \sum_{j=1}^d \int (R_j f)g_j.$$

By Proposition 7.14, there exists h_j such that $\|h_j\|_{BMO} \leq c\|g_j\|_\infty$ and $\int (R_j f)g_j = \int fh_j$ for all $f \in H^1$. If we take $g = g_0 + \sum_{j=1}^d h_j$, then $g \in BMO$ and $\varphi(f) = \int fg$. $\qquad\square$

(7.16) Lemma. *If M_t is a martingale and $L = \limsup s^d \mathbb{E}^{(0,s)} M^* < \infty$, then the function $f(x) = \lim \mathbb{E}_x^{(0,s)} M_\tau$ is in H^1 with $\|f\|_{H^1} \leq cL$.*

Proof. By the duality theorem, it suffices to show that $\int f(x)g(x)dx \leq cL\|g\|_{BMO}$ if $g \in BMO$. Now

$$\int f(x)g(x)dx = \lim s^d \mathbb{E}^{(0,s)}[f(X_\tau)g(X_\tau)] = \lim s^d \mathbb{E}^{(0,s)}[M_\tau g(X_\tau)]$$

by Proposition III.2.7, where we denote the harmonic extension of g by g again. By the probabilistic version of Fefferman's inequality (Proposition 7.4),

$$\lim s^d \mathbb{E}^{(0,s)}[M_\tau g(X_\tau)] \leq (\limsup s^d \mathbb{E}^{(0,s)} M^*)\|g(X_\tau)\|_{BMO} \leq cL\|g\|_{BMO}.$$

$\square$

(7.17) Corollary. *R_j is a bounded operator on H^1 and on BMO.*

Proof. By duality we need only show that R_j is a bounded operator on H^1. (Actually we need to be a little careful: although $(H^1)^* = BMO$, it is not true that $(BMO)^* = H^1$. Nevertheless a duality argument works (see Exercise 29).) If $f \in H^1$, then by Theorem 6.2 $M_t = u(Z_{t \wedge \tau}) = \int_0^{t \wedge \tau} \nabla u(Z_r) \cdot dZ_r$ satisfies the hypotheses of Lemma 7.16 with $L \leq c\|f\|_{H^1}$. So

$$\limsup s^d \mathbb{E}^{(0,s)}\left[\left(\int_0^\tau |\nabla u(Z_r)|^2 dr\right)^{1/2}\right] \leq cL.$$

If A is the matrix given in Corollary 3.7,

$$\limsup s^d \mathbb{E}^{(0,s)}\left[\left(\int_0^\tau |A\nabla u(Z_r)|^2 dr\right)^{1/2}\right] \leq cL,$$

or $\int_0^{t \wedge \tau} A\nabla u(Z_r) \cdot dZ_r = \int_0^{t \wedge \tau} A\nabla u(Z_r)dY_r$ satisfies the hypotheses of Lemma 7.16 with L replaced by cL. By Lemma 7.16, the function

$$\lim \mathbb{E}_x^{(0,s)}\left[\int A\nabla u(Z_r)dY_r\right]$$

is a H^1 function with H^1 norm bounded by cL. By Corollary 3.7, this function is $cR_j f$. $\square$

Singular integrals

We can also prove the H^1 boundedness of a wide class of singular integrals.

(7.18) Proposition. *Suppose K satisfies (7.11), (7.13), and (5.11). Then K is a bounded operator on H^1 and on BMO.*

Proof. By duality we need only to show that K is a bounded operator on H^1. By Theorem 6.2, we need to show that if $f \in H^1$, then Kf and $R_j Kf$, $j = 1, \ldots, d$, are all in L^1. Since R_j and K are both translation invariant, they commute, so we need to show Kf and $KR_j f$ are all in L^1. Also, $f \in H^1$ implies $R_j f \in H^1$, and $K : L^\infty \to BMO$ implies by duality that $K : H^1 \to L^1$. Hence $KR_j f \in L^1$. $\qquad\square$

Another application of Proposition 7.10 is to show $K : f \to K * f$ is a bounded operator on L^p, $p \in (1, \infty)$. In proving this fact we do not use the H^1-BMO duality. Let

$$f^\#(x) = \sup_{x \in Q} \frac{1}{|Q|} \int_Q |f(x) - f_Q| dx,$$

where the supremum is over all cubes Q and $f_Q = |Q|^{-1} \int_Q f(x) dx$. Of course, $f^\# \in L^\infty$ if and only if $f \in BMO$. We will show that if $f^\# \in L^p$ (and $f \in L^1$), then $|\{|f| > \alpha\}| \leq c \|f^\#\|_p^p / \alpha^p$. Actually, $\|f\|_p \leq c \|f^\#\|_p$ (see Fefferman and Stein [1] and Garsia [1]), but we do not need this stronger statement.

(7.19) Lemma. *Suppose $K \geq 1$, $p \in (1, \infty)$, and M_n is a discrete-time martingale satisfying $|M_n| \leq K|M_{n-1}|$, a.s., for every n. Suppose*

$$\left\| \sup_n \mathbb{E}\left[|M_\infty - M_n| \,\big|\, \mathcal{F}_n\right] \right\|_p \leq 1.$$

Then there exists c such that

$$\mathbb{P}(M_\infty^* > \alpha) \leq c/\alpha^p.$$

Proof.

$$(7.28) \qquad \mathbb{P}(M_\infty^* \geq 4K\alpha) \leq \mathbb{P}(M_\infty^* \geq 4K\alpha, |M_\infty| \leq 2K\alpha) \\ + \mathbb{P}(|M_\infty| > 2K\alpha).$$

Let $N_1 = \inf\{n : |M_n| \geq 3K\alpha\}$. Then the first term on the right in (7.28) is bounded by

$$\mathbb{P}(|M_\infty - M_{N_1}| \geq K\alpha, N_1 < \infty)$$
$$= \mathbb{E}\left[\mathbb{P}(|M_\infty - M_{N_1}| \geq K\alpha|\mathcal{F}_{N_1}); N_1 < \infty\right]$$
$$\leq \frac{1}{K\alpha}\mathbb{E}\left[\mathbb{E}\left[|M_\infty - M_{N_1}| \,\big|\, \mathcal{F}_{N_1}\right]; N_1 < \infty\right]$$
$$\leq \frac{1}{K\alpha}\mathbb{P}(N_1 < \infty)^{1/q} \leq \frac{1}{K\alpha}\mathbb{P}(M_\infty^* \geq 3K\alpha)^{1/q},$$

using Hölder's inequality, where q is the conjugate exponent to p.

Let $N_2 = \inf\{n : |M_n| \geq \alpha\}$. Then $|M_{N_2-1}| < \alpha$, so $\alpha \leq |M_{N_2}| \leq K\alpha$. Thus the second term on the right in (7.28) is bounded by $\mathbb{P}(|M_\infty - M_{N_2}| \geq K\alpha, N_2 < \infty)$, and similarly to the above this is less than $(1/K\alpha)\mathbb{P}(M_\infty^* \geq \alpha)^{1/q}$. So

$$(7.29) \qquad \mathbb{P}(M_\infty^* \geq 4K\alpha) \leq \frac{2}{K\alpha}\left(\mathbb{P}(M_\infty^* \geq \alpha)\right)^{1/q}.$$

By Exercise 30, (7.29) implies the lemma. $\qquad\square$

(7.20) Proposition. *If $1 < p < \infty$, there exists c such that $\|f^\#\|_p \leq c\|f\|_p$ and*

$$|\{x : |f(x)| \geq \alpha\}| \leq \frac{c\|f^\#\|_p^p}{\alpha^p}.$$

Proof. If $x \in Q$, let Q' be the smallest cube centered at x containing Q. Then

$$\frac{1}{|Q|}\int_Q |f(x) - f_Q| \leq 2f_Q \leq 2\frac{|Q'|}{|Q|}\frac{1}{|Q'|}\int_{Q'} |f(y)|dy \leq cMf(x).$$

The first inequality follows immediately.

For the second , it suffices to show that if Q_0 is a cube, then

$$|\{x \in Q_0 : |f(x)| \geq \alpha\}| \leq \frac{c\|f^\#\|_p^p}{\alpha^p},$$

where c does not depend on the cube Q_0. Let us normalize so that $\|f^\#\|_p = 1$. Let $\mathcal{F}_n$ be the dyadic decomposition of Q_0 and let f_n be the corresponding dyadic martingale. Then

$$\mathbb{E}\left[|f - f_n| \,|\mathcal{F}_n\right](x) = \frac{1}{|Q|}\int_Q |f - f_Q| \leq f^\#(x)$$

if Q is the cube in $\mathcal{F}_n$ containing x. As in Lemma I.4.17, $|f_n| \leq 2^d|f_{n-1}|$. Since $|f| \leq \sup_n |f_n|$, the result now follows by Lemma 7.19. $\qquad\square$

We say that an operator T is weak type (p,p) if $|\{|Tf(x)| \geq \alpha\}| \leq c\alpha^{-p}\int |f(x)|^p dx$ for all α. Saying T is sublinear means that $|T(f + g)| \leq |T(f)| + |T(g)|$. If T is defined on L^p and L^q and $p < r < q$, we can define T on L^r by letting $Tf = T(f1_{(|f|<1)}) + T(f1_{(|f|\geq 1)})$.

(7.21) Theorem. (Marcinkiewicz interpolation theorem) *Suppose T is sublinear and weak type (p,p) and weak type (q,q) for some $1 \leq p \leq q \leq \infty$. If $p < r < q$, then T is a bounded operator on L^r.*

Proof. We will do the case when $q < \infty$. For the case $q = \infty$, see Exercise 33 and the proof of Theorem I.4.7.

Let $\alpha > 0$ and let $f_1 = f1_{(|f(x)|>\alpha)}$, $f_2(x) = f(x)1_{(|f(x)|\leq\alpha)}$. Note that if $f \in L^r$, then $f_1 \in L^p$ and $f_2 \in L^q$. Since

$$\{|Tf(x)| > \alpha\} \subseteq \{|Tf_1(x)| > \alpha/2\} \cup \{|Tf_2(x)| > \alpha/2\},$$

then

$$|\{|Tf(x)| > \alpha\}| \leq |\{|Tf_1(x)| > \alpha/2\}| + |\{|Tf_2(x)| > \alpha/2\}|$$

$$\leq \frac{c}{(\alpha/2)^p} \int |f_1|^p + \frac{c}{(\alpha/2)^q} \int |f_2|^q$$

$$\leq c\alpha^{-p} \int_{|f|>\alpha} |f|^p + c\alpha^{-q} \int_{|f|\leq\alpha} |f|^q.$$

So

$$\int |Tf(x)|^r dx = \int_0^\infty r\alpha^{r-1} |\{|Tf(x)| > \alpha\}| d\alpha$$

$$\leq c \int_0^\infty \alpha^{r-p-1} \int_{|f|>\alpha} |f|^p d\alpha + c \int_0^\infty \alpha^{r-q-1} \int_{\alpha\leq|f|} |f|^q d\alpha$$

$$= c \int |f|^p \int_0^{|f|} \alpha^{r-p-1} d\alpha + c \int |f|^q \int_{|f|}^\infty \alpha^{r-q-1} d\alpha$$

$$\leq c \int |f|^p |f|^{r-p} + c \int |f|^q |f|^{r-q} \leq c \int |f|^r.$$

$\square$

(7.22) Theorem. *If K satisfies (7.11), (7.13), and (5.11), then $f \to K * f$ is a bounded operator on L^p, $p \in (1, \infty)$.*

Proof. Since $\widehat{K}$ is bounded by Proposition 7.11, the result follows for $p = 2$ by Plancherel's theorem:

$$\|Kf\|_2 = c\|\widehat{Kf}\|_2 \leq c\|\widehat{f}\|_2 = c\|f\|_2.$$

By duality, we obtain the result for $p < 2$ once we have it for $p > 2$, so let us do the case $p > 2$. Let $p_0 \in (p, \infty)$.

Define $Tf = K * f$ and $Uf = (Tf)^\#$. Since T is linear, it is easy to see that U is sublinear. By Proposition 7.10, if $f \in L^\infty$ with compact support, then $Tf \in BMO$, or $Uf \in L^\infty$, with L^∞ norm bounded by $c\|f\|_\infty$. Note that if $f \in L^2$, then

$$\|Uf\|_2 \leq c\|Tf\|_2 \leq c\|f\|_2.$$

By Exercise 33, $\|Uf\|_{p_0} \leq c\|f\|_{p_0}$ if $f \in L^{p_0}$ with compact support. We can extend U to a bounded operator on L^{p_0}. By Proposition 7.20,

$$|\{|Tf(x)| > \alpha\}| \leq \frac{c\|(Tf)^\#\|_{p_0}^{p_0}}{\alpha^{p_0}} \leq \frac{c\|f\|_{p_0}^{p_0}}{\alpha^{p_0}},$$

or T satisfies a weak $(p_0\text{-}p_0)$ inequality. By the Marcinkiewicz interpolation theorem again, T is a bounded linear operator on L^p. □

Extensions

Another characterization of BMO functions is that their harmonic extensions give rise to Carleson measures. See Fefferman and Stein [1].

In this chapter we have been exclusively concerned with operators arising from convolution with some function. What about operators of the form $Tf(x) = \int K(x,y)f(y)dy$, where $K(x,y)$ is not of the form $K(x-y)$? The main difficulty here is the L^2 boundedness of T, and a very useful result is the $T1$ theorem (David and Journé [1]). Certain boundedness and smoothness properties of the kernel K are necessary and the hypotheses that T and its adjoint map the function 1 into BMO.

The dual of H^1 is BMO, and it turns out that the dual of the spaces H^p, $p < 1$, are the spaces C^α of Sect. 3, where α depends on p and the dimension d (Fefferman and Stein [1]). It would be nice to see a probabilistic proof of this fact. This is likely to be a bit tricky, since in general, $\mathcal{M}^p$ does not have a dual if $p < 1$. See Durrett [1] and Herz [1].

One can interpolate between various H^p spaces as well as between BMO and L^p spaces. One nice result is that if T is a bounded operator on H^p for some $p < 1$ and on L^q for some $q > 1$, then T satisfies a weak (1-1) inequality (Folland and Stein [1]).

8 Exercises and further results

Exercise 1. Show that if f is supported in $B(0,1)$ and $\int_{B(0,1)} |f|(1 + \log^+ |f|) < \infty$, then $\int_{(B(0,1))} Mf < \infty$. (Recall $\log^+ x = 0 \vee \log x$.)

Exercise 2. Show that if $f \in L^p$ and $1 \le p < \infty$, then

$$\lim_{r \to 0} |B(0,1)|^{-1} \int_{B(x,r)} |f(y) - f(x)|^p dy = 0$$

for almost every x.

Exercise 3. Show there exists c such that c_p in Theorem 1.1(b) can be taken less than $c(1 \vee (p-1)^{-1})$ for $1 < p < \infty$.

Exercise 4. Show that under the hypotheses of Exercise 1, we have

$$\int_{B(0,1)} |Hf| < \infty.$$

Exercise 5. Show there exists c such that the c_p in Theorem 2.2(a) can be taken less than $c/(p-1)$ if $1 < p \le 2$ and less than cp if $2 \le p < \infty$.

Exercise 6. Show that if f is bounded with support in $B(0,1)$, then there exists α such that $\int_{B(0,1)} e^{\alpha |Hf|} < \infty$.

Hint: Expand $e^{\alpha |Hf|}$ in a Taylor series and use Exercise 5.

Exercise 7. Show that if $f \in C_K^1$, there exists c such that

$$|Hf(x)| \le c/(1 + |x|).$$

Exercise 8. Show that if $f \in C_K^1$, u is the harmonic extension of f, and v is the harmonic extension of Hf, then $s(v(0,s))^2 \to 0$ as $s \to \infty$ and $s|u(0,s)| \le c\|f\|_1$.

Exercise 9. Find an $f \in L^1$ such that $Hf \notin L^1$. (Such an f is in $L^1 - H^1$.) Find an $f \in L^\infty$ such that $Hf \notin L^\infty$.

Exercise 10. Show that if $\alpha > 0$ and $z \le 1$, then $1 - (1-z)^\alpha \le \alpha z$.

Exercise 11. Show that if S_y, h, and u are as in Proposition 3.6, then

$$\mathbb{E}^{(0,s)} \int_0^{S_y} \partial_y h(Z_r) \partial_y u(Z_r)\, dr \to \mathbb{E}^{(0,s)} \int_0^T \partial_y h(Z_r) \partial_y u(Z_r) dr$$

and

$$\mathbb{E}_x^{(0,s)} \int_0^{S_y} \partial_y u(Z_r) dY_r \to \mathbb{E}_x^{(0,s)} \int_0^T \partial_y u(Z_r) dY_r$$

as $y \to 0$.

Hint: Get a bound on $\mathbb{E}^{(0,s)} \int_{S(2^{-n})}^{S(2^{-n-1})}$ and similarly with $\mathbb{E}^{(0,s)}$ replaced by $\mathbb{E}_x^{(0,s)}$.

Exercise 12. Show that if f is in the Schwartz class, then so is $\hat{f}$.

Exercise 13. Let $\beta = \alpha - d/p$. Prove that if $f \in L^p$ and $1 > \beta > 0$, then $I_\alpha f \in C^\beta$.

Hint: Let $K_1(x) = I_\alpha(x) \wedge A$, $K_2(x) = I_\alpha(x) - K_1(x)$ for suitable A. K_1 is differentiable and $K_2 f(x)$ is small. (Cf. the proof of Theorem II.3.16.)

Exercise 14. Prove Corollary 3.15(c).

Exercise 15. Show the following: to prove the inequality $\|g(f)\|_p \le c\|f\|_p$, it suffices to prove the inequality for $f \in C_K^1$ with $f \ge 0$. Show the same is true if $g(f)$ is replaced by $g_\lambda^*(f)$ or $A(f)$.

Exercise 16. Show there exists c_1 and c_2 such that $\|g_\lambda^*(f)\|_2 = c_1\|f\|_2$ and $\|g_y(f)\|_2 = c_2\|f\|_2$.

Exercise 17. Show (4.14).

Exercise 18. Suppose that $K : \mathbb{R} \to \mathbb{R}$ is an odd function and $|K(x)| \le c/|x|$, $x \ne 0$. Show $\lim_{\varepsilon \to 0, N \to \infty} \int_{N > |x| > \varepsilon} K(t) P_y(t - x) \, dt$ exists for each x.

Exercise 19. Extend the proof of Theorem 5.3 to cover the case when $d > 1$. Assume $K : \mathbb{R}^d \to \mathbb{R}$ satisfies (7.11), (7.13), and $|\nabla K(x)| \le c|x|^{-d-1}$, $x \ne 0$.

Exercise 20. Prove the second inequality in (5.15).

Exercise 21. Suppose $f \in C_K^\infty$. Let $S_{(-\infty, u]}$ be defined by (5.14). Show that there exist jointly measurable functions $g(u, x)$ such that $g(u, x) = S_{(-\infty, u]} f(x)$, a.e., for each u and $g(u, x)$ is continuous in u for almost every x.

Hint: Use Theorem I.3.11.

Exercise 22. Show that if $\|m\|_\infty = \infty$, then the operator T_m is not bounded on L^2.

Exercise 23. Show Theorem 5.5 when $1 < p < 2$.

Hint: Use duality.

Exercise 24. Show that if $f \in H^1$, then $\int f(x) \, dx = 0$.

Exercise 25. Show that if $f \in H^1$, then $P_\varepsilon f$ converges to f in H^1 norm as $\varepsilon \to 0$. Show the same for any approximation to the identity satisfying the hypotheses of Theorem 1.6.

Exercise 26. Prove Lemma 6.7.

Exercise 27. Prove that if m is a symmetric matrix, P is unitary, and $n = P^t m P$, then $\sum_{i,j} m_{ij}^2 = \sum_{i,j} n_{ij}^2$ and (6.10) holds.

Exercise 28. Prove the function $(\log x) 1_{(x > 0)}$ is in BMO.

Exercise 29. Show that if T is an operator given by $\widehat{Tf} = \widehat{T}\widehat{f}$ for $f \in C_K^\infty$, then T is a bounded operator on H^1 if and only if T is a bounded operator on BMO.

Exercise 30. Suppose $p \in (1, \infty)$ and q is the exponent conjugate to p. Show that if X is a nonnegative random variable satisfying

$$\mathbb{P}(X \ge R\lambda) \le \frac{c_1}{R\lambda} \left(\mathbb{P}(X \ge \lambda) \right)^{1/q}$$

for all λ, then there exists c_2 depending on R, c_1, and p such that $\mathbb{P}(X \ge \lambda) \le c_2/\lambda^p$.

Hint: Let $\lambda = 2c_1$ and $R = 2$. By induction, show

$$\mathbb{P}(X \ge R^n \lambda) \le R^{-a_n} 2^{-b_n},$$

where

$$a_n = n + (n-1)/q + \cdots + 1/q^{n-1}, \quad b_n = 1 + 1/q + \cdots + 1/q^{n-1}.$$

Conclude that $\mathbb{P}(X \geq R^n \lambda) \leq c/R^{np}$.

Exercise 31. If H is the Hilbert transform, show $HP_y(x) = cx/(x^2 + y^2)$.

Exercise 32. Prove Proposition 2.5.

Exercise 33. Let $1 < p < q < \infty$. Suppose T is a sublinear operator that is weak type (p, p) and there exists c_1 such that $\|Tf\|_\infty \leq c_1 \|f\|_\infty$ for all $f \in L^\infty$ with compact support. Show there exists c_2 such that $\|Tf\|_q \leq c_2 \|f\|_q$ for all $f \in L^q$ with compact support.

Notes

Analytical approaches to the first five sections can be found in Stein [1]. Proposition 1.3 is due to Burkholder and Gundy [2]. For more on Theorem 1.5 see Durrett [1].

The probabilistic proofs of Theorem 2.2(a) are due to Burkholder, Gundy, and Silverstein [1]. For Theorem 2.2(b) see Burkholder [1].

The characterization of the Riesz transforms in terms of martingale transforms is due to Gundy and Silverstein [1] and Gundy and Varopoulos [1]. The proofs of the boundedness of the Riesz transforms via probability are due to Bennett [1] and Bañuelos [1].

The probabilistic aspects of Sect. 4 are from Meyer [2, 3].

Theorems 5.2 and 5.3 are slightly different from but are similar to what can be found in Stein [1], Chap. 4. The rest of Sect. 5 follows Stein [1] closely.

The probability results in Sect. 6 are from Burkholder and Gundy [2]. The remainder is from Stein [1] and Fefferman and Stein [1].

For Sect. 7 we followed Durrett [1] and Fefferman and Stein [1].

V

ANALYTIC FUNCTIONS

1 Conformal invariance

Lévy's theorem

The connection between two-dimensional Brownian motion and complex analysis comes about through the result of Lévy that an analytic function composed with a two-dimensional Brownian motion is a time change of another Brownian motion. After describing some notation we will begin with this theorem.

Let $\mathbb{C}$ denote the complex plane, $\mathbb{D}$ the unit disk $\{z : |z| < 1\}$, and H the half-plane $\{z : \operatorname{Im} z > 0\}$. We will let Z_t denote two-dimensional Brownian motion. Z_t can be considered either as $\operatorname{Re} Z_t + i \operatorname{Im} Z_t$ or as $(\operatorname{Re} Z_t, \operatorname{Im} Z_t)$, depending on the context. We will write just τ for $\tau_\mathbb{D}$ and τ_r for $\tau_{B(0,r)}$. ∂_x denotes $\partial/\partial x$ and similarly ∂_y denotes $\partial/\partial y$.

We will split Lévy's theorem into two pieces. First, suppose that f is not constant and that f is entire, that is, analytic on all of $\mathbb{C}$.

(1.1) Theorem. *Let*

$$(1.1) \qquad A_s = \int_0^s |f'(Z_r)|^2 dr \qquad and \qquad \sigma_t = \inf\{s : A_s \geq t\}.$$

Then $f(Z_{\sigma_t})$ is a two-dimensional Brownian motion.

Proof. If a function is analytic in a domain and its zeroes have a cluster point in the domain, then the function must be identically 0 (Ahlfors [1],

p. 127). Since f is not identically constant, f' has at most countably many zeroes. Because two-dimensional Brownian motion does not hit a countable set (Proposition I.5.8), A_s is strictly increasing. Hence σ_t is continuous in t, a.s., and $f(Z_{\sigma_t})$ is a continuous process. By Exercise 1, $A_s \to \infty$, a.s., as $s \to \infty$.

Suppose $f = u + iv$. By Itô's formula,

$$(1.2) \qquad u(Z_t) = u(Z_0) + \int_0^t \nabla u(Z_s) \cdot dZ_s$$

and

$$(1.3) \qquad v(Z_t) = v(Z_0) + \int_0^t \nabla v(Z_s) \cdot dZ_s.$$

In particular, $u(Z_t)$ and $v(Z_t)$ are martingales. By the Cauchy-Riemann equations we have $|\nabla u(z)|^2 = |\nabla v(z)|^2$ and $\nabla u(z) \cdot \nabla v(z) = 0$ for all z. Hence $\langle u(Z) \rangle_t = \langle v(Z) \rangle_t = A_t$ and $\langle u(Z), v(Z) \rangle_t = 0$.

Therefore, $f(Z_{\sigma_t})$ is a continuous two-dimensional process, each of its components is a martingale, $\langle u(Z_{\sigma.}) \rangle_t = \langle v(Z_{\sigma.}) \rangle_t = t$, and $\langle u(Z_{\sigma.}), v(Z_{\sigma.}) \rangle_t = 0$. By Theorem I.5.10, $f(Z_{\sigma_t})$ is a two-dimensional Brownian motion. □

We next give the version of Lévy's theorem for functions that are analytic in a domain D.

(1.2) Theorem. *Suppose f is analytic in a domain D. Let A_s and σ_t be as above. If $F_t = f(Z(\sigma_{t \wedge \tau_D}))$, then $F_{t \wedge \tau(f(D))}$ is a two-dimensional Brownian motion stopped on exiting $f(D)$.*

Proof. As in the proof of Theorem 1.1, F_t is a continuous martingale with $\langle \operatorname{Re} F \rangle_t = \langle \operatorname{Im} F \rangle_t = t \wedge \tau_D$ and $\langle \operatorname{Re} F, \operatorname{Im} F \rangle_t = 0$. Let W_t be a two-dimensional Brownian motion started at 0 and independent of Z_t and define

$$F_t' = \begin{cases} F_t, & t \le \tau_D; \\ F_{\tau_D} + W_{t - \tau_D}, & t > \tau_D. \end{cases}$$

It is easy to check that $\langle \operatorname{Re} F' \rangle_t = \langle \operatorname{Im} F' \rangle_t = t$ and $\langle \operatorname{Re} F', \operatorname{Im} F' \rangle_t = 0$, and so F_t' is a two-dimensional Brownian motion. It is not hard to see that $f(Z_t)$ first hits $\partial f(D)$ either at the same time or before Z_t first hits ∂D. So $F_{t \wedge \tau(f(D))}$ is two-dimensional Brownian motion stopped at the hitting time of $\partial f(D)$. □

If f is one to one, then in fact the hitting time of $\partial f(D)$ by $f(Z_t)$ is equal to the hitting time of ∂D by Z_t.

Fundamental theorem of algebra

For a first application, let us give a probabilistic proof of the fundamental theorem of algebra.

(1.3) Theorem. *Let $P(z)$ be a nonconstant polynomial in z. Then there is a least one point z_0 such that $P(z_0) = 0$.*

Proof. If $P(z) = a_n z^n + a_{n-1} z^{n-1} + \cdots + a_0$ with $a_n \neq 0$, then

$$|P(z)| = |a_n| \, |z|^n \left| 1 + \frac{a_{n-1}}{a_n} z^{-1} + \cdots \frac{a_0}{a_n} z^{-n} \right| \to \infty$$

as $|z| \to \infty$. Pick R such that $|P(z)| \geq 1$ if $|z| \geq R$.

Suppose $P(z)$ never equals 0. Since $P(\overline{B(0,R)})$ is compact, hence closed, there is a neighborhood V of 0 that is disjoint from it. By taking V to have diameter less than $1/2$, $V \cap P(\mathbb{C}) = \emptyset$. However, $P(Z_t)$ is a time change of two-dimensional Brownian motion. Since P is nonconstant, Exercise 1 says that $\langle P(Z) \rangle_t \to \infty$ as $t \to \infty$. That means $P(Z_t)$ hits every neighborhood infinitely often; in particular it hits V, a contradiction to $V \cap P(\mathbb{C}) = 0$. $\qquad\square$

Exit distributions

As we saw in Chap. II, if we know the exit distribution of Brownian motion for a domain, we can give an explicit solution to the Dirichlet problem. From Lévy's theorem, we can compute the exit distribution of two-dimensional Brownian motion in a number of domains. In principle, it can be done for any simply connected region by virtue of the Riemann mapping theorem (see Theorem 1.14), but in practice it is not easy except for certain simple domains.

(1.4) Proposition. *Let W_α be the wedge $\{re^{i\theta} : r > 0, 0 < \theta < \alpha\}$, $0 < \alpha < 2\pi$, and suppose $z_0 \in W_\alpha$. Then*

$$(1.4) \qquad \mathbb{P}^{z_0}\left(Z_{\tau(W_\alpha)} \in (dt, 0)\right) = \alpha^{-1} t^{\pi/\alpha - 1} \frac{\mathrm{Im}\left(z_0^{\pi/\alpha}\right)}{|z_0^{\pi/\alpha} - t^{\pi/\alpha}|^2} \, dt.$$

Proof. The mapping $f(z) = z^{\pi/\alpha}$ maps W_α onto the half-space $H = \{\mathrm{Im}\, z > 0\}$. Since $0 \notin W_\alpha$, f is analytic on W_α. If $C_u = [0, u] \times \{0\}$, then

$$\mathbb{P}^{z_0}\left(Z_{\tau(W_\alpha)} \in C_u\right) = \mathbb{P}^{f(z_0)}\left(f(Z_{\tau(W_\alpha)}) \in f(C_u)\right).$$

Since $f(Z_t)$ is the time change of a two-dimensional Brownian motion, the place where $f(Z_t)$ exits the domain H and where a two-dimensional Brownian motion exits H are the same. So we need to compute

$$\mathbb{P}^{z_0^{\pi/\alpha}}(Z'_{\tau_H} \in [0, u^{\pi/\alpha}] \times \{0\}),$$

where Z'_t is another two-dimensional Brownian motion. By (II.1.17), this is

$$\int_0^{u^{\pi/\alpha}} \frac{1}{\pi} \frac{\operatorname{Im} z_0^{\pi/\alpha}}{|w - z_0^{\pi/\alpha}|^2} dw.$$

Differentiating in u gives (1.4). □

The mapping $f(z) = \exp(z)$ maps the strip $S = \{0 < \operatorname{Im} z < \pi/2\}$ onto the half-plane H. The same method of proof gives the exit distribution of a strip (Exercise 3).

Sometimes a combination of conformal mapping and other techniques is useful. For example, to find the exit distribution of $W_\alpha \cap B(0, M)$ or $\{0 < \operatorname{Im} z < \pi/2, \operatorname{Re} z < M\}$, conformal mapping reduces the problem to finding the exit distribution of $H \cap B(0, N)$ for suitable N. This is easily done by the reflection principle. For later, we only need the following bound.

(1.5) Proposition. *There exists c independent of N such that if z_0 is in $H \cap B(0, N/2)$, then*

$$\mathbb{P}^{z_0}(Z_{\tau(H \cap B(0,N))} \in \partial B(0, N)) \leq c|z_0|/N.$$

Proof.

$$h_1(z) = \mathbb{P}^z(Z_{\tau(B(0,N))} \in H \cap \partial B(0, N))$$

is harmonic in $B(0, N)$. If $z = x + iy$ and $h_2(z) = h_1(x - iy)$, then by symmetry considerations and the fact that $\overline{Z_t}$, the complex conjugate of Z_t, is another Brownian motion,

$$h_2(z) = \mathbb{P}^{x+iy}(\overline{Z}_{\tau(B(0,N))} \in H^c \cap B(0, N))$$

is harmonic in $B(0, N)$. (This is the reflection principle in disguise.) The difference $h = h_1 - h_2$ is harmonic in $B(0, N)$, takes boundary value 1 on $H \cap \partial B(0, N)$ and takes the value 0 on $\partial H \cap B(0, N)$. Therefore $h(z) = \mathbb{P}^z(Z_{\tau(H \cap B(0,N))} \in \partial B(0, N))$ for $z \in H \cap B(0, N)$. By Corollary II.1.4, $|\nabla h|$ is bounded by c/N in $H \cap B(0, N/2)$. Since $h(0) = 0$, the proposition follows. □

Maximum modulus principle

If u is harmonic in D and continuous on $\overline{D}$, and D is a bounded domain, then $\sup_{\overline{D}} u = \sup_{\partial D} u$. The same is true for analytic functions. Recall from the remarks following Definition II.6.6 that f is subharmonic if $-f$ is superharmonic.

(1.6) Proposition. (Maximum modulus principle) *If D is a bounded domain and f is analytic in D and continuous in $\overline{D}$, then $\sup_{\overline{D}} |f| = \sup_{\partial D} |f|$.*

Proof. If $f = u + iv$ where u and v are harmonic, then $|f|^2 = u^2 + v^2$ is subharmonic, since $\Delta(u^2) = 2|\nabla u|^2 \geq 0$ and the same for $\Delta(v^2)$. So $|f(z)|^2 \leq \mathbb{E}^z |f(Z_{\tau_D})|^2 \leq \sup_{\partial D} |f|^2$ for $z \in D$. Taking square roots gives our result. $\qquad\square$

If D is not bounded, this need not be true at all. For example, let $D = H$ and $f(z) = e^{-iz} = e^{-ix+y}$. This is bounded on ∂H, but not on H. The Phragmén-Lindelöf theorem say that if f also satisfies a growth condition, the maximum modulus result still holds.

First we need a lemma. Although one has to be careful when defining $\log f(z)$ so as to avoid the zeroes of f, $|f(z)|$ is a nonnegative real number and there is no problem defining $\log^+ |f(z)|$, where $\log^+ x = 0 \vee \log x$.

(1.7) Lemma. *If f is analytic in D, then $\log^+ |f(z)|$ is subharmonic and continuous in D.*

Proof. If $|f(z_0)| \leq 1/2$ for some $z_0 \in D$, then by the continuity of f, $|f(z)| < 1$ in a neighborhood of z_0 and so $\log^+ |f(z)| = 0$ in a neighborhood of z_0. If $|f(z_0)| > 1/2$ for some $z_0 \in D$, the same will be true for z in a neighborhood of z_0. A direct calculation shows that $\log |f(z)|$ is harmonic in this neighborhood (cf. Sect. II.3), and so $\log^+ |f(z)|$ is subharmonic and continuous. Therefore $\log^+ |f(z)|$ is subharmonic and continuous in a neighborhood of every point in D. $\qquad\square$

(1.8) Theorem. *Suppose f is analytic in H, continuous on $\overline{H}$, and bounded by K on ∂H. If for some $\alpha < 1$,*

$$|f(z)| \leq c \exp(|z|^\alpha),$$

then $|f(z)| \leq K$ for all $z \in H$.

Proof. By multiplying f by a constant, we may suppose that $K \geq 1$. Fix $z_0 \in H$. Let $\varepsilon > 0$. Let $u(z) = \log^+ |f(z)|$. By our growth assumption on f, there exists $M > 2|z_0|$ large such that $u(z) \leq \varepsilon |z|$ if $|z| \geq M$. Then by Proposition 1.5

$$
\begin{aligned}
u(z_0) &\leq \mathbb{E}^{z_0} u(Z_{\tau(B(0,M)\cap H)}) \\
&\leq (\log K)\mathbb{P}^{z_0}(Z_{\tau(B(0,M)\cap H)} \in \partial H) \\
&\quad + \varepsilon M \mathbb{P}^{z_0}(Z_{\tau(B(0,M)\cap H)} \in \partial B(0,M) \cap H) \\
&\leq \log K + c\varepsilon.
\end{aligned}
$$

Since ε is arbitrary, $u(z_0) \leq \log K$. $\qquad\square$

As a corollary, if S is the strip $\{0 < \operatorname{Im} z < \pi/2\}$, we have

(1.9) Corollary. *Suppose f is analytic in a strip S and continuous on $\overline{S}$. Suppose $|f|$ is bounded by K on ∂S. If for some $\alpha < 1$, $|f(z)| \leq c \exp(|\exp(\alpha z)|)$ for all $z \in S$, then $|f|$ is bounded by K in S.*

Proof. Let $F(z) = \log z$. Then $|f(F(z))|$ is bounded by K on H and $|f(F(z))| \leq c \exp(|z|^{\alpha})$. By Theorem 1.8, $|f(F(z))| \leq K$ on ∂H, which implies $|f|$ bounded by K in S. $\square$

One application of the Phragmén-Lindelöf result is to the interpolation of linear operators. We use Theorem 1.9 to prove the "three lines lemma," which could also be derived as a consequence of the maximum modulus principle.

(1.10) Lemma. *Suppose φ is bounded on the strip $\{0 < \operatorname{Re} z < 1\}$ and analytic in the interior. If $|\varphi| \leq M_0$ on $\{\operatorname{Re} z = 0\}$, $|\varphi| \leq M_1$ on $\{\operatorname{Re} z = 1\}$, and $t \in (0, 1)$, then $|\varphi| \leq M_0^{1-t} M_1^t$ on $\{\operatorname{Re} z = t\}$.*

Proof. Let $\psi(z) = \varphi(z) M_0^{z-1} M_1^{-z}$. Since $M_0^z = e^{z \log M_0}$, then $|M_0^z| = e^{\operatorname{Re}(z \log M_0)} \leq c$ on the strip. Similarly M_1^{-z} is bounded on the strip, hence ψ is bounded and continuous on the strip and analytic in the interior. When $\operatorname{Re} z = 0$, $|\psi(z)| \leq 1$ and the same is true when $\operatorname{Re} z = 1$. By a rotation, scaling, and Corollary 1.9, $|\psi|$ is bounded by 1 on the strip. Therefore $|\varphi(z)| \leq |M_0^{1-z}| |M_1^z| = M_0^{1-t} M_1^t$ if $\operatorname{Re} z = t$. $\square$

This leads to the Riesz-Thorin interpolation theorem.

(1.11) Theorem. *Suppose T is a bounded linear functional on L^p and on L^q. Suppose $1 \leq p < r < q \leq \infty$. Then T is a bounded linear functional on L^r and $\|T\|_r \leq \|T\|_p^{1-t} \|T\|_q^t$ if $1/r = (1-t)/p + t/q$.*

Proof. Let $M_0 = \|T\|_p$, $M_1 = \|T\|_q$. The key step is to show

$$(1.5) \qquad \int (Tf)g \leq M_0^{1-t} M_1^t$$

if $f \in L^r$ with $\|f\|_r = 1$ and $g \in L^{r'}$ with $\|g\|_{r'} = 1$, where r' is the conjugate exponent to r (that is, $1/r + 1/r' = 1$), and f and g are simple functions. For then, if (1.5) holds, taking the supremum over such simple functions g implies that $\|Tf\|_r \leq M_0^{1-t} M_1^t$.

To prove (1.5), let us write $f = \sum_{j=1}^m a_j 1_{A_j}$ and $g = \sum_{k=1}^n b_k 1_{B_k}$, where the A_js are disjoint and so are the B_ks. If $a_j = |a_j| e^{i\alpha_j}$ and $b_k = |b_k| e^{i\beta_k}$, let

$$f_z = \sum |a_j|^{s(z)/s(t)} e^{i\alpha_j} 1_{A_j}$$

and

$$g_z = \sum |b_k|^{(1-s(z))/(1-s(t))} e^{i\beta_k} 1_{B_k},$$

where

$$s(z) = \frac{1-z}{p} + \frac{z}{q}$$

and $s(t)$ denotes $s(t + 0\mathrm{i})$. Let

$$\varphi(z) = \int (Tf_z)g_z$$
$$= \sum_{j,k} |a_j|^{s(z)/s(t)} |b_k|^{(1-s(z))/(1-s(t))} e^{\mathrm{i}(\alpha_j + \beta_k)} \int (T1_{A_j}) 1_{B_k}.$$

φ is an entire function of z. Note $\varphi(t) = \int (Tf)g$. Also, the exponents of $|a_j|$ and $|b_k|$ have bounded real parts, hence φ is bounded on the strip $\{0 \le \operatorname{Re} z \le 1\}$.

If we write $z = x + \mathrm{i}y$, then when $\operatorname{Re} z = 0$,

$$s(z) = \frac{1 - \mathrm{i}y}{p} + \frac{\mathrm{i}y}{q} = \frac{1}{p} + \mathrm{i}y\left(\frac{1}{q} - \frac{1}{p}\right)$$

and so $|f_{\mathrm{i}y}| = |f|^{(1/p)/s(t)} = |f|^{r/p}$. Therefore

$$\|f_{\mathrm{i}y}\|_p = \left(\int |f_{\mathrm{i}y}|^p\right)^{1/p} = \left(\int |f|^r\right)^{1/p} = 1$$

by our normalization. Similarly, if p' is the conjugate exponent to p, $\|g_{\mathrm{i}y}\|_{p'} = 1$. By Hölder's inequality,

$$|\varphi(\mathrm{i}y)| \le \|Tf_{\mathrm{i}y}\|_p \|g_{\mathrm{i}y}\|_{p'} \le M_0 \|f_{\mathrm{i}y}\|_p \|g_{\mathrm{i}y}\|_{p'} \le M_0.$$

So φ is bounded by M_0 on $\{\operatorname{Re} z = 0\}$. Similarly, it is bounded by M_1 on $\{\operatorname{Re} z = 1\}$. By Lemma 1.10, $\varphi(z)$ is bounded on $\{\operatorname{Re} z = t\}$ by $M_0^{1-t} M_1^t$. $\qquad\square$

The Marcinkiewicz interpolation theorem (Theorem IV.7.21) says that if T is sublinear, i.e., $|T(f+g)| \le |T(f)| + |T(g)|$ and $|T(cf)| = |c| \, |T(f)|$, and T is weak type (p, p) and weak type (q, q), then T is strong type (r, r). Weak type (p, p) means that $|\{x : |Tf(x)| \ge \lambda\}| \le c\|f\|_p^p/\lambda^p$ for all $\lambda > 0$, while strong type (r, r) simply means that T is a bounded linear functional on L^r. By Chebyshev's inequality, if an operator is strong type (p, p), it is also weak type (p, p). Note the Marcinkiewicz interpolation theorem has weaker assumptions than the Riesz-Thorin theorem. However the bound on $\|T\|_r$ from the Riesz-Thorin theorem is much better.

Both interpolation theorems can be modified to handle the case where $T : L^{p_1} \to L^{p_2}$ and $T : L^{q_1} \to L^{q_2}$ to conclude $T : L^{r_1} \to L^{r_2}$ if $p_1, p_2, q_1, q_2, r_1,$ and r_2 are chosen appropriately. Both theorems have been extended to quite abstract settings, and the Riesz-Thorin theorem and the Marcinkiewicz theorem are the basis for the complex and real methods of interpolation, respectively; see Bergh and Löfström [1].

Riemann mapping theorem

We now use some ideas from Chap. II to give a proof of the Riemann mapping theorem, which says that any simply connected domain other than the whole plane can be mapped one to one onto the unit disk. One nice feature of the construction we give here is that it is somewhat constructive.

First we recall the argument principle: if γ is a simple closed curve enclosing a domain D, f is analytic in a domain D' containing $\overline{D}$, and f has no zeroes on γ, then

$$\frac{1}{2\pi i} \int_{\gamma} \frac{f'(z)}{f(z)} dz$$

is equal to the number of zeroes of f in D, counted according to their multiplicity (see Ahlfors [1], p. 151).

We will need the following corollary.

(1.12) Corollary. (Rouché's theorem) *Let γ and D be as above. Suppose f and g are analytic in a domain D' containing $\overline{D}$ and $|f(z) - g(z)| < |f(z)|$ on γ. Then f and g have the same number of zeroes in D.*

Proof. Let $F(z) = g(z)/f(z)$. The hypothesis implies that $|F(z) - 1| < 1$. By the maximum modulus principle applied to $F - 1$, $|F - 1| < 1$ on D, and so F has no zeroes in D. Therefore F'/F is analytic in D. Applying Cauchy's integral theorem and the argument principle, we see F has no zeroes in D. Since F is analytic in D, that implies that f and g must have zeroes at the same places with the same multiplicity. $\square$

Since most of the results in Sect. II.3 were for d-dimensional Brownian motion with $d \geq 3$, we take a few moments to present some facts about Green functions for two-dimensional Brownian motion. For simplicity, we will only consider domains D such that every point of ∂D is regular for D^c and such that D is contained in some half-plane (see Port and Stone [1] for the full story). By a change of coordinate systems, we may suppose $D \subseteq H$.

Let

(1.6) $\qquad u_H(x, y) = -\pi^{-1} \log|x - y| + \pi^{-1} \log|\overline{x} - y|, \qquad x, y \in H.$

Since $|\overline{x} - y| = |x - \overline{y}|$, u_H is symmetric in x and y. If we let $\widehat{Z}_t$ be two-dimensional Brownian motion killed on exiting H, then by Exercise II.8.44,

$$\mathbb{E}^x \int_0^\infty f(\widehat{Z}_t) dt = \int_H u_H(x, y) f(y) \, dy$$

if f is nonnegative and measurable. Set

(1.7) $\qquad g_D(x, y) = u_H(x, y) - \mathbb{E}^x u_H(\widehat{Z}_{\tau_D}, y), \qquad x, y \in D,$

and $g_D(x, y) = 0$ if x or y is in D^c. We call g_D the Green function for D.

(1.13) Proposition. *Suppose $D \subseteq H$ and every point of ∂D is regular for D^c.*

(a) *If $f \geq 0$, then $\int g_D(x,y)f(y)\,dy = \mathbb{E}^x \int_0^{\tau_D} f(\widehat{Z}_t)\,dt$.*

(b) *$g_D(x,y)$ differs from $-\pi^{-1}\log|x-y|$ by a function that is harmonic in D.*

(c) *$g_D(x,y)$ is symmetric in x and y.*

Proof. (a) follows by the argument used in equations (II.3.14) through (II.3.16) with u replaced by u_H. For each y, $\mathbb{E}^x u_H(\widehat{Z}_{\tau_D}, y)$ is harmonic in x on the set D. Hence g_D differs from u_H by a function that is harmonic in D. (b) follows, since $-\pi^{-1}\log|x-\overline{y}|$ is harmonic in x on the set H. To prove (c), note that the transition density of $\widehat{Z}_t$ is symmetric in x and y by Exercise II.8.43. So just as in the proof of Proposition II.4.1,

$$\int_C \mathbb{P}^x(\widehat{Z}_t \in A, \tau_D \geq t)\,dx = \int_A \mathbb{P}^x(\widehat{Z}_t \in C, \tau_D \geq t)\,dx.$$

If we integrate this over t from 0 to ∞, on the left we obtain

$$\int_C \mathbb{E}^x \int_0^{\tau_D} 1_A(\widehat{Z}_t)dt\,dx = \int_C \int_A g_D(x,y)dy\,dx.$$

Similarly, on the right we have $\int_A \int_C g_D(x,y)dy\,dx$, from which we conclude that $g_D(x,y) = g_D(y,x)$ for almost every pair (x,y). Also, $g_D(x,y)$ is continuous in x on the set $D - \{y\}$ since it is harmonic there, and it is continuous in y on the set $D - \{x\}$ by the continuity of u_H in y. (c) follows. $\qquad \square$

We now prove the Riemann mapping theorem. Recall that a domain D is simply connected if D^c consists of a single unbounded component.

(1.14) Theorem. *If D is a simply connected domain other than the entire plane and $z_0 \in D$, then there exists f analytic in D mapping D one to one onto the unit disk with $f(z_0) = 0$ and $f'(z_0) > 0$.*

It is not hard to show that f must be unique; see Ahlfors [1], p. 222.

Proof. We first show that we may take D to be bounded. Suppose D is unbounded. If D is not the entire complex plane, there is at least one point not in D. Let us suppose that point is 0. Since D is simply connected, we can define a single-valued branch of $\log z$, namely, we connect a fixed point z_0 in D to z by a curve γ and let $\log z = \int_\gamma dz/z + \log|z_0|$. We can define $f_1(z) = z^{1/2} = e^{(1/2)\log z}$ as a single-valued analytic function in D. f_1 is one to one, since it has an inverse, namely, z^2. Let $D' = f_1(D)$.

There exists a point z_1 and a neighborhood U of z_1 contained in D'. We claim $-U$ is disjoint from D'. For if $z \in -U$ and $z \in D'$, then z^2 and $(-z)^2$ would both be in D, contradicting the fact that f_1 is one to one.

Therefore the complement of D' contains some ball, say $B(z_2, \varepsilon)$. Let f_2 be inversion in the ball $B(z_2, \varepsilon)$; we set $f_2(z) = \varepsilon^2/(z - z_2)$. f_2 is clearly one to one and analytic and now $D'' = f_2(D') \subseteq B(z_2, \varepsilon)$. We have thus reduced the problem to the case of D bounded.

Now suppose D is bounded. We proceed to define f. By Proposition II.1.14, every point of ∂D is regular for D^c. Let $g(z) = g_D(z_0, z)$ be the Green function for D. Then $g(z) = 0$ if $z \in \partial D$. What we would like to do is to let $\widetilde{g}$ be the conjugate harmonic function to g on $D - \{z_0\}$. Then we would let $f(z) = \exp(-\pi[g(z) + i\widetilde{g}(z)])$. Since $g(z) \geq 0$, then $|f(z)| \leq 1$ and f maps D into the unit disk. The difficulty is that g is harmonic in $D - \{z_0\}$, not in D, and $D - \{z_0\}$ is not simply connected. So a single-valued conjugate harmonic function need not exist in $D - \{z_0\}$ (e.g., $\log|z|$ is harmonic in $\mathbb{D} - \{0\}$ without a single-valued conjugate harmonic function in that domain).

What we do is fix $z_1 \in D - \{z_0\}$ and if $z \in D$, we connect z_1 to z by a curve ψ lying in $D - \{z_0\}$. Then let

$$(1.8) \qquad \widetilde{g}(z) = \int_\psi (-\partial_y g + i\partial_x g)\, dz, \qquad f(z) = \exp(-\pi[g(z) + i\widetilde{g}(z)]).$$

Although $\widetilde{g}$ is not uniquely defined, f is nevertheless single valued (see Exercise 4).

We show f is analytic in D. f is clearly analytic if $z \neq z_0$. By Proposition 1.13, $|g(z) - (-\pi^{-1}\log|z - z_0|)|$ remains bounded in a neighborhood of z_0. Therefore f is bounded in a neighborhood of z_0. It follows that z_0 is a removable singularity for f. Since $\lim_{z \to z_0} f(z) = 0$ by (1.8), $f(z_0) = 0$ as we wished. Since $g(z) \to \infty$ only if $z \to z_0$, in fact z_0 is the only zero of f.

Note $g(z) = -\pi^{-1}\log|z - z_0|$ plus a harmonic function, say u. Since f has a zero at z_0, it follows that $f(z) = (z - z_0)h(z)$, where h is analytic and $|h(z)| = e^{-\pi u(z)}$. Since u is harmonic at z_0, it is differentiable there, which implies $f'(z_0) \neq 0$. So the zero at z_0 is a simple one. If we follow f by the application of the appropriate linear fractional transformation, we can arrange $f'(z_0)$ to be real and positive.

Let us show f is one to one. Suppose $f(z_1) = w$. Then $g(z_1) = -\pi^{-1}\log|w|$. Since $g(z) \to \infty$ as $z \to z_0$ and $g(z) \to 0$ as z tends to the boundary, $\{z : g(z) = w\}$ is a positive distance from ∂D and from z_0. For each $z \in D - \{z_0\}$, there is a neighborhood of z in which $\widetilde{g}$, the conjugate harmonic function to g, is single-valued. $g + i\widetilde{g}$ is analytic in this neighborhood, hence the derivative is analytic. Since g cannot be constant in a subdomain (Exercise III.6.25), ∇g is zero for only countably many points inside this neighborhood. Therefore there are only countably many points in $D - \{z_0\}$ at which ∇g is zero. Let $\varepsilon > 0$ be small and let γ be the set $\{z : g(z) = -\pi^{-1}\log|w| - \varepsilon\}$. We can choose ε so that γ contains no zero of ∇g. By the implicit function theorem, γ is a smooth curve. We claim γ is a simple closed curve. It is closed, because if not, we could find a curve ψ connecting z_0 to ∂D avoiding γ; this is impossible because $g(z_0) = \infty$,

$g = 0$ on ∂D, and g is continuous in $D - \{z_0\}$. Suppose γ were not simple. Then there is a piece of γ that encloses a subdomain D', with $z_0 \notin D'$. Since g is constant on γ, hence on $\partial D'$, then by the maximum principle, g is constant in D'. Then $\nabla g = 0$ in D', contradicting what we had earlier.

Let us now apply Rouché's theorem to the curve γ and the functions f and $f - w$. On γ,

$$|f - (f - w)| = |w|.$$

On γ, $|f(z)| = e^{-\pi g(z)} = e^{\pi \varepsilon}|w| > |w|$. So by Rouché's theorem, f and $f - w$ have the same number of zeroes inside γ. Since f is zero only at z_0, this says $f - w$ has only one zero.

That f is onto the unit disk comes from the same proof. Pick w in the unit disk and define γ as above. Since $f - w$ has exactly one zero in D, there must exist a point z_1 such that $f(z_1) = w$. □

It is not necessarily true that f can be extended continuously to $\overline{D}$ (suppose D is the slit disk). However if ∂D is a simple closed curve, then f can be extended continuously to $\overline{D}$ (Pommerenke [1]). If ∂D is even smoother, one can ask about further smoothness of f, for example, does f' exist? See Sect. 5.

2 Range of analytic functions

In this section we give probabilistic proofs of Picard's little theorem and the Koebe distortion theorem. The first says that the range of an entire function is either all of $\mathbb{C}$ or else $\mathbb{C}$ minus at most one point. The second says that the range of a one to one analytic function on $\mathbb{D}$ contains a ball of a certain size.

Picard's theorem

Picard's little theorem says the following.

(2.1) Theorem. *If f is an entire function and nonconstant, then the range of f consists of $\mathbb{C}$ minus at most one point.*

The analytic proof of this fact takes some doing. We give an elegant proof due to Davis [2].

Recall that two closed curves γ_1 and γ_2 are homotopic in a domain D if γ_1 can be stretched continuously to γ_2. More precisely, suppose γ_1 and γ_2 are continuous mappings of $[0, 1]$ into a domain D with $\gamma_i(0) = \gamma_i(1)$, $i = 1, 2$. γ_1 and γ_2 are homotopic if there exists F mapping $[0, 1]^2$ into D such that F is continuous, $F(s, 0) = \gamma_1(s)$ for all $s \in [0, 1]$, $F(s, 1) = \gamma_2(s)$ for all $s \in [0, 1]$, and $F(0, t) = F(1, t)$ for all $t \in [0, 1]$. The set of curves homotopic to each other is easily seen to be an equivalence class. Also, if γ_1 and γ_2 are homotopic to each other and f is continuous, then $f(\gamma_1)$ and $f(\gamma_2)$ are homotopic to each other. In a simply connected domain each closed curve is homotopic to the curve γ_0 that is a single point in the domain.

Proof of Theorem 2.1. If $f(\mathbb{C})$ omits more than one point of $\mathbb{C}$, we can assume without loss of generality that two of the points it omits are $-a$ and b, where a and b are positive reals, and that $f(0) = 0$. Since f is entire, $|f'| > 0$ except for a countable set, and by the recurrence of Brownian motion, $\int_0^t |f'(Z_s)|^2 ds \to \infty$ as $t \to \infty$ (cf. Exercise 1). Hence $f(Z_t)$ is the time change of a Brownian motion (not killed or stopped).

Let ε be small so that if $|z| < \varepsilon$, $f(z)$ can be connected to 0 by a curve lying in $f(\mathbb{C}) \cap B(0, (a \wedge b)/2)$. Infinitely often $Z_t \in B(0, \varepsilon)$. Let L_t be the straight line segment connecting Z_t to 0. The curve consisting of adding the line segment L_t to the end of Z_t is a closed curve homotopic to the single point 0. So the curve consisting of adding $f(L_t)$ to the end of $f(Z_t)$ must also be homotopic to a single point. By Proposition 2.2, for t sufficiently large, this curve is not homotopic to a single point, a contradiction. $\qquad\square$

(2.2) Proposition. *Let L_t be as in the proof above. With probability one, there exists t_0 (depending on ω) such that if $t > t_0$, the curve formed by adding $f(L_t)$ to the end of $f(Z_t)$ is not homotopic to a single point.*

Warning: Although this proof is obviously modeled after the account in Durrett [1], there are significant differences. For example, the definition of reduced sequences is different.

Proof. Let
$$A_0 = \{-a < x < b, y = 0\},$$
$$A_1 = \{x < -a, y = 0\},$$
$$A_2 = \{x > b, y = 0\},$$
$$A_3 = \{x = -a \text{ or } x = b, y > 0\},$$
$$A_4 = \{x = -a \text{ or } x = b, y < 0\}.$$

Let $T_1 = 0$. Let $b_i \in \{0, 1, 2, 3, 4\}$ be such that $f(Z_{T_i}) \in A_{b_i}$, and
$$T_{i+1} = \inf\{t > T_i : f(Z_t) \in \cup_{j=0}^4 A_j - A_{b_i}\}.$$

Thus the T_is are the times to hit one of the sets A_j different from the last one hit.

We form the sequence $0b_1 b_2 \ldots b_n$, and then form the reduced sequence as follows:

(a) If our sequence ends in 030 or 040, delete the last two entries.

(b) If the last entry is the same as the second from the last and the next to the last entry is not 0, delete the last two entries. (So $\ldots 0424$ becomes $\ldots 04$, but $\ldots 3404$ does not become $\ldots 34$.)

(c) If the sequence ends in a 0 and has the form $\ldots 0b_{i_1} \ldots b_{i_j} 0b_{i_j} \ldots b_{i_1} 0$ (i.e., the piece between the next to the last 0 and the end is the same as the piece from the second from the last 0 and the next to the last 0 in reverse order), eliminate everything after the second to the last 0. For example, $\ldots 0423140413240$ becomes $\ldots 0$.

We apply the rules (a), (b), (c) until the sequence cannot be reduced any further, and that is our reduced sequence.

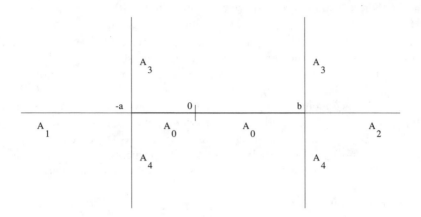

FIGURE 2.1. The sets A_i, $i = 0, \ldots, 4$.

The reduced sequence completely describes the homotopy class for $f(Z_t)$ with $f(L_t)$ added on whenever $Z_t \in B(0, \varepsilon)$, and for this curve to be homotopic to a point, it is necessary that the reduced sequence consist of the single entry 0. So to prove our proposition, it suffices to prove that the number of 0s in the reduced sequence tends to ∞, a.s. Define a block to be a portion of a reduced sequence starting and ending with a 0 and with no 0s in the middle.

Suppose our last block in the reduced sequence is $\ldots 03140$. In order to reduce the number of 0s, the next block must be 04130. By symmetry about the $y = 0$ line, we are equally likely to have 03140, which increases the number of 0s by one. Also, by the support theorem, there is positive probability bounded away from 0 that the next block will be 03240 or 04230, each of which increases the number of 0s. A similar symmetry argument can be given no matter what the last block is, and so there exists $\delta > 0$ such that
$$\mathbb{P}(\text{number of 0s increases by 1}) \geq 1/2 + \delta$$
and
$$\mathbb{P}(\text{number of 0s decreases by 1}) \leq 1/2 - \delta.$$
By Exercise 5, the number of 0s increases to ∞, a.s. □

Distortion theorem

We now give a probabilistic proof of what is known as the distortion theorem or Koebe's 1/4 theorem.

(2.3) Theorem. *Suppose f is analytic and one to one on $\mathbb{D}$ with $f(0) = 0$ and $|f'(0)| = 1$. Then there exists a (not depending on f) such that $B(0, a) \subseteq f(\mathbb{D})$.*

The best constant possible is $a = 1/4$, which explains the name of the theorem. The Koebe function $f(z) = z/(1 - z)^2$ shows that the constant is attained. We will give a probabilistic proof of Theorem 2.3; the constant we obtain is not as good as the one obtained from an analytic proof.

First we need an elementary fact about one to one analytic functions.

(2.4) Lemma. *If f is one to one on a domain D, then f' never vanishes on D.*

Proof. Suppose f' vanishes at some point of D. Without loss of generality, let us suppose that the point is $0 \in D$, and by adding a constant, we may assume $f(0) = 0$. Then in a neighborhood of 0, $f(z) = z^n g(z)$ where n is an integer greater than or equal to 2 and g is analytic in this neighborhood with $g(0) \neq 0$. By continuity, g is not zero in a neighborhood of 0, and so we can define $g^{1/n}$ as a single-valued analytic function there. Let $h(z) = zg(z)^{1/n}$, and we have $f(z) = [h(z)]^n$ in a neighborhood of 0. Note $h(0) = 0$ and $h'(0) = g(0)^{1/n} \neq 0$. Since $h'(0) \neq 0$, the inverse of h exists in a neighborhood of $h(0)$ by the inverse function theorem. Thus if ε is a sufficiently small positive real and $\omega = e^{i2\pi/n}$, there exist distinct points z_1 and z_2 such that $h(z_1) = \varepsilon$ and $h(z_2) = \varepsilon\omega$. Then $f(z_1) = f(z_2) = \varepsilon^n$, contradicting the fact that f is one to one. $\square$

Next we have the following lemma.

(2.5) Lemma. *Suppose $|f'(0)| = 1$. There exists δ independent of f such that*

$$(2.1) \qquad \mathbb{P}^0\left(\int_0^\tau |f'(Z_s)|^2 ds > \delta \right) \geq \delta.$$

Proof. Let

$$g(r) = \frac{1}{2\pi} \int_0^{2\pi} |f'(re^{i\theta})|^2 d\theta.$$

Since $|f'|^2 = |\partial_x u|^2 + |\partial_y u|^2$ and $\partial_x u$ and $\partial_y u$ are both harmonic, then $|f'|^2$ is subharmonic. So

$$1 = |f'(0)|^2 \leq \mathbb{E}^0 |f'(Z_{\tau_r})|^2 = \frac{1}{2\pi} \int_0^{2\pi} |f'(re^{i\theta})|^2 d\theta = g(r).$$

Define $F(z) = |f'(z)|^2/g(|z|)$, where we set $g(0) = 1$. Then $F(0) = 1$, $F(z) \le |f'(z)|^2$, and $(2\pi)^{-1} \int_0^{2\pi} F(re^{i\theta})d\theta = 1$.

The Green function for $\mathbb{D}$ is given by Exercise II.8.45. Hence

$$\mathbb{E}^0 \int_0^\tau F(Z_s)ds = \int_{\mathbb{D}} g_{\mathbb{D}}(0, z)F(z)\,dz \ge \int_{B(0,1/4)} g_{\mathbb{D}}(0, z)F(z)\,dz$$

$$\ge c \int_0^{1/4} \log(1/r) r \int_0^{2\pi} F(re^{i\theta})d\theta\,dr \ge c_1.$$

On the other hand, if $w \in \mathbb{D}$,

$$g_{\mathbb{D}}(w, z) \le -c\log(|w - z|/2) \le -c\log((|w| - |z|)/2),$$

and so there exists c_2 such that for all $w \in \mathbb{D}$,

$$\mathbb{E}^w \int_0^\tau F(Z_s)ds = \int_{\mathbb{D}} g_{\mathbb{D}}(w, z)F(z)dz$$

$$\le -c \int_0^1 r\log((r - |w|)/2) \int_0^{2\pi} F(re^{i\theta})d\theta\,dr \le c_2.$$

If $A_t = \int_0^{t \wedge \tau} F(Z_s)ds$, then

$$\mathbb{E}^w[A_\infty - A_t | \mathcal{F}_t] = \mathbb{E}^{Z_t} A_\infty \le c_2.$$

By Theorem I.6.10 with $B = c_2$, there exists c_3 such that

$$(2.2) \qquad \sup_w \mathbb{E}^w \left[\int_0^\tau F(Z_s)ds \right]^2 \le c_3.$$

Then

$$c_1 \le \mathbb{E}^0 \int_0^\tau F(Z_s)ds = \mathbb{E}^0 \left[\int_0^\tau F(Z_s)ds; \int_0^\tau F(Z_s)ds \le c_1/2 \right]$$

$$+ \mathbb{E}^0 \left[\int_0^\tau F(Z_s)ds; \int_0^\tau F(Z_s)ds > c_1/2 \right]$$

$$\le c_1/2 + \left(\mathbb{E}^0 \left[\int_0^\tau F(Z_s)ds \right]^2 \right)^{1/2} \left(\mathbb{P}^0 \left(\int_0^\tau F(Z_s)ds > c_1/2 \right) \right)^{1/2}.$$

Using (2.2),

$$\mathbb{P}^0 \left(\int_0^\tau F(Z_s)ds > c_1/2 \right) \ge \frac{(c_1/2)^2}{c_3}.$$

Since $|f'(z)|^2 \ge F(z)$, this proves the lemma. $\qquad\square$

Now we are ready to prove Theorem 2.3. The idea is that by the support theorem, $f(Z_t)$ will make a loop around the outside of the ball of radius a about 0. Since $f(\mathbb{D})$ is simply connected, this means the ball of radius a must be contained in $f(\mathbb{D})$. Here are the details.

Proof of Theorem 2.3. Let W_t be two-dimensional Brownian motion. Let δ be as in Lemma 2.2. We will show

(2.3)   *There exists $a > 0$ such that W_t makes a loop around $B(0, a)$ before time δ with probability at least $1 - \delta/2$.*

By making a loop around $B(0, a)$, we mean the graph of $\{W_s : 0 \leq s \leq \delta\}$ contains a closed curve γ with $B(0, a)$ contained in one of the bounded components of γ^c.

Before showing (2.3), let us see how (2.3) implies the theorem. Let Z_t be two-dimensional Brownian motion in $\mathbb{D}$ and let W_t be a two-dimensional Brownian motion that agrees with the time change of $f(Z_t)$ up to the first exit of D. By Lemma 2.2 and Theorem 1.1, W_t will not have exited D by time δ with probability at least δ. So by (2.3), with probability at least $\delta/2$ the graph of $\{W_s : 0 \leq s \leq \tau_D(W)\}$ will make a loop around $B(0, a)$. This graph is the same as the graph of $\{f(Z_s) : 0 \leq s \leq \tau_{\mathbb{D}}(Z)\}$. (Here $\tau_D(W)$ is the exit time of D for W and $\tau_{\mathbb{D}}(Z)$ is the exit time of $\mathbb{D}$ for Z.) It follows that $f(\mathbb{D})$ contains a closed curve containing $B(0, a)$. Since f is one to one, f' is never 0 by Lemma 2.1 and f^{-1} is analytic. Since f and f^{-1} are both continuous, $f(\mathbb{D})$ is simply connected. If there were $z \in B(0, a)$ that was not in $f(\mathbb{D})$, that would contradict the fact that $f(D)$ is simply connected. Therefore $B(0, a) \subseteq f(\mathbb{D})$.

We now show (2.3). There exists r small such that

(2.4) $$\mathbb{P}^0(\tau_r > \delta) < \delta/4.$$

Let $\psi(t)$ be a curve that starts at 1, goes around the circumference of $B(0, 1)$ twice, then goes from 1 to 2 along the positive real axis, and finally from 2 to $1/4$ along the positive real axis. If $\varepsilon = 1/16$, by the support theorem (Theorem I.6.6), there exists $\eta > 0$ such that the graph of Brownian motion started at 1 will, with probability at least η, contain a closed curve that has $B(0, 1/4)$ in its interior.

Let

$$A_k = \{T_{\partial B(0, 16^{-k}r/2)} < T_{\partial B(0, 8 \cdot 16^{-k}r)} \text{ and the graph of}$$
$$\{W_s, s \in [0, T_{\partial B(0, 16^{-k}r/2)} \wedge T_{\partial B(0, 8 \cdot 16^{-k}r)})\}$$
contains a closed curve containing $B(0, 16^{-k}r/4)\}.$

By scaling, if $|z| = 16^{-k}r$, then $\mathbb{P}^z(A_k) \geq \eta$, or $\mathbb{P}^z(A_k^c) \leq 1 - \eta$. By the strong Markov property at time $T_{\partial B(0, 16^{-4k}r)}$,

$$\mathbb{P}^0(A_4^c \cap A_8^c \cap \cdots \cap A_{4k}^c) \leq (1 - \eta)\mathbb{P}^0(A_4^c \cap \cdots \cap A_{4k-4}^c).$$

By induction,
$$\mathbb{P}^0(A_4^c \cap A_8^c \cap \cdots \cap A_{4k}^c) \leq (1 - \eta)^k.$$

If we take k large enough so that $(1 - \eta)^k \leq \delta/4$ and combine with (2.4), we have (2.3) with $a = 16^{-4k}r/4$. $\square$

(2.6) Corollary. *Suppose f is one to one and analytic on $\mathbb{D}$, $z \in \mathbb{D}$. Then*

$$\operatorname{dist}\left(f(z), \partial f(\mathbb{D})\right) \le a^{-1}|f'(z)|(1-|z|).$$

Proof. Let $r = a^{-1}|f'(z)|(1-|z|)$ and suppose $r < \operatorname{dist}\left(f(z), \partial f(\mathbb{D})\right)$. Apply the distortion theorem to f^{-1} and $B(f(z), r)$. By scaling,

$$B\left(z, ar|(f^{-1})'(f(z))|\right) \subseteq f^{-1}\left(B(f(z), r)\right),$$

or

$$f\left(B(z, ar/|f'(z)|)\right) \subseteq B(f(z), r).$$

So we have $f(B(z, 1-|z|)) \subseteq B(f(z), r)$. If $w \to z/|z|$ with $w \in B(z, 1-|z|)$, then $f(w) \to \partial\mathbb{D}$ yet $f(w) \in B(f(z), r))$. This contradicts $r < \operatorname{dist}\left(f(z), \partial\mathbb{D}\right)$. $\qquad\square$

3 Boundary behavior of analytic functions

Nontangential limits

We are going to look at the boundary behavior of functions analytic in the unit disk $\mathbb{D}$. First of all, a great deal can be said simply by looking at the real and imaginary parts of f. Both are harmonic functions and the results of Chap. IV can be easily modified to deal with the case $d = 2$ and the unit disk instead of the half-space. One can go through the proofs and make the appropriate modifications. Or, since we are looking at analytic functions, we can simply use a conformal mapping argument.

However, in preparation for describing some of the behavior peculiar to analytic functions, we will need to know that a harmonic function u converges nontangentially the same places $u(Z_t)$ converges. Let C_θ be the convex hull of $\{e^{i\theta}\} \cup B(0, 1/4)$, the Stolz domain with vertex at $e^{i\theta}$, and let

$$(3.1) \qquad\qquad L_u = \{\theta : \lim_{z \in C_\theta, z \to e^{i\theta}} u(z) \text{ exists}\}.$$

For the probabilistic analog, let

$$(3.2) \qquad\qquad \mathcal{L}_u = \{\theta : \lim_{t \to \tau} u(Z_t) \text{ exists } \mathbb{P}_\theta^0 - \text{a.s.}\}.$$

We are assuming u is harmonic, but not necessarily positive and $\mathbb{P}_\theta^0$ is the law of Brownian motion conditioned to exit $\mathbb{D}$ at $e^{i\theta}$, or phrased another way, the law of Brownian motion h-path transformed by the Poisson kernel with pole at $e^{i\theta}$.

First we show $\mathcal{L}_u \subseteq L_u$ up to a set of measure 0. The analog of this for $d > 2$ is not true; see Durrett [1].

(3.1) Proposition. *Almost every $\theta \in \mathcal{L}_u$ is in L_u. Moreover for almost every $\theta \in \mathcal{L}_u$,*

$$\lim_{t \to \tau} u(Z_t) = \lim_{z \in C_\theta, z \to e^{i\theta}} u(z), \qquad \mathbb{P}_\theta^0 - a.s.$$

Proof. Fix $\theta \in \mathcal{L}_u$. By the zero-one law (Theorem III.2.9), $\lim_{t \to \tau} u(Z_t)$ is a constant, say ℓ, $\mathbb{P}_\theta^0$ a.s. Suppose there exist $z_1, z_2, \ldots$ in C_θ converging to $e^{i\theta}$ with $u(z_n) - \ell > \varepsilon$ for some $\varepsilon > 0$. We will show there exists c independent of n such that

$$(3.3) \qquad \mathbb{P}_\theta^0(Z_t \text{ makes a loop around } z_n) \geq c.$$

As in Sect. V.2, by "Z_t makes a loop around z_n," we mean that the path of Z_t, $t < \tau$, contains a closed curve γ_n with z_n inside the region enclosed by γ_n. Before proving (3.3), let us show that (3.3) implies the proposition. Since $u(Z_t)$ converges to ℓ, we can take δ small enough so that $u(Z_t) \leq \ell + \varepsilon/2$ if $t > \tau - \delta$ (γ_n and δ are random, of course). Since $Z_t \to e^{i\theta}$, $\mathbb{P}_\theta^0$-a.s., γ_n is contained in the graph of $\{Z_t : \tau - \delta < t < \tau\}$ if n is large enough. By the maximum principle, since z_n is in the region enclosed by γ_n, $u(z_n) \leq \sup_{\gamma_n} u$, which will be less than $\ell + \varepsilon/2$ if n is large enough, a contradiction. Therefore $\limsup_n u(z_n) \leq \ell$. The lim inf is treated similarly.

So we need to prove (3.3). This is a consequence of the support theorem. Let us suppose for simplicity that $r_n = 1 - |z_n| < 1/4$, the other case being easier, and suppose $\theta = 0$.

Let $h_\theta(z)$ be the Poisson kernel for $\mathbb{D}$ with pole at 0 (Theorem II.1.17). Let

$$L_n = \{(1 - r_n)e^{i\theta} : -\pi r_n/16 < \theta < \pi r_n/16\}.$$

Then

$$(3.4) \qquad \mathbb{P}_0^0(Z_{\tau(1-r_n)} \in L_n) = \frac{\mathbb{E}^0[h_0(Z_{\tau(1-r_n)}); Z_{\tau(1-r_n)} \in L_n]}{h_0(0)}$$

$$\geq cr_n^{-1}\mathbb{P}^0(Z_{\tau(1-r_n)} \in L_n) \geq c.$$

Fix $z \in L_n$. Let ψ_n be the curve that goes at constant speed from z toward the origin a distance $r_n/2$, moves clockwise along $\partial B(0, 1 - 3r_n/2)$ an angle $\pi r_n/4$, moves radially away from the origin a distance r_n, moves counterclockwise along $\partial B(0, 1 - r_n/2)$ an angle $\pi r_n/2$, moves toward the origin a distance $3r_n/4$, and then moves clockwise along $\partial B(0, 1 - 5r_n/4)$ an angle πr_n. Let $\varepsilon_n = r_n/16$. Let $A_n = \{\sup_{s \leq \sqrt{r_n}} |Z_s - \psi_n(s)| < \varepsilon_n\}$. By the support theorem (Theorem I.6.6) and scaling,

$$\mathbb{P}^z(Z_t \text{ makes a loop around } z_n) \geq \mathbb{P}^z(A_n) \geq c.$$

In fact, since A_n is in $\mathcal{F}_{\tau(1-r_n/4)\wedge T(\partial B(0,1-2r_n))}$ and $h_0(w)/h_0(v)$ is bounded above for $w, v \in B(0, 1 - r_n/4) - B(0, 1 - 2r_n)$,

$$(3.5) \qquad \mathbb{P}_0^z(Z_t \text{ makes a loop around } z_n) \geq \mathbb{P}_0^z(A_n) \geq c.$$

By the strong Markov property, (3.4) and (3.5),

$$\mathbb{P}_0^0(Z_t \text{ makes a loop around } z_n)$$
$$\geq \mathbb{E}_0^0[\mathbb{P}_0^{Z_{\tau(1-r_n)}}(Z_t \text{ makes a loop around } z_n); Z_{\tau(1-r_n)} \in L_n]$$
$$\geq c.$$

This proves (3.3). $\square$

The other direction, that L_u is contained in $\mathcal{L}_u$ almost everywhere works for any dimension. First we need a lemma due to Marcinkiewicz. Let σ be normalized surface measure on $\partial \mathbb{D}$.

(3.2) Lemma. *Suppose F is a closed subset of $\partial \mathbb{D}$. Then for almost every $x \in F$,*

$$(3.6) \qquad I(x) = \int_{\partial \mathbb{D}} \frac{\text{dist}\,(y, F)}{|x - y|^2} \sigma(dy) < \infty.$$

The integral $I(x)$ is known as the integral of Marcinkiewicz, and the analog of this lemma holds in the half-space for any d.

Proof. We show $I(x)$ is finite a.e. by showing $\int_{\partial \mathbb{D} \cap F} I(x)\sigma(dx) < \infty$. By Fubini's theorem, this is equal to

$$\int \int_F \frac{\text{dist}\,(y, F)}{|x - y|^2} \sigma(dx)\, \sigma(dy).$$

Note dist $(y, F) = 0$ unless $y \notin F$. If $x \in F$ and $y \notin F$, then $|x - y|$ must be greater than dist (y, F). So the double integral is bounded by

$$2 \int \int_{\text{dist}\,(y,F)}^{\infty} \frac{dz}{z^2} \text{dist}\,(y, F)\, \sigma(dy) = 2 \int \sigma(dy) < \infty.$$

$\square$

(3.3) Proposition. *Almost every θ in L_u is also in $\mathcal{L}_u$.*

Proof. Fix M and let $E = E_M = \{e^{i\theta} : \sup_{z \in C_\theta} |u(z)| \leq M\}$. Let $G = \cup_{e^{i\theta} \in E} C_\theta$. Then $|u| \leq M$ in G. G is a sawtooth domain (cf. Fig. IV.6.1). Because $\sup_{z \in C_\theta} |u(z)|$ is a lower semicontinuous function, E is closed. The key step is to show:
    If $e^{i\theta} \in E$ and

(3.7)
$$\int \frac{\text{dist}\,(e^{ix}, E)}{|e^{i\theta} - e^{ix}|^2} \sigma(dx) < \infty,$$

then

(3.8)
$$\mathbb{P}_\theta^0(\limsup_{t<\tau} |u(Z_t)| \le M) = 1.$$

To prove (3.8) we first make four observations. By the Harnack inequality, there exists c such that

(3.9)
$$h_\theta(\rho e^{i\phi}) \le c h_\theta(s e^{i\alpha})$$

if $\rho e^{i\phi}, s e^{i\alpha} \in \partial G - \partial \mathbb{D}$ with $|\phi - \alpha| \le (1-\rho)/2$. Second, by the support theorem, there exists c such that

(3.10)
$$\mathbb{P}^{\rho e^{i\phi}}(|Z_\tau - e^{i\phi}| \le (1-\rho)/2) \ge c.$$

Third, note that

(3.11)
$$h_\theta(s e^{i\alpha}) \le \frac{c(1-s)}{|s e^{i\alpha} - e^{i\theta}|^2} \le \frac{c\,\text{dist}\,(e^{i\alpha}, E)}{|e^{i\alpha} - e^{i\theta}|^2}$$

if $s e^{i\alpha} \in \partial G - \partial \mathbb{D}$. Fourth, if $z \in C_\theta$,

(3.12)
$$\frac{h_\alpha(z)}{h_\theta(z)} = \frac{|z - e^{i\theta}|^2}{|z - e^{i\alpha}|^2} \le c.$$

If $e^{i\phi} \notin E$, $\rho e^{i\phi} \in \partial G - \partial \mathbb{D}$, and $|\alpha - \phi| \le (1-\rho)/2$, then $e^{i\alpha} \notin E$. For such α choose s_α such that $s_\alpha e^{i\alpha} \in \partial G - \partial \mathbb{D}$. Combining (3.9), (3.10), and (3.11),

(3.13)
$$h_\theta(\rho e^{i\phi}) \le c h_\theta(\rho e^{i\phi}) \mathbb{P}^{\rho e^{i\phi}}(|Z_\tau - e^{i\phi}| \le (1-\rho)/2)$$
$$\le c h_\theta(\rho e^{i\phi}) \int_{\phi-(1-\rho)/2}^{\phi+(1-\rho)/2} h_\alpha(\rho e^{i\phi})d\alpha$$
$$\le c \int_{\phi-(1-\rho)/2}^{\phi+(1-\rho)/2} h_\theta(s_\alpha e^{i\alpha}) h_\alpha(\rho e^{i\phi})d\alpha$$
$$\le c \int_{\phi-(1-\rho)/2}^{\phi+(1-\rho)/2} \frac{\text{dist}\,(e^{i\alpha}, E)}{|e^{i\alpha} - e^{i\theta}|^2} h_\alpha(\rho e^{i\phi})d\alpha.$$

If $r > 0$ and $|\rho e^{i\phi} - e^{i\theta}| < r$, then $\phi+(1-\rho)/2 \le \theta+cr$ and $\phi-(1-\rho)/2 \ge \theta - cr$. So using (3.12), (3.13), and the fact that $h_\theta(Z_{t\wedge\tau})$ is a martingale, if $z \in C_\theta$,

$$(3.14) \quad \mathbb{P}_\theta^z(\tau_G < \tau, |Z_{\tau_G} - e^{i\theta}| < r)$$

$$= \mathbb{E}^z\left[h_\theta(Z_{\tau_G}); \tau_G < \tau, |Z_{\tau_G} - e^{i\theta}| < r\right]/h_\theta(z)$$

$$\leq c\mathbb{E}^z \int_{\theta-cr}^{\theta+cr} \frac{\text{dist}\,(e^{i\alpha}, E)}{|e^{i\alpha} - e^{i\theta}|^2} \frac{h_\alpha(Z_{\tau_G})}{h_\theta(z)} d\alpha$$

$$= c \int_{\theta-cr}^{\theta+cr} \frac{\text{dist}\,(e^{i\alpha}, E)}{|e^{i\alpha} - e^{i\theta}|^2} \frac{h_\alpha(z)}{h_\theta(z)} d\alpha$$

$$\leq c \int_{\theta-cr}^{\theta+cr} \frac{\text{dist}\,(e^{i\alpha}, E)}{|e^{i\alpha} - e^{i\theta}|^2} d\alpha.$$

Since θ satisfies (3.7), there exists r such that the right-hand side of (3.14) is less than $1/2$. Taking $z \in C_\theta$ with z sufficiently close to $e^{i\theta}$,

$$\mathbb{P}_\theta^z(|Z_{\tau_G} - e^{i\theta}| > r) \leq 1/4.$$

Hence $\mathbb{P}_\theta^z(\tau_G < \tau) \leq 3/4$. If $\tau_G = \tau$, then $u(Z_t)$ does not hit G^c before $\partial\mathbb{D}$, and $\sup_{t<\tau} |u(Z_t)| \leq M < \infty$. So $\mathbb{P}_\theta^z(\limsup_{t\to\tau} |u(Z_t)| \leq M) \geq 1/4$. The zero-one law (Theorem III.2.9) implies (3.8).

To complete the proof, by Lemma 3.2 and (3.8), for almost every $\theta \in E_M$ we have

$$(3.15) \qquad\qquad \mathbb{P}_\theta^0(\sup_{t<\tau} |u(Z_t)| < \infty) = 1.$$

Since clearly $L_u \subseteq \cup_{M=1}^\infty E_M$, (3.15) holds for almost every $\theta \in L_u$. By Proposition III.2.7,

$$\mathbb{P}^0(\sup_{t<\tau} |u(Z_t)| < \infty, Z_\tau \in L_u) = \mathbb{P}^0(Z_\tau \in L_u).$$

By Exercise 8,

$$\mathbb{P}^0(Z_\tau \in L_u) = \mathbb{P}^0(\lim_{t\to\tau} u(Z_t) \text{ exists}, Z_\tau \in L_u),$$

or for almost every $\theta \in L_u$, $\mathbb{P}_\theta^0(\lim_{t\to\tau} u(Z_t) \text{ exists}) = 1$. This is the assertion of the proposition. $\qquad\square$

The above proposition has analogs in the half-space and for any dimension d. We took our Stolz domains to be the convex hull of $\{e^{i\theta}\} \cup B(0, 1/4)$. Of course, the $1/4$ could be replaced by any number between 0 and 1, and so our C_θ could have any aperture angle we desired between 0 and π.

Privalov theorem

We can now easily prove some interesting results about the boundary behavior of analytic functions.

(3.4) Theorem. (Privalov) *(a) If* $\lim_{z \in C_\theta, z \to e^{i\theta}} f(z) = 0$ *on a set of θs of positive Lebesgue measure, then f is identically 0.*

(b) If $\lim_{z \in C_\theta, z \to e^{i\theta}} |f(z)| = \infty$ *on a set of θs of positive Lebesgue measure, then f is identically infinite.*

Proof. (a) Suppose f is not identically 0. Then $f(Z_t)$ is a time change of two-dimensional Brownian motion. Since with probability one two-dimensional Brownian motion never hits 0 and does not tend to 0 as $t \to \infty$, then $\mathbb{P}^0(\lim_{t \to \tau} f(Z_t) = 0) = 0$. So $\{\theta : \mathbb{P}^0_\theta(\lim_{t \to \tau} f(Z_t) = 0) = 1\}$ must have Lebesgue measure 0. By Proposition 3.3, $\{\theta : \lim_{z \in C_\theta, z \to e^{i\theta}} f(z) = 0\}$ must also have Lebesgue measure 0.

(b) is very similar, observing that two-dimensional Brownian motion does not tend to infinity in modulus as $t \to \infty$ (Proposition I.5.8).   □

There is an interesting open problem concerning the analog of this in higher dimensions. Suppose u is harmonic in $B(0,1)$ (or in a half-space) and C^2 (or even C^∞) in $\overline{B(0,1)}$. Let

$$(3.16) \qquad A = \Big\{\theta : \lim_{z \in B(0,1), z \to e^{i\theta}} u(z) = 0, \lim_{z \in B(0,1), z \to e^{i\theta}} \nabla u(z) = 0\Big\}.$$

If $|A| > 0$, must u be identically 0? Amazingly enough this need not be true if we only require $u \in C^{1+\alpha}(\overline{B(0,1)})$ for some $\alpha > 0$ (see Bourgain and Wolff [1]). The question is still open for $u \in C^2$ or even $u \in C^\infty$.

Plessner's theorem

After some preliminary lemmas we can also prove Plessner's theorem.

(3.5) Lemma. *For almost every ω, either $\lim_{t \to \tau} f(Z_t(\omega))$ exists or for every ε, the set $\{f(Z_t(\omega)), t \in [\tau(\omega) - \varepsilon, \tau(\omega))\}$ is dense in $\mathbb{C}$.*

Proof. Let $U_t = \operatorname{Re} f(Z_{t \wedge \tau})$, $V_t = \operatorname{Im} f(Z_{t \wedge \tau})$. By the Cauchy-Riemann equations, $\langle U \rangle_\tau = \langle V \rangle_\tau$. For almost every ω for which $\langle U \rangle_\tau < \infty$, we know $\lim_{t \to \tau} U_t$ and $\lim_{t \to \tau} V_t$ exist by Exercise 8. If ω is such that $\langle U \rangle_\tau = \langle V \rangle_\tau = \infty$, $f(Z_t)$ is the time change of a two-dimensional Brownian motion path. Since two-dimensional Brownian motion is neighborhood recurrent, $f(Z_t)$ hits every neighborhood in $\mathbb{C}$ infinitely often.   □

(3.6) Lemma. *If A is a Borel subset of $\mathbb{D}$,*

$$\mathbb{P}^z\big(Z_\tau \in A, Z_t \in \cup_{e^{i\theta} \in A} C_\theta \text{ for all } t \in [\tau - \varepsilon, \tau)\big) \to \mathbb{P}^z(Z_\tau \in A)$$

as $\varepsilon \to 0$.

Proof. Let $h(z) = \mathbb{P}^z(Z_\tau \in A)$. If $w \notin \cup_{e^{i\theta} \in A} C_\theta$, there exists $\delta > 0$ such that $\mathbb{P}^w(Z_\tau \notin A) > \delta$ by the support theorem (cf. proof of Proposition IV.6.3).

So $h(z) \leq 1 - \delta$ on $(\cup_{e^{i\theta} \in A} C_\theta)^c$. Since $h(z) = \mathbb{E}^z 1_A(Z_\tau)$ is harmonic, $h(Z_t) \to 1_A(Z_\tau)$, a.s., by the martingale convergence theorem. Then

$$(Z_\tau \in A, Z_t \notin \cup_{e^{i\theta} \in A} C_\theta \text{ i.o.}) \subseteq (Z_\tau \in A, h(Z_t) < 1 - \delta \text{ i.o.})$$
$$\subseteq (Z_\tau \in A, \lim_{t \to \tau} h(Z_t) < 1 - \delta),$$

and the last event has probability 0.   □

(3.7) Theorem. (Plessner) *For almost every θ, either f has a nontangential limit at $e^{i\theta}$ or $f(C_\theta)$ is dense in $\mathbb{C}$.*

Proof. Let $L_f = \{\theta : \text{the nontangential limit of } f \text{ exists at } e^{i\theta}\}$. Suppose $|L_f^c| > 0$ and $A \subseteq L_f^c$ with $|A| > 0$. Then for a.e. θ with $e^{i\theta} \in A$, $\theta \notin \mathcal{L}_f$, or

$$\mathbb{P}_\theta^0 \left(\lim_{t \to \tau} f(Z_t) \text{ exists} \right) = 0.$$

By Proposition III.2.7, $\mathbb{P}^0(\lim f(Z_t) \text{ exists}, Z_\tau \in A) = 0$. Then by Lemma 3.1,

$$\mathbb{P}^0 \left(\{f(Z_t), t \in [\tau - \varepsilon, \tau)\} \text{ is not dense in } \mathbb{C} \text{ for all } \varepsilon, Z_\tau \in A \right)$$
$$\leq \mathbb{P}^0 \big(\{f(Z_t), t \in [\tau - \varepsilon, \tau)\} \text{ is not dense in } \mathbb{C} \text{ for all } \varepsilon,$$
$$\lim f(Z_t) \text{ does not exist} \big)$$
$$+ \mathbb{P}^0 (\lim f(Z_t) \text{ exists}, Z_\tau \in A) = 0.$$

By Lemma 3.6, on the set $(Z_\tau \in A)$ we know that Z_t will be in $\cup_{e^{i\theta} \in A} C_\theta$ for all t sufficiently close to τ. Hence $f(\cup_{e^{i\theta} \in A} C_\theta)$ must be dense in $\mathbb{C}$.

This is true for all $A \subseteq L_f^c$. Let $D_\theta = \overline{f(C_\theta)}$. Let $\{O_i\}$ be the collection of balls with rational radii and centers that have rational real and imaginary parts. So $\{O_i\}$ is a basis for $\mathbb{C}$. If $|\{\theta \in L_f^c : D_\theta \neq \mathbb{C}\}| > 0$, then for some i_0, $|\{\theta \in L_f^c : D_\theta \cap O_{i_0} = \emptyset\}| > 0$. If we let $A = \{\theta \in L_f^c : D_\theta \cap O_{i_0} = \emptyset\}$, then $f(\cup_{e^{i\theta} \in A} C_\theta)$ does not intersect O_{i_0}, a contradiction. Hence for almost every $\theta \in L_f^c$, $f(C_\theta)$ is dense in $\mathbb{C}$.   □

A stronger statement is McMillan's twist theorem. This says that for almost every θ either f' has a nontangential limit at $e^{i\theta}$ or else

$$(3.17) \qquad \limsup_{z \in C_\theta, z \to e^{i\theta}} \arg(f(z) - f(e^{i\theta})) = \infty \qquad \text{and}$$

$$\liminf_{z \in C_\theta, z \to e^{i\theta}} \arg(f(z) - f(e^{i\theta})) = -\infty$$

($\arg w$, the argument of w, is the imaginary part of $\log w$); see Pommerenke [2] for a proof.

Nevanlinna class

If $\sup_{r<1} \int_0^{2\pi} |f(re^{i\theta})|^p d\theta < \infty$ for some $p \in (0,\infty)$, then by analogy to the half-space case, we say that $f \in H^p$. If $f \in H^1$, then by Theorem IV.6.10, f then has nontangential limits a.e. However, in two dimensions, less is needed. If $\sup_{r<1} \int_0^{2\pi} \log^+ |f(re^{i\theta})| d\theta < \infty$, we say that f is in the Nevanlinna class. The most interesting property of such fs is a decomposition into simpler analytic functions (see Garnett [1]). A consequence of this decomposition is that fs in the Nevanlinna class have nontangential limits. We give a probabilistic proof.

(3.8) Proposition. *If f is in the Nevanlinna class, then f has nontangential limits for almost every θ.*

Proof. By Lemma 1.7, $\log^+ |f(z)|$ is subharmonic, so $\log^+ |f(Z_t)|$ is a nonnegative submartingale. It is uniformly bounded in L^1 since

$$\mathbb{E}^0 \log^+ |f(Z_{t\wedge\tau})| \leq \sup_{r<1} \mathbb{E}^0 \log^+ |f(Z_{t\wedge\tau_r})| \leq \sup_{r<1} \mathbb{E}^0 \log^+ |f(Z_{\tau_r})|$$

by Fatou's lemma and optional stopping.

$$\sup_{r<1} \mathbb{E}^0 \log^+ |f(Z_{\tau_r})| = \sup_{r<1} \frac{1}{2\pi} \int_0^{2\pi} \log^+ |f(re^{i\theta})| d\theta$$

is finite since f is in the Nevanlinna class, and we have the uniform boundedness in L^1. Therefore $\log^+ |f(Z_t)|$ converges almost surely, and in particular, $\sup_{t<\tau} |f(Z_t)| < \infty$, a.s.

For each M, let $G_M = \{\cup C_\theta : \sup_{z \in C_\theta} |f(z)| \leq M\}$. By the finiteness of $\sup |f(Z_t)|$, $\mathbb{P}^0(\tau_{G_M} < \tau) \to 0$ as $M \to \infty$. If $f = u + iv$, then u and v are both bounded on G_M, so $u(Z_t)$ and $v(Z_t)$ both converge as $t \to \tau \wedge \tau_{G_M}$. With the above this implies that $u(Z_t)$ and $v(Z_t)$ converges a.s. as $t \to \tau$. By Proposition 3.1, this implies the result. □

LIL for Bloch functions

A Bloch function is a function f that is analytic in $\mathbb{D}$ with

$$(3.18) \qquad \|f\|_B = |f(0)| + \sup_{z \in \mathbb{D}} (1 - |z|)|f'(z)| < \infty.$$

Thus, these are analytic functions whose derivative at z grows no faster than the distance from z to the boundary. By Corollary II.1.4, bounded analytic functions are Bloch functions. Another example of a Bloch function is the lacunary series $f(z) = \sum_{k\geq 0} z^{2^k}$ (Exercise 9). (This series is called lacunary because of the "lacuna" or "holes" in the sequence of exponents of z.)

The important members of the Bloch class for us are given by the following.

(3.9) Proposition. *If f is one to one on $\mathbb{D}$, then $\log f'$ is a Bloch function.*

Proof. Let us normalize f so that $f(0) = 0$ and $f'(0) = 1$. By the distortion theorem, $f(\mathbb{D}) \supseteq B(0, a)$, so f^{-1} is one to one on $B(0, a)$. By applying the distortion theorem to f^{-1} and $B(0, a)$, $f^{-1}(B(0, a)) \supseteq B(0, a^2)$, or $f(B(0, a^2)) \subseteq B(0, a)$. This can be rephrased by saying f is bounded by a on $B(0, a^2)$. By Corollary II.1.4 applied to the real and imaginary parts of $f(z)$,

$$|f''(0)| \le \frac{c}{(a^2)^2} \sup_{z \in B(0, a^2)} |f(z)| \le \frac{c}{a^3} = c.$$

By our normalization,

$$f(z) = z + b_2 z^2 + b_3 z^3 + \cdots .$$

So what we have shown is that $|b_2| \le c$, c independent of f.

The mapping $z \to (z + z_0)/(1 + \overline{z_0} z)$ is one to one from $\mathbb{D}$ to itself and maps 0 to z_0. Hence

$$h(z) = \frac{f((z + z_0)/(1 + \overline{z_0} z)) - f(z_0)}{(1 - |z_0|^2) f'(z_0)}$$

is one to one on $\mathbb{D}$ and

$$h(z) = z + \left(\frac{1}{2}(1 - |z_0|^2) \frac{f''(z_0)}{f'(z_0)} - \overline{z_0} \right) z^2 + \cdots .$$

Therefore,

$$\left| \frac{1}{2}(1 - |z_0|^2) \frac{f''(z_0)}{f'(z_0)} - \overline{z_0} \right| \le c.$$

Since $|z_0| \le 1$,

$$(1 - |z_0|) |f''(z_0)/f'(z_0)| \le c,$$

c independent of f. This is the same as saying $\log f'$ is in the Bloch class. $\qquad\square$

It is not hard to show that if f is one to one on $\mathbb{D}$ with $f(0) = 0$, $f'(0) = 1$, then $|f''(0)| \le 2$. The Bieberbach conjecture, proved by deBranges [1], is the assertion that the nth coefficient in the Taylor expansion of f about 0 is bounded in absolute value by n.

We will show Makarov's law of the iterated logarithm for Bloch functions. In the next section we will use it to settle a question about harmonic measure.

(3.10) Theorem. *Suppose f is a Bloch function with $\|f\|_B \le 1$. There exists β not depending on f such that*

$$(3.19) \qquad \limsup_{r \to 1} \frac{|f(re^{i\theta})|}{\sqrt{\log(1/(1-r)) \log \log \log(1/(1-r))}} \le \beta$$

for almost all θ.

By being a little more careful than we will be, one can show the result for $\beta = 2$. There is no lower bound for the lim sup; if f is bounded, for example, the lim sup is zero.

The proof will be broken up into several steps. In $\mathbb{D}$ the appropriate substitute for the g_λ^* function of (IV.4.2) is

$$g^*(f)(\theta) = \left(\mathbb{E}_\theta^0 \left[\int_0^\tau |f'(Z_s)|^2 ds \right] \right)^{1/2}.$$

(3.11) Proposition. *Let f be analytic with $g^*(f)$ bounded. Then*

$$\frac{1}{2\pi} \int_0^{2\pi} e^{\lambda(f(e^{i\theta}) - f(0))} d\theta \leq e^{\lambda^2 \|g^*(f)\|_\infty^2 / 2}.$$

Proof. Let us normalize so that $f(0) = 0$ and $\|g^*(f)\|_\infty = 1$. Since Z_τ has a uniform distribution on $\partial \mathbb{D}$ under $\mathbb{P}^0$,

$$\frac{1}{2\pi} \int_0^{2\pi} e^{\lambda f(e^{i\theta}) - \lambda^2 (g^*(f)(\theta))^2 / 2} d\theta$$

$$= \mathbb{E}^0 \exp \left(\lambda f(Z_\tau) - \frac{\lambda^2}{2} \mathbb{E}_{Z_\tau}^0 \left[\int_0^\tau |f'(Z_s)|^2 ds \right] \right)$$

$$= \mathbb{E}^0 \exp \left(\mathbb{E}_{Z_\tau}^0 \left[\lambda f(Z_\tau) - \frac{\lambda^2}{2} \int_0^\tau |f'(Z_s)|^2 ds \right] \right).$$

By Jensen's inequality, this is less than

$$\mathbb{E}^0 \mathbb{E}_{Z_\tau}^0 \left[\exp \left(\lambda f(Z_\tau) - \frac{\lambda^2}{2} \int_0^\tau |f'(Z_s)|^2 ds \right) \right].$$

By Proposition III.2.7, this is equal to

$$\mathbb{E}^0 \left[\exp \left(\lambda f(Z_\tau) - \frac{\lambda^2}{2} \int_0^\tau |f'(Z_s)|^2 ds \right) \right]$$

$$= \mathbb{E}^0 [\exp(\lambda f(Z_\tau) - \lambda^2 \langle f(Z) \rangle_\tau / 2)] = 1.$$

Since $g^*(f)$ is bounded by 1,

$$\frac{1}{2\pi} \int_0^{2\pi} e^{\lambda f(e^{i\theta})} d\theta \leq e^{\lambda^2 / 2}.$$

$\square$

Next suppose f is a Bloch function with $\|f\|_B \leq 1$. Let $f_r(e^{i\theta}) = f(re^{i\theta})$, denote the extension of f_r by f_r also, and let

$$f_r^*(\theta) = \sup_{0 \le s \le r} |f_r(e^{i\theta})|.$$

We have the following.

(3.12) Proposition. *There exist c_1 and c_2 such that*

$$|\{\theta : f_r^*(\theta) > \alpha\}| \le c_1 \exp\left(-\frac{\alpha^2}{c_2 \log(1/(1-r))}\right).$$

Proof. Let us first calculate $g^*(f_r)$.

$$
\begin{aligned}
(g^*(f_r)(\theta))^2 &= \mathbb{E}_\theta^0\left[\int_0^\tau |f_r'(Z_s)|^2 ds\right] \\
&= r^2 \mathbb{E}_\theta^0\left[\int_0^\tau |f'(rZ_s)|^2 ds\right] \\
&\le r^2 \mathbb{E}_\theta^0\left[\int_0^\tau \frac{ds}{(1-r|Z_s|)^2}\right]
\end{aligned}
$$

since f is a Bloch function. Since $h_\theta(z)$, the Poisson kernel with pole at $e^{i\theta}$, is constant in θ when $z = 0$ and $h_\theta(Z_s)$ is a martingale, this is

$$cr^2 \mathbb{E}^0\left[\int_0^\tau \frac{ds}{(1-r|Z_s|)^2} h_\theta(Z_\tau)\right] = cr^2 \mathbb{E}^0\left[\int_0^\tau \frac{h_\theta(Z_s)}{(1-r|Z_s|)^2} ds\right].$$

Using the Green function for $\mathbb{D}$, this is equal to

$$
\begin{aligned}
cr^2 \int_\mathbb{D} \log(1/|z|) h_\theta(z) \frac{dz}{(1-r|z|)^2} \\
&= cr^2 \int_0^1 \int_0^{2\pi} s \log(1/s) \frac{1}{(1-rs)^2} h_\theta(se^{i\phi}) d\phi\, ds \\
&= cr^2 \int_0^1 s \log(1/s) \frac{ds}{(1-rs)^2},
\end{aligned}
$$

where in the last equality we used the fact that $\int_0^{2\pi} h_\theta(se^{i\phi}) d\phi$ is constant in θ by rotational invariance. We have

$$cr^2 \int_0^1 \frac{s \log(1/s) ds}{(1-rs)^2} \le cr \int_0^1 \frac{ds}{1-rs} = c \log\frac{1}{1-r}.$$

So $\|g^*(f_r)\|_\infty^2 \le c \log(1/(1-r))$.

Next $\|f_r^*\|_p^p \le c^p \|f_r\|_p^p$ by Exercise IV.8.3. Therefore

$$\int_0^{2\pi} e^{\lambda f_r^*(\theta)} d\theta = \sum_{p=0}^\infty \int_0^{2\pi} \frac{\lambda^p}{p!} |f_r^*(\theta)|^p d\theta$$

$$\le \sum_{p=0}^\infty \int_0^{2\pi} \frac{\lambda^p c^p}{p!} |f_r(e^{i\theta})|^p d\theta$$

$$\le \int_0^{2\pi} e^{\lambda c |f_r(e^{i\theta})|} d\theta.$$

By Proposition 3.11, this is less than

$$\exp\left(\lambda^2 c^2 \|g^*(f_r)\|_\infty^2 / 2\right) = \exp\left(\lambda^2 c^2 \log(1/(1-r))/2\right).$$

Now use Chebyshev's inequality:

$$|\{\theta : f_r^*(\theta) > \alpha\}| = |\{\theta : e^{\lambda f_r^*(\theta)} > e^{\lambda\alpha}\}| \le e^{-\lambda\alpha} \int_0^{2\pi} e^{\lambda f_r^*(\theta)} d\theta.$$

Letting $\lambda = \alpha/c^2 \log(1/(1-r))$, we obtain our result. $\qquad\square$

Proof of Theorem 3.10. This is standard (cf. Exercise I.8.4). Let $r_n = 1 - \exp(-2^n)$, let $\ell_n = \log(1/(1-r_n)) = 2^n$, let c_2 be as in Proposition 3.12, and let

$$B_n = \{\theta : f_{r_n}^*(e^{i\theta}) > 2(c_2\ell_n \log\log\ell_n)^{1/2}\}.$$

By Proposition 3.12, $|B_n| \le cn^{-4}$. By the Borel-Cantelli lemma, $|(B_n \text{ i.o.})| = 0$. If $|f_r(e^{i\theta})| > 4(c_2 \log(1/(1-r)) \log\log\log(1/(1-r)))^{1/2}$ for some r, then for some r_n, $\theta \in B_n$. The result follows with $\beta = 4c_2^{1/2}$. $\qquad\square$

There are generalizations of Theorem 3.10 to harmonic functions where the distance to the boundary is replaced by expressions involving the area functional. See Bañuelos, Klemeš, and Moore [1, 2] and Bañuelos and Moore [1, 2].

4 Harmonic measure

Beurling's projection theorem

Recall harmonic measure is the hitting distribution of Brownian motion: if $E \subseteq \partial D$ and $x_0 \in D$,

(4.1) $$\omega^{x_0}(E) = \mathbb{P}^{x_0}(X_{\tau_D} \in E).$$

By the Harnack inequality, harmonic measure relative to x_0 and harmonic measure relative to another point $x_1 \in D$ are mutually absolutely continuous and the density is bounded above and below by positive constants (see Theorem III.2.6).

We start by giving a proof of Beurling's projection theorem. The circular projection of a set $E \subseteq \mathbb{D}$ is

$$(4.2) \qquad \gamma(E) = \{|z| : z \in E\}.$$

$\gamma(E)$ is a subset of the positive real axis. Beurling's projection theorem says that Brownian motion killed on exiting $\mathbb{D}$ is more likely to hit E than $\gamma(E)$. Let $a = -1/2$.

(4.1) Theorem. *Suppose E is a closed subset of $\mathbb{D}$.*

$$\mathbb{P}^a(T_E < \tau) \geq \mathbb{P}^a(T_{\gamma(E)} < \tau).$$

The proof involves the reflection principle, and we start by seeing the effect of reflection across a line. Let $\rho(A) = \{x - iy : x + iy \in A\}$, reflection across the x-axis.

(4.2) Lemma. *Suppose $E \subseteq \mathbb{D}$. Let*

$$E^+ = E \cap \{\operatorname{Im} z \geq 0\}, \qquad E^- = E \cap \{\operatorname{Im} z < 0\},$$

and

$$\zeta_0(E) = E^- \cup \rho(E^+).$$

Then if $\operatorname{Im} w \geq 0$,

$$\mathbb{P}^w(T_E < \tau) \geq \mathbb{P}^w(T_{\zeta_0(E)} < \tau).$$

Proof. If $H = E^+ \cup (E^- \rho(E^+))$, then $H \subseteq E$ and $\zeta_0(H) = \zeta_0(E)$. So we may assume $\rho(E^+) \cap E^- = \emptyset$.

Let R be the real axis, and suppose for now that E is open and a positive distance from $R \cup \partial \mathbb{D}$. Let $U_1 = \inf\{t : Z_t \in E \cup \rho(E)\} \wedge \tau$, and for $i \geq 1$,

$$V_i = \inf\{t > U_i : Z_t \in R\} \wedge \tau, \qquad U_{i+1} = \inf\{t > V_i : Z_t \in E \cup \rho(E)\} \wedge \tau.$$

By the continuity of the paths of Z_t,

$$(4.3) \quad \mathbb{P}^w(T_{\zeta_0(E)} < \tau)$$
$$= \mathbb{P}^w(Z_{U_1} \in \zeta_0(E)) + \mathbb{P}^w(Z_{U_1} \notin \zeta_0(E), Z_{U_2} \in \zeta_0(E))$$
$$+ \mathbb{P}(Z_{U_1} \notin \zeta_0(E), Z_{U_2} \notin \zeta_0(E), Z_{U_3} \in \zeta_0(E)) + \ldots.$$

We also have

$$(4.4) \qquad \mathbb{P}^w(T_E < \tau)$$
$$\geq \mathbb{P}^w(Z_{U_1} \in E) + \mathbb{P}^w(Z_{U_1} \notin E, Z_{U_2} \in E)$$
$$+ \mathbb{P}(Z_{U_1} \notin E, Z_{U_2} \notin E, Z_{U_3} \in E) + \ldots,$$

since the sets on the right are disjoint and contained in the set on the left.

By symmetry, $\mathbb{P}^z(Z_{U_1} \in E) = \mathbb{P}^z(Z_{U_1} \in \zeta_0(E))$ if $z \in R$. So the nth term on the right-hand side of (4.3) is equal to the nth term on the right-hand side of (4.4), using the strong Markov property $n - 1$ times.

We remove the assumption that E is open and a positive distance from $R \cup \partial D$ by a limiting argument. $\qquad \square$

Proof of Theorem 4.1. By the proof of Proposition I.2.7, we may assume E is open. Let

$$(4.5) \qquad \zeta_{\theta_0}(A) = e^{i\theta_0}(\zeta_0(e^{-i\theta_0}A)).$$

ζ_{θ_0} is just the rotation of the operation $A \to \zeta_0(A)$ by the angle θ_0. Let $E_1 = \zeta_\pi(E)$. Then let $E_2 = \zeta_{\pi/2}(E_1)$, $E_3 = \zeta_{\pi/4}(E_2)$, $E_4 = \zeta_{\pi/8}(E_3)$, etc. Using Lemma 4.2 repeatedly with the lines $\{\theta = \pi\}$, $\{\theta = \pi/2\}$, $\{\theta = \pi/4\}$, $\{\theta = \pi/8\}$, etc., we deduce

$$\mathbb{P}^a(T_E < \tau) \geq \mathbb{P}^a(T_{E_1} < \tau) \geq \mathbb{P}^a(T_{E_2} < \tau) \geq \cdots .$$

It is easy to see $T_{E_n} \to T_{\gamma(E)}$, a.s., hence $\mathbb{P}^a(T_E < \tau) \geq \mathbb{P}^a(T_{\gamma(E)} < \tau)$. $\quad \square$

Hall's lemma

Very similar techniques can be used to prove Hall's lemma (see Øksendal [1]), but for variety we give a potential theory proof. Let

$$(4.6) \qquad \sigma(E) = \{z/|z| : z \in E\}$$

be the radial projection of E onto ∂D.

(4.3) Theorem. *Suppose E is a closed subset of $\mathbb{D}$. There exists c not depending on E such that*

$$(4.7) \qquad \mathbb{P}^0(T_E < \tau) \geq c\mathbb{P}^0(T_{\sigma(E)} < \tau).$$

Hall's lemma holds in dimensions greater than two as well as in two dimensions. It is not known what the best value of c is, or even if one can choose c independent of the dimension d.

Proof. Suppose first that E consists of finitely many circular segments $\{r_k e^{i\theta} : a_k \leq \theta \leq b_k\}$, with $0 \leq a_k \leq b_k \leq 2\pi$, the (a_k, b_k) disjoint, and $r_k \leq 1$. Define

$$(4.8) \qquad U(z) = \int_E \frac{g_{\mathbb{D}}(z, w)}{\log(1/|w|)} d\mu,$$

where $d\mu$ refers to angular measure: $\mu(A) = |\sigma(A)|$. We claim there exist c_1 and c_2 such that

(4.9) $$U(0) \geq c_1 \mathbb{P}^0(Z_\tau \in \sigma(E))$$

and

(4.10) $$U(z) \leq c_2, \qquad z \in E.$$

Suppose (4.9) and (4.10) have been established. If $h(z) = \mathbb{P}^z(T_E < \tau)$, $h(z)$ is harmonic in $\mathbb{D} - E$. $U(z)$ is also harmonic off E. Both U and h are 0 on $\partial\mathbb{D}$ and $h = 1$ on E. So by (4.10), $c_2 h(z) - U(z)$ is nonnegative on $\partial\mathbb{D} \cup \partial E$ and harmonic in $\mathbb{D} - E$. By the maximum principle, $c_2 h(z) - U(z) \geq 0$ on $\mathbb{D} - E$, hence on $\mathbb{D}$. Then by (4.9),

$$c_1 \mathbb{P}^0(Z_\tau \in \sigma(E)) \leq U(0) \leq c_2 h(0) = c_2 \mathbb{P}^0(T_E < \tau),$$

which is what we wanted to show for E of the above form.

(4.9) is easy to see.

$$\mathbb{P}^0(Z_\tau \in \sigma(E)) = c \int_{\sigma(E)} d\theta = c \int_E 1 \, d\mu,$$

while

$$U(0) = \int_E \frac{g_\mathbb{D}(0, w)}{\log(1/|w|)} \, d\mu.$$

Note, however, that $g_\mathbb{D}(0, w)/\log(1/|w|)$ is a positive constant in $\mathbb{D}$.

(4.10) is a bit more complicated. Without loss of generality, let us suppose $z = \rho > 0$. For each θ we bound

$$g_D(z, y) = \pi^{-1} \log(|y| |z - y^*| / |z - y|)/(1 - |y|)$$

for $y = b e^{i\theta}$ and $y^* = y/|y|^2$. We leave the cases $\rho \leq 1/2$ and $b \leq 1/2$ for Exercise 10, so we suppose both b and $\rho \geq 1/2$. Since $\log(1/b) \geq 1 - b$, we need to bound $g_\mathbb{D}(z, y)/(1 - |y|)$.

Substituting for y and z, we see

$$
\begin{aligned}
(4.11) \quad g_D(z, y)/(1 - |y|) &= c \log\left(\frac{[b\rho - \cos\theta]^2 + [\sin\theta]^2}{[\rho - b\cos\theta]^2 + [b\sin\theta]^2}\right)/(1 - b) \\
&= c \log\left(\frac{b^2\rho^2 - 2b\rho\cos\theta + 1}{\rho^2 - 2b\rho\cos\theta + b^2}\right)/(1 - b) \\
&= c \log\left(1 + \frac{(1 - b^2)(1 - \rho^2)}{\rho^2 - 2b\rho\cos\theta + b^2}\right)/(1 - b).
\end{aligned}
$$

If either $|\theta| \geq 1 - \rho$ or else $|\theta| \leq 1 - \rho$ and $|b - \rho| \geq (1 - \rho)/2$, then there exists c such that $|\rho - be^{i\theta}| \geq c|\rho - e^{i\theta}|$. Then the right-hand side of (4.11) is bounded by

$$(4.12) \quad c\left(\frac{(1 - b^2)(1 - \rho^2)}{|\rho - be^{i\theta}|^2}\right)/(1 - b) = c(1 + b)\frac{1 - \rho^2}{|\rho - be^{i\theta}|^2} \leq \frac{c(1 - \rho^2)}{|\rho - e^{i\theta}|^2}.$$

The ratio on the right is just a constant times the Poisson kernel for $\mathbb{D}$ (see Theorem II.1.17).

If $|\theta| \le 1 - \rho$ and $|b - \rho| \le (1 - \rho)/2$, then $1 - \rho \le 2(1 - b)$ and $1 - b \le 3(1 - \rho)/2$. In this case, the expression in (4.11) is bounded by

$$(4.13) \qquad c \log \left(1 + \frac{c(1 - b)(1 - \rho)}{(\rho - b)^2 + 2b\rho(1 - \cos\theta)}\right) / (1 - b)$$

$$\le c \log \left(1 + \frac{c(1 - \rho)^2}{\theta^2}\right) / (1 - \rho).$$

Note

$$(4.14) \qquad \int_{-(1-\rho)}^{1-\rho} \log\left(1 + \frac{c(1 - \rho)^2}{\theta^2}\right) \frac{d\theta}{1 - \rho} = \int_{-1}^{1} \log(1 + c/s^2)\,ds < \infty.$$

Combining (4.11), (4.12), (4.13), and (4.14),

$$(4.15) \qquad \sup_{1/2 \le \rho < 1} \int_0^{2\pi} \sup_{1/2 \le b < 1} \frac{g_{\mathbb{D}}(\rho, be^{i\theta})}{1 - b}\,d\theta \le c < \infty.$$

Combining (4.8), (4.15), and Exercise 10 implies (4.10).

This completes the proof in the case E is the finite union of circular arcs. Let a polar rectangle be a set of the form $I = \{(r, \theta) : a_1 \le r \le a_2, b_1 \le \theta \le b_2\}$. Since $J_I = \{re^{i\theta} : r = a_1, b_1 \le \theta \le b_2\}$ is contained in I, the probability Brownian motion hits J_I before τ is less than the probability it hits I before τ. So if E is the finite union of polar rectangles, say $\cup_k I_k$, let $F = \cup_k J_{I_k}$. By what we have shown above,

$$(4.16) \quad \mathbb{P}^0(T_E < \tau) \ge \mathbb{P}^0(T_F < \tau) \ge c\mathbb{P}^0(T_{\sigma(F)} < \tau) = c\mathbb{P}^0(T_{\sigma(E)} < \tau).$$

By taking a limit, (4.16) also holds for the countable union of polar rectangles, and since any compact set can be approximated from above by countable unions of polar rectangles, it follows that (4.16) also holds for any compact E. That suffices to complete the proof. $\qquad\square$

The Besicovitch covering lemma

We will need a covering lemma due to Besicovitch.

(4.4) Theorem. *Let A be a bounded set in $\mathbb{R}^d$. Suppose for each $x \in A$ there exists a cube $Q(x)$ centered at x. We can select a sequence of cubes from $\{Q(x) : x \in A\}$, say $Q_1, Q_2, \ldots$, such that*
(a) $A \subseteq \cup_k Q_k$;
(b) No point of $\mathbb{R}^d$ is in more than N_1 of the Q_k (N_1 depends only on d); and

(c) $\{Q_k\}$ can be split into N_2 disjoint collections such that each Q_k is in exactly one of the N_2 subcollections, and the cubes in any subcollection have pairwise disjoint interiors. (N_2 depends only on d.)

Proof. For simplicity we consider the case $d = 2$, although only minor changes are needed for $d \neq 2$.

Let $a_1 = \sup\{|Q(x)| : x \in A\}$. If $a_1 = \infty$, we pick a square larger than $2 \operatorname{diam} A$, and we are done. So let us suppose $a_1 < \infty$. Pick $x_1 \in A$ such that $|Q(x_1)| \geq (9/10)a_1$, and let that be Q_1. Let $a_2 = \sup\{|Q(x)| : x \in A - Q_1\}$, pick $x_2 \in A - Q_1$ such that $|Q(x_2)| \geq (9/10)a_2$, let that be Q_2, let $a_3 = \sup\{|Q(x)| : x \in A - (Q_1 \cup Q_2)\}$, etc.

Let $s_i = (|Q_i|)^{1/2}$, the side length of Q_i. If $i < j$, note $|Q_j| \leq (10/9)|Q_i|$ (or else $|Q(x_i)| < (9/10)a_i$), so $s_j \leq (10/9)^{1/2}s_i$. If $i < j$, then $x_j \notin Q_i$ by construction, so $|x_j - x_i| \geq s_i/2$.

Let $\widetilde{Q}_i$ be the square with the same center as Q_i but with sides $1/4$ as long. We claim the $\widetilde{Q}_i$ are disjoint. For if $\widetilde{Q}_i$ intersects $\widetilde{Q}_j$, let y be a point in the intersection and suppose that $i < j$. Then

$$s_i/2 \leq |x_j - x_i| \leq |x_j - y| + |x_i - y|$$
$$\leq (\sqrt{2}/8)s_j + (\sqrt{2}/8)s_i \leq (9/20)s_i,$$

a contradiction.

Let us now prove (a), (b) and (c). If the process of constructing the Q_is stops after finitely many steps, (a) must hold, or else we would continue the construction. Suppose there are infinitely many Q_i. Since the $\widetilde{Q}_i$ are disjoint and each is contained in the set of points that are within $2a_1$ of A, we must have $|Q_i| \to 0$. If there exists $y \in A - \cup_i Q_i$, then $|Q(y)| > 2|Q_i|$ for some i large, a contradiction to the way Q_i was selected. Therefore $A \subseteq \cup_i Q_i$ and (a) is proved.

We next prove (b). Let y be any point in $\mathbb{R}^2$. By a change of coordinates, let us suppose $y = 0$. Look at all the Q_i centered at points x_i in the wedge $W = \{re^{i\theta} : r \geq 0, 0 \leq \theta \leq \pi/16\}$ that contain 0, and let i_0 be the smallest subscript. If Q_i is one of these squares, we have $x_i \notin Q_{i_0}$, $s_{i_0} \geq (9/10)^{1/2}s_i$, and $\widetilde{Q}_{i_0}$ and $\widetilde{Q}_i$ are disjoint. Some simple geometry shows this is not possible for any $i \neq i_0$, and therefore there is only one square centered at a point in W containing 0. We repeat the argument for $e^{\pi k/16}W$, $k = 1, \ldots, 32$, and conclude (b) with $N_1 = 32$.

Finally we prove (c). Let $S = [-1/2, 1/2]^2$ and let T be the eight points consisting of the corners of S and the midpoints of the sides of S. Let $T_i = x_i + s_i T$, the corresponding eight points on Q_i. Fix i. If $k < i$ and Q_k intersects Q_i, it is easy to see that since $s_k \geq (9/10)^{1/2}s_i$ and $x_i \notin Q_k$, then Q_k must contain at least one of the eight points in T_i. Since each of the points of T_i is in at most $N_1 = 32$ of the Q_i, we see that at most $N_2 = 256$ of the Q_k with $k < i$ intersect Q_i.

We now form 256 subcollections of Q_i, say $I_1, \ldots, I_{256}$. We start by putting Q_i in I_i, $i = 1, \ldots, 256$. Q_{257} must be disjoint from at least one of the $Q_1, \ldots, Q_{256}$, say Q_{i_0}. We then put Q_{257} in I_{i_0}. Again, Q_{258} can intersect at most 256 of the $Q_1, \ldots, Q_{257}$, so it must be disjoint from all the cubes in at least one of the I_i. Put Q_{258} in the first of the I_i for which this is so. If we continue in this way, (c) is proved.

For $d \neq 2$, the only changes necessary are to change the quantities $9/10$, N_1, N_2, replace wedges by cones, and to increase the number of points in T in a suitable way. $\qquad \square$

In the application we will make, A is actually a subset of $\partial \mathbb{D}$. We have the following corollary.

(4.5) Corollary. *Suppose $A \subseteq \partial \mathbb{D}$ with $|\partial \mathbb{D} - A| = 0$. Suppose for each $z \in A$ there is an arc L_z centered at z. We can select a countable sequence of these arcs $L_1, L_2, \ldots$, such that $A \subseteq \cup_i L_i$ and $\sum_i |L_i| < \infty$.*

The proof of Corollary 4.5 is Exercise 12.

Support of harmonic measure

We give here Makarov's proof that harmonic measure for a simply connected domain is concentrated on a set of Hausdorff dimension 1. (We will give definitions in a moment.)

Consider the von Koch snowflake (see Fig. 4.1).

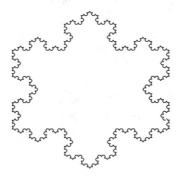

FIGURE 4.1. The von Koch snowflake.

Brownian motion does not hit points, so it does not hit any of the corners. Yet the Hausdorff dimension of the boundary is larger than 1, and "most" of the boundary points are not hit. The support of harmonic measure, i.e., the smallest closed set whose complement has measure 0, is actually all of ∂D. We are interested in the essential support: a set on which ω^{x_0} is concentrated.

The construction of Lebesgue measure on the line starts out by defining

$$m(A) = \inf\Big\{\sum_j |I_j| : I_j \text{ is a countable collection of intervals}$$

$$\text{whose union contains } A\Big\},$$

where $|I_j|$ is just the length of I_j. Similarly we start out defining the Lebesgue measure of $A \subseteq \mathbb{R}^2$ by replacing I_j in the above by balls or squares.

If $\varphi(t)$ is a continuous and increasing function with $\varphi(0) = 0$, define

$$(4.17) \quad \Lambda_\varphi(A) = \lim_{\delta \to 0} \inf\Big\{\sum_j \varphi(\text{diam } B_j) : B_j \text{ is a collection of balls,}$$

$$\text{each with radius less than } \delta, \text{ whose union contains } A\Big\}.$$

$\Lambda_\varphi(A)$ is called the Hausdorff measure of A with respect to the measure function φ. Except for the restriction that the radii must be less than δ and the taking of the limit as $\delta \to 0$, this is entirely analogous to the definition of Lebesgue measure.

When $\varphi_\alpha(t) = t^\alpha$ and $A \subseteq \mathbb{R}^d$, it is easy to see (Exercise 13) that $\Lambda_{\varphi_\alpha}(A)$ will be 0 for all α bigger than some α_0, and infinite for all α less than α_0. When $\alpha = \alpha_0$, it could be infinite, 0, or some positive number. The number $\alpha = \inf\{\alpha : \Lambda_{\varphi_\alpha}(A) = 0\}$ is called the Hausdorff dimension of A. We will denote the measure Λ_{φ_α} simply by Λ_α.

The first theorem we prove is the following.

(4.6) Theorem. *Suppose D is a simply connected domain other than $\mathbb{C}$. There exists $H \subset \partial D$ such that the Hausdorff dimension of H is 1 and $\omega^{x_0}(\partial D - H) = 0$.*

Thus ω^{x_0} is concentrated on a set of Hausdorff dimension at most 1. Let us say that a subset A of D is a dominating subset of D if

$$\|f\|_\infty = \sup_{z \in A} |f(z)|$$

for every f that is analytic and bounded in D. If almost every boundary point $e^{i\theta}$ of ∂D is such that $[\mathbb{D} - B(0, r)] \cap A \cap C_\theta \neq \emptyset$ for all r close to 1, then A is a dominating subset for $\mathbb{D}$ (Exercise 14). If A is a dominating subset of $\mathbb{D}$, then $f(A)$ is a dominating subset of $f(\mathbb{D})$ (Exercise 14).

Let f be a one to one analytic function mapping $\mathbb{D}$ to D.

(4.7) Lemma. *Let $\varepsilon > 0$. There exists a dominating subset $\widetilde{A}$ of D such that*

$$\sum_{z \in \widetilde{A}} [\text{dist }(z, \partial D)]^{1+\varepsilon} < \infty.$$

Proof. First, we assert that for almost every $\theta \in [0, 2\pi)$,

$$(4.18) \qquad \liminf_{z \in C_\theta, z \to e^{i\theta}} |f'(z)|(1 - |z|)^{\varepsilon/(1+\varepsilon)} = 0.$$

For if not, $\lim_{z \in C_\theta, z \to e^{i\theta}} |f'(z)| = \infty$ on a set of positive measure. By Theorem 3.4(b), f' is identically infinite, a contradiction.

Let $\eta > 0$. For each θ not in the null set, there exists $z_\theta \in C_\theta$ such that $1 - |z_\theta| < \eta$ and $|f'(z_\theta)| \leq \eta(1 - |z|)^{-\varepsilon/(1+\varepsilon)}$. Let L_θ be a subarc of $\partial \mathbb{D}$, centered at $e^{i\theta}$, with length $1 - |z_\theta|$. By Corollary 4.5, we can find a countable subcollection of the set of arcs $\{L_\theta\}$ such that $\sum_j (1 - |z_{\theta_j}|) < \infty$ and $\partial \mathbb{D} - \cup_j L_{\theta_j}$ has 0 Lebesgue measure.

Let $A_\eta = \{f(z_{\theta_j})\}$. By Corollary 2.6,

$$\operatorname{dist}(f(z), \partial D) \leq c|f'(z)|(1 - |z|).$$

Hence

$$
\begin{aligned}
[\operatorname{dist}(f(z_{\theta_j}), \partial D)]^{1+\varepsilon} &\leq c\Big[\eta(1 - |z_{\theta_j}|)^{-\varepsilon/(1+\varepsilon)}(1 - |z_{\theta_j}|)\Big]^{1+\varepsilon} \\
&= c\eta^{1+\varepsilon}(1 - |z_{\theta_j}|).
\end{aligned}
$$

Therefore

$$\sum_j [\operatorname{dist}(f(z_{\theta_j}), \partial D)]^{1+\varepsilon} \leq c\eta^{1+\varepsilon}.$$

Let $\widetilde{A} = \cup_{k>0} A_{2^{-k}}$. By Exercise 14, A is a dominating subset for D. $\qquad \square$

Proof of Theorem 4.6. For each $z \in \widetilde{A}$, let $E_z = \partial D \cap B(z, 2\operatorname{dist}(z, \partial D))$. Since D is simply connected, by the proof of Proposition II.1.14, $\mathbb{P}^z(Z_{\tau_D} \in E_z) \geq c > 0$. Let $E = \cup_{z \in \widetilde{A}} E_z$.

We claim $\omega^{z_0}(E) = 1$. If not, there exists $F \subseteq \partial D - E$ with $\omega^{z_0}(F) > 0$. Let $h(z) = \mathbb{P}^z(Z_{\tau_D} \in F)$, let $\widetilde{h}$ be the conjugate harmonic function of h, and let $f = e^{h + i\widetilde{h}}$. Then $\|f\|_\infty = e^{\|h\|_\infty}$, and by taking a suitable sequence $z \in D$ converging to a point in F, $\|h\|_\infty = 1$. If $z \in \widetilde{A}$, $|f(z)| = e^{h(z)}$, and $h(z) \leq 1 - c$, so $|f(z)| \leq e^{1-c}$. This contradicts the fact that $\widetilde{A}$ is a dominating subset. Therefore $\omega^{z_0}(E) = 1$.

Label the points of $\widetilde{A}$ as $z_1, z_2, \ldots$, and for each m, let $\widetilde{A}_m = \{z_{k_m}, z_{k_m+1}, \ldots\}$, where k_m is chosen large so that

$$\Big[\sum_{j=k_m}^\infty \operatorname{dist}(z_j, \partial D)\Big]^{1+\varepsilon} < 2^{-m}$$

and $\operatorname{dist}(z_j, \partial D) < 2^{-m}$ if $j \geq k_m$. Let $E_m = \cup_{z \in \widetilde{A}_m} E_z$. Then $\widetilde{A}_m$ is still a dominating subset of D, and so $\omega^{z_0}(E_m) = 1$. If $G_\varepsilon = \cap_m E_m$, then $\omega^{z_0}(G_\varepsilon) = 1$, $G_\varepsilon \subseteq E_m$ for all m, and $\{B(z_j, 2\operatorname{dist}(z_j, \partial D)); z_j \in \widetilde{A}_m\}$ is a cover of G_ε by balls with

$$\sum_{z_j \in \widetilde{A}_m} (\operatorname{diam} B(z_j, 2\operatorname{dist}(z_j, \partial D)))^{1+\varepsilon} \leq 2^{1+\varepsilon} 2^{-m}.$$

Hence $\Lambda_{1+\varepsilon}(G_\varepsilon) = 0$. Finally, let $H = \cap_{\ell > 0} G_{2-\ell}$. Then $\omega^{z_0}(H) = 1$ and the dimension of H is less than or equal to 1. $\square$

We now present the result that harmonic measure cannot be concentrated on a set of Hausdorff dimension strictly less than 1. We suppose f is one to one and continuous on $\mathbb{D}$ and analytic on $\mathbb{D}$.

(4.8) Lemma. *Let $\varepsilon > 0$, $\delta \in (0, \varepsilon)$, $r \in [1/2, 1)$, and $A \subseteq \partial \mathbb{D}$. If* diam $f(A) \leq \varepsilon$,

$$(4.19) \qquad\qquad |f(rz) - f(z)| \leq \varepsilon, \qquad z \in A,$$

and

$$(4.20) \qquad\qquad (1 - r)|f'(rz)| \geq \delta, \qquad z \in A,$$

then there exists $c > 2$ such that A can be covered by at most $c(\varepsilon/\delta)^2$ sets of diameter $1 - r$.

Proof. Consider a grid of squares of mesh size $b\delta$, b to be chosen later. Let Q_k be the squares that contain points of $f(rA)$. Since diam $f(A) \leq \varepsilon$, (4.19) implies diam $f(rA) \leq 3\varepsilon$. If m is the number of squares, then

$$(b\delta)^2 m = |Q_1 \cup \cdots \cup Q_m| \leq c\varepsilon^2.$$

Thus it suffices to show that $A_k = \{z \in A : f(rz) \in Q_k\}$ has diameter less than $1 - r$ if b is chosen appropriately.

Suppose $z_1, z_2 \in A_k$ with $|z_1 - z_2| > 1 - r$. Then $|rz_1 - rz_2| > r(1-r)$. By Exercise 15 and (4.20),

$$(4.21) \qquad \delta < (1 - r^2)|f'(rz_1)| \leq c|f(rz_1) - f(rz_2)| \leq c_1 b\delta,$$

since $f(rz_1), f(rz_2) \in Q_k$. This is a contradiction if we choose $b = 1/2c_1$. $\square$

(4.9) Theorem. *Suppose f is one to one and continuous on $\overline{\mathbb{D}}$ and analytic in $\mathbb{D}$. Let $D = f(\mathbb{D})$. Suppose $\alpha < 1$, $C \subseteq \partial \mathbb{D}$, and $|C| > 0$. Then $\Lambda_\alpha(f(C)) > 0$.*

Proof. Let

$$(4.22) \qquad \psi(r) = 2\beta \Big(\log(1/(1 - r)) \log \log \log(1/(1 - r)) \Big)^{1/2},$$

where β is the constant given in Theorem 3.10. Since $\log f'$ is a Bloch function, there exist r_0 and $A \subseteq C$ such that $|A| > 0$ and

$$-\psi(r) \le \log|f'(rz)| \le \psi(r)$$

if $z \in A$ and $r \in [r_0, 1)$.
Hence

$$|f(z) - f(rz)| \le \int_r^1 |f'(sz)|ds \le \int_r^1 e^{\psi(s)}ds,$$

$f(z)$ existing since the integral is finite.
We have the inequality

$$\int_r^1 e^{\psi(s)}ds \le 2(1-r)e^{\psi(r)}$$

if r is close to 1. To see this, $g(r) = \int_r^1 e^{\psi(s)}ds - 2(1-r)e^{\psi(r)}$ has $g'(r) > 0$ for r close to 1 and $\lim_{r \to 1-} g(r) = 0$.

Hence, increasing r_0 if necessary,

$$|f(z) - f(rz)| \le 2(1-r)e^{\psi(r)},$$

if $r \in [r_0, 1)$, $z \in A$. Let $\{B_k\}$ be any collection of sets that are the intersection of open balls with ∂D such that $\cup_k B_k$ contains $f(A)$, and let $A_k = \{z \in A : f(z) \in B_k\}$. Let $\varepsilon_k = \operatorname{diam} B_k$, define r_k so that $\varepsilon_k = 2(1 - r_k)e^{\psi(r_k)}$, and let $\delta_k = (1 - r_k)e^{-\psi(r_k)}$. By Lemma 4.8, A_k can be covered by at most $c(\varepsilon_k/\delta_k)^2$ sets of diameter $1 - r_k$. So

$$|A| \le \sum_k |A_k| \le c\sum_k (\varepsilon_k/\delta_k)^2(1 - r_k)$$

$$= c\sum_k (1 - r_k)e^{4\psi(r_k)} \le c\sum_k \varphi_\alpha(\varepsilon_k).$$

Hence $\Lambda_\alpha(f(A)) \ge |A|/c > 0$. □

(4.10) Corollary. *Suppose $B \subseteq \partial D$ has positive harmonic measure. Then the Hausdorff dimension of B is at least 1.*

Proof. By Lévy's theorem (Theorem 1.2), $C = f^{-1}(B)$ has positive harmonic measure in $\mathbb{D}$, hence $|C| > 0$. Theorem 4.9 says that $\Lambda_\alpha(B) = \Lambda_\alpha(f(C)) > 0$ for all $\alpha < 1$, which implies that B has Hausdorff dimension at least 1. □

Both for the lower and upper bounds stronger statements can be made. Rather than bounds on Hausdorff dimensions, one can even describe which Hausdorff measure functions give 0 or infinite mass to the set on which harmonic measure is concentrated (see Makarov [1] or Pommerenke [2]; see also Wolff [2]).

If D is multiply connected, the upper bound on the Hausdorff dimension still holds; see Jones and Wolff [1]. There is no lower bound in this case; given any $\alpha < 1$, sets of Hausdorff dimension α can be shown to be hit by two-dimensional Brownian motion (Exercise 27), and so if D is the complement of one of these sets, the support of harmonic measure will have Hausdorff dimension α.

When $d > 2$ it is not true that the support of harmonic measure lies in a set of Hausdorff dimension of at most $d - 1$ (see Wolff [1]). However, there exists ε depending only on d such that the Hausdorff dimension of the support of harmonic measure is at most $d - \varepsilon$ (see Bourgain [2]).

5 The angular derivative problem

Angular derivatives

In this section we give an introduction to the angular derivative problem. We have seen in Theorem 1.14 that any simply connected domain can be mapped one to one onto the unit disk. We are now interested in: when is this mapping smooth? When does it have a derivative? More precisely, if $f : \mathbb{D} \to D$ is one to one and analytic, what are geometric conditions on D such that f' has a nontangential limit at a given boundary point? That is, does $\lim_{z \in C_\theta, z \to e^{i\theta}} f'(z)$ exist and is it nonzero? This limit, when it exists, is called the angular derivative.

Sometimes instead of the above, one sees the definition of angular derivative as

$$(5.1) \qquad \lim_{z \in C_\theta, z \to e^{i\theta}} \frac{f(z) - f(e^{i\theta})}{z - e^{i\theta}} = a,$$

where $a \neq \infty$ and $f(e^{i\theta})$ is the radial limit of f. The radial limit of f at $e^{i\theta}$ is $\lim_{r \to 1} f(re^{i\theta})$.

These definitions are equivalent, since we have the following.

(5.1) Proposition. *f has angular derivative a at $e^{i\theta}$ if and only if f has a radial limit at $e^{i\theta}$ and*

$$\lim_{z \in C_\theta, z \to e^{i\theta}} \frac{f(z) - f(e^{i\theta})}{z - e^{i\theta}} = a.$$

Proof. Suppose f has an angular derivative at $e^{i\theta}$. The radial limit exists since $f(re^{i\theta}) = e^{i\theta} \int_0^r f'(se^{i\theta}) ds$ converges as $r \to 1$. Since $z(1-t) + te^{i\theta} \in C_\theta$ if $t \in (0,1)$ and $z \in C_\theta$ by the convexity of C_θ, then

$$\frac{f(e^{i\theta}) - f(z)}{e^{i\theta} - z} = \int_0^1 f'(z + (e^{i\theta} - z)t) dt \to a$$

as $z \to e^{i\theta}$ with $z \in C_\theta$.

To prove the converse, suppose $z_n \in C_\theta, z_n \to e^{i\theta}$. Let

$$g_n(s) = \frac{1}{e^{i\theta} - z_n}\left[f(z_n + (e^{i\theta} - z_n)s) - f(e^{i\theta})\right].$$

As $n \to \infty$, $g_n(s) \to a(s-1)$ uniformly over $|s| \le 1/2$. Hence $g_n'(0) \to a$. However, $g_n'(0) = f'(z_n)$. $\qquad\square$

Since the disk $\mathbb{D}$ and the half-space $H = \{\text{Im } z > 0\}$ are mapped into each other by a smooth mapping, we may as well work with mappings from H into D. We will investigate the existence of the angular derivative at the boundary point 0 and we assume $f : H \to D$ is such that $f(0) = 0$.
Set

(5.2) $$V_0 = \{x + iy \in H : |x| < y\}.$$

We want to know when $\lim_{z \in V_0, z \to 0} f(z)/z$ exists and is a finite positive real number. We will make some nontrivial simplifying assumptions in the remainder of the section:

(5.3) ∂D has ∂H as a tangent line at 0, V_0 is contained in D, $-V_0$ is contained in D^c, and f has a continuous extension to $\overline{H}$.

Let us make some comments about these assumptions. Saying ∂D has a tangent at 0 means that

$$\lim_{z \in \partial D, z \to 0-} \arg f(z) = \pi \qquad \text{and} \qquad \lim_{z \in \partial D, z \to 0+} \arg f(z) = 0,$$

where $\arg w$ is the imaginary part of $\log w$. A necessary and sufficient condition for f to have an extension to $\overline{H}$ is that ∂D be a Jordan curve (see Pommerenke [1] for a proof); here we make this our assumption. Finally, there exist f that have angular derivatives at 0 but for which ∂D does not have a tangent (Exercise 17).

(5.2) Proposition. *If D has a tangent at 0, then $\lim_{z \in H, z \to 0} \arg(f(z)/z)$ exists.*

Proof. Let $w(z) = \arg(f(z)/z) = \arg f(z) - \arg z$. The assumption that D has a tangent at 0 implies that w restricted to ∂H is continuous at 0. Since f has a continuous extension to ∂H, w restricted to ∂H is continuous away from 0. Since $V_0 \subseteq D$ and $-V_0 \subseteq D^c$, w is bounded. Since w is harmonic in H, it follows by Theorem II.1.15 that $w(z)$ is continuous at 0 over $z \in H$. $\qquad\square$

Since we are assuming D has a tangent at 0, we need to look at the nontangential limit of $|f(z)|/|z|$. It is actually enough to look at the radial limit. Let

(5.4) $$I = \{iy : y > 0\}.$$

(5.3) Proposition. *Suppose*

$$\lim_{z\in I,z\to 0}|f(z)/z| = a \in (0,\infty) \qquad and \qquad \arg_{z\in H,z\to 0}\arg(f(z)/z) = 0.$$

Then $\lim_{z\in V_0,z\to 0}|f(z)/z| = a.$

Proof. Let $V_1 = \{x + iy : |x| < 2y\}$. Let $\varepsilon > 0$. Since f is one to one and nonzero in H, then $g(z) = \log(f(z)/z)$ is a single-valued analytic function in H. Since $\operatorname{Im} g(z) \to 0$ as $z \to 0$, then $|\operatorname{Im} g(z)| < \varepsilon$ for $z = x + iy \in V_1$ provided y is sufficiently small. By Corollary II.1.4, $|\nabla \operatorname{Im} g(x + iy)| \le c\varepsilon/y$ if $x + iy \in V_0$ and y is sufficiently small. Since $|\nabla \operatorname{Re} g| = |\nabla \operatorname{Im} g|$ by the Cauchy-Riemann equations,

$$|\operatorname{Re} g(x + iy) - \log a| \le |\operatorname{Re} g(iy) - \log a| + |x|c\varepsilon/y$$
$$\le |\operatorname{Re} g(iy) - \log a| + c\varepsilon$$

if $x + iy \in V_0$ and y is sufficiently small. Since ε is arbitrary and $\operatorname{Re} g(iy) \to \log a$ as $y \to 0$, the result follows. $\qquad\square$

For a generalization to the case where $\arg(f(z)/z)$ does not tend to 0, see Pommerenke [1].

We next convert the question of radial limits of $|f(z)/z|$ to one about the normal derivative of the Green function.

(5.4) Proposition. *Suppose D is as above. Fix $w_0 \in D$. Then D has an angular derivative at 0 if and only if the normal derivative of $g_D(y, w_0)$ at 0 is positive and finite.*

Proof. Let $\Gamma = f^{-1}(I)$, where I is defined in (5.4). A very similar argument to the proof of Proposition 5.3 shows that

$$(5.5) \qquad \lim_{z\in V_0,z\to 0}|f(z)|/|z| = \lim_{z\in\Gamma,z\to 0}|f(z)|/|z|.$$

Since $f'(0)$ is real (because $\arg f'(0) = 0$) and positive, $\lim_{z\in\Gamma,z\to 0}\operatorname{Im} z/|z| = 1$. It is easy, by Proposition II.3.8, to see that $\lim_{z\to 0,z\in H} g_H(z, y_0)/\operatorname{Im} z$ exists and is positive for every $y_0 \in H$. So the expression in (5.5) is equal to

$$\lim_{z\in\Gamma,z\to 0} c|f(z)|/g_H(z, f^{-1}(w_0))$$

$$= \lim_{w\in I,w\to 0} c|f(f^{-1}(w))|/g_H(f^{-1}(w), f^{-1}(w_0))$$

$$= \lim_{w\in I,w\to 0} c|w|/g_D(w, w_0),$$

since $g_H(z, w) = g_D(f(z), f(w))$ by Exercise 18. $\qquad\square$

Ratios of harmonic functions

The following proposition will allow us to prove the existence of a normal derivative by proving upper and lower bounds on $g_D(iy, w_0)/y$, $y > 0$. See Doob [2] for generalizations.

(5.5) Proposition. *Suppose $d = 2$, $N > 1$, D is the intersection of the region above the graph of a Lipschitz function with $B(0, N)$, $0 \in \partial D$, and $w_0 \in D$. Suppose u is positive and harmonic in D. Then*

$$(5.6) \qquad \lim_{y \downarrow 0} u(iy)/g_D(iy, w_0) = \liminf_{z \in D, z \to 0} u(z)/g_D(z, w_0).$$

Proof. Let z, z_1 be points in $D \cap B(0, |w_0|/2)$. By Exercise 21, there exists c_1 such that

$$\frac{g_D(z, w_0) g_D(w_0, z_1)}{g_D(z, z_1)} \le c \big[1 + \log^+(1/|z - w_0|) + \log^+(1/|z_1 - w_0|) \big] \le c_1.$$

If we let $z_1 \to 0$ in D, we see that

$$(5.7) \qquad g_D(z, w_0)/M(z, 0) \le c_1,$$

where $M(z, 0)$ is the Martin kernel with pole at 0. As usual, we abbreviate $\mathbb{P}^w_{M(\cdot, 0)}$ by $\mathbb{P}^w_0$.

Let $\ell = \liminf_{z \in D, z \to 0} u(z)/g_D(z, w_0)$. Our result is trivial if $\ell = \infty$, so let us suppose $\ell < \infty$. Let $\varepsilon > 0$ and

$$A = \{ z \in D \cap B(0, |w_0|/2) : u(z) - (\ell + \varepsilon) g_D(z, w_0) > 0 \}.$$

Since $g_D(\cdot, w_0)$ and u are harmonic in $D \cap B(0, |w_0|/2)$, A is open. Let

$$(5.8) \qquad L_t = \big[u(Z_t) - (\ell + \varepsilon) g_D(Z_t, w_0) \big] / M(Z_t, 0).$$

It is easy to see that L_t is a martingale with respect to $\mathbb{P}^w_0$ for each $w \in D$, and by (5.7), L_t is bounded below by $-c_1(\ell + \varepsilon)$ when Z_t is in $D \cap B(0, |w_0|/2)$.

If we use the definition of ℓ to choose w close to 0 with $u(w)/g_D(w, w_0) < \ell + \varepsilon/2$, we see by the optional stopping theorem (Theorem I.4.4) and dominated convergence that

$$\mathbb{E}^w_0 L_{T_A \wedge \tau_D} < -\frac{\varepsilon g_D(w, w_0)}{2M(w, 0)}.$$

So $\mathbb{P}^w_0(T_A < \tau_D) < 1$. By the zero-one law (Proposition III.2.7),

$$\mathbb{P}^w_0(\limsup_{t \to \tau_D} L_t > 0) = 0,$$

or

$$\mathbb{P}_0^w(\limsup_{t\to\tau_D} u(Z_t)/g_D(Z_t,w_0) > \ell+\varepsilon) = 0.$$

Clearly, $\liminf_{t\to\tau_D} u(Z_t)/g_D(Z_t,w_0) \geq \ell$ under $\mathbb{P}_w^0$, and so the limit of $u(Z_t)/g_D(Z_t,w_0)$ exists and equals ℓ, $\mathbb{P}_w^0$ a.s.

Let y_n be any sequence tending to 0 with $u(iy_n)/g_D(iy_n,w_0) \to a < \infty$. By the Harnack inequality applied to $g_D(z,w_0)$ and $u(z)$, there exists c such that $|u(z)/g_D(z,w_0)-a| < \varepsilon$ if $z \in B(iy_n, cy_n)$ for some n. By the argument of Theorem III.4.3 or Proposition 3.1, we must have $\ell = a$. A similar argument holds for sequences y_n tending to 0 with $u(iy_n)/g_D(iy_n,w_0) \to \infty$. The proposition follows. □

Domains contained in a half-space

We will look at the cases when $D \subseteq H$, $D \supseteq H$, and more general domains separately. Let us here suppose $D \subseteq H$ and let us further suppose that

(5.9) *D is the region above the graph of a Lipschitz function F that is bounded by 1 and has Lipschitz constant at most 1.*

For domains satisfying (5.9) we can give a complete answer to the angular derivative problem.

(5.6) **Theorem.** *Suppose D satisfies (5.3) and (5.9). Let $w_0 \in D$. $g_D(\cdot,w_0)$ has a normal derivative at 0 (and hence the angular derivative exists) if and only if $\int_{-\infty}^{\infty} F(x)/x^2\,dx < \infty$.*

To begin the proof of Theorem 5.6, we first want to reduce to the case where $F(x)$ is 0 if $|x| \geq 8$. Let

$$F_1(x) = \begin{cases} 0 & \text{if } |x| \geq 8, \\ 1 & \text{if } |x| \leq 7, \\ 8-|x| & \text{if } 7 < |x| < 8 \end{cases}$$

and let D_1 be the union of D and the region above the graph of F_1. D_1 still satisfies the conditions (5.3) and (5.9). By the boundary Harnack principle in D with the harmonic functions $u = g_{D_1}(\cdot,w_0)$ and $v = g_D(\cdot,w_0)$, we see $g_D(\cdot,w_0)/g_{D_1}(\cdot,w_0)$ is bounded above and below by positive constants in a neighborhood of 0. So the normal derivative of g_D at 0 will be finite and positive if and only if the normal derivative of g_{D_1} is positive and finite at 0. Hence without loss of generality we may assume that $F(x) = 0$ if $|x| \geq 8$.

Proof of Theorem 5.6. Since $D \subseteq H$, then $g_D(z,w) \leq g_H(z,w)$ for all $z, w \in D$. An easy calculation with Proposition II.3.8 shows that $g_H(iy,w_0)/y$ converges to a finite positive limit as $y \downarrow 0$ for each w_0, hence $\lim_{y\downarrow 0} g_D(iy,w_0)/y < \infty$. By Proposition 5.5 in $D\cap B(0,1)$ with $u(z) = \text{Im } z$, it suffices to show $\liminf_{y\downarrow 0} g_D(iy,w_0)/y > 0$.

Let $U = \{x + iy : y = 10\}$. In $D \cap B(0,1)$, $g_D(z, w_0)$ and $\mathbb{P}^z(T_U < \tau_D)$ are both positive harmonic functions that vanish on ∂D. So by the boundary Harnack principle these two functions are comparable near 0. It therefore suffices to show

$$(5.10) \qquad \frac{\mathbb{P}^{iy}(T_U < \tau_D)}{y} \geq c$$

for some positive constant c and for y close to 0.

Let $h(x + iy) = y$. This is harmonic and $(\mathbb{P}^z_h, X_t)$ is Brownian motion in H conditioned to hit U before ∂H.

$$\mathbb{P}^{iy}(T_U < \tau_D)/y = c\mathbb{P}^{iy}_h(T_U < \tau_D).$$

Now Brownian motion in H h-path transformed by this particular h still has independent components. X_t is still a Brownian motion and (see Exercise 19) Y_t has the same law as the modulus of a three-dimensional Brownian motion. Let

$$A = \{(x^1, x^2, x^3, x^4) : ((x^2)^2 + (x^3)^2 + (x^4)^2)^{1/2} \geq F(x^1)\}.$$

Let W_t be a four-dimensional Brownian motion. What we look at is $\mathbb{P}^0(T_A < \tau_V)$, where $B(0, 10)$ is the ball of radius 10 in $\mathbb{R}^3$, $V = \mathbb{R} \times B(0, 10)$, and T_A and τ_V are defined relative to W_t.

Let $A_k = \{x \in A : 2^{-(k+1)} < |x| \leq 2^{-k}\}$. By Lemma 5.7, $\int_{-8}^{8} F(x)/x^2 dx = \infty$ if and only if $\sum 2^{2k} C(A_k) = \infty$, where $C(A_k)$ is the capacity (relative to W_t) of A_k. If $\sum 2^{2k} C(A_k) = \infty$, then by Wiener's test (Theorem II.5.15), 0 is regular for A, hence $\mathbb{P}^0(T_A < \tau_V) = 1$. In this case $\mathbb{P}^y(T_A > \tau_V) \to 0$ as $|y| \to 0$, and so the normal derivative of g_D is 0 at 0. On the other hand, $\sum 2^{2k} C(A_k) < \infty$ implies $\mathbb{P}^0(T_A = 0) = 0$. Then $\mathbb{P}^0(T_A > \tau_V) > 0$. Therefore (Exercise 20) $\mathbb{P}^y(T_A > \tau_V)$ is bounded away from 0 for y in the intersection of I with a neighborhood of 0, which implies (5.10). $\qquad \square$

(5.7) Lemma. $\sum 2^{2k} C(A_k) = \infty$ if and only if $\int_{-8}^{8} F(x)/x^2 dx = \infty$.

Proof. Let I be the interval $[0, 1]$ and let

$$B = \{(x^1, x^2, x^3, x^4) : x^1 \in I, ((x^2)^2 + (x^3)^2 + (x^4)^2)^{1/2} \leq F(x^1)\}.$$

We show there exist constants c_1 and c_2 such that

$$(5.11). \qquad c_1 \int_I F(x) dx \leq C(B) \leq c_2 \int_I F(x) dx.$$

The lemma will follow by scaling and summing over the collections of intervals $(2^{-(k+1)}, 2^{-k}]$ and $[-2^{-k}, 2^{-(k+1)})$.

Let us form a collection of disjoint intervals. Divide I into two equal subintervals. If there exists $x \in [0, 1/2)$ such that $F(x) \geq 1$, we select

$[0, 1/2)$ as one of our intervals; otherwise we divide $[0, 1/2)$ into two equal subintervals and continue. We also perform this procedure on $[1/2, 1]$. We select an interval J_i if there exists a point x in it for which $F(x) \geq 2|I_i|$. More precisely, an interval $J = [k/2^m, (k+1)/2^m)$ is one of our intervals if there exists $x \in J$ with $F(x) \geq 2|J|$ and J is not contained in any larger dyadic interval J' for which there exists $y \in J'$ with $F(y) \geq 2|J'|$.

Since F is Lipschitz with Lipschitz constant at most 1, $F \geq |J_i|$ on the interval J_i in our sequence. Since J_i is not contained in any larger interval in the sequence, $F \leq 2|J_i|$ on J_i. Note $F = 0$, a.e., on $I - \cup_i J_i$.

Let $B_i = B \cap \{(x^1, x^2, x^3, x^4) : x^1 \in J_i\}$. By the estimate of Proposition II.5.16 for the capacity of a cylinder, there exists c such that

$$(5.12) \qquad c^{-1}|J_i|^2 \leq C(B_i) \leq c|J_i|^2.$$

Therefore

$$C(B) \leq \sum_i C(B_i) \leq c \sum_i |J_i|^2$$

$$\leq c \sum_i \int_{J_i} F(x)dx \leq c \int_I F(x)dx.$$

If $J_i = [k/2^m, (k+1)/2^m)$, let $\tilde{J}_i = [(k-1/2-\varepsilon)/2^m, (k+1/2+\varepsilon)/2^m]$, where ε will be chosen later and let $\tilde{B}_i = B \cap \{(x^1, x^2, x^3, x^4); x^1 \in \tilde{J}_i\}$. By Propositions II.5.16 and II.5.11, there exists c such that

$$(5.13) \qquad |J_i|^2 \leq cC(\tilde{B}_i).$$

By Proposition II.5.8,

$$(5.14) \qquad C(B) = c \int \mathbb{P}^y(T_B < \infty)\sigma(dy),$$

where σ is normalized surface measure on $\partial B(0, 6)$ and T_B is defined in terms of W_t, a four-dimensional Brownian motion. If $|y| = 6$,

$$(5.15) \quad \mathbb{P}^y(T_B < \infty) \geq \sum_i \mathbb{P}^y(T_{\tilde{B}_i} < \infty) - \sum_{i \neq j} \mathbb{P}^y(T_{\tilde{B}_i} < \infty, T_{\tilde{B}_j} < \infty).$$

As in the proof of Theorem II.5.15,

$$(5.16) \quad \mathbb{P}^y(T_{\tilde{B}_i} < \infty, T_{\tilde{B}_j} < \infty) \leq \mathbb{P}^y(T_{\tilde{B}_i} < \infty, T_{\tilde{B}_j} \circ \theta_{T(\tilde{B}_i)} < \infty)$$
$$+ \mathbb{P}^y(T_{\tilde{B}_j} < \infty, T_{\tilde{B}_i} \circ \theta_{T(\tilde{B}_j)} < \infty)$$

and

$$(5.17) \qquad \mathbb{P}^y(T_{\tilde{B}_i} < \infty, T_{\tilde{B}_j} \circ \theta_{T(\tilde{B}_i)} < \infty)$$
$$= \mathbb{E}^y\left[\mathbb{P}^{W_{T(\tilde{B}_i)}}(T_{\tilde{B}_j} < \infty); T_{\tilde{B}_i} < \infty\right].$$

Let x_i be the center of J_i and let $r_{ij} = |x_i - x_j|$. If $w \in \widetilde{B}_i$, then by Proposition I.5.8

$$(5.18) \qquad \mathbb{P}^w(T_{\widetilde{B}_j} < \infty) \leq c(\varepsilon)|J_j|^2 r_{ij}^{-2}$$

where $c(\varepsilon)$ is a constant that goes to 0 as $\varepsilon \to 0$. Substituting in (5.16) and (5.17),

$$\mathbb{P}^y(T_{\widetilde{B}_i} < \infty, T_{\widetilde{B}_j} < \infty) \leq c(\varepsilon)|J_j|^2 r_{ij}^{-2}\mathbb{P}^y(T_{\widetilde{B}_i} < \infty)$$
$$+ c(\varepsilon)|J_i|^2 r_{ij}^{-2}\mathbb{P}^y(T_{\widetilde{B}_j} < \infty).$$

Putting this in (5.14) and (5.15),

$$(5.19) \quad C(B) \geq \sum_i C(\widetilde{B}_i) - c(\varepsilon) \sum_{i \neq j}\{|J_j|^2 r_{ij}^{-2}C(\widetilde{B}_i) + |J_i|^2 r_{ij}^{-2}C(\widetilde{B}_j)\}$$
$$\geq \sum_i C(\widetilde{B}_i) - c(\varepsilon) \sum_{i \neq j} C(\widetilde{B}_i)C(\widetilde{B}_j)r_{ij}^{-2},$$

using (5.13).

If $y \in \widetilde{J}_j$, $|y - x_i| \leq cr_{ij}$. Suppose $|J_j| \leq |J_i|$. By (5.12),

$$C(\widetilde{B}_j)r_{ij}^{-2} \leq c|J_j|^2 r_{ij}^{-2} \leq c|J_j| \int_{\widetilde{J}_j} \frac{dy}{|y - x_i|^2} \leq c|J_i| \int_{\widetilde{J}_j} \frac{dy}{|y - x_i|^2}.$$

Then

$$\sum_{\{j:|J_j| \leq |J_i|, i \neq j\}} C(\widetilde{B}_j)r_{ij}^{-2} \leq c|J_i| \int_{J_i^c} \frac{dy}{|y - x_i|^2} \leq c|J_i||J_i|^{-1} \leq c.$$

This gives us

$$c(\varepsilon) \sum_{i,j} C(\widetilde{B}_i)C(\widetilde{B}_j)r_{ij}^{-2} \leq 2c(\varepsilon) \sum_{|J_i| \geq |J_j|, i \neq j} C(\widetilde{B}_i)C(\widetilde{B}_j)r_{ij}^{-2}$$
$$\leq c(\varepsilon) \sum_i C(\widetilde{B}_i) \leq \frac{1}{2}\sum_i C(\widetilde{B}_i)$$

if ε is small enough. Finally, substituting in (5.19),

$$C(B) \geq \sum_i C(\widetilde{B}_i) - (1/2)\sum_i C(\widetilde{B}_i) \geq c\sum_i |J_i|^2$$
$$\geq c\sum_i \int_{J_i} F(x)dx = c\int_I F(x)dx.$$

$\square$

Domains containing a half-plane

Next we turn to the case $D \supseteq H$. Let us suppose D is the region above $-F(x)$, where $F(x)$ is a nonnegative Lipschitz function bounded by 1 with Lipschitz constant at most 1 and that D satisfies (5.3). Let $w_0 \in D$. As above, there is no loss of generality in assuming $F(x) = 0$ if $|x| \geq 8$.

(5.8) Theorem. *Suppose D is as above. Then the normal derivative of $g_D(\cdot, w_0)$ at 0 is positive and finite (and hence the angular derivative exists) if and only if $\int_{-\infty}^{\infty} F(x)/x^2 dx < \infty$.*

We prove this theorem in a number of steps. Let $U = \{x + iy : y = 10\}$.

(5.9) Proposition. *The normal derivative of $g_D(\cdot, w_0)$ is positive and finite if and only if*

$$\int_{-8}^{8} \frac{\mathbb{P}^x(T_U < \tau_D)}{|x|^2} dx < \infty.$$

Proof. Since $H \subseteq D$, $g_D(z, w) \geq g_H(z, w)$ for all z and w. As in Theorem 5.6,

$$\liminf_{y \downarrow 0} g_D(iy, w_0)/y \geq \lim_{y \downarrow 0} g_H(iy, w_0)/y > 0.$$

We will show

(5.20) $$\limsup_{y \downarrow 0} g_D(iy, w_0)/y < \infty.$$

Since $g_D(\cdot, w_0)$ is harmonic in $H - \{w_0\}$,

$$\frac{g_D(iy, w_0)}{y} = \frac{1}{y} \int_{\partial H} \frac{1}{\pi} \frac{y}{x^2 + y^2} g_D(x, w_0) \, dx = \int_{\partial H} \frac{1}{\pi} \frac{g_D(x, w_0)}{x^2 + y^2} dx.$$

Because this increases as $y \downarrow 0$, convergence along a subsequence $y_n \downarrow 0$ implies the existence of the limit as $y \downarrow 0$.

As in the proof of Theorem 5.6, the boundedness of (5.20) is equivalent to the boundedness of $\mathbb{P}^{iy}(T_U < \tau_D)/y$ near 0.

$$\mathbb{P}^{iy}(T_U < \tau_D) = \mathbb{E}^{iy}\left[\mathbb{P}^{X_T(\partial H)}(T_U < \tau_D)\right] = \int_{-\infty}^{\infty} P_H(iy, x)\mathbb{P}^x(T_U < \tau_D) dx,$$

where P_H is the Poisson kernel for H. We have

$$P_H(iy, x) = \frac{cy}{y^2 + x^2} \leq \frac{cy}{x^2},$$

so

$$\frac{\mathbb{P}^{iy}(T_U < \tau_D)}{y} \leq c \int_{-\infty}^{\infty} \mathbb{P}^x(T_U < \tau_D)/x^2 dx.$$

On the other hand,

$$\frac{\mathbb{P}^{iy}(T_U < \tau_D)}{y} \to c \int_{-\infty}^{\infty} \mathbb{P}^x(T_U < \tau_D)/x^2 dx$$

as $y \to 0$ by the monotone convergence theorem.

To complete the proof, we observe that $F(x) = 0$ and so $\mathbb{P}^x(T_U < \tau_D) = 0$ if $|x| \geq 8$. Hence we can replace the integral $\int_{-\infty}^{\infty}$ by the integral $\int_{-8}^{8}$. $\square$

(5.10) Lemma. *There exists c such that if $x \in \partial H$, then $\mathbb{P}^x(T_U < \tau_D) \geq cF(x)$.*

Proof. Let $x \in \partial H$. Since F is Lipschitz with constant at most 1, $Q(x) \subseteq D$, where $Q(x)$ is the rectangle $[x - F(x)/4, x + F(x)/4] \times [-F(x)/4, 6F(x)]$. Let $V(x)$ be the upper boundary of $Q(x)$. By scaling and the support theorem, $\mathbb{P}^x(T_{V(x)} < \tau_D) \geq c$. If $y \in V(x)$,

$$\mathbb{P}^y(T_U < \tau_D) \geq \mathbb{P}^y(T_U < T_{\partial H}) \geq cF(x)$$

by Proposition I.4.9. So by the strong Markov property, $\mathbb{P}^x(T_U < \tau_D) \geq cF(x)$. $\square$

The converse of Lemma 5.10 is not true and a slightly different argument is needed.

Proof of Theorem 5.8. If $\int_{-8}^{8} F(x)/x^2 dx = \infty$, then by Lemma 5.10, $\int_{-8}^{8} \mathbb{P}^x(T_U < \tau_D)/x^2 dx = \infty$, and hence the normal derivative of $g_D(\cdot, w_0)$ is infinite by Proposition 5.9.

So let us suppose $\int_{-8}^{8} F(x)/x^2 dx < \infty$. Set $\bar{\mu}(dx) = 1_{[1,2]}(x)dx$, a measure on ∂H, and let $\bar{R} = D \cap \{x + iy : 1/2 \leq x \leq 4\}$. We claim

$$(5.21) \qquad \int \mathbb{P}^x(T_U < \tau_{\bar{R}})\bar{\mu}(dx) \leq c \int_{1/2}^{4} F(x)dx.$$

Let us show how (5.21) is used to prove the theorem. Let $\mu(dx) = 1_{[1,2]}(|x|)dx$ and $R = D \cap \{x + iy : 1/2 \leq |x| \leq 4\}$. By symmetry and an addition, (5.21) implies that

$$\int \mathbb{P}^x(T_U < \tau_R)\mu(dx) \leq c \int_{1/2 \leq |x| \leq 4} F(x)dx.$$

Let $\mu_k(dx) = 1_{[2^{-k}, 2^{-k+1}]}(|x|)dx$ and $R_k = D \cap \{x + iy : 2^{-k-1} \leq |x| \leq 2^{-k+2}\}$. By scaling,

$$\int \mathbb{P}^x(T_U < \tau_{R_k}) \frac{\mu_k(dx)}{\mu_k(\mathbb{R})} \leq c2^k \int_{2^{-k-1} \leq |x| \leq 2^{-k+2}} F(x)dx.$$

Note $\mu_k(\mathbb{R}) = c2^{-k}$. Since x^{-2} is comparable to 2^{2k} for $|x| \in [2^{-k-1}, 2^{-k+2}]$, taking the sum over k leads to

$$\int_{-8}^{8} \mathbb{P}^x(T_U < \tau_D)/x^2 dx \le c \sum_k 2^{2k} \int \mathbb{P}^x(T_U < \tau_D) \mu_k(dx)$$

$$\le c \sum_k \int_{2^{-k-1} \le |x| \le 2^{-k+2}} F(x)/x^2 dx$$

$$\le c \int_{-8}^{8} F(x)/x^2 dx.$$

To obtain the last inequality we used the facts that for any x, $|x|$ is in at most eight of the intervals $[2^{-k-1}, 2^{-k+2}]$ and that $F(x) = 0$ if $|x| \ge 8$. Proposition 5.9 then completes the proof.

So we must show (5.21). Let $S_k = \inf\{t : \operatorname{Im} Z_t = -2^{-k}\}$, $L_k = \{x+iy : y = -2^{-k}\}$, and let $K = \inf\{k : S_k < \tau_D\}$.

$$(5.22) \qquad \mathbb{P}^x(T_U < \tau_{\overline{R}}) = \sum_{k=0}^{\infty} \mathbb{P}^x(T_U < \tau_{\overline{R}}, K = k).$$

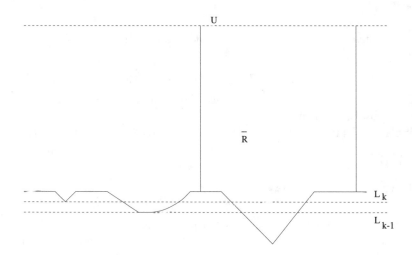

FIGURE 5.1. Diagram for Theorem 5.8.

If $T_U < \tau_{\overline{R}}$ and $K = k$, this means Z_t hits $L_k \cap \overline{R}$ and then hits U before hitting L_{k-1}. By the strong Markov property, this is less than or equal to

$$(5.23) \quad \mathbb{P}^x(Z_{S_k} \in L_k \cap \overline{R}, T_U \circ \theta_{S_k} < S_{k-1})$$
$$= \mathbb{E}^x\left[\mathbb{P}^{Z(S_k)}(T_U < S_{k-1}); Z_{S_k} \in L_k \cap \overline{R}\right].$$

If $y \in L_k$, $\mathbb{P}^y(T_U < S_{k-1}) \le c2^{-k}$. Now

$$(5.24) \int \mathbb{P}^x(Z_{S_k} \in L_k \cap \overline{R})\overline{\mu}(dx) = c \int_1^2 \int_{L_k \cap \overline{R}} \frac{2^{-k}}{2^{-2k} + |x - w|^2} \, dw \, dx$$

$$\leq c \int_{L_k \cap \overline{R}} dx.$$

Substituting (5.23) and (5.24) in (5.22) and summing over k,

$$(5.25) \qquad \mathbb{P}^x(T_U < \tau_{\overline{R}}) \leq c \sum_k 2^{-k} |L_k \cap \overline{R}|.$$

By Exercise 22, the right-hand side is less than or equal to $c \int_{1/2}^4 F(x) \, dx$. This gives (5.21). $\qquad\qquad\qquad\qquad\qquad\qquad\qquad\qquad\qquad\qquad\quad \square$

More general domains

Our main result of the section is the following.

(5.11) Theorem. *Suppose h is a Lipschitz function bounded by 1 with Lipschitz constant at most 1 and $h(0) = 0$. Suppose $h'(0)$ exists and is equal to 0. Let D be the region above the graph of h. If $J^- = \int_{-\infty}^{\infty} h^-(x)/x^2 \, dx < \infty$, then D has an angular derivative at 0 if and only if $J^+ = \int_{-\infty}^{\infty} h^+(x)/x^2 \, dx < \infty$.*

Here h^+ and h^- are the positive and negative parts of h. Note that the finiteness of J^+ and J^- implies $h(0) = 0$ and $h'(0) = 0$.

It can also be shown (Burdzy [1]) that under the above hypotheses on h, if $J^- = \infty$ and $J^+ < \infty$, then the angular derivative does not exist at 0.

Proof. Let $D_1 = D \cap H$, $D_2 = D \cup H$. Suppose first that both integrals J^- and J^+ are finite. Let $w_0 \in D$. By Theorems 5.6 and 5.8, $g_{D_1}(iy, w_0)/y$ and $g_{D_2}(iy, w_0)/y$ both converge as $y \downarrow 0$ to finite nonzero constants. Since $D_1 \subseteq D \subseteq D_2$, the lim sup and lim inf of $g_D(iy, w_0)/y$ are positive and finite. By Proposition 5.5 in $D_1 \cap B(0, 1)$ with $u(z) = g_D(z, w_0)$, we see that $g_D(iy, w_0)/g_{D_1}(iy, w_0)$ converges to a positive finite limit as $y \downarrow 0$. Multiplying by $g_{D_1}(iy, w_0)/y$ implies that $g_D(iy, w_0)/y$ has a positive finite limit. By Proposition 5.4, the angular derivative exists.

Now suppose $J^- < \infty$ but $J^+ = \infty$. Let $U = \{x + iy : y = 10\}$, let $h(z) = \mathbb{P}^z(T_U < \tau_{D \cup H})$, and let $k(z) = \mathbb{P}^z(T_U < \tau_H) = c \operatorname{Im} z$. Since $H \subseteq H \cup D$, then $h(z) \geq k(z)$. Since $J^- < \infty$, then $h(iy)/k(iy) = ch(iy)/y$ is bounded above and below for y near 0 by Theorem 5.8.

For each n,

$$\mathbb{P}_h^{iy}(T_{(H-D)\cap B(0,1/n)} < T_U)$$

$$\geq \mathbb{E}^{iy}[h(Z_{T_U}); T_{(H-D)\cap B(0,1/n)} < T_U < \tau_{D\cup H}]/h(iy)$$

$$\geq c\mathbb{E}^{iy}[k(Z_{T_U}); T_{(H-D)\cap B(0,1/n)} < T_U < \tau_H]/k(iy)$$

$$\geq c\mathbb{P}_k^{iy}(T_{(H-D)\cap B(0,1/n)} < T_U),$$

c independent of n. Hence by Exercise 20

(5.26) $\mathbb{P}_h^0(T_{(H-D)\cap B(0,1/n)} < T_U) \geq c\mathbb{P}_k^0(T_{(H-D)\cap B(0,1/n)} < T_U).$

Since $J^+ = \infty$, it follows from the proof of Theorem 5.6 that

$$\mathbb{P}_k^0(T_{(H-D)\cap B(0,1/n)} = 0) > 0$$

for each n. The conclusion of the zero-one law (Corollary I.3.6) still holds for $(\mathbb{P}_k^0, Z_t)$ because the same proof works. Hence

$$\mathbb{P}_k^0(T_{(H-D)\cap B(0,1/n)} = 0) = 1$$

for each n. Using (5.26),

$$\mathbb{P}_h^0(T_{(H-D)\cap B(0,1/n)} < T_U) \geq c,$$

where c is independent of n. By the continuity of the paths of Z_t under $\mathbb{P}_h^0$, the Borel-Cantelli lemma, and the fact that Z_t is never equal to 0 for $t > 0$ under $\mathbb{P}_h^0$, we see that $\mathbb{P}_h^0(T_{H-D} = 0) \geq c$. By the conclusion of the zero-one law (Corollary I.3.6) for $(\mathbb{P}_h^0, Z_t)$ as above, $\mathbb{P}_h^0(T_{H-D} = 0) = 1$. It follows (Exercise 20) that if $\varepsilon > 0$, then for y sufficiently small, $\mathbb{P}_h^{iy}(\tau_D > T_U) \leq \varepsilon$, hence

$$\mathbb{P}^{iy}(T_U < \tau_D) \leq c\varepsilon y.$$

Finally, by the boundary Harnack principle in D, $g_D(z, w_0)$ and $\mathbb{P}^z(T_U < \tau_D)$ are comparable for z near 0. Since ε is arbitrary, $g_D(iy, w_0)/y \to 0$ as $y \to 0$. By Proposition 5.4, the angular derivative does not exist at 0. $\square$

A comparison theorem holds for the angular derivative: if $D_1 \subseteq D \subseteq D_2$ and D_1 and D_2 both have angular derivatives at 0, then so does D. The existence of the angular derivative is a local property: if D_1 has an angular derivative at 0 and $D_1 \cap B(0,r) = D_2 \cap B(0,r)$ for some r, then D_2 has an angular derivative at 0. These two facts, the proofs of which can be found in Burdzy [1], together with Theorem 5.11, allow one to settle the question of the existence of the angular derivative for many interesting domains; see Burdzy [1] for details.

Analytic proofs of Burdzy's results can be found in Rodin and Warschawski [1], Carroll [1], Gardiner [1], and Sastry [1]. In the analysis literature, the angular derivative problem is sometimes converted to a problem concerning the behavior at ∞ of a domain contained in a strip by means of the mapping $z \to (-1/\pi)\log z$.

The proofs given above for the finiteness and positivity of the normal derivative of the Green function can be easily modified to cover the case when the dimension is greater than 2; see Burdzy [1].

There are still many cases where the angular derivative problem has not been solved. Even the case of domains above Lipschitz functions is still open.

6 The corona problem

Maximal ideals

In this section we give a proof of the famous corona theorem. There are several proofs; the proof we give is due to Varopoulos [2]. The first thing we want to do is to reduce the statement of the corona theorem to a concrete problem concerning the existence of a solution to a certain functional equation (Theorem 6.2). We then prove the probabilistic analog of this equation (Theorem 6.7), and finally deduce the analytic result from the probabilistic one.

Let H^∞ be the set of functions that are analytic on $\mathbb{D}$ and bounded on $\overline{\mathbb{D}}$. H^∞ is an algebra under the operation of pointwise multiplication. M is an ideal in this algebra if M is a subring of H^∞ and $f \in M$ and $g \in H^\infty$ implies $fg \in M$. Any ideal containing the constant function 1 is, of course, equal to H^∞ itself. If $z \in \mathbb{D}$, $M_z = \{f \in H^\infty : f(z) = 0\}$ is an example of an ideal, and in fact, M_z is a maximal ideal. That is, it is a proper subset of H^∞ and is not properly contained in any other proper ideal (see Exercise 24). If $\mathcal{M}$ denotes the set of maximal ideals, $z \to M_z$ is an embedding of $\mathbb{D}$ into $\mathcal{M}$. In a moment we will discuss a topology on $\mathcal{M}$. The corona theorem is the assertion that in this topology, $\overline{\mathbb{D}} = \mathcal{M}$. $\mathcal{M} - \overline{\mathbb{D}}$ is what is left over if the face of the sun ($\mathcal{M}$) is obscured by the shadow of the moon ($\overline{\mathbb{D}}$) during a full eclipse, the corona of the sum. The corona theorem says that the corona is empty.

To put a topology on $\mathcal{M}$, we need to very briefly discuss the Gel'fand mapping.

(6.1) Proposition. *If $M \in \mathcal{M}$, then H^∞/M is isomorphic to $\mathbb{C}$. Let $f(M) = f + M$ denote the mapping from H^∞ to $\mathbb{C}$. Then the mapping is linear, $fg(M) = f(M)g(M)$, $|f(M)| \leq \|f\|_\infty$, $1(M) = 1$, and $f(M) = 0$ if and only if $f \in M$.*

Here $|f(M)| = \|f + M\| = \inf_{g \in M} \|f + g\|_\infty$.

Proof. Let us first show that M is closed. Since the closure of an ideal is also an ideal, and M is maximal, either $M = \overline{M}$ or $\overline{M} = H^\infty$. Let us show that the latter case cannot happen. If $1 \in \overline{M}$, there must exist $g \in M$ with

$\|1 - g\| < 1$. Then the series $\sum_{n=0}^{\infty}(1 - g)^n$ converges absolutely, and is equal to g^{-1}, since

$$g \sum_{n=0}^{\infty}(1 - g)^n = [1 - (1 - g)] \sum_{n=0}^{\infty}(1 - g)^n = 1.$$

So $g^{-1} \in H^{\infty}$ and therefore $1 = gg^{-1} \in M$, contradicting the fact that M is proper. Therefore M is closed.

Let $B = H^{\infty}/M$, the quotient algebra. Since M is maximal, B is a field (Exercise 23). We want to show it is isomorphic to $\mathbb{C}$. Suppose it is not. Then there exists $f \in B$ such that $f \neq \lambda + M$ for any $\lambda \in \mathbb{C}$. $(f - \lambda)$ is an element of B that is not 0, hence it has an inverse. Thus $(f - \lambda)^{-1} \in B$ for all $\lambda \in \mathbb{C}$. Clearly $f^{-1} \neq 0$, and by the Hahn-Banach theorem, there exists a bounded linear functional on B such that $\|L\| = 1$ and $L(f^{-1}) \neq 0$.

Provided $|\lambda - \lambda_0| < 1/\|(f - \lambda_0)^{-1}\|$,

$$(f - \lambda)^{-1} = \sum_{n=0}^{\infty}(\lambda - \lambda_0)^n((f - \lambda_0)^{-1})^{n+1}.$$

Then defining $F(\lambda) = L((f - \lambda)^{-1})$,

$$F(\lambda) = \sum_{n=0}^{\infty} L(((f - \lambda_0)^{-1})^{n+1})(\lambda - \lambda_0)^n.$$

Since $\|L\| = 1$, the series converges uniformly, and still provided $|\lambda - \lambda_0| < 1/\|(f - \lambda_0)^{-1}\|$, F is analytic. So F is analytic in some neighborhood of any point λ_0, and therefore F is entire.

Now if $|\lambda|$ is sufficiently large,

$$(f - \lambda)^{-1} = \frac{1}{\lambda} \sum \frac{f^n}{\lambda^n},$$

and

$$|F(\lambda)| = |L((f - \lambda)^{-1})| \leq \frac{1}{|\lambda|} \sum \frac{\|f\|^n}{|\lambda|^n} \leq \frac{c}{|\lambda|}.$$

By the maximum modulus theorem, for each N, $\sup_{|\lambda| \leq N} |F(\lambda)| \leq c/N$, from which we conclude $F \equiv 0$. We know, however, that $L(f^{-1}) \neq 0$, a contradiction. Therefore B is isomorphic to $\mathbb{C}$.

The only remaining part of the proposition that takes some work is the assertion that $|f(M)| \leq \|f\|_{\infty}$. Suppose $f \in H^{\infty}$ with $\|f\|_{\infty} < 1$ but $|f(M)| = 1$. By looking at $e^{i\theta}f$ for suitable θ, we may assume $f(M) = 1$. Then $(1 - f)^{-1}$ exists and equals $\sum_{n=0}^{\infty} f^n$. Then

$$1 = 1(M) = (1 - f)^{-1}(M)(1(M) - f(M)) = 0,$$

a contradiction. $\qquad\square$

For $f \in H^\infty$, define the Gel'fand transform $f \to \widehat{f}$ by letting $\widehat{f}$ map $\mathcal{M}$ into $\mathbb{C}$ with $\widehat{f}(M) = f(M)$. Define a topology on $\mathcal{M}$ by defining a basic neighborhood of $M' \in \mathcal{M}$ to be a set of the form

$$V = \{M \in \mathcal{M} : |\widehat{f_j}(M) - \widehat{f_j}(M')| < \varepsilon, 1 \le j \le n\}$$

for some $\varepsilon > 0$ and $f_1, \ldots, f_n \in H^\infty$.

The main project in this section is to prove the following.

(6.2) Theorem. *Suppose* $\delta > 0$, $n > 1$, *and* $f_1, \ldots, f_n \in H^\infty$ *with*

$$(6.1) \qquad \max_{1 \le j \le n} |f_j(z)| \ge \delta, \qquad z \in \mathbb{D}.$$

Then there exist $g_1, \ldots, g_n \in H^\infty$ *such that*

$$(6.2) \qquad f_1(z)g_1(z) + \cdots + f_n(z)g_n(z) = 1, \qquad z \in \mathbb{D}.$$

The f_j are called corona data and the g_j corona solutions. As a corollary, we have the corona theorem.

(6.3) Corollary. $\mathbb{D}$ *is dense in* $\mathcal{M}$.

Proof of the corollary. Suppose $\{M_z; z \in \mathbb{D}\}$ is not dense in $\mathcal{M}$. Then there exists $N \in \mathcal{M}$ and a neighborhood V of N of the form $V = \{M \in \mathcal{M} : |\widehat{h_j}(M) - \widehat{h_j}(N)| < \delta, j = 1, \ldots, n\}$ that contains no M_z, where $h_1, \ldots, h_n \in H^\infty$. Let $f_j = h_j - \widehat{h_j}(N) \cdot 1$. Then $f_j \in H^\infty$ and $V = \{M \in \mathcal{M} : |f_j(M)| < \delta, j = 1, \ldots, n\}$. Moreover, $\widehat{f_j}(N) = \widehat{h_j}(N) - \widehat{h_j}(N) = 0$, so each f_j is in N. If $z \in \mathbb{D}$, then since $M_z \notin V$, $|f_j(z)| \ge \delta$ for at least one j.

Since N is an ideal, $1 = f_1 g_1 + \cdots + f_n g_n \in N$, where the $g_1, \ldots, g_n$ are given by Theorem 6.2. This contradicts N being proper. $\square$

Holomorphic martingales

Let us return to the situation where Z_t is two-dimensional Brownian motion. In this section we need to distinguish between complex-valued Brownian motion and $\mathbb{R}^2$-valued Brownian motion, so we write $Z_t = \operatorname{Re} Z_t + i \operatorname{Im} Z_t$ and we let $\widehat{Z}_t$ denote the vector $(\operatorname{Re} Z_t, \operatorname{Im} Z_t)$.

(6.4) Definition. *A complex-valued process* M_t *is a holomorphic martingale if* $\sup_t |M_t| \in L^p$ *and* $M_t = M_0 + \int_0^t H_s dZ_s$ *for some predictable complex-valued process* H_s.

As one would expect, $\int_0^t H_s \, dZ_s$ is defined to be

$$(6.3) \qquad \int_0^t (\operatorname{Re} H_s + i \operatorname{Im} H_s)\, d(\operatorname{Re} Z_s + i \operatorname{Im} Z_s)$$

$$= \int_0^t (\operatorname{Re} H_s\, d\operatorname{Re} Z_s - \operatorname{Im} H_s d\operatorname{Im} Z_s)$$

$$+ i \int_0^t (\operatorname{Re} H_s d\operatorname{Im} Z_s + \operatorname{Im} H_s d\operatorname{Re} Z_s).$$

The prototypes for holomorphic martingales are $M_t = f(Z_t)$ where f is analytic. If we write $f = u + iv$, then u and v are harmonic, and by Itô's formula,

$$u(Z_t) = u(Z_0) + \int_0^t \big(\partial_x u(Z_s) d\operatorname{Re} Z_s + \partial_y u(Z_s) d\operatorname{Im} Z_s\big),$$

and similarly for $v(Z_t)$. So by the Cauchy-Riemann equations

$$(6.4) \qquad f(Z_t) = (u + iv)(Z_t)$$

$$= f(Z_0) + \int_0^t (\partial_x u + i\partial_x v)(Z_s) d\operatorname{Re} Z_s$$

$$+ \int_0^t (\partial_y u + i\partial_y v)(Z_s) d\operatorname{Im} Z_s$$

$$= f(Z_0) + \int_0^t H_s dZ_s,$$

where $H_s = (\partial_x u)(Z_s) + i(\partial_x v)(Z_s) = f'(Z_s)$.

Just as with analytic functions, the real and imaginary parts of holomorphic martingales are closely tied together.

(6.5) Lemma. *Suppose*

$$(6.5) \qquad \operatorname{Re} M_t = \operatorname{Re} M_0 + \int_0^t K_s \cdot d\widehat{Z}_s$$

for some K_s predictable. Then M_t is a holomorphic martingale if and only if

$$(6.6) \qquad \operatorname{Im} M_t = \operatorname{Im} M_0 + \int_0^t A K_s \cdot d\widehat{Z}_s,$$

where

$$(6.7) \qquad A = \begin{pmatrix} 0 & -1 \\ 1 & 0 \end{pmatrix}.$$

Proof. We may suppose without loss of generality that $M_0 = 0$. Suppose M_t is holomorphic martingale, so that $M_t = \int_0^t H_s dZ_s$. Then

$$\mathrm{Re}\, M_t = \int_0^t [\mathrm{Re}\, H_s \, d\mathrm{Re}\, Z_s - \mathrm{Im}\, H_s \, d\mathrm{Im}\, Z_s]$$

and

$$\mathrm{Im}\, M_t = \int_0^t [\mathrm{Im}\, H_s \, d\mathrm{Re}\, Z_s + \mathrm{Re}\, H_s \, d\mathrm{Im}\, Z_s].$$

Hence $K_s = \begin{pmatrix} \mathrm{Re}\, H_s \\ -\mathrm{Im}\, H_s \end{pmatrix}$ and $AK_s = \begin{pmatrix} \mathrm{Im}\, H_s \\ \mathrm{Re}\, H_s \end{pmatrix}$, and so indeed $\mathrm{Im}\, M_t = \int_0^t AK_s \cdot d\widehat{Z}_s$.

Conversely, suppose (6.5) and (6.6) hold. If $K_s = \begin{pmatrix} B_s \\ C_s \end{pmatrix}$, we let $H_s = B_s - iC_s$, and see that $M_t = \int_0^t H_s dZ_s$. $\qquad\square$

We define the space $\mathcal{H}^p$, $1 \le p \le \infty$, to be the set of holomorphic martingales M_t such that $\sup_{t<\infty} |M_t| \in L^p$. We are especially interested in the spaces $\mathcal{H}^\infty$. We saw that if $f \in H^\infty$, then $f(Z_{t \wedge \tau_D}) \in \mathcal{H}^\infty$. One of the key ideas is how to construct an analytic function from an $\mathcal{H}^\infty$ martingale.

(6.6) Proposition. *Suppose $M \in \mathcal{H}^\infty$. Let f be a function whose real and imaginary parts are harmonic in $\mathbb{D}$ and f has boundary values $f(e^{i\theta}) = \mathbb{E}_\theta^0 M_\tau$. Then $f \in H^\infty$.*

This could be expressed as $f(e^{i\theta}) = \mathbb{E}^0[M_\tau | Z_\tau = e^{i\theta}]$. Saying that f has boundary values $f(e^{i\theta})$ means that the nontangential limits of the real and imaginary parts of f are equal to the real and imaginary parts of $f(e^{i\theta})$ for almost every θ.

Proof. It is clear that f is bounded. What we need to show is that f is analytic in $\mathbb{D}$. Write $u = \mathrm{Re}\, f$ and $v = \mathrm{Im}\, f$. By Lemma 6.5, $\mathrm{Re}\, M_\tau = \int_0^\tau K_s d\widehat{Z}_s$ and $\mathrm{Im}\, M_\tau = \int_0^\tau AK_s d\widehat{Z}_s$ where A is the matrix given by (6.7) and K_s is some predictable process.

Suppose h is a L^2 function on $\partial \mathbb{D}$ and let h also denote its harmonic extension. Let $\widetilde{h}$ denote the conjugate harmonic function to h. By Exercise 25, the nontangential limit of $\widetilde{h}$ exists at almost every point of the boundary; if we denote the boundary values by $\widetilde{h}$ also, then $\widetilde{h} \in L^2(\partial \mathbb{D})$. Then

$$\frac{1}{2\pi} \int_0^{2\pi} v(e^{i\theta}) h(e^{i\theta}) d\theta = \mathbb{E}^0 [v(Z_\tau) h(Z_\tau)]$$

$$= \mathbb{E}^0 [(\mathrm{Im}\, M_\tau) h(Z_\tau)]$$

by Proposition III.2.7. Since $\mathrm{Im}\, M_t$ and $h(Z_t)$ are both martingales, the integration by parts formula shows this is equal to

$$\mathbb{E}^0 \langle \mathrm{Im}\, M, h(Z) \rangle_t = \mathbb{E}^0 \int_0^\tau AK_s \cdot \nabla h(Z_s) ds$$

$$= \mathbb{E}^0 \int_0^\tau (A^t \nabla h)(Z_s) \cdot K_s \, ds.$$

Now $A^t \nabla h = -\nabla \widetilde{h}$. So reversing the above calculation, this is equal to

$$-\mathbb{E}^0 \int_0^\tau (\nabla \widetilde{h})(Z_s) \cdot K_s ds = -\mathbb{E}^0[(\operatorname{Re} M_\tau)\widetilde{h}(Z_\tau)] = -\mathbb{E}^0[u(Z_\tau)\widetilde{h}(Z_\tau)].$$

Let u be the harmonic extension of u and let $\widetilde{u}$ be the conjugate harmonic function to u. Then similarly, since $u(Z_t)$ and $v(Z_t)$ are martingales,

$$\mathbb{E}^0[\widetilde{u}(Z_\tau)h(Z_\tau)] = \mathbb{E}^0\langle \widetilde{u}(Z), h(Z)\rangle_\tau$$
$$= E^0 \int_0^\tau \nabla \widetilde{u}(Z_s) \cdot \nabla h(Z_s) ds$$
$$= \mathbb{E}^0 \int_0^\tau (A\nabla u)(Z_s) \cdot \nabla h(Z_s) ds$$
$$= \mathbb{E}^0 \int_0^\tau \nabla u(Z_s) \cdot (A^t \nabla h)(Z_s) ds$$
$$= -\mathbb{E}^0 \int_0^\tau \nabla u(Z_s) \cdot \nabla \widetilde{h}(Z_s) ds.$$

Reversing the calculation, we see that this is $-\mathbb{E}^0[u(Z_\tau)\widetilde{h}(Z_\tau)]$.

We conclude $\int_0^{2\pi} v(e^{i\theta})h(e^{i\theta})d\theta = \int_0^{2\pi} \widetilde{u}(e^{i\theta})h(e^{i\theta})d\theta$ whenever h is in $L^2(\partial \mathbb{D})$. It follows that $v = \widetilde{u}$ a.e., which shows that the harmonic extension of v is the conjugate harmonic function to u, or $u + iv$ is analytic. $\square$

We can now state the corona theorem for martingales.

(6.7) Theorem. *Suppose $F_1, \ldots, F_n \in \mathcal{H}^\infty$ and there exists $\delta > 0$ such that*

$$\max_{1 \leq j \leq n} |F_j(t)| > \delta, \qquad \text{a.s. },$$

for each t. Then there exist $G_1, \ldots, G_n \in \mathcal{H}^\infty$ such that

$$F_1 G_1 + \cdots + F_n G_n = 1, \qquad \text{a.s. },$$

for all t.

A first guess would be to let $G_j(t) = \overline{F}_j(t)/\sum |F_j(t)|^2$, where $\overline{F}_j$ is, of course, the complex conjugate. There is no reason to expect G_j to be holomorphic, and what we do instead is look at $\overline{F}_j(t)/\sum |F_j(t)|^2 + \sum_{k=1}^n W_{jk}(t)F_k(t)$ for suitable W.

Let us set $F(t) = (F_1(t), \ldots, F_n(t))$ and let $\overline{F}(t) = (\overline{F}_1(t), \ldots, \overline{F}_n(t))$ be the complex conjugates. $F_j(t)$ is holomorphic, and we denote by $F'_j(t)$ the predictable process such that $F_j(t) = F_j(0) + \int_0^t F'_j(t) dZ_t$. $F'(t)$ and $\overline{F'}(t)$ are the corresponding vectors. $\|F(t)\|$ will denote $(\sum_{j=1}^n |F_j(t)|^2)^{1/2}$. Let

(6.8) $$L_j(t) = \overline{F}_j(t)/\|F(t)\|^2.$$

Note since $\operatorname{Re} Z_s$ and $\operatorname{Im} Z_s$ are independent, $d\langle \operatorname{Re} Z, \operatorname{Im} Z\rangle_t = 0$, or

(6.9) $$d\langle Z, Z\rangle_t = 0, \qquad d\langle Z, \overline{Z}\rangle_t = 2\,dt.$$

We need some lemmas.

(6.8) Lemma.

$$L_j(t) = L_j(0) + \int_0^t A_j(t)dZ_t + \int_0^t B_j(t)d\overline{Z}_t + \int_0^t C_j(t)dt,$$

where

$$A_j(t) = -\overline{F}_j(t)\frac{F(t)\cdot F'(t)}{\|F(t)\|^4},$$

$$B_j(t) = \frac{\overline{F}'_j(t)}{\|F(t)\|^2} - \overline{F}_j(t)\frac{F(t)\cdot\overline{F}'(t)}{\|F(t)\|^4},$$

and

$$C_j(t) = \overline{F}_j(t)\frac{4|F(t)\cdot\overline{F}'(t)|^2 - 2\|F(t)\|^2\|F'(t)\|^2}{\|F(t)\|^6} - 2\overline{F}'_j(t)\frac{\overline{F}(t)\cdot F'(t)}{\|F(t)\|^4}.$$

Proof. It is easier to compute $\overline{L}_j(t)$ and then take complex conjugates. Let

$$H(t) = \|F(t)\|^2 = \sum |F_j(t)|^2 = \sum F_j(t)\overline{F}_j(t).$$

By Itô's formula and (6.9),

(6.10) $$dH(t) = \sum F_j(t)d\overline{F}_j(t)$$
$$+ \sum \overline{F}_j(t)dF_j(t) + \sum \langle F_j, \overline{F}_j\rangle_t$$
$$= \sum F_j(t)\overline{F}'_j(t)d\overline{Z}_t + \sum \overline{F}_j(t)F'_j(t)dZ_t$$
$$+ 2\sum F'_j(t)\overline{F}'_j(t)\,dt.$$

Then

(6.11) $$d\langle H\rangle_t = 2(F(t)\cdot\overline{F}'(t))(\overline{F}(t)\cdot F'(t))\,dt.$$

If $K = H^{-1}$,

(6.12) $$dK(t) = -H(t)^{-2}dH(t) + 2H(t)^{-3}d\langle H\rangle_t.$$

Also,

(6.13) $$d\overline{L}_j(t) = F_j(t)dK(t) + K(t)dF_j(t) + d\langle F_j, K\rangle_t.$$

Combining (6.10), (6.11), (6.12), and (6.13) and taking complex conjugates proves the lemma. $\square$

A random variable U is said to be in $\mathcal{BMO}$ if the martingale $\mathbb{E}\left[U|\mathcal{F}_t\right]$ is in $\mathcal{BMO}$.

(6.9) Lemma. *Suppose*

$$X_t = X_0 + \int_0^t A_s\, d\mathrm{Re}\, Z_s + \int_0^t B_s\, d\mathrm{Im}\, Z_s$$

is in $\mathcal{BMO}$. *Suppose* H_s *is predictable and* $\sup_s |H_s| \le c < \infty$, *a.s.*
(a) The martingales

$$\int_0^\tau H_s A_s d\mathrm{Re}\, Z_s, \quad \int_0^\tau H_s B_s d\mathrm{Re}\, Z_s, \quad \int_0^\tau H_s A_s d\mathrm{Im}\, Z_s, \quad and$$

$$\int_0^\tau H_s B_s d\mathrm{Im}\, Z_s$$

are all in $\mathcal{BMO}$.
(b) The random variables

$$\int_0^\tau H_s A_s^2 ds, \quad \int_0^\tau H_s A_s B_s ds, \quad and \quad \int_0^\tau H_s B_s^2 ds$$

are all in $\mathcal{BMO}$.

Proof. All the stochastic integral terms in (a) are similar. Let us look at

$$Y_t = \int_0^t H_s A_s d\mathrm{Re}\, Z_s.$$

Then

$$\mathbb{E}\left[|Y_\tau - Y_t|^2|\mathcal{F}_t\right] = \mathbb{E}\left[\int_t^\tau H_s^2 A_s^2 ds|\mathcal{F}_t\right] \le c\mathbb{E}\left[\int_t^\tau [A_s^2 + B_s^2]ds|\mathcal{F}_t\right]$$
$$= c\mathbb{E}\left[|X_\tau - X_t|^2|\mathcal{F}_t\right] \le c\|X\|_{\mathcal{BMO}}^2.$$

The three terms in (b) are similar. Let us let $U_\tau = \int_0^\tau H_s A_s B_s ds$ and $U_t = \mathbb{E}\left[U_\tau|\mathcal{F}_t\right]$. Then if $t < \tau$,

$$\mathbb{E}\left[|U_\tau - U_t| |\mathcal{F}_t\right] = \mathbb{E}\left[\left|\int_t^\tau H_s A_s B_s ds - \mathbb{E}\left[\int_t^\tau H_s A_s B_s\, ds|\mathcal{F}_t\right]\right| |\mathcal{F}_t\right]$$

$$\le 2\mathbb{E}\left[\int_t^\tau |H_s A_s B_s|ds\, |\mathcal{F}_t\right]$$

$$\le c\mathbb{E}\left[\int_t^\tau [A_s^2 + B_s^2]ds|\mathcal{F}_t\right].$$

As above, this is bounded by $c\|X\|_{\mathcal{BMO}}^2$. By Theorem I.6.11 and Exercise 26, $U_\tau \in \mathcal{BMO}$. $\qquad\square$

(6.10) Lemma. *Suppose* $N \in \mathcal{BMO}$ *and* $\mathbb{E}[NM_\infty] = 0$ *for all* $M \in \mathcal{H}^2$. *Then* $N \in \mathcal{H}^2$.

Proof. Since $N \in \mathcal{BMO}$, then $N \in \mathcal{M}^2$, and we need to show that N is holomorphic. Let $N_t = \mathbb{E}[N|\mathcal{F}_t]$. By the martingale representation theorem (Corollary I.5.14), $N_t = N_0 + \int_0^t a_s d\mathrm{Re}\, Z_s + \int b_s d\mathrm{Im}\, Z_s$ for some a_s, b_s predictable and complex-valued. If we set $A_s = (a_s - ib_s)/2$ and $B_s = (a_s + ib_s)/2$, then

$$N_t = N_0 + \int_0^t A_s dZ_s + \int_0^t B_s d\overline{Z}_s.$$

Let $M_t = \int_0^t S_s dZ_s$ where S_s is predictable, bounded, and 0 after some fixed time t_0. Then $M_t \in \mathcal{H}^2$ and

$$0 = \mathbb{E}[NM_\infty] = \mathbb{E}\langle N, M\rangle_\infty = 2\int_0^\infty B_s S_s ds,$$

using (6.9). Since this holds for all such S_s, then $B_s = 0$, a.s. for almost every s, or $N_t = \int_0^t A_s dZ_s$ is holomorphic. □

Proof of Theorem 6.7. Let

$$(6.14) \qquad H_{jk}(t) = \frac{\overline{F}_j(t)\overline{F}'_k(t) - \overline{F}_k(t)\overline{F}'_j(t)}{\|F(t)\|^4}$$

and

$$(6.15) \qquad K_{jk}(t) = 4\frac{\overline{F}(t)\cdot F'(t)}{\|F(t)\|^6}[\overline{F}_k(t)\overline{F}'_j(t) - \overline{F}_j(t)\overline{F}'_k(t)].$$

Note both H_{jk} and K_{jk} are antisymmetric.

Let

$$U_{jk}(t) = \int_0^t H_{jk}(s)d\overline{Z}_s + \int_0^t K_{jk}(s)ds.$$

By Lemma 6.9, $\mathbb{E}[U_{jk}(\infty)|\mathcal{F}_t] \in \mathcal{BMO}$. By Fefferman's inequality (Proposition IV.7.4),

$$L(M) = \mathbb{E}[U_{jk}(\infty)M_\infty]$$

is a bounded linear functional on $\mathcal{M}^1$. Since $\mathcal{M}^1 \subseteq L^1(dP)$, we can extend L to a bounded linear functional on $L^1(dP)$. There thus exists a complex-valued random variable $V_{jk} \in L^\infty(dP)$ such that

$$L(M) = \mathbb{E}[V_{jk}M_\infty]$$

for all $M \in \mathcal{H}^1$, and hence

$$\mathbb{E}[U_{jk}(\infty)M_\infty] = \mathbb{E}[V_{jk}M_\infty]$$

for all $M \in \mathcal{H}^1$.

Recall U_{jk} is antisymmetric. We may assume without loss of generality that V_{jk} is also antisymmetric, for if we replace V_{jk} by $V'_{jk} = (V_{jk} - V_{kj})/2$,

$$\mathbb{E}\,[V'_{jk}M_\infty] = (\mathbb{E}\,[V_{jk}M_\infty] - \mathbb{E}\,[V_{kj}M_\infty])/2$$
$$= (\mathbb{E}\,[U_{jk}(\infty)M_\infty] - \mathbb{E}\,[U_{kj}(\infty)M_\infty])/2 = \mathbb{E}\,[U_{jk}(\infty)M_\infty].$$

By Lemma 6.10 and the fact that $\mathcal{H}^2 \subseteq \mathcal{H}^1$, we see that $\mathbb{E}\,[U_{jk}(\infty) - V_{jk}|\mathcal{F}_t]$ is holomorphic.

Next we define

$$(6.16) \quad W_{jk}(t) = -\mathbb{E}\,[U_{jk}(\infty) - V_{jk}|\mathcal{F}_t] + \int_0^t H_{jk}(s)d\overline{Z}_s + \int_0^t K_{jk}(s)ds.$$

Then $W_{jk}(\infty) = V_{jk} \in L^\infty$.

Finally, let

$$G_j(t) = L_j(t) + \sum_{k=1}^n W_{jk}(t)F_k(t).$$

Clearly $G_j \in L^\infty$. Since W_{jk} is antisymmetric, $\sum_{j,k} F_j(t)W_{jk}(t)F_k(t) = 0$, and so $\sum_{j=1}^n F_j(t)G_j(t) = 1$. All that remains is to show that the $G_j(t)$ are holomorphic. Making the substitutions, we have

$$(6.17) \quad dG_j(t) = dL_j(t) + \sum_{k=1}^n [W_{jk}(t)dF_k(t) + F_k(t)dW_{jk}(t) + d\langle F_k, W_{jk}\rangle_t].$$

In view of (6.9) and the fact that $F_k(t)$ and $\mathbb{E}\,[U_{jk}(\infty) - V_{jk}|\mathcal{F}_t]$ are holomorphic,

$$d\langle F_k, W_{jk}\rangle_t = 2F'_k(t)H_{jk}(t)\,dt.$$

Substituting this, (6.14), (6.15), and (6.17) in (6.16) shows $G_j(t)$ is holomorphic. □

We can now prove Theorem 6.2.

Proof of Theorem 6.2. Let f_j be in H^∞ with $\max_{1\leq j\leq n}|f_j(z)| \geq \delta > 0$. Let $F_j(t) = f_j(Z_{t\wedge\tau})$. By Theorem 6.7, there exists G_j holomorphic in $\mathcal{H}^\infty$ such that $\sum_{j=1}^n F_j(t)G_j(t) = 1$. Let $g_j(e^{i\theta}) = \mathbb{E}_\theta^0 G_j(\tau)$. By Proposition 6.6, each function g_j equals the boundary values of an analytic function in H^∞; let us denote that function by g_j also. Since $\sum F_j(\tau)G_j(\tau) = 1$, then

$$1 = \sum \mathbb{E}_\theta^0[F_j(\tau)G_j(\tau)] = \sum \mathbb{E}_\theta^0[f_j(e^{i\theta})G_j(\tau)] = \sum f_j(e^{i\theta})g_j(e^{i\theta})$$

for almost all θ. Hence $\sum f_j(z)g_j(z) = 1$ for all z in the interior of $\mathbb{D}$. □

The corona theorem has also been proved for many other domains in $\mathbb{C}$. See Behrens [1, 2], Jones and Marshall [1], and Garnett and Jones [1]. It is still an open problem whether the corona is empty for every planar domain. (Where the proof breaks down is in Proposition 6.6; in multiply connected domains there is not a single-valued conjugate harmonic function.) There does exist a Riemann surface in which the analog of Theorem 6.2 fails; see Gamelin [1], Chap. 4.

The corona theorem was first proved by Carleson as a consequence of his solution of an interpolation problem; see Carleson [1].

7 Exercises and further results

Exercise 1. Suppose f is entire and nonconstant. Show $\langle f(Z)\rangle_t \to \infty$ a.s. as $t \to \infty$.

Hint: Use the recurrence of two-dimensional Brownian motion.

Exercise 2. State and prove a version of the Phragmén-Lindelöf theorem for W_α, the wedge of aperture α.

Exercise 3. If $S = \{0 < \operatorname{Im} z < \pi/2\}$ and $z \in S$, find $\mathbb{P}^z(Z_{\tau_S} \in dt)$ for $t \in \{\operatorname{Im} z = 0\}$.

Exercise 4. Show that the f defined in the proof of Theorem 1.14 is single-valued.

Hint: For $\delta < \operatorname{dist}(z_0, \partial D)/2$, let γ_δ consist of the curve that starts at $z_0 + i\delta$ and goes once around $B(z_0, \delta)$ clockwise. Let Δ_δ be the difference between the starting value and the ending value of $f(z)$ as z follows γ_δ once around. Argue that it suffices to show that $\Delta_\delta = 0$. By connecting γ_δ to γ_ε by a straight line segment that is traversed once in each direction, show that $\Delta_\delta = \Delta_\varepsilon$ for all $\varepsilon < \delta$. Finally, use estimates of g in small neighborhoods of z_0 to show $\lim_{\varepsilon \to 0} \Delta_\varepsilon = 0$.

Exercise 5. Suppose $\delta > 0$. Suppose X_n is an integer-valued process such that $X_{n+1} - X_n = 1$ or -1 with probability 1 and $\mathbb{P}(X_{n+1} - X_n = 1|\mathcal{F}_n) \geq 1/2 + \delta$, a.s., if $X_n > 0$. Show that $X_n \to \infty$ a.s.

Hint: Let $P_n = \mathbb{P}(X_{n+1} - X_n = 1|\mathcal{F}_n)$. Let U_n be a sequence of independent identically distributed random variables that are independent of the Xs and have a uniform distribution on $[0, 1]$, that is $\mathbb{P}(U_n < a) = a$ if $a \in [0, 1]$. Define $Y_{n+1} = Y_n + 1$ if $X_{n+1} - X_n = +1$ and $U_n < (1/2 + \delta)/P_n$. Otherwise let $Y_{n+1} = Y_n - 1$. Show Y_n is the sum of independent identically distributed random variables that take the value $+1$ with probability $1/2 + \delta$ and the value -1 with probability $1/2 - \delta$. Conclude from Exercise I.8.18 that $Y_n \to \infty$ a.s. Show that $X_n \geq Y_n$, a.s.

Exercise 6. Suppose that D is a Jordan domain (so that by Pommerenke [1] D is the image of $\mathbb{D}$ under a one to one function f that is analytic in $\mathbb{D}$ and continuous on $\partial\mathbb{D}$). Use conformal mapping to show that the Martin boundary of D may be identified with the Euclidean boundary of D. Show that the boundary Harnack principle holds in D.

Exercise 7. Suppose u is a positive harmonic function in $\mathbb{D}$ that converges radially at $e^{i\theta}$. Show u converges nontangentially at $e^{i\theta}$.

Exercise 8. Show that if M_t is a continuous martingale, the two sets $(\lim_{t\to\infty} M_t$ exists$)$ and $(\langle M\rangle_\infty < \infty)$ differ by a set of probability 0.

Exercise 9. Show that $f(z) = \sum_{k\geq 0} z^{2^k}$ is a Bloch function.

Exercise 10. Complete the proof of (4.10) in Hall's lemma (that is, consider the cases $\rho \leq 1/2$ and $b \leq 1/2$).

Exercise 11. Prove a version of Theorem 4.4 where instead of cubes $Q(x)$ centered at points x there are balls $B(x)$ centered at x.

Exercise 12. Prove Corollary 4.5.

Exercise 13. In the notation of (4.17), show that if $A \subseteq \mathbb{R}^d$, there exists α_0 such that $\Lambda_{\varphi_\alpha}(A) = 0$ if $\alpha > \alpha_0$ and $\Lambda_{\varphi_\alpha}(A) = \infty$ if $0 < \alpha < \alpha_0$.

Exercise 14. Show that if almost every boundary point $e^{i\theta}$ of $\partial\mathbb{D}$ is such that $[\mathbb{D} - B(0,r)] \cap A \cap C_\theta \neq \emptyset$ for all r close to 1, then A is a dominating subset of $\mathbb{D}$. Show that if A is a dominating subset of $\mathbb{D}$ and f is a one to one analytic mapping of $\mathbb{D}$ onto D, then $f(A)$ is a dominating subset of D.

Exercise 15. Justify the second inequality in (4.21).

Hint: Use the distortion theorem.

Exercise 16. Let D be a domain in $\mathbb{R}^d$ whose Martin boundary coincides with its Euclidean boundary, $w_0 \in D$, $z \in \partial D$. A function f has a minimal fine limit a at z if $\lim_{t\to\tau_D} f(X_t) = a$ almost surely with respect to $\mathbb{P}^{w_0}_{M(\cdot,z)}$, where $M(x,z)$ is the Martin kernel for D with pole at z. A set A is minimal thin at z if 1_A has minimal fine limit 0 at z. Show that A is minimal thin at z in D if and only if

$$\limsup_{x\to z, x\in D-A} g_{D-A}(x, w_0)/g_D(x, w_0) > 0.$$

Exercise 17. Find a domain D at which the angular derivative at 0 exists, but ∂D does not have a tangent at 0.

Exercise 18. Show that if f is a one to one map of H onto D, then $g_H(z, w) = g_D(f(z), f(w))$.

Exercise 19. Let $\mathbb{P}^x_1$ be the law of $|X_t|$, where X_t is three-dimensional Brownian motion started at $(x, 0, 0)$. Let Y_t be one-dimensional Brownian motion killed on hitting 0, and let P^x_2 be the law of the h-path transform of Y_t, where $h(x) = x$. Show $\mathbb{P}^x_1 = \mathbb{P}^x_2$ for all $x \geq 0$.

Hint: Show that the same stochastic differential equation is satisfied by both processes and use Exercise I.8.33.

Exercise 20. Let $D \subseteq \mathbb{C}$ be the domain above the graph of a Lipschitz function with $0 \in \partial D$. Let $A \subseteq D$ be the region between two Lipschitz functions Γ_1 and Γ_2 with $\Gamma_1(0) = \Gamma_2(0) = 0$. Let h be a positive harmonic function in D that vanishes on $\partial D \cap B(0,1)$. Show that if $x_0 \in D$ and ε is sufficiently small, then

$$\mathbb{P}_h^{iy}(T_A < T_{B(x_0, \varepsilon)}) \to \mathbb{P}_h^0(T_A < T_{B(x_0, \varepsilon)})$$

as $y \downarrow 0$ (see Exercise III.6.14 for the definition of $\mathbb{P}_h^0$).

Exercise 21. If D is a bounded Lipschitz domain in $\mathbb{R}^2$, show there exists c such that

$$\frac{g_D(x,y)g_D(y,z)}{g_D(x,z)} \le c\left[1 + \log^+(1/|x-y|) + \log^+(1/|y-z|)\right].$$

Hint: Cf. Theorem III.3.6.

Exercise 22. If F is a nonnegative Lipschitz function bounded by 1, D is the region above $-F$, and L_k is the line $\{\operatorname{Im} z = -2^{-k}\}$, show

$$\sum_{k=0}^{\infty} 2^{-k}|L_k \cap [0,1]| \le c\int_0^1 F(x)\, dx.$$

Exercise 23. If M is a maximal ideal in H^∞, show the quotient algebra H^∞/M is a field.

Exercise 24. If $M_z = \{f \in H^\infty : f(z) = 0\}$, show using elementary methods (that is, not using the corona theorem) that M_z is a maximal ideal in H^∞.

Exercise 25. Let h be a function in $L^2(\partial\mathbb{D})$. Denote the harmonic extension of h by h also and let $\widetilde{h}$ be the conjugate harmonic function. Show $\widetilde{h}$ has nontangential limits, a.s. If we denote the nontangential limit by $\widetilde{h}$ also, show $\widetilde{h} \in L^2(\partial\mathbb{D})$.

Hint: Imitate the proof of the boundedness of the Hilbert transform on $L^2(\mathbb{R})$.

Exercise 26. Suppose there exists c such that $\mathbb{E}\left[|U_\infty - \mathbb{E}[U|\mathcal{F}_t]|\,|\mathcal{F}_t\right] \le c$, a.s., for each t. Show $U \in \mathcal{BMO}$.

Exercise 27. Suppose $\varepsilon > 0$ and A is a Borel set whose Hausdorff dimension is greater than ε. Show $\mathbb{P}^0(T_A < \infty) = 1$.

Notes

Much of Sect. 1 is adapted from Durrett [1]. Theorem 1.3 is taken from Revuz and Yor [1]. Our approach to the Phragmén-Lindelöf theorem is new. The proof of Theorem 1.11 is from Folland [1]. We learned this proof of Theorem 1.14 from D. Marshall.

The proof of Picard's theorem using Brownian motion is due to Davis [2]. We modified a version given in Durrett [1]. A discussion of Picard's "big" theorem and an alternate proof of Theorem 2.3 are in Davis [3]. See Davis [1] for Lemma 2.5. Lemma 2.4 is from Ahlfors [1].

The results in Sect. 3 up through Theorem 3.7 are taken from Durrett [1]. A probabilistic proof of Proposition 3.8 was given in Durrett [1]; our proof is slightly different. For a discussion of Bloch functions, see Pommerenke [2]. Our proof of Theorem 3.10 follows Bañuelos [2].

The proof of Theorem 4.1 given here is due to Øksendal [1]. Our proof of Hall's lemma is an adaptation of one given in Duren [1]. For Theorem 4.4, we followed the presentation in deGuzmán [1]. The proof of Theorem 4.6 follows Makarov [1], while the proof of Theorem 4.9 follows the presentation in Pommerenke [2].

The approach given in Sect. 5 follows Burdzy [1], except that we removed the use of excursion theory. The general results about angular derivatives (Propositions 5.1, 5.2, and 5.3) were from Pommerenke [1]. Proposition 5.5 is a special case of a result from Doob [2], 1 XII 14.

The discussion of maximal ideals is from Duren [1]. The remainder of the section follows Varopoulos [2].

BIBLIOGRAPHY

L. Ahlfors
[1]   *Complex Analysis*, 3d ed. McGraw-Hill, New York, 1979.

A. Ancona
[1]   Principe de Harnack à la frontière et théorème de Fatou pour
      un opérateur elliptique dans un domaine lipschitzien. *Ann. Inst.
      Fourier, Grenoble* **28** (1978) 169–213.
[2]   Negatively curved manifolds, elliptic operators, and the Martin
      boundary. *Ann. Math.* **125** (1987) 495–535.

R. Bañuelos
[1]   Martingale transforms and related singular integrals. *Trans. Amer.
      Math. Soc.* **293** (1986) 547–563.
[2]   Brownian motion and area functions. *Indiana Univ. Math. J.* **35**
      (1986) 643–668.
[3]   A sharp good-λ inequality with an application to Riesz transforms.
      Michigan Math. J. **35** (1988) 117–125.
[4]   Intrinsic ultracontractivity and eigenvalue estimates for Schrödinger
      operators. *J. Functional Anal.* **100** (1991) 181–206.
[5]   Lifetime and heat kernel estimates in non-smooth domains. In: *Par-
      tial Differential Equations with Minimal Smoothness and Applica-
      tions*, 37–48. Springer, New York, 1992.

R. Bañuelos, R.F. Bass, and K. Burdzy
[1]   Hölder domains and the boundary Harnack principle. *Duke Math.
      J.* **64** (1991) 195–200.

R. Bañuelos and J. Brossard

[1]   The area integral and its density for BMO and VMO functions. *Ark. Mat.* **31** (1993) 175–196.

R. Bañuelos and B. Davis

[1]   Heat kernel, eigenfunctions, and conditioned Brownian motion in planar domains. *J. Functional Anal.* **84** (1989) 188–200.

[2]   A geometrical characterization of intrinsic ultracontractivity for planar domains with boundaries given by the graphs of functions. *Indiana Math. J.* **41** (1992) 885–913.

R. Bañuelos, I. Klemeš, and C.N. Moore

[1]   An analogue for harmonic functions of Kolmogorov's law of the iterated logarithm. *Duke Math J.* **57** (1988) 37–68.

[2]   Lower bounds in the law of the iterated logarithm for harmonic functions. *Duke Math. J.* **60** (1990) 689–715.

R. Bañuelos and C.N. Moore

[1]   Some results in analysis related to the law of the iterated logarithm. In: *Analysis in Urbana*, Vol. 1, 47–80. Cambridge Univ. Press, Cambridge, 1989.

[2]   Laws of the iterated logarithm, sharp good-λ inequalities and L^p estimates for caloric and harmonic functions. *Indiana Math. J.* **38** (1989) 315–344.

[3]   Sharp estimates for the nontangential maximal function and the Lusin area function in Lipschitz domains. *Trans. Amer. Math. Soc.* **312** (1989) 641–662.

[4]   Distribution function inequalities for the density of the area integral. *Ann. Inst. Fourier, Grenoble* **41** (1991) 137–171.

M.T. Barlow and M. Yor

[1]   (Semi)Martingale inequalities and local times. *Zeit. f. Wahrsch.* **55** (1981) 237–254.

R.F. Bass

[1]   A probabilistic approach to the boundedness of singular integral operators. In: *Séminaire de Probabilités XXIV*, 15–40. Springer, New York, 1990.

[2]   The Doob-Meyer decomposition revisited. Preprint, 1994.

R.F. Bass and K. Burdzy

[1]   A probabilistic proof of the boundary Harnack principle. In: *Seminar on Stochastic Processes, 1989*, 1–16. Birkhäuser, Boston, 1990.

[2]   A boundary Harnack principle for twisted Hölder domains. *Ann. Math.* **134** (1991) 253–276.

[3]   Lifetimes of conditioned diffusions. *Probab. Theory & Rel. Fields* **91** (1992) 405–444.

[4]   The Martin boundary in non-Lipschitz domains. *Trans. Amer. Math. Soc.* **337** (1993) 361–378.

[5]   The boundary Harnack principle for non-divergence form operators. *J. London Math. Soc.* **50** (1994) 157–169.

[6]   Conditioned Brownian motion in planar domains. *Probab. Theory & rel. Fields*, to appear.

R.F. Bass and P. Hsu

[1]   Some potential theory for reflecting Brownian motion in Hölder and Lipschitz domains. *Ann. Probab.* **19** (1991) 486–508.

R.F. Bass and D. Khoshnevisan

[1]   Local times on curves and uniform invariance principles. *Probab. Theory & Rel. Fields* **92** (1992) 465–492.

M. Behrens

[1]   On the corona problem for a class of infinitely connected domains. *Bull. Amer. Math. Soc.* **76** (1970) 387–391.

[2]   The maximal ideal space of algebras of bounded analytic functions on infinitely connected domains. *Trans. Amer. Math. Soc.* **161** (1971) 358–380.

A.G. Bennett

[1]   Probabilistic square functions and a priori estimates. *Trans. Amer. Math. Soc.* **291** (1985) 159–166.

A. Benveniste and J. Jacod

[1]   Systèmes de Lévy des processus de Markov. *Inventiones Math.* **21** (1973) 183–198.

J. Bergh and J. Löfström

[1]   *Interpolation Spaces*. Springer, Heidelberg, 1976.

P. Billingsley

[1]   *Convergence of Probability Measures*. Wiley, New York, 1968.

[2]   *Probability and Measure*, 2d ed. Wiley, New York, 1986.

C.J. Bishop

[1]   Some questions concerning harmonic measure. In: *Partial Differential Equations with Minimal Smoothness and Applications*, 89–97. Springer, New York, 1992.

R.M. Blumenthal and R.K. Getoor

[1]   *Markov Processes and Potential Theory*. Academic Press, New York, 1968.

J. Bourgain

[1]   Some remarks on Banach spaces in which martingale difference sequences are unconditional. *Ark. Mat.* **21** (1983) 163–168.

[2]   On the Hausdorff dimension of harmonic measure in higher dimensions. *Invent. Math.* **87** (1987) 477–483.

J. Bourgain and T.H. Wolff
[1]   A remark on gradients of harmonic functions in dimensions ≥ 3. *Colloq. Math.* **60/61** (1990) 253–260.

L. Breiman
[1]   *Probability.* Addison-Wesley, Reading, MA, 1968.

M. Brelot
[1]   *Eléments de la Théorie Classique du Potentiel*, 4th ed. Centre du Documentation Universitaire Paris, Paris, 1969.

J. Brossard and L. Chevalier
[1]   Classe $L \log L$ et densité de l'intégrale d'aire dans $\mathbb{R}^{n+1}_+$. *Ann. Math.* **128** (1988) 603–618.

K. Burdzy
[1]   *Multidimensional Brownian Excursions and Potential Theory.* Longman, Harlow, Essex, 1987.

D.L. Burkholder
[1]   Distribution function inequalities for martingales. *Ann. Probab.* **1** (1973) 19–42.
[2]   A geometric condition that implies the existence of certain singular integrals of Banach-space-valued functions. In: *Conference on Harmonic Analysis in Honor of Antoni Zygmund*, Vol. 1, 270–286. Wadsworth, Belmont, CA, 1983.

D.L. Burkholder and R.F. Gundy
[1]   Distribution function inequalities for the area integral. *Studia Math.* **44** (1972) 527–544.
[2]   Boundary behaviour of harmonic functions in a half-space and Brownian motion. *Ann. Inst. Fourier, Grenoble* **23** (1973) 195–212.

D.L. Burkholder, R.F. Gundy, and M.L. Silverstein
[1]   A maximal function characterization of H^p. *Trans. Amer. Math. Soc.* **157** (1971) 137–153.

L. Caffarelli, E. Fabes, S. Mortola, and S. Salsa
[1]   Boundary behavior for nonnegative solutions of elliptic operators in divergence form. *Indiana Math. J.* **30** (1981) 621–640.

L. Carleson
[1]   Interpolation by bounded analytic functions and the corona problem. *Ann. Math.* **76** (1962) 547–559.
[2]   On convergence and growth of partial sums of Fourier series. *Acta Math.* **116** (1966) 135–157.

T.F. Carroll
[1] A classical proof of Burdzy's theorem on the angular derivative. *J. London Math. Soc.* **38** (1988) 423–441.

H. Cartan
[1] Théorie du potentiel Newtonien: Énergie, capacité, suites de potentiels. *Bull. Soc. Math. France* **73** (1945) 74–106.

K.L. Chung
[1] *A Course in Probability Theory*, 2d ed. Academic Press, New York, 1974.

K.L. Chung and J.B. Walsh
[1] To reverse a Markov process. *Acta Math.* **123** (1969) 225–251.

R.R. Coifman
[1] A real-variable characterization of H^p. *Studia Math.* **51** (1974) 269–274.

R.R. Coifman and C. Fefferman
[1] Weighted norm inequalities for maximal functions and singular integrals. *Studia Math.* **51** (1974) 241–250.

R.R. Coifman, P.W. Jones, and S. Semmes
[1] Two elementary proofs of the L^2 boundedness of Cauchy integrals on Lipschitz curves. *J. Amer. Math. Soc.* **2** (1989) 553–564.

M. Cranston
[1] Lifetime of conditioned Brownian motion in Lipschitz domains. *Zeit. f. Wahrsch.* **70** (1985) 335–340.
[2] Conditional Brownian motion, Whitney squares, and the conditional gauge theorem. In: *Seminar on Stochastic Processes, 1988*, 109–119. Birkhäuser, Boston, 1989.
[3] A probabilistic approach to gradient estimates. *Canadian J. Math.* **35** (1992) 46–55.

M. Cranston, E. Fabes, and Z. Zhao
[1] Conditional gauge and potential theory for the Schrödinger operator. *Trans. Amer. Math. Soc.* **307** (1988) 171–194.

M. Cranston and T.R. McConnell
[1] The lifetime of conditioned Brownian motion. *Zeit. f. Wahrsch.* **65** (1983) 1–11.

M. Cranston and T.S. Salisbury
[1] Martin boundaries of sectorial domains. *Ark. Mat.* **31** (1993) 27–49.

B. Dahlberg
[1] Estimates of harmonic measure. *Arch. Rat. Mech. Anal.* **65** (1977) 275–288.

G. David and J.-L. Journé

[1]   A boundedness criterion for generalized Calderón-Zygmund opera-
      tors. *Ann. Math.* **120** (1984) 371–397.

E.B. Davies and B. Simon

[1]   Ultracontractivity and the heat kernel for Schrödinger operators and
      Dirichlet Laplacians. *J. Functional Anal.* **59** (1984) 335–395.

B. Davis

[1]   An inequality for the distribution of the Brownian gradient function.
      Proc. Amer. Math. Soc. **37** (1973) 189–194.

[2]   Picard's theorem and Brownian motion. *Trans. Amer. Math. Soc.*
      213 (1975) 353–362.

[3]   Brownian motion and analytic functions. *Ann. Probab.* **7** (1979)
      913–932.

[4]   Intrinsic ultracontractivity and the Dirichlet Laplacian. *J. Func-
      tional Anal.* **100** (1991) 163–180.

L. deBranges

[1]   A proof of the Bieberbach conjecture. *Acta Math.* **154** (1985) 137–
      152.

M. deGuzmán

[1]   *Differentiation of Integrals in $\mathbb{R}^n$.* Springer, New York, 1975.

C. Dellacherie and P.-A. Meyer

[1]   *Probabilités et Potentiel*, Vol. 1. Hermann, Paris, 1975.

[2]   *Probabilités et Potentiel: Théorie des Martingales.* Hermann, Paris,
      1980.

[3]   *Probabilités et Potentiel: Théorie Discrète du Potentiel.* Hermann,
      Paris, 1983.

C. Doléans-Dade and P.-A. Meyer

[1]   Inegalités de normes avec poids. In: *Séminaire de Probabilités XIII*,
      313–331. Springer, New York, 1979.

J.L. Doob

[1]   Conditioned Brownian motion and the boundary limits of harmonic
      functions. *Bull. Math. Soc. France* **85** (1957) 431–458.

[2]   *Classical Potential Theory and Its Probabilistic Counterpart.*
      Springer, New York, 1984.

R.M. Dudley

[1]   *Real Analysis and Probability.* Brooks/Cole, Pacific Grove, CA,
      1989.

P.L. Duren

[1]   *Theory of H^p spaces.* Academic Press, New York, 1970.

R. Durrett
  [1]   *Brownian Motion and Martingales in Analysis*. Wadsworth, Belmont, CA, 1984.
  [2]   *Probability: Theory and Examples*. Wadsworth, Belmont, CA, 1991.

A. Erdélyi
  [1]   *Bateman Manuscript Project: Tables of Integral Transforms*, Vol. 1. McGraw-Hill, New York, 1954.

E. Fabes, M. Jodeit, and N. Rivière
  [1]   Potential techniques for boundary-value problems on C^1 domains. *Acta Math.* **141** (1978) 165–186.

C. Fefferman
  [1]   The multiplier problem for the ball. *Ann. Math* **94** (1971) 330–336.

C. Fefferman and E.M. Stein
  [1] H^p spaces of several variables. *Acta Math.* **129** (1972) 137–193.

W. Feller
  [1]   *An Introduction to Probability Theory and Its Applications*, Vol. 1, 3d ed. Wiley, New York, 1968.
  [2]   *An Introduction to Probability Theory and Its Applications*, Vol. 2, 2d ed. Wiley, New York, 1971.

G.B. Folland
  [1]   *Real Analysis: Modern Techniques and Their Applications*. Wiley, New York, 1984.
  [2]   *Fourier Analysis and Its Applications*. Brooks/Cole, Pacific Grove, CA, 1992.

G.B. Folland and E.M. Stein
  [1]   *Hardy Spaces on Homogeneous Groups*. Princeton Univ. Press, Princeton, 1982.

M. Fukushima
  [1]   *Dirichlet Forms and Markov Processes*. North Holland/Kodansha, Amsterdam, 1980.

P. Gao
  [1]   The boundary Harnack principle for some degenerate elliptic operators. *Comm. in PDE* **18** (1993) 2001–2022.
  [2]   The lifetime of conditioned diffusions associated with some degenerate elliptic operators. Preprint, 1993.

T.W. Gamelin
  [1]   *Uniform Algebras and Jensen Measures*. Cambridge Univ. Press, Cambridge, 1978.

S.J. Gardiner
[1]   A short proof of Burdzy's theorem on the angular derivative. *Bull. London Math. Soc.* **23** (1991) 575–579.

J.B. Garnett
[1]   *Bounded Analytic Functions.* Academic Press, New York, 1981.

J.B. Garnett and P.W. Jones
[1]   The corona theorem for Denjoy domains. *Acta Math.* **155** (1985) 37–40.

A.M. Garsia
[1]   *Martingale Inequalities.* Benjamin, Reading, MA, 1973.

D. Gilbarg and N.S. Trudinger
[1]   *Elliptic Partial Differential Equations of Second Order,* 2d ed. Springer, New York, 1983.

R.F. Gundy
[1]   *Some Topics in Probability and Analysis.* Amer. Math. Soc., Providence, 1989.

R.F. Gundy and M.L. Silverstein
[1]   On a probabilistic interpretation of the Riesz transforms. In: *Functional Analysis in Markov Processes,* 199–203. Springer, New York, 1982.
[2]   The density of the area integral in $\mathbb{R}^{n+1}_+$. *Ann. Inst. Fourier, Grenoble* **35** (1985) 215–229.

R.F. Gundy and N. Th. Varopoulos
[1]   Les transformations de Riesz et les intégrales stochastiques. *C.R. Acad. Sci. Paris A* **289** (1979) 13–16.

L.L. Helms
[1]   *Introduction to Potential Theory.* Wiley, New York, 1969.

C. Herz
[1] H_p spaces of martingales, $0 < p \leq 1$. *Zeit. f. Wahrsch.* **28** (1974) 189–205.

R.A. Hunt
[1]   On the convergence of Fourier series. In: *Orthogonal Expansions and Their Continuous Analogues,* 235–255. Southern Illinois Univ. Press, Carbondale, 1968.

R.A. Hunt and R.L. Wheeden
[1]   On the boundary values of harmonic functions. *Trans. Amer. Math. Soc.* **132** (1968) 307–322.
[2]   Positive harmonic functions on Lipschitz domains. *Trans. Amer. Math. Soc.* **132** (1970) 507–527.

N. Ikeda and S. Watanabe
[1] *Stochastic Differential Equations and Diffusion Processes.* North Holland/Kodansha, Amsterdam, 1981.

D.S. Jerison and C.E. Kenig
[1] The Dirichlet problem in non-smooth domains. *Ann. Math.* **113** (1981) 367–382.
[2] Boundary value problems on Lipschitz domains. In: *Studies in Partial Differential Equations*, 1–68. Math. Assoc. Amer., Washington, DC, 1982.
[3] Boundary behavior of harmonic functions in non-tangentially accessible domains. *Adv. in Math.* **46** (1982) 80–147.

P.W. Jones and D.E. Marshall
[1] Critical points of Green's function, harmonic measure, and the corona problem. *Ark. Mat.* **23** (1985) 281–314.

P.W. Jones and T.H. Wolff
[1] Hausdorff dimension of harmonic measures in the plane. *Acta Math.* **161** (1988) 131–144.

I. Karatzas and S.E. Shreve
[1] *Brownian Motion and Stochastic Calculus.* Springer, New York, 1988.

R.H. Latter
[1] A decomposition of $H^p(\mathbb{R}^n)$ in terms of atoms. *Studia Math.* **62** (1978) 92–101.

T. Lindvall and L.C.G. Rogers
[1] Coupling of multi-dimensional diffusions by reflection. *Ann. Probab.* **14** (1986) 860–872.

M. Loève
[1] *Probability Theory*, 4th ed. Springer, New York, 1977.

N.G. Makarov
[1] On the distortion of boundary sets under conformal mappings. *Proc. London Math. Soc.* **51** (1985) 369–384.

M. Marias
[1] Littlewood-Paley-Stein theory and Bessel diffusions. *Bull. Sci. Math.* **111** (1987) 313–331.

J.E. Marsden and A.J. Tromba
[1] *Vector Calculus.* W.H. Freeman, San Francisco, 1976.

T.R. McConnell
[1] On Fourier multiplier transformations of Banach-valued functions. *Trans. Amer. Math. Soc.* **285** (1984) 739–757.

[2]   A conformal inequality related to the conditional gauge theorem. *Trans. Amer. Math. Soc.* **318** (1990) 721–733.

P.-A. Meyer
[1]   *Probability and Potentials.* Blaisdell, Waltham MA, 1966.
[2]   Démonstration probabiliste de certaines inégalités de Littlewood-Paley. In: *Séminaire de Probabilités X*, 125–183. Springer, New York, 1976.
[3]   Retour sur la théorie de Littlewood-Paley. In: *Séminaire de Probabilités XV*, 151–166. Springer, New York, 1981.
[4]   Note sur les processus d'Ornstein-Uhlenbeck en dimension infinie. In: *Séminaire de Probabilités XVI*, 95–132. Springer, New York, 1982.
[5]   Transformations de Riesz pour les lois gaussiennes. In: *Séminaire de Probabilités XVIII*, 179–193. Springer, New York, 1984.

M. Ohtsuka
[1]   An elementary introduction of Kuramochi boundary. *J. Sci. Hiroshima* **28** (1964) 271–299.

B. Øksendal
[1]   Projection estimates for harmonic measure. *Ark. Mat.* **21** (1983) 191–203.

R.R. Phelps
[1]   *Lectures on Choquet's Theorem.* D. Van Nostrand, Princeton, 1966.

G. Pisier
[1]   Riesz transforms: A simpler analytic proof of P.A. Meyer's inequality. In: *Séminaire de Probabilités XXII*, 485–501. Springer, New York, 1988.

Ch. Pommerenke
[1]   *Univalent Functions.* Vandenhoeck & Ruprecht, Göttingen, 1975.
[2]   *Boundary Behaviour of Conformal Maps.* Springer, New York, 1992.

S.C. Port and C.J. Stone
[1]   *Brownian Motion and Classical Potential Theory.* Academic Press, New York, 1978.

P.E. Protter
[1]   *Stochastic Integration and Differential Equations: A New Approach.* Springer, New York, 1990.

K.M. Rao
[1]   On the decomposition theorem of Meyer. *Math. Scand.* **24** (1969) 66–78.

D. Revuz and M. Yor
[1]   *Continuous Martingales and Brownian Motion.* Springer, New York, 1991.

B. Rodin and S.E. Warschawski
  [1]   Remarks on a paper of K. Burdzy. *J. d'Analyse Math.* **46** (1986) 251–260.

J.L. Rubio de Francia
  [1]   A Littlewood-Paley inequality for arbitrary intervals. *Rev. Mat. Iberoamer.* **1** (1985) 1–13.

W. Rudin
  [1]   *Real and Complex Analysis*, 2d ed. McGraw-Hill, New York, 1974.
  [2]   *Principles of Mathematical Analysis*, 3d ed. McGraw-Hill, New York, 1976.

S. Sastry
  [1]   Existence of angular derivatives for a class of strip domains. *Proc. Amer. Math. Soc.*, to appear.

M.J. Sharpe
  [1]   *General Theory of Markov Processes.* Academic Press, Boston, 1988.

E.M. Stein
  [1]   *Singular Integrals and Differentiability Properties of Functions.* Princeton Univ. Press, Princeton, 1970.
  [2]   *Topics in Harmonic Analysis Related to the Littlewood-Paley Theory.* Princeton Univ. Press, Princeton, 1970.
  [3]   *Harmonic Analysis: Real-Variable Methods, Orthogonality, and Oscillatory Integrals.* Princeton Univ. Press, Princeton, 1993.

E.M. Stein and G.Weiss
  [1]   *Introduction to Fourier Analysis on Euclidean Spaces.* Princeton Univ. Press, Princeton, 1971.

D.W. Stroock and S.R.S. Varadhan
  [1]   *Multidimensional Diffusion Processes.* Springer, New York, 1979.

N. Th. Varopoulos
  [1]   Aspects of probabilistic Littlewood-Paley theory. *J. Functional Anal.* **38** (1980) 25–60.
  [2]   The Helson-Szegö theorem and A_p functions for Brownian motion and several variables. *J. Functional Anal.* **39** (1980) 85–121.

G.C. Verchota
  [1]   Layer potentials and regularity for the Dirichlet problem for Laplace's equation in Lipschitz domains. *J. Functional Anal.* **59** (1984) 572–611.

J. Wermer
  [1]   *Potential Theory*, 2d ed. Springer, New York, 1981.

D. Williams

[1] *Diffusions, Markov Processes and Martingales, Vol.1: Foundations.*
     Wiley, New York, 1979.

T. Wolff

[1] Counterexamples with harmonic gradients in $\mathbb{R}^3$. Preprint, 1987.
[2] Plane harmonic measure lives on sets of σ-finite length. *Ark. Mat.*
     31 (1993) 137–172.

J.-M. G. Wu

[1] Comparison of kernel functions, boundary Harnack principle, and
     relative Fatou theorem on Lipschitz domains. *Ann. Inst. Fourier,
     Grenoble* **28** (1978) 147–167.

J. Xu

[1] The lifetime of conditioned Brownian motion in planar domains of
     infinite area. *Probab. Theory & Rel. Fields* **87** (1991) 469–487.

K. Yosida

[1] *Lectures on Differential and Integral Equations.* Interscience, New
     York, 1960.

Z. Zhao

[1] Green functions and conditioned gauge theorem for a two dimen-
     sional domain. In: *Seminar on Stochastic Processes, 1987*, 283–294.
     Birkhäuser, Boston, 1988.

INDEX